Practical Construction Equipment Maintenance Reference Guide

The McGraw-Hill Engineering Reference Guide Series

This series makes available to professionals and students a wide variety of engineering information and data available in McGraw-Hill's library of highly acclaimed books and publications. The books in the series are drawn directly from this vast resource of titles. Each one is either a condensation of a single title or a collection of sections culled from several titles. The Project Editors responsible for the books in the series are highly respected professionals in the engineering areas covered. Each Editor selected only the most relevant and current information available in the McGraw-Hill library, adding further details and commentary where necessary.

Hicks · CIVIL ENGINEERING CALCULATIONS REFERENCE GUIDE

Hicks · MACHINE DESIGN CALCULATIONS REFERENCE GUIDE

Hicks · PLUMBING DESIGN AND INSTALLATION REFERENCE GUIDE

Hicks · POWER GENERATION CALCULATIONS REFERENCE GUIDE

Hicks · POWER PLANT EVALUATION AND DESIGN REFERENCE GUIDE

Higgins · PRACTICAL CONSTRUCTION EQUIPMENT MAINTENANCE REFERENCE GUIDE

Johnson & Jasik · ANTENNA APPLICATIONS REFERENCE GUIDE

Markus and Weston · CLASSIC CIRCUITS REFERENCE GUIDE

Merritt · CIVIL ENGINEERING REFERENCE GUIDE

Perry · BUILDING SYSTEMS REFERENCE GUIDE

Rosaler and Rice · INDUSTRIAL MAINTENANCE REFERENCE GUIDE

Rosaler and Rice · PLANT EQUIPMENT REFERENCE GUIDE

Woodson · HUMAN FACTORS REFERENCE GUIDE FOR ELECTRONICS AND COMPUTER PROFESSIONALS

Woodson · HUMAN FACTORS REFERENCE GUIDE FOR PROCESS PLANTS

Practical Construction Equipment Maintenance Reference Guide

Lindley R. Higgins, P.E. Consulting Engineer
Senior Editor, *Construction Contracting*
(Formerly *Construction Methods and Equipment*)
McGraw-Hill, Inc., New York, N.Y.

Tyler G. Hicks Project Editor

McGraw-Hill Book Company
New York St. Louis San Francisco Auckland Bogotá
Hamburg London Madrid Mexico Milan
Montreal New Delhi Panama Paris São Paulo
Singapore Sydney Tokyo Toronto

Library of Congress Cataloging-in-Publication Data

Practical construction equipment maintenance reference guide.

(McGraw-Hill engineering reference guide series)
"Condensation of Handbook of construction equipment maintenance, Lindley R. Higgins, ed."—T.p. verso
Includes index.
1. Construction equipment—Maintenance and repair.
I. Higgins, Lindley R.   II. Hicks, Tyler Gregory, 1921-   .   III. Handbook of construction equipment maintenance.   IV. Series.
TH900.P67   1987   621.8′16   87-3410
ISBN 0-07-028772-4

Copyright © 1987 by McGraw-Hill, Inc. All rights reserved. Printed in the United States of America. Except as permitted under the United States Copyright Act of 1976, no part of this publication may be reproduced or distributed in any form or by any means, or stored in a data base or retrieval system, without the prior written permission of the publisher.

1234567890   DOC/DOC   89210987

The material in this volume has been published previously in the *Handbook of Construction Equipment Maintenance*, by Lindley R. Higgins (ed.), Copyright © 1979 by McGraw-Hill, Inc. All rights reserved.
Printed and bound by R. R. Donnelley & Sons Company.

The pronouns "he" and "his" and terms such as "foreman" have been used in a purely generic sense in this book to accommodate the text to the limitations of the English language and avoid awkward grammatical constructions.

Contents

Contributors vii
Preface ix

Section 1. Basic Maintenance Technology

1. Corrosion Control — 1-1
2. Equipment Parts Cleaning — 1-5
3. Lubrication — 1-11
4. Centralized Lubrication Systems — 1-21
5. Arc Welding in Maintenance — 1-27
6. Gas Welding in Maintenance — 1-79
7. Metal Resurfacing — 1-107
8. Maintenance of Plain Bearings — 1-125
9. Maintenance of Rolling Bearings — 1-135
10. Maintenance of Mechanical Power Transmission Equipment — 1-155
11. Maintenance of Scaffolds and Ladders — 1-171
12. Chain Hoists and Chain Slings — 1-187
13. Maintenance of Belt Conveyors and Conveying Equipment — 1-203
14. Maintenance of Hydraulic Hose and Fittings — 1-223
15. Steam and Hot Water Cleaning — 1-245

Section 2. Maintenance of Power Systems

1. Maintenance of Electrical Power Systems — 2-1
2. Maintenance of Diesel Power Systems — 2-9
3. Maintenance of Gasoline Power Systems — 2-33
4. Principles and Maintenance of Hydraulic Systems — 2-39

5. Maintenance of Electric Motors — 2-53

Section 3. Maintenance of Ground Contact Elements
1. Upkeep and Maintenance of Tires for Construction Equipment — 3-1
2. Wear and Maintenance of the Undercarriage — 3-15
3. Dozer Moldboard and Ripper Tooth Maintenance — 3-23
4. Maintenance of Earthmoving Buckets and Bucket Teeth — 3-31
5. Maintenance of Vibratory Compactator Drums — 3-43

Index Follows Section 3

Contributors

ANDERSON, K. D. *Terex Division, General Motors Corporation, Hudson, Ohio* (SECTION 3, CHAPTER 2)

ARBORE, LOUIS A. *Manager, Michelin Tire Corporation, Technical Group, Earthmoving Dept., Lake Success, N.Y.* (SECTION 3, CHAPTER 1)

ARNOLD, R. E. *Manager, Product Engineering, General Electric Company, Industrial Motor Division, Schenectady, N.Y. Mr. Arnold is now retired.* (SECTION 2, CHAPTER 5)

AXELSON, WILLIAM *Vice-President, The Hotsey Corporation, Englewood, Colo.* (SECTION 1, CHAPTER 15)

BISHOP, A. *Product Manager, Mounted Bearings, Dodge Division, Reliance Electric Company, Mishawaka, Ind.* (SECTION 1, CHAPTER 10)

BOEYER, G. *Product Manager, Shaft Mounted Speed Reducers, Flexible Couplings, Dodge Division, Reliance Electric Company, Mishawaka, Ind.* (SECTION 1, CHAPTER 10)

BREWER, ALLEN F. *Consultant in Lubrication, St. Lucie County, Fla.* (SECTION 1, CHAPTER 3)

BURROWS, LINCOLN *Sperry Vickers, Division of Sperry Rand Corporation, Troy, Mich.* (SECTION 2, CHAPTER 4)

BURTON, H. *Product Manager, Dry Fluid Drums and Couplings, Dodge Division, Reliance Electric Company, Mishawaka, Ind.* (SECTION 1, CHAPTER 10)

CALLAHAN, JAMES J. *Vice-President, Research and Engineering, Trabor Lubricating Systems, Lubriquip Division, Houdaille Industries, Inc., Cleveland, Ohio.* (SECTION 1, CHAPTER 4)

DEICHEL, ARNOLD *Manager, Product Services, Ingersoll-Rand Company, Compactor Division, Shippensburg, Pa.* (SECTION 3, CHAPTER 5)

D'HOORE, M. *Product Specialist, Babbitted and Bronze Sleeve Type Bearings, Dodge Division, Reliance Electric Company, Mishawaka, Ind.* (SECTION 1, CHAPTER 10)

ELSON, R. *Product Specialist, Mounted Bearings, Dodge Division, Reliance Electric Company, Mishawaka, Ind.* (SECTION 1, CHAPTER 10)

ESCO CORPORATION *Engineering Staff, Portland, Ore.* (SECTION 3, CHAPTER 4)

FRICKE, C. L. *Engine Service Manager, Briggs & Stratton Corporation, Milwaukee, Wisc.* (SECTION 2, CHAPTER 3)

GLEEKMAN, LEWIS W., DR. *Materials and Corrosion Engineering Services, Smithfield, Mich.* (SECTION 1, CHAPTER 1)

Contributors

GOLDENBOGEN, W. N. *Supervisor, Material Quality, Gould, Inc., Clevite Engine Parts Division. Cleveland, Ohio.* (SECTION 1, CHAPTER 8)

GOOD, W. RALPH *SKF Industries, King of Prussia, Pa. Mr. Good is now retired.* (SECTION 1, CHAPTER 9)

HAGELIN, VERNON D. *Technical Services, Deere & Company, Moline, Ill.* (SECTION 2, CHAPTER 2)

HINKEL, J. E. *Manager, Application Engineering, The Lincoln Electric Company, Cleveland, Ohio.* (SECTION 1, CHAPTER 5)

HOLLIDAY, H. E. *Service Manager, CM Hoist Division, Columbus McKinnon Corporation, Tonawanda, N.Y. Mr. Holliday is now deceased.* (SECTION 1, CHAPTER 12)

HOPPENRATH, R. A. *Product Application Manager, Conveyors, Barber-Greene Company, Aurora, Ill.* (SECTION 1, CHAPTER 13)

HUEMMER, C. *Product Manager, V-Belt Drives, Tapered Bushings, Dodge Division, Reliance Electric Company, Mishawaka, Ind.* (SECTION 1, CHAPTER 10)

HULLMANN, HERMAN A. *Technical Services Representative, Fleet Services, Fiat-Allis Construction Machinery Inc., Springfield, Ill.* (SECTION 3, CHAPTER 3)

JOHNSON, JESS *Kohler Company, Kohler, Wisc.* (SECTION 2, CHAPTER 1)

KELLEY, K. N. *Manager, Marketing Communications, Stoody Company, Industry, Calif. Mr. Kelley is now retired.* (SECTION 1, CHAPTER 7)

LOUCKS, CHARLES M. *Consulting Chemist, Arlington Heights, Ill.* (SECTION 1, CHAPTER 2)

OLSON, ROBERT L., P.E. *Product Application Department, The Gates Rubber Co., Englewood, Colo.* (SECTION 1, CHAPTER 14)

SAFIER, LEONARD *President, Patent Scaffolding Company, A Division of Harsco Corporation, Ft. Lee, N.J.* (SECTION 1, CHAPTER 11)

SAREEN, A. *Product Specialist, Dry Fluid Drives and Couplings, Dodge Division, Reliance Electric Company, Mishawaka, Ind.* (SECTION 1, CHAPTER 10)

SHERMAN, MARVIN *Sperry Vickers, Division of Sperry Rand Corporation, Troy, Mich.* (SECTION 2, CHAPTER 4)

STREJC, W. *Product Specialist, Chain Drives, Dodge Division, Reliance Electric Company, Mishawaka, Ind.* (SECTION 1, CHAPTER 10)

UNION CARBIDE CORPORATION *Engineering Staff, Linde Division, New York, N.Y.* (SECTION 1, CHAPTER 6)

ZAJAC, J. D. *Product Standards and Service, CM Hoist Division, Columbus McKinnon Corporation, Tonawanda, N.Y.* (SECTION 1, CHAPTER 12)

ZAMBELAS, GEORGE *Engineer, Michelin Tire Corporation, Technical Group, Earthmoving Department, Lake Success, N.Y.* (SECTION 3, CHAPTER 1)

Preface

This is a concise presentation of practical construction equipment maintenance techniques for managers, maintenance supervisors, and mechanics. It is comprised of key material selected from the Higgins—*Handbook of Construction Equipment Maintenance* (McGraw-Hill, 1979).

The coverage begins with basic maintenance technology and discusses the following: corrosion control, equipment parts cleaning, lubrication, centralized lubrication systems, arc welding, gas welding, metal resurfacing, plain bearings, rolling bearings, mechanical power transmission equipment, scaffolds and ladders, chain hoists and chain slings, belt conveyors and conveying equipment, hydraulic hose and fittings, and steam and hot-water cleaning.

Next, maintenance of power systems is considered. Topics in this section include: electrical power, diesel power, gasoline power, hydraulic systems, and electric motors.

The final section of this reference guide covers maintenance of ground contact elements, including: upkeep and maintenance of tires, wear and maintenance of the undercarriage, dozer moldboard and ripper teeth, earthmoving buckets and bucket teeth, and vibratory compactor drum maintenance.

Each section is authored by outstanding specialists in the field, and presents useful information that can help solve day-to-day on-the-job problems for readers. Amongst those who will benefit from the information presented are construction managers, maintenance supervisors, maintenance mechanics, diesel engine operators and maintenance personnel, mechanical-power transmission supervisors and mechanics, and electrical maintenance workers. With hundreds of procedures and practical tips, the book will prove to be a handy daily companion that will save both money and time on any type of construction job.

Tyler G. Hicks, P.E.

Practical Construction Equipment Maintenance Reference Guide

Section 1

Basic Maintenance Technology

Chapter 1

Corrosion Control

DR. LEWIS W. GLEEKMAN
Materials and Corrosion Engineering Services, Southfield, Mich.

To the construction equipment maintenance engineer, corrosion is best defined as the deterioration of any material in contact with its surroundings. This applies not only to metal parts of equipment but also to plastics and protective coatings. While the cost of corrosion damage to equipment and accessories (estimated at $6 billion per year) is huge, more critical costs are involved in lost man-hours caused by equipment failure.

Learning to identify various types of corrosion and knowing what to do about them are important parts of the job for today's construction equipment maintenance engineer or foreman.

Corrosion may be uniform or localized. *Uniform corrosion* means that a piece of equipment is being attacked by wet or dry chemical or electrochemical forces over most or all of its surface. *Localized corrosion* can be divided into two forms, one too small to see and the other visible to the human eye; both usually affect only part of a piece of construction equipment.

TYPES OF CORROSION

The following are basic types of corrosion: galvanic, erosion, crevice corrosion, pitting, selective leaching, exfoliation, intergranular corrosion, and stress cracking.

Galvanic Galvanic corrosion occurs when two different metals are in contact in the presence of a conductive solution. In galvanic corrosion, the more active metal deteriorates while the less active or noble metal is protected. To minimize galvanic corrosion it is best to avoid having dissimilar metals in contact. A second method places insulation between dissimilar metals.

Erosion A selective type of corrosion, erosion, is caused by the motion of a corroding solution over a metal surface. Such motion removes the protective film, usually by mechanical wear. Erosion appears as smooth-bottomed shallow pits with a very directional appearance. It is frequently seen on threaded areas and elbows.

Cavitation, a special form of erosion, is caused by the formation and collapse of vapors at the metal surfaces. The change from high to low pressure disturbs the base metal by removing its normal protective film. *Fretting,* another form of erosion, occurs when metal slides over metal and causes mechanical damage to either one or both of the metals. Fretting is most commonly caused by vibration.

1-2   Basic Maintenance Technology

Crevice corrosion   Crevices normally exist at lap joints, gaskets, or around bolts and rivets. Crevices are also created by deposits of corrosion products on a surface or scratches in the paint film. Acidity changes in the crevice, lack of oxygen, detrimental metallic-ion buildup, and depletion of a corrosion inhibitor are other causes.

Materials such as stainless steel and titanium, which depend on an oxide film to achieve corrosion resistance, are more susceptible to crevice corrosion than others.

Pitting   A highly localized type of attack, pitting can be prevented in many cases by the use of an inhibitor such as sodium or potassium dichromate. Surface cleanliness and selection of materials known to be resistant to pitting are the best ways of avoiding the problem.

Selective leaching   Sometimes called *parting corrosion*, selective leaching is the removal of one element from an alloy. It occurs in many alloys, though it is most frequently found in copper-based alloys. The addition of small amounts of arsenic, antimony, or tellurium inhibits this form of corrosion in brass, but the best method of prevention is the use of nonsusceptible alloys.

Exfoliation   A variation of selective leaching is exfoliation, which is surface corrosion spreading below the surface. It differs from pitting in that the attack has a laminated appearance with whole layers of material eaten away in the form of a flaky or blistered surface.

Intergranular corrosion   This is an attack concentrated at the grain boundaries without appreciable corrosion evident on the grains themselves. Austenitic stainless steels, high-nickel alloys, and aluminum alloys are most frequently involved.

Stress cracking   This cracking results from residual or applied stresses plus corrosion and usually occurs without notable loss of metal in the form of uniform corrosion. Most alloys are susceptible to this problem, but the number of combinations of alloy and corrosion that cause it are relatively few. Stresses that cause cracking arise from residual cold work and thermal stresses on contraction after welding or after other thermal treatments; they may also be applied externally during use of the equipment.

METHODS OF STOPPING CORROSION

The following methods are the key to corrosion prevention: change material, change the environment, protect the material.

Change material   The practice of using a more corrosion-resistant material can mean a change in alloying or switching to nonmetallic parts—plastics, with or without reinforcement, for example. The matter of changing material is not merely a function of selecting a material that has improved corrosion resistance; other factors such as thermal and electrical properties, ease of fabrication, strength, availability, and cost must be considered.

Change the environment   To reduce corrosion, the easiest and most obvious method is to lower the temperature. Corrosion processes are chemical reactions, and every 18°F decrease in temperature reduces the reaction rate by half. Other environmental changes involve agitation, aeration, and velocity.

Protect the material   Isolate the metallic surface from the corrosive environment with either organic or metallic coatings on the surface. Organic coatings can be thick or thin, paint film or solid linings, or plastic in the form of tape, sheet, or powder fused to the surface. Metallic coatings can be applied as electroplated materials or deposited by chemical means in an electrodeless deposition similar to silvering glass to make a mirror. A metal part may be coated at moderate temperatures by diffusion as from applying zinc in galvanizing or aluminum in aluminizing or metallizing of the surface by spray application of partially melted materials.

Design factors   In addition to these methods, choosing equipment designed to minimize corrosion is recommended. The following design factors should be considered when buying equipment:

1. Butt joints should be used where possible.
2. Equipment that must be washed and drained should have an effective system that allows prompt attention to this detail.
3. Means of access for inspection and maintenance should be provided at all necessary points.

4. Dissimilar metals should not be used in contact with each other where such use can be avoided.

5. Localized turbulence and areas of high velocity at feed and drain connections of fittings and lines should be minimized where possible.

6. Equipment should not be allowed to rest in liquid or damp material (including damp earth) and should be promptly cleaned of such material. Porous material should be waterproofed.

NONMETALLIC MATERIALS

Use of nonmetallic materials for some equipment parts has proven very successful in lowering corrosion.

Plastics Although limited in use in the past, plastics are becoming more useful for construction equipment parts with the development of reinforced and thermoset materials. Some plastics are now capable of withstanding thousands of pounds of pressure and high temperatures. Polyethylene, polypropylene, polyvinylchloride, the styrene–synthetic rubber blends, the acrylics, and the fluorocarbons are thermoplastics which now have relatively high thermal distortion temperatures. Thermoset (materials not softened by heat) reinforced plastics include polyesters, epoxy and furane resins, and phenolic and epoxy resins. Polyvinyl-chloride, polypropylene, and reinforced polyesters are beginning to replace stainless steel, lead, and galvanized steel in some equipment parts—particularly pumps, blowers, and fan wheels. Nylon is becoming more common in bearings and other parts. Plastics are generally not subject to pitting, stress corrosion cracking, or other forms of corrosion common to metal. They also do not require protective painting. However, some plastics do present stress problems when used in combination with metal parts.

A chemically resistant plastic, tetrafluoroethylene (Teflon or Halon), which withstands temperatures to 500°F, is being used for gaskets, "O" rings, seals, and other relatively small molded items. Loose linings, including nozzle linings, also use this material. Polyethylene, a lower-cost plastic, is being used for piping and tubing.

Fire-retardant grades of most plastics are now available and should be used in construction equipment parts where required. This is particularly true of plastics such as urethane and epoxy foam used as insulation.

Rubber and elastomers For many years, these materials have been used where chemical corrosion is common. Natural rubber compounds resist a wide variety of chemical solutions, yet are readily attacked by strong oxidizing acids. The temperature limit for soft rubber compounds is about 140°F and about 180°F for hard rubber compounds. Elastomer compounds extend these ranges, yet still provide natural rubber's corrosion-resistant qualities.

Wood, concrete, and glass are other materials that offer good resistance to corrosion in some cases, but their application in construction equipment is limited.

PROTECTIVE COATINGS

Protective coatings are probably the most widely used and, at the same time, the most controversial material used to minimize corrosion of metal and certain other materials. While protective coating and painting provide protection, such protection is limited by conditions under which the equipment is used. Any coating will have some pinholes or holidays, and these points require continuous maintenance. Faults in a coating can also be minimized by adding extra coats of the protection or by increasing the degree of the coating's cure. Baked phenolics, baked epoxies, air-dried epoxy, vinyl, and neoprene are some of the coatings commonly used.

INHIBITORS

The corrosion of iron and other metals can frequently be minimized or inhibited by the addition of soluble chromates, phosphates, molybdates, silicates, amines, and other chemicals singly or in combination. These are most useful for equipment used in a closed system requiring constant immersion of construction equipment.

REFERENCES

Ailor, W. H.: "Handbook of Corrosion Testing and Evaluation," sponsored by Electrochemical Society, John Wiley & Sons, New York, 1971.
"Corrosion Data Survey, IV Edition," NACE, Houston, Texas, 1968.
Fontana, M. G., and Greene, N. D.: "Corrosion Engineering," McGraw-Hill Book Company, New York, 1967.
LaQue, F. D., and Copson, H. R.: "Corrosion Resistance of Metals and Alloys," 2d ed., Reinhold Publishing Co., New York, 1963.
Seymour, R. B., and Steiner, R. H.: "Corrosion Resistant Plastics," Reinhold Book Corporation, New York, 1955.
Uhlig, H. H. (ed.): "The Corrosion Handbook," sponsored by the Electrochemical Society, John Wiley & Sons, Inc., New York, 1958.

Chapter **2**

Equipment Parts Cleaning

CHARLES M. LOUCKS
Consulting Chemist, Arlington Heights, Ill.

When fouling of construction equipment has occurred because of corrosion and water scaling, chemical solvents may be used for the necessary cleaning. The methods are discussed in this chapter.

A large construction firm may have facilities for cleaning by means of tanks, jets and steam jennies, vapor degreasers, electrolytic cleaners, or ultrasonic transducers. Such facilities are used mainly for removing soil or rust from external surfaces of metal parts.

Quite another problem is the cleaning of inaccessible interior surfaces when equipment items are too big to be handled or disassembled. Such to-be-cleaned construction equipment becomes the containing vessel when solvents are pumped in by means of special truck-mounted tanks, pumps, mixers, and heaters.

Cleaning materials The removal of corrosion products and/or scale from large construction equipment requires large volumes of cleaning solvents, perhaps 1000 to 100,000 gal.

Acids and alkalis Much of such cleaning can be done with dilute solutions of relatively inexpensive acids and alkalis. Solutions of soda ash, caustic soda, phosphates, or silicates, plus synthetic detergents for better wetting and emulsifying, will remove oil, grease, and general soil when applied with heat and turbulent movement. Alkalis are also used after acids have been applied for scale removal. This assures that acid residues have been neutralized.

The most common acid solvent is inhibited muriatic acid. This acid is inhibited, as all acid solvents must be, to reduce chemical attack on metal surfaces to an acceptable level. Muriatic acid is cheap and effective. It forms reaction products that are generally water-soluble so they are removed in the used solvent. Sulfuric acid is seldom used for the opposite reason, insoluble reaction products. Nitric acid cannot be prevented from attacking carbon steels or copper alloys. It has certain special applications where the substrate metal is stainless steel or aluminum. Of special interest for small-scale construction equipment maintenance cleaning is sulfamic acid, mainly because it is a dry, solid product that is safely handled. It has acid properties only after it is dissolved in water. Solid inhibitors and wetting additives can be premixed in the packaged product.

Sequestrants Recently a class of alkaline salts called *sequestrants* have come into general use for prevention of scale formation and for periodic removal of both water scale and corrosion products.[1]* The most useful examples are derived from an organic acid called ethylene-diamine-tetra-acetic acid (EDTA). The sodium salt dissolves water-hardness scale. The ammonium salt is used to remove iron oxides and copper.

Another use for sequestrants such as EDTA and sodium gluconate is in the alkaline

*Superior numbers refer to references at the end of the chapter.

rinse used after a conventional acid stage.[2] The sequestrants prevent the precipitation of dissolved metal ions by the alkali. Less rinsing saves time and rinse water.

Thiourea and its derivatives can form acid-soluble complexes that are used to prevent dissolved copper from plating out of an acid solution onto steel surfaces.

Synthetic detergents and acid inhibitors Although they serve different purposes, both synthetic detergents and acid inhibitors are large organic molecules that are attracted to surfaces. Synthetic detergents are attracted to oil-water interfaces where they promote wetting, emulsion formation, detergency, and foam. Acid inhibitors are attracted to metal surfaces where they interfere with the chemical reaction of acid on metal.

Organic solvents Certain relatively small-volume cleaning jobs require nonwater solvents for removing oil and grease. Shop cleaning of engine parts may be done with Stoddard solvent, kerosene, or diesel fuel. On-site degreasing of larger parts is more likely to employ the chlorinated solvents trichlor or perchlor, which are nonflammable but have toxic vapors. Freons, with both chlorine and fluorine atoms in the molecules, are nonflammable and relatively nontoxic, but the cleaning uses are limited because of cost. They are especially recommended for cleaning electric motors.

Specialty cleaning products There are many packaged proprietary products used for construction equipment maintenance cleaning and housekeeping. The quantities required each day are small. A purchaser buys the special formulations and convenient packaging provided by the vendor.

For cleaning large equipment the vendor is likely to be an outside contractor who handles bulk chemicals by the truckload and does his own formulating.

METHODS OF APPLICATION

Fill and empty In cleaning the intricate internal surfaces of tank trucks or other carrier and storage vessels where exposure to water has led to formation of corrosion products and scale, the equipment can generally be filled with a liquid solvent and, at the proper time, emptied by opening a drain valve.

A vessel is pumped full of hot inhibited acid, for example. The acid reacts with corrosion products and scale; then the drain valve is opened and the solvent is removed. Rinsing and neutralizing solutions are handled in the same way. The mechanical requirements are simple. There are tank trucks to haul a liquid acid, such as inhibited concentrated muriatic acid, to the site. A water line to the truck position supplies water. A steam line furnishes steam for heating as the acid is diluted and pumped into the vessel through a temporary pipeline from the temporary acid pump. Simple connections will be needed at the fill line and, if necessary, on the vessel. A vent line allows air and gases, including hydrogen, to escape from the high point on the equipment tank. A gauge glass will indicate the final solvent level unless the vessel is to be filled to overflowing at the vent.

During the reaction time the service engineer in charge takes samples, runs analyses, records temperatures, and decides when the reaction period is finished. The solvent is drained to waste through another temporary line provided for that purpose. It may have been decided to blanket the internal surfaces with nitrogen gas during the draining and rinsing operations. In that case, nitrogen is admitted at the vent connection as draining proceeds.

Other decisions will have been made regarding the rinsing and neutralizing. It may be agreed to keep the rinse water slightly acid to prevent precipitation of metal salts. This can be done with very small amounts of any acid. Citric acid is often used. Or it may have been decided to reduce the number of water rinses by using immediately an alkaline solution containing the neutralizing agent such as soda ash plus sequestrants of the EDTA, gluconate types. Use of an alkaline solution will also prevent precipitation of metal ions remaining in the acid residuals. The vessel will eventually be opened for inspection and for the removal of any loose, undissolved debris that normally remains after a chemical cleaning operation.

Flow-through vessels Many vessels such as pipelines were designed for mass flow-through circulation rather than for filling and draining. Such a situation presents a problem to the chemical cleaning engineer. The ideal solution to the problem would be to clean the system while operating it in a manner for which it was designed. The present state of the technology, however, may not allow this ideal solution, so cleaning engineers and equipment owners have improvised other methods that work rather well.

One solution is to provide large temporary circulating pumps that, according to the design engineer, will provide flow through all parallel paths. Instead of draining, each fluid is replaced by the next until solvents, rinses, and neutralizers have been put in and then completely removed.

The foamed-acid technique[3] has also been useful because the flow characteristics of foam allow it to enter one end of a pipe from a water box, fill the pipe completely, and emerge into an empty water box on the opposite end.

Some awkward pipeline problems have been approached by using a flow of steam[4] adequate to carry cleaning reagents and loose debris over the ups and downs. Otherwise, each high and low would need to be provided with vents and drain lines. Or solvents, rinses, and neutralizers have been put in lines and held in position by rubber plugs as the train moves along under the pressure of fluid pumped in behind.

When circulating such a system the natural corrosiveness of the fluid used becomes important. Inhibited muriatic acid may be replaced by inhibited organic acids. Higher temperatures are generally used with the organic acids but control of the temperature becomes easier and less critical than with muriatic acid.

Large hollow vessels Vessels of large volume and limited surface area to be cleaned are not adapted to fill-and-empty or flow-through methods. Cleaning reagents have been applied in the form of a gel. Or interior surfaces may be cleaned by using automated spray devices that do not require personnel to remain inside the vessel. In some instances, reactants (both alkalis and acids) have been put into the hollow space by means of steam and allowed to condense on the interior walls. Cleaning tank exteriors chemically has not met with success.

New ideas The chemical cleaning business has traditionally been conservative, partly because cleaning engineers are not at liberty to take chances with contractors' equipment and partly because a common inertia exists which favors using tomorrow the same methods that were used yesterday. Perhaps the most significant trend has been toward the use of reactants such as EDTA-type chelants rather than cheap, aggressive muriatic acid. Use of chelants, plus reducing agents, for both scale and corrosion and ammoniated EDTA for removing iron oxide and copper represent real progress. The cost of such expensive materials can be offset by savings elsewhere. Time is the largest potential saving; indeed, the ultimate goal should be to return the serviced equipment to the construction site as quickly as possible.

The use of noncondensed phases as carriers, the use of chelants to save rinse time and water, better analytical control during cleaning, and the search for entirely new ideas are only a few of the interesting possibilities for future development.

Pigs, plugs, balls, and jets Mechanical devices often are used alone or in connection with chemical solvents. In pipe cleaning, tools can be made to travel through the line by the force of a fluid behind them. The fluid may be water, oil, gas, or a chemical cleaning solution. The tool may have rubber disks that fit the inside of the pipe, the disks being attached to a central shaft to which scrapers and brushes may also be attached. Even a radioactive capsule may be attached to assist in showing the location of the tool if the radiation can be detected outside the line. Sometimes rubber stoppers are used to separate a slug of one fluid from the next. In this manner, cleaning, rinsing, and neutralizing solutions have been put through pipes in proper sequence to avoid filling entirely with first one fluid, then the next. There are rubber balls of any diameter, with chain mesh to fit them when inflated. The balls move with the fluid flow as do the pigs and plugs.

Water jets powered by pumps up to 10,000 psi have become very useful cleaning tools where chemical removal of deposits is not the best method.

Disposal problems In recent years, the problem of disposing of large volumes of cleaning solvents has come to require serious attention.[5] The problem may indeed dictate what reactants will be permitted. For years laboratory tests have been run to determine a degree of dilution which might avoid fish kill. Acid has been dumped onto sludge from the water softener or into pits filled with crushed limestone or neutralized with caustic soda, soda ash, or powdered lime or limestone. Solvents containing ammonia, oxidizing agents, and dissolved copper have been disposed of by dilution. Eventually, more sophisticated and expensive methods will be required to control not only pH but all dissolved solids as well. Methods involving ion exchange, reverse osmosis, electrodialysis, or evaporation suggest themselves.

SUMMARY OF DATA AND DECISIONS

It is necessary to know the nature of deposit phases. Chemical cleaning involves chemical reactions between substances that foul the equipment (corrosion products and scale) and chemicals that are chosen to correct the fouling condition. As a reaction goes on, new substances will be formed. Knowledge of deposit phases reveals what has been going on in a system to cause fouling. Such knowledge helps in selecting the solvent to be used for cleaning and in anticipating the identity of the reaction products, which must be soluble in the cleaning solution to be removed from the system. Reaction products that are flammable, toxic, or corrosive must be anticipated. In addition to identifying deposit ingredients, it is common practice to use deposit samples for solvent trials and corrosion tests.

It is necessary to know engineering materials and design. What engineering materials will be exposed to the cleaning solvent? What are the design details? Both materials and design are described in prints and descriptive bulletins furnished by equipment vendors. If there is any doubt about the chemical properties of a material versus solvents to be proposed, laboratory trials should be made. Specimens of materials can be exposed to solvents and deposits under conditions that simulate the cleaning conditions that are being planned.

After securing the information indicated, there are decisions to be made. If an outside service agency is employed, the decisions may be left to them or they may be reached by consultation between the agency and the contractor. If the contractor's own men do the job, someone must make the following decisions:

1. What reactants, inhibitors, surfactants, or neutralizers should be used?

2. What reactant concentrations and temperatures are necessary? What mixing and heating facilities will provide the chosen conditions?

3. What precautions are necessary to protect people and equipment? This involves safety instructions, clothing and special protective devices for workmen. It involves isolating the equipment to prevent the solvent escaping through forgotten connections or valves that fail to hold. It may mean roping off the work area and posting safety signs. Perhaps most important, it means providing for reaction products that may be hazardous. Hydrogen gas always is anticipated when cleaning involves acids and steel; providing a suitable vent for hydrogen should be routine. The possibility of toxic gases such as hydrogen sulfide or chlorine must be anticipated, either to prevent the reaction or to dispose of the product. Solvents and rinses must be disposed of in a way that avoids pollution problems.

4. What supervision is needed during the operation? What people are needed and what engineering and chemical data so that people can follow the progress of the cleaning process? Someone must decide when each step has been completed and it is time to go to the next until everything has been done and equipment is once again ready for use.

5. What records should be kept and by whom? There are data and records to be kept during planning, performing, and evaluating of the results.

CHEMICAL VERSUS ALTERNATIVE METHODS

To compare costs of chemical methods versus alternatives, consider cleaning time, lost equipment usage time, man-hours, tools, equipment, materials needed, degree of restored efficiency to be expected, and safety to equipment and personnel.

In any case, the objective is to get the most cleaning per dollar of cost. If construction equipment is too large or complex to allow mechanical methods to be used, the answer is to use chemical methods. Still the question is: What is the best way to get the most cleaning for the money? Can equipment maintenance personnel do the work, or is it better to seek outside sources? Are outside services bought by bids? On what are the bids based? Who specifies procedures, materials, results? What assurance is there of competent planning and performance? Some answers depend on whether the purchaser has men who know chemical cleaning technology well enough to furnish specifications describing what is being purchased and to judge the competence of the service that is offered for hire. Without such knowledge, the purchaser is in the position of a layman seeking advice from medical experts. Bids and lowest prices are not necessarily the greatest bargain. The experience and qualifications of whoever will be responsible for the service are most important.

REFERENCES

1. Blake, D. M., J. P. Engle, and C. A. Lesinski: The Use of Chelating Agents in Chemical Cleaning, Proceedings of the 23d International Water Conference, Engineers Society of Western Pennsylvania, pp. 135–142, 1962.
2. U.S. Patent No. 3,067,070.
3. (a) Carroll, D. B., C. L. Eddington, and J. P. Engle: Chemical Cleaning with Foamed Solvents, Proceedings of the 22d Annual Water Conference, Engineers Society of Western Pennsylvania, pp. 35–40, 1961. (b) U.S. Patent No. 3,037,887.
4. (a) Loucks, C. M.: Something New in Chemical Cleaning, *Power Engineering*, vol. 65, pp. 58, 59, June 1961. (b) U.S. Patent No. 3,084,076.
5. Bell, W. E., and E. D. Escher: Disposal of Chemical Cleaning Waste Solvents, *Materials Protection and Performance*, vol. 9, pp. 15–18, December 1970.

Chapter **3**

Lubrication

ALLEN F. BREWER
Consultant in Lubrication, St. Lucie County, Fla.

Lubrication is of interest to the construction equipment maintenance engineer because it has an influence on the costs he must charge to maintenance. Any machine will operate most dependably when it is properly lubricated. Under such conditions the maintenance engineer has only to note that lubrication is being properly maintained and that lubricants most suited to the machinery are being used. This leads to minimum cost of maintenance, fewer headaches for the maintenance engineer, and lower-cost construction.

Conventional tests For lubrication oils such tests include viscosity, flash and fire points, pour points, carbon-residue content, emulsification and demulsibility, acidity or neutralization number, and saponification number.

For greases, the base, penetration, and dropping- or melting-point tests are the most significant.

Viscosity As an indication of the relative fluidity of any lubricating oil viscosity is discussed regardless of the service. Machinery builders develop their lubrication recommendations around viscosity; it is the number one test when purchasing. Viscosity, however, does not denote quality; it indicates simply how the oil will flow at the temperatures under which lubrication must be maintained.

Flash and Fire Points These points are customarily quoted in listing the characteristics of a lubricating oil; but, unless the oil is to be used under very high temperature conditions where vaporization could be a factor, the flash and fire points are of little interest to the construction equipment maintenance engineer.

Pour-Point Test Here is another temperature test that should be considered with respect to the viscosity of an oil. It indicates how fluid an oil will be at very low temperatures. With straight mineral oils (without use of pour-depressant additives or special dewaxing) naphthenic-base oils will show lower pour tests (for the same viscosity) than paraffin-base oils. The pour test is most useful when oils are to be used in circulating systems which may be exposed to low atmospheric temperatures. In small-diameter piping, an oil of inadequate pour test could become so congealed during a cold overnight shutdown as to result in starved lubrication and the need for bearing replacement.

Carbon-Residue Content The carbon-residue content of an oil is a factor in internal combustion engine service. Being an indication of residual matter, it becomes allied with the lubricating ability of the oil. For clean engine performance, it should be as low as possible. With modern detergent and dispersant types of heavy-duty engine oils, engine cleanliness is more positively assured by the quality and additive make-up of the oil than by the carbon-residue content of the base oil. In internal combustion engine maintenance it is very important to watch the cleanliness of the air intake and the water temperature. Dirty air or condensation of moisture in the crankcase can contribute far more to cause a

1-11

dirty engine and need for frequent overhaul than the fractional percent of carbon residue which may be indicated by the laboratory test.

Emulsification and Demulsibility   These are tests which definitely relate to lubricating ability when an oil is to be used in hydraulic operations. Emulsification indicates the tendency of an oil to mix intimately with water to form a more or less stable emulsion. Demulsibility indicates the readiness with which subsequent separation will occur. Best performance of turbine-grade oils occurs when the equipment operates at temperatures of 150 to 160°F or less, and when an oil-reconditioning system is operating to remove water and contaminants.

Neutralization Number   This number is related to acidity. It measures the number of milligrams of potassium hydroxide required to neutralize 1 g of oil, normally less than 0.10. An abnormal rise in oil used in turbine or hydraulic equipment can indicate oil oxidation.

Saponification Number   Saponification is a chemical reaction involving the action of an alkali on a fat or fatty acid; the resultant combination is called a soap. Saponification will rarely occur in a well-refined mineral oil and then only where fatty acids may have resulted because of oxidation or chemical breakdown. In effect, the principle of saponification is the basis for the manufacture of certain greases.

The tendency which a petroleum lubricating oil may have to saponify is determined by neutralization and is measured by the equivalent amount of caustic potash required to react with or neutralize 1 g of oil under test. In terms of milligrams of caustic the resultant figure is called the saponification number of the oil. Obviously this should be as low as possible. Increase in the tendency toward saponfication may have a like effect upon the tendency toward oxidation and gum formation in the oil.

Lubricating-grease characteristics—tests

Base.   Lubricating greases are classified broadly according to the type of soap used in their manufacture. The more conventional products include the calcium (lime) base or general-purpose greases usable to around 160°F, sodium base products usable up to at least 250°F, and soda-lime (mixed base) greases which are so widely used for service where the combined features of their respective elements are enhanced by inclusion of antioxidation additives.

In addition there are the more recently developed multipurpose greases of primarily lithium, barium, or strontium base, as well as the aluminum and lead soap products required along with high-temperature durability up to 275°F (plus).

Among the nonpetroleum materials which are applicable to grease service are the bentones, which can be compounded with certain petroleum products, and molybdenum disulfide, which has excellent high-temperature stability, good film tenacity, and a low coefficient of friction. Table 3-1 shows in detail the conventional types of lubricating greases, their characteristics and, as will be noted, the base of each.

Penetration   This word is used to describe the consistency of a grease and to some extent its texture. The construction equipment maintenance engineer is concerned with grease consistency because the grease must continually reach the parts to be lubricated. In a pressure system serving a considerable number of bearings, a grease too heavy for the diameter or pumping ability of the pump could cause clogging of the lines and, again, inadequate lubrication, which could mean only parts renewal later.

Dropping or Melting Point   The melting point is a temperature measurement that indicates the tendency of a grease to soften with increase in temperature. The percentage and type of soap used, the viscosity of the mineral oil, and the type of alkali affect the dropping point of the finished grease. As a rule sodium- and lithium-soap greases show higher dropping points than those of calcium base. To some extent this indicates that the former are better suited for higher-temperature equipment, although the dropping point should not be assumed to indicate the maximum usable temperature.

TYPES OF LUBRICANTS

Petroleum lubricants are broadly classified according to the service for which they are most widely used. Some are virtually specialties; others can be successfully applied to such a wide variety of equipment as to become multipurpose in nature.

The maintenance engineer is interested in the following classifications:
  Circulating oils
  Gear oils
  Machine or engine oils
  Steam-cylinder oils
  Wire-rope lubricants
  Greases of calcium, sodium, aluminum, lithium, or barium base
  Synthetic and solid lubricants

TABLE 3-1 Classification Chart for Greases

Worked penetration range at 77°F by ASTM Method D 217	NLGI° consistency No.	Nature of grease according to consistency	Typical means of application	Typical types of grease within adopted ranges
400–430	00	Semifluid	Packed in special felt bag of good porosity	Nonmelting inorganic base containing MoS_2
355–385	0	Semifluid	Brush	Sodium or calcium
310–340	1	Very soft	Pin-type cup	Sodium, calcium, or aluminum
265–295	2	Soft	Pressure gun or centralized pressure system	Sodium, calcium, lithium, or mixed (sodium-calcium); or inorganic thickened
220–250	3	Light cup		
175–205	4	Medium cup		
130–160	5	Heavy cup	Pressure gun or by hand	Sodium or calcium
85–115	6	Block type	Hand applied, cut to fit lubricator or bearing pocket	Sodium

° National Lubricating Grease Institute.

NOTE: In addition to these conventional products, there are those multipurpose greases prepared with inorganic thickeners to impart distinctive temperature-consistency (nonmelting) relationship, dependable mechanical stability, marked load-carrying ability, excellent water resistance; usable up to 350° plus. A. Gordon Brewer and P. C. Jarvis, "Inorganic Thickened Greases in the Steel Industry," Shell Oil Company, discussed at the AISE Annual Meeting in 1963, published in *Iron and Steel Engineer*, March 1964.

Circulating oils These are probably the highest-quality lubricants available today. They are obtained over a comparatively wide range of viscosities, that is, from about 21 to 550 centistokes or from 100 to 2500 Saybolt Seconds Universal (SSU) viscosity at 100°F. In this category are included steam-turbine-grade oils, hydraulic oils, heavy-duty internal combustion engine oils.

Circulating oils may have a paraffin or naphthenic base according to equipment. For turbine, hydraulic, and similar equipment, the former predominates. Either naphthenic- or paraffin-base oils are used for heavy-duty engine service. The viscosity range at 100°F is given in Table 3-2.

Circulating oils contain additives. Turbine and hydraulic oils are fortified to enable them to resist oxidation and to retard rusting in the system; they also usually contain a foam dispersant.

The modern heavy-duty internal combustion engine oil is specifically refined to function under high engine temperatures and bearing loads. These oils are highly resistant to oxidation and are fortified with detergent and dispersant additives.

Gear oils These may be straight mineral oils of widely varying viscosity or compounded oils containing extreme-pressure additives to improve the film strength and load-carrying ability.

Straight mineral gear oils range normally from SAE 80 to 250. The lower viscosity grades are used for low-temperature equipment; the heavier grades, SAE 140 or 250, are selected for equipment which will normally range above 100°F.

Where gears of the above type are exposed and bath or hand lubricated, the viscosity of the lubricant must be increased to enable the film to resist throwoff. Exposed gears generally do not run too fast, but, because of their location, they may be exposed to wide temperature ranges such as in the swing gear on a power shovel. For such gears a straight mineral residual petroleum product is used which may be compounded with a small percentage of pine tar to improve the adhesiveness. Some such lubricants are cut back with solvents to facilitate application. These thinners later evaporate from the film, but, as some solvents may be flammable, it is well not to use such lubricants in enclosed areas. The question of toxicity also is important.

TABLE 3-2 Viscosity Range of Circulating Oils

Equipment	Viscosity at 100°F	
	Centistokes	SSU
Steam turbine:		
Direct-connected	32–40	150–185
Geared	65–110	300–500
Hydraulic:		
Light-service	21–54	100–250
Machine tools	30–121	140–550
Heavy-duty	Up to 154	Up to 700
Internal combustion engines:		
Heavy-duty SAE 30 to SAE 60	110–370	500–1700

Machine or engine oils The straight mineral red oils come under this classification. They came into usage for general lubrication of external operating parts of engines, pumps, compressors, and general equipment when unit lubrication by oil can or oil cup was practiced. Later they were adapted to ring oilers, but on modern equipment the higher-quality turbine-grade oils are used.

The average so-called machine or engine oil is a good lubricant for once-through lubrication. However, since the resistance to oxidation is lower than that of modern premium-grade oils, ordinary machine oils are not recommended for equipment where formation of sludge or gummy residues could add to the troubles of the maintenance engineer.

Steam-cylinder oils The necessity for lubricating steam cylinders with something more dependable than the time-honored tallow pot became evident when steam engines were operated on high-pressure steam and when multistaging or expansion was adopted. By that time the petroleum industry had perfected methods of refinement and residual lubricating stocks were available so that compounding with a small percentage of fatty (animal) oil such as lard oil, tallow, or wool fat to improve the wetting ability of the finished oil became standard practice. The principle remains the same today, although the petroleum chemist has isolated certain base stocks such as bright stock and the fire stocks dewaxed more or less. These, together with steam-refined stock, make available a variety of products for compounding according to the nature of the steam (i.e., pressure, temperature, moisture content), the utilization of the exhaust, and whether or not very rapid atomization is desired.

Being residual in nature, steam-cylinder oils are necessarily of higher viscosity than distilled oils such as turbine oils. They can be grouped into three broad classifications according to viscosity:

    Light—100 to 120 SSU at 210°F
    Medium—120 to 150 SSU at 210°F
    Heavy—150 and above SSU at 210°F

Steam-cylinder oils are used by injection into the steam line by a hydrostatic or mechanical forced lubricator. When a suitable injection quill is installed in the line, the steam atomizes the drops of oil as they pass onto the quill and thereby carry a so-called fog or mist of oil to all parts of the cylinder walls, pistons, valves, and valve seats.

Wire-rope lubricants Lubrication of wire rope has undergone quite a transition. At one time it was felt that the best protection resulted from using a comparatively heavy residual-type lubricant similar to a heavy gear lubricant. Today the idea of using lighter-bodied oils is popular, applying them by spray to assure better penetration between the rope strands. For this purpose a specially prepared fluid lubricant of about 600 SSU viscosity is adaptable. Inclusion of a small percentage of pine tar gives it added stickiness and also penetrative ability. This type of lubricant is especially suited for wire rope which must be exposed to the weather and to low temperatures as on aerial tramways.

For less severe service a somewhat heavier straight mineral product is satisfactory; this type of lubricant can be applied by drip, brush, or split box.

The construction equipment maintenance engineer is particularly concerned with good wire-rope lubrication because strand breakage can require removing the rope from service and installing a new rope. Safety precautions require rigid inspection.

Greases The American Society for Testing and Materials (ASTM) defines a lubricating grease as "a combination of a petroleum product, and a soap or mixture of soaps, suitable for certain lubrication applications." The metal used in making the metallic soap constituent of a grease denotes its base, for example, calcium, sodium, aluminum, lithium, or barium. In addition a mixture of calcium plus sodium produces what is called a mixed-base grease. Table 3-1 indicates the features and serviceability of modern greases.

Greases are further identified by the type and viscosity of the petroleum oil used in their make-up, by their degree of plasticity, and by their dropping or melting point. A combination of these factors scientifically worked out will produce a grease of remarkable stability and endurance over a wide range of temperature conditions. A multipurpose grease is the ideal, as it reduces the possibility of misapplication and is a factor in storage.

There used to be a popular conception that greases were chosen for a job when a fluid oil could not be retained because of the housing or inadequacy of seals. Modern design has relegated this idea to the past. Today, precision manufacture is so fine and seals are so perfect that lifetime lubrication by just a few grams of grease in a bearing is practicable. Furthermore, greases are available which will function over very wide temperature ranges. But when any such bearing is to be disassembled from the other parts of a piece of equipment during overhaul procedure, the sealed ball or roller bearings must be carefully handled. An effective seal is effective only as long as it is not abused. Careless handling or soaking in solvents may lead subsequently to entry of abrasive dust.

In maintenance work the protection of a grease lubrication system is just as important as protection of the bearings or other parts being lubricated. Careless use of tools around fittings, control outlets, lengths of pipe, or inadvertent striking while moving a beam or scaffold might render one or more outlets inoperative, because of stricture or grease leakage.

Synthetic and solid lubricants Names and characteristics of nonpetroleum lubricants are given in Tables 3-3 and 3-4.

ADDITIVES

Additives serve a variety of useful purposes; accordingly, there are a variety of additives.

Pour-point depressants These depressants are added to lubricating oils to enable them to flow at and be usable at lower temperatures than would be possible with the base oil. Such additives retard or change the action of formation of wax crystals at low temperatures so that they do not interfere with the fluidity of the oil.

Viscosity-index improvers Viscosity-index improvers are added to an oil to improve the viscosity-temperature relationship or, in other words, to obtain as little change in viscosity as possible over the expected service-temperature range.

Antioxidants Antioxidants, often called *inhibitors*, are widely used to fortify steam-turbine, hydraulic, and circulating oils against oxidation when subject to oxidizing conditions. Oxidation of lubricants vitally concerns the engineer because the resulting gums and sludges generally call for considerable expensive equipment overhaul and cleaning of the working parts.

TABLE 3-3 Characteristics of Synthetic Lubricants*

	Low-temperature flow properties	Inhibitor susceptibility	Resistance to oxidation	Water solubility	Thermal stability	Lubricating ability	Viscosity index
Hydrocarbons	Fair to 0°F, although some are OK to −40°F	Poor to good according to type	Fair to good according to type	Immiscible	Poor to good according to type	Comparable with equivalent petroleum oils	Slightly above petroleum paraffin-base oils
Polyalkylene oxides and polyethers (glycols)	Generally good to −40°F	Generally fair	Low	Mostly water-soluble to insoluble	Suitable for high-temperature work	Good where high-temperature usage requires vaporization with minimum residue	Up to 150
Esters	Generally good to −70°F	Good	Fair	Mostly immiscible	Have high flash point		Up to 150
Silicones	Good to −70°F	Poor	Good up to 390°F; above this they oxidize rapidly	Immiscible	Good	Most effective only when one surface is nonferrous. Additives improve	High
Fluorocarbons	Fair	....	Excellent	Immiscible	Excellent but volatility may be high	Not too well known	Below 100

*From Allen F. Brewer, "Effective Lubrication," Robert E. Krieger Publishing Co., Inc, Huntington, N.Y., 1973.

TABLE 3-4 Solid Lubricants*

Product	Name and characteristics	Lubrication service adaptability
Bentones	Produced by reacting hydrous magnesium aluminum silicate (montmorillonite) or bentone clay with an ammonium salt	Effective in compound with petroleum greases. Prepared by a gelling process. No soap is involved
	Features are stability at high temperatures, water resistance; do not liquefy	Well suited for high-temperature service and extreme water conditions
Boron nitride	Sometimes referred to as white graphite owing to unctuous nature. Produced as a light fluffy white powder in an arc furnace at very high temperature. Has excellent stability at high temperatures. Insoluble in water but decomposes in most acids. Has disadvantage of inability (by itself) to adhere to metal surfaces	Usable as a component with silicone-type lubricants in range of 5 to 25% concentration. Carrier serves chiefly to carry the boron nitride to the surfaces to be lubricated. Film of this material is very durable at high temperatures even after carrier has been dissipated
Fuller's earth	Silica base—finely divided	Can be used dry or mixed with water, light oil, or grease
		Effective in retarding fretting corrosion. High-temperature resistant up to around 700°F
Graphite	Produced from coke or anthracite coal. Milled to obtain colloidal graphite usable for lubrication	Can be used dry or mixed with oil or grease. Its chemical inertness enables its use where high thermal stability is required. Maximum usable temperature is around 1500°F. Not too effective in preventing corrosion when used dry
	The flake nature in form of sheets piled on top of each other imparts the lubricating effect as these sheets slide over one another in motion	
Molybdenum disulfide (MoS_2)	Stable at high temperatures. Good film tenacity. Low coefficient of friction	Effective in reducing friction at high sliding velocities. May be mixed with a solvent for application to parts to be lubricated. To obtain best results from a chemically active lubricant of this type, the metal surfaces should be *clean*
Mica	A natural mineral which is ground very finely	Can be used like talc as a lapping material to obtain high surface finish of machine parts. Sometimes added as a filler or thickener in certain lubricants
Talc	Powdered soapstone	Suitable as a lapping material for finishing or working in machine parts
Zinc oxide (ZnO_2)	White in color. Particle size very small—requires no milling. Has low coefficient of friction	Usable as a component with mineral oil for lubrication of parts where perishable products are being produced as in food handling and meat processing

*From Allen F. Brewer, "Effective Lubrication," Robert E. Krieger Publishing Company, Inc., Huntington, N.Y., 1973.

Foam depressants These additives are useful in turbine and circulating oils to prevent foaming when they are agitated with air; foam depressants also accelerate foam dispersion when it has once formed.

Anticorrosion additives and rust preventives These serve a very useful purpose especially when added to circulating oils and to some types of greases. They retard metal corrosion and rusting when the surfaces are exposed to moist air or to water.

Extreme-pressure additives Extreme-pressure (EP) additives are most widely known probably because they are so closely associated with the automotive hypoid gear. An extreme-pressure additive is a chemical compound which increases the load-carrying ability of the lubricating film when subjected to high rubbing speeds. Extreme-pressure additives reduce friction and wear. When metal-to-metal contact occurs between meshing gear teeth, so-called spot welding develops between the surfaces under the extremely high spot temperatures which prevail. This friction can lead to serious tearing away of surface metal and to serious malfunctioning of the gear set. Then a maintenance problem results, usually calling for gear replacement. An extreme-pressure additive in the gear oil imposes a sufficiently protective and easily sheared lubricating film between the teeth.

Engine-cleanliness additives Detergents and dispersants are included in this category. They are used in the make-up of modern heavy-duty motor and engine oils.

Detergents are cleaners. They ensure most satisfactory performance of a circulating oil by preventing residual nonlubricating matter, such as sludges which result from oil decomposition and fuel combustion, from accumulating around piston rings, in bearing clearances, and elsewhere on the engine parts. Regardless of the original degree of purity of the oil, ultimately it will get dirty because of entry of road dust via the crankcase breather or air cleaner, condensation water, and the natural results of service under high temperatures. The oil filter removes some of these contaminants, but some will remain to develop the sticky, gummy substances which ultimately bake onto the hot metal surfaces. A suitable detergent in the oil prevents this buildup by virtue of its continual dissolving and cleansing action. *Dispersants* are companion additives to assist in this function. A dispersant is included to keep these finely divided insoluble materials dispersed in the oil until drain-out. No harm to the engine results from this action because after filtration the dispersed material is virtually nonabrasive.

CONSTRUCTION EQUIPMENT LUBRICATION MAINTENANCE

While lubrication can retard wear, it cannot entirely prevent it. Wear can result from dust contamination or failure of the lubricating system to maintain a proper film on moving parts. Furthermore, the load on a lubricant under severe operating conditions is equally heavy on the equipment. Adequate maintenance and proper design, however, can control the effects of these loads.

A general inspection should be made once a month. At every third such inspection, the foreman should invite an engineer from the lubricant supplier to accompany the maintenance people on their tour. By working together they can determine where the use of an improved bearing or lubricating device would make the equipment run more smoothly, longer, and more economically; where a more suitable grade of lubricant would reduce the frequency of lubrication; or where housing of overexposed gears would improve safety, reduce fire hazards, and cut the cost of removing dripped lubricant.

Equipment operators have a responsibility in equipment lubrication, too. They should not squirt oil carelessly or turn down grease cups with a wrench. Critical parts may be lubricated, but lubricant is wasted and bearing seals may be broken.

Lubricant Protection

Protection of lubricating oils and greases in service is just as important as selecting products with the right characteristics. Protection is a requirement which is often neglected. Premium-grade products are purchased and then stored in a dirty storeroom or even out-of-doors, drums on end, to accumulate water and dirt. Subsequently, it is almost impossible to draw oil or scoop out grease from such a container without some contamination. Obviously, the answer is to provide a special location *indoors* for storing lubricants and to plan a definite schedule for taking stock, refilling containers or lubricating systems, and cleaning, with assigned personnel responsible for the schedule.

Contractors are taking lubricant protection more and more seriously. Many companies have experienced lubrication engineers on their operating staffs. The lubricating-equipment people, in turn, have developed devices and procedures for handling and distributing lubricants which are equally progressive.

Location and personnel A clean, well-lighted room or building is advisable with provisions for heating in cold weather. It should be specifically kept for lubricant storage and reserve lubricating equipment. In this way the responsibility for cleanliness and proper location of lubricant containers can be assigned to one or two individuals who, in reality, become assistant lubrication engineers. They can be trained by the maintenance engineer to appreciate the problem should a bearing or gear set fail as a result of contaminated lubricant. Likewise, they can be schooled in appreciating the value of quality lubricants and the reasons why such products are virtually specialties for the service to be performed.

Fire protection The possibility of fire in a well-planned lubricant-storage area is remote, assuming that *no smoking* rules are observed, that casual visits from other people are prohibited, that oil drip is prevented or cleaned up promptly, that waste or wiping rags are stored in metal containers and in minimum quantity, that sparking or arcing tools are used only under conditions of good ventilation. Even so, insurance regulations will require installation of suitable fire-extinguishing equipment. The accepted foam-type device for smothering oil fires is best. In a small storeroom one or two hand units may suffice. In a larger area a multiple-gallon foam cart with adequate hose may be required.

Endurance Value of Petroleum Lubricants

The petroleum chemist goes beyond the laboratory procedures in considering tests to predict lubricating or endurance value. He considers those factors which will denote the wetting ability of the product, its film-forming ability, its behavior when exposed to water, and its tendency to form sludge or saponifiable by-products. Service considerations will dictate the relative importance of each of these factors.

Wetting ability The wetting ability of the lubricating film is regarded as its most important function if dependable and protective lubrication is to be maintained. In other words, the extent to which effective lubrication can be expected depends upon the extent to which the lubricating film actually wets the surfaces of the metal parts between which motion is taking place.

Wetting ability as a function of adhesion can be illustrated by wetting steel-strip surfaces with oils which are to be compared. The surface which is wet or coated with a satisfactory oil will retain this film when dipped in water. When a steel strip is coated with an oil of poor wetting ability, the water will displace the oil film. Increase in temperature (use of warm water) hastens this displacement effect.

Surface tension Surface tension in a liquid involves the cohesive action of the component particles. It is related directly to viscosity, temperature, and emulsion-forming tendency. Inasmuch as it is an indication of the relative strength of the lubricating film, higher-viscosity oils can be expected to produce films of greater strength at the same temperature. As the temperature is increased, the surface tension will be reduced.

Interfacial tension Interfacial tension in petroleum oils is affected by oxidation. The compounds formed during this reaction tend to reduce interfacial tension. As these compounds usually have an emulsifying attraction for water, a low interfacial tension may well indicate its presence.

Adhesion Adhesion already has been mentioned as being associated with wetting ability in a lubricating film. If good wetting ability prevails, one may assume that the adhesion property is good. To some extent, however, this will depend on the surface finish of the contact metals and the way in which the lubricant has been refined. Too much polish or surface finish is not conducive to good wetting ability or adhesion. The same holds true for overrefinement of an oil.

Saponfication and emulsification These conditions, as they relate to the lubricating value and utilization of petroleum lubricants, have already been discussed as to test procedure. While saponification is a characteristic which is relatively negligible in petroleum lubricating oils, organic acids may exist or may develop to react with an alkali. An oil with a comparatively high saponification number may be susceptible to emulsification, an effect that would be undesirable in a hydraulic system.

Lubrication Procedures

Lubrication procedure involves the means provided for lubrication, timing according to the nature of the lubricants and the requirements of the construction equipment, training of personnel as to resultant benefits, arrangement of records, and analysis of failures which may be traced to faulty lubrication.

Hand-pressure grease guns Modern hand-pressure grease guns provide application pressures as high as 10,000 to 12,000 psi. Such guns can be used either with or without hose connections, according to the fitting and the location of the part to be lubricated. Pressure can be applied either before or after attaching the gun to the fitting. The usual method of developing pressure in a hand gun is to force a plunger against the grease in the barrel, the stem of the plunger being threaded to enable a screw-down action when the handle is turned.

When pressure is to be developed before attachment of the gun to the fitting, a check valve is installed in the tip. In such a gun the act of attachment opens this valve and permits grease to be forced automatically to the bearing.

The purpose of designing rigid-connection guns with check valves is to eliminate the necessity of relieving the pressure before detaching the gun from the fitting and to enable pressure to be raised before attachment; this eliminates the possibility of twisting off the fitting. Direct connection also reduces the possibility of leaks which might develop in flexible hose.

Power guns Where a considerable amount of equipment is involved with numerous grease fittings, the portable power gun is suggested. It hold up to 100 lb of grease. Some of the latest designs work directly from the grease container.

Smaller power guns develop 2000 to 3000 psi and are powered by a pump handle or lever. Later models are electric- and air-powered.

Timing Timing begins with the establishment of a suitable schedule. The schedule would require study of the equipment, the extent to which its parts are housed or protected to conserve lubricant, the speed of various parts, the possibility of lubrication contamination, and the ability of the lubricant to act as a flushing agent.

Lubrication personnel There is a decided trend to pay more attention to the status of lubrication personnel. Organization and training are key factors in developing a lubrication-minded staff. The idea of a so-called grease monkey must be completely discarded. A lubrication mechanic's job is as important as that of any crane operator, for, without the mechanic, the crane could not continue to operate effectively.

Lubrication failures Lubrication failures are most often caused by one of the following:
1. Unsuitable grade or type of lubricant.
2. The lubrication system is not suited to the design of the equipment.
3. The lubricant is contaminated by dust, dirt, water, or dilution by fuel.
4. A suitable lubrication schedule is not followed.

Chapter **4**

Centralized Lubrication Systems

JAMES J. CALLAHAN
Vice President, Research and Engineering, Trabon Lubricating
Systems, Lubriquip Division, Houdaille Industries, Inc.,
Cleveland, Ohio

As labor and equipment replacement costs continue to rise, more and more heavy-duty vehicle owners are looking for ways to improve equipment performance and reduce maintenance costs. One way to accomplish this is through the installation of centralized lubrication systems.

A centralized lubrication system, properly designed and installed, will meter the proper amount of lubricant to all points, with the exception of such rotating points as universal joints. This measurement will minimize wear, even with heavier loads and poor operating environments, resulting in longer equipment life and less downtime.

A centralized system also minimizes problems of lubricant contamination and allows for more economical use of grease, since each lube point receives only its precise requirement. Lubrication also can be centrally monitored, so that the operator will be alerted if a lubrication point becomes damaged or blocked by dirt. Another frequently overlooked advantage is that personnel safety is increased because it is no longer necessary for an operator to climb over a machine to reach lubrication points (Fig. 4-1).

The basic system A basic centralized lubrication system for mobile equipment can be broken down into three functions: (1) lubricant reservoir and pump actuation, (2) proportioning network, and (3) lubrication points.

Of primary concern in the proper design of a system is the quantity and size of the various lubrication points. First, the equivalent area of each point must be calculated. This value is then multiplied by a film thickness to determine the volumetric requirement of each point. The equivalent area for bearings is determined by the following equations:
Plain bearings

$$A = LD$$

where L = length, in.
 D = diameter, in.
Antifriction bearings

$$A = D^2 R$$

where D = shaft diameter, in.
 R = number of rows

For heavy equipment, the film thickness requirement is 0.001 to 0.002 in. of grease per hour. This value multiplied by the equivalent area will yield the volume of lubricant required by the bearing in 1 hr. By knowing the relative sizes of all bearings, the designer

1-22 Basic Maintenance Technology

can then determine the proportioning network. This consists of feeders which operate progressively to divide the lubricant. Once the master feeder and secondary feeders are selected, the pump can be selected and the reservoir size determined.

Each time the pump is actuated, either by hand or automatically, lubricant is delivered to the system. The master feeder, or first feeder, divides the lubricant into amounts required by the secondary feeder assemblies. These, in turn, serve any number of lubrication points. The pump will develop only the pressure that is needed.

Since a centralized lubrication system works on a progressive piston-displacement basis, every piston must complete its stroke before the next piston can cycle. If any one

Fig. 4-1 Centralized lubrication lines have access to all key points on this track system.

piston in the system fails to complete its cycle, pressure will build up and warn the operator of a malfunction. When high pressure occurs, the blockage can be traced by checking the indicators on the feeder assemblies.

On automatic systems, the power supply of the lubrication pump can be air, hydraulic, or electric. Selection of power take-off is generally worked out with the maintenance supervisor before designing the system in order to be certain that other service areas are not affected.

In order to better illustrate how a typical centralized system works, let's take a look at two specific examples.

Front end loader A centralized lubrication system on a wheel-type articulated front end loader consists of a grease reservoir, an air-operated pump, and a feeder assembly (Fig. 4-2). Special lubrication points, other than those common to most heavy equipment, include bucket hinge pins, cylinder pivots, and lift arm pins. This particular system is very flexible and both small and large bearings can be automatically lubricated by correct selection of feeder piston size.

Both bearing size and manufacturer's suggested lubrication intervals are considered when designing a proportioning network of this type. Special protection also is given to fittings and components that are subjected to severe use.

Heavily loaded points such as the bucket pivot and tilt bearings are given greater lubrication quantities to help keep out dirt and other foreign matter. High-pressure hose or nylon tubing is used where flexing is required. Feeders are mounted on the vehicle in such a way that flex lines are minimized and only short lines to bearings are required. Use of short lines reduces the possibility of line breakage and also helps keep system pressure low.

The lubrication pump is actuated each time the operator depresses the vehicle's brake pedal. Other take-off points may include a compressor governor unloader valve, or in the case of a double-acting hydraulic pump, opposite ends of a hydraulic cylinder. The pump is adjustable, so that lubricant delivery can be changed to suit all applications. Where necessary, a timer can be used in conjunction with a solenoid valve for added control of lube quantity.

Controlling the frequency of pump actuations under varying conditions often is a problem in installations of this type. The physical environment in which a vehicle

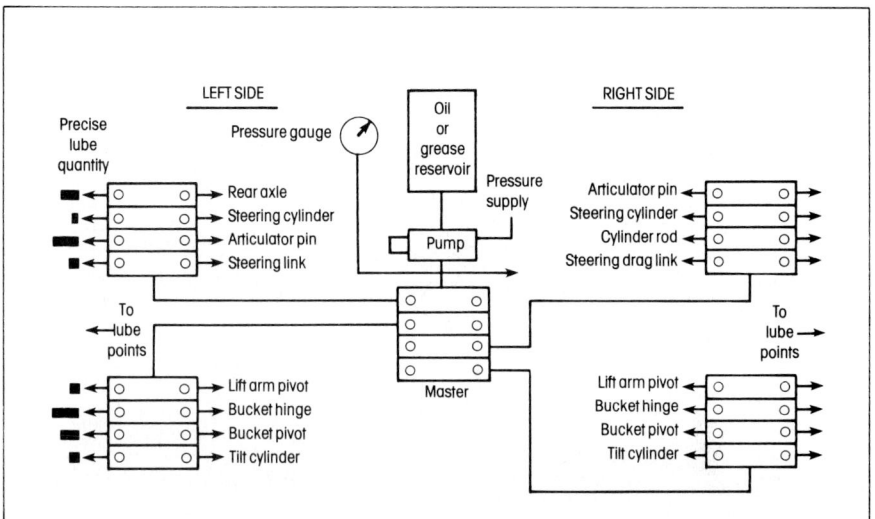

Fig. 4-2 Flow schematic for centralized lubrication system applied to a front end loader.

operates and the operator's habits, such as brake usage and loading procedures, can affect the amount of lubricant being delivered to lubrication points.

A rugged, economical solid state dc program timer has been developed that can be used in conjunction with a solenoid valve to eliminate this problem when it occurs. The time can be adjusted to any one of 12 frequencies per hr—from a minimum of 1 to a maximum of 60 cycles.

By setting the pump at a predetermined output and using the program timer, the designer can precisely control the amount of lubricant used in a given period. This permits an accurate estimate of reservoir-filling frequency based on machine operating hours. The timer was specially designed and tested for heavy-duty mobile equipment.

Feeders on this front end loader are mounted out of the way. However, the master feeder is generally located within sight of the operator so that he can observe the system's performance from the cab. A cycle indicator—an extension of one of the feeder pistons—is included to indicate that the system is working properly. Also, a pressure gauge mounted in the cab alerts the operator when unusually high pressure occurs as a result of a blocked bearing so that he can take corrective action.

The lubricant used in the system should be selected on the basis of prevailing mean ambient temperatures in the geographical area where the vehicle operates.

Large-drill or shovel system Because of the capital outlay required for very large equipment such as a drill or shovel, an automatic, centralized lubrication system is really a necessity. This ensures that all points are lubricated regularly, thereby prolonging the life and improving the performance of the vehicle. The system specified, however, is basically the same as that designed for smaller vehicles. The only difference is the degree of sophistication applied in monitoring and operating the various components.

Since electric power is generally available on larger equipment and because a large

1-24 Basic Maintenance Technology

quantity of lubricant is needed, barrel pumps are generally employed. This type of system can service hundreds of lubrication points on even the largest off-the-road equipment.

The particular system illustrated in Fig. 4-3 is a large Schramm rotary blast-hole drill. The automatic grease system for the main frame, crawlers, proper bearings on the upper works, main machinery, revolving frame, and boom are served by a common grease reservoir. The system provides continuous lubrication to the lubrication points on the crawlers and propel bearings when the machine is being moved. A diverter valve is used to send the lubricant to different master feeders. Once the machine is stationary and the operator begins drilling, a solenoid valve is energized and lubrication is delivered to the

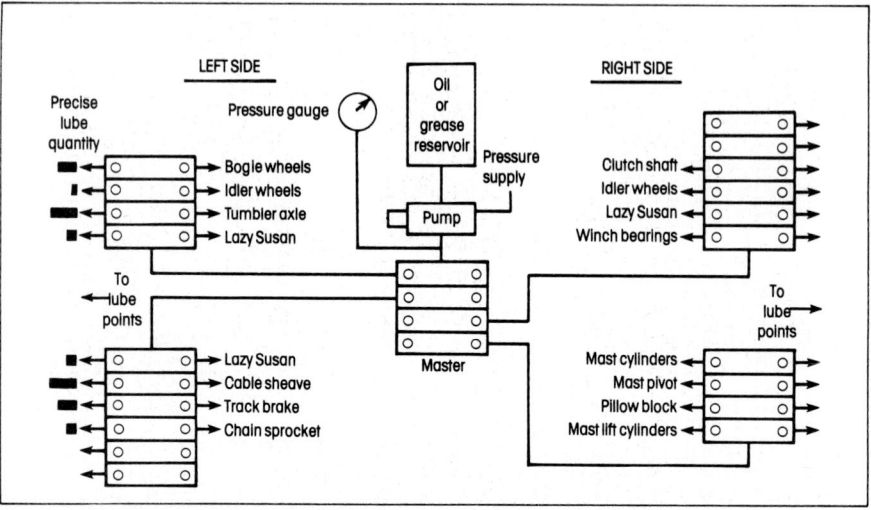

Fig. 4-3 Flow schematic for centralized lubrication system applied to a drill unit.

main-group master feeder. One of the outlets on the master supplies part of the lubricant to the propel-group master, thus keeping the crawlers and other drive points constantly lubricated.

Another microswitch on the master, actuated by piston movement, permits centralized monitoring of system operation. The quantity of grease is controlled by an adjustable timer-counter located in the operator's cab. If, for any reason, the microswitch is not actuated in a predetermined time interval, a red light will go on and warn the operator of a malfunction. The exact location of the difficulty can be traced by using various types of performance indicators. This particular system, therefore, is truly centralized and automatic from the standpoint of both operating and monitoring.

On large equipment of this type, grease is used at the rate of approximately 15 cu in./hr during drilling or digging and at about three times this amount when the machine is moving and lubrication is continuous.

On vehicles with open gears, a separate system that sprays lubricant onto the gear teeth may be installed.

Dollar savings accrue The experience of the G Company with centralized lubrication systems is typical of many heavy-equipment operations. The firm began using automatic lubricators on their mobile equipment in 1963; the results have been impressive.

For example, a 16-yd Auto-Car, used primarily for stocking stone, was equipped with a Trabon system that services 68 lubrication points. All the system's components are in the cab, and it is actuated every time the brakes are applied. The truck uses approximately 2 qt of #90 gear oil per week.

The direct savings in man-hours and materials on this truck, after installation of automatic lubricators, amount to $250 per year. Since the total cost of labor and materials for installing the system was $730, the company reached the breakeven point in only 3 years.

In another instance, a 5-yd 88-B Bucyrus-Erie shovel was equipped with a semiautomatic Trabon lubrication system that services 130 lubrication points. The system consists of three zones: one for the propel mechanism, one for the machinery on the revolving frame, and one for the boom and hook rollers.

Each zone is manually serviced with an air-operated pressure gun, actuated by the shovel air compressor. The gun is held on the master valve until the entire block of feeder valves has cycled the proper number of times. The exact number of cycles is determined by experience. This procedure is then repeated for the other two zones.

This system allows the shovel to be serviced in 15 min as opposed to the usual 70 to 90 min. Direct savings in man-hours and materials amounts to $1375 per year, so that the breakeven point was reached in about 2 years.

As other heavy-equipment companies are discovering, the industry demand for larger, faster, and more heavily loaded pieces of equipment is increasing. As it does, lubrication at the right time and in the correct quantity is becoming more important in order to protect larger investments and to ensure that the equipment will operate up to its rated capacity.

A centralized system offers maintenance personnel a powerful tool in keeping equipment on the job and out of the shop.

Chapter **5**

Arc Welding in Maintenance

J. E. HINKEL
Manager, Application Engineering, The Lincoln Electric Company, Cleveland, Ohio

Among the more important uses of welding in maintenance are repairing and making machinery and equipment. In this respect, welding is an indispensable tool without which operations would soon shut down. Fortunately, welding machines and electrodes have been developed to the point where reliable welding can be accomplished under the most adverse circumstances. Frequently, welding must be done under something less than ideal conditions, and therefore equipment and men for maintenance welding should be the best.

Besides the quick on-the-spot repairs of broken machinery parts, welding offers maintenance a means of making many items needed to meet a particular demand in a required minimum of time. Broken castings, when new ones are no longer available, can be replaced with steel weldments fashioned out of standard shapes and plates. Special machine tools for specific operations can be designed and made for a fraction of the cost that might be needed to buy a standard machine that would have to be adapted to do the job.

The almost infinite variety of this type of welding makes it impossible to do more than suggest what can be done. Figures 5-1 to 5-5 show what the imaginations of some maintenance men have accomplished in this field. As for the welding involved, it should present no particular problems if the welders have the necessary training and background to provide them with a knowledge of the many welding techniques that can be used.

With welding, a maintenance crew can fabricate and erect many of the structures required, even to the extent of making structural steel. Welding can be done either in the maintenance department or at the site. Structures must, of course, be adequately designed to be able to withstand the loads to which they will be subjected. Such loads will vary from those of wind and snow in a simple shed to dynamic loads of several tons where a crane is involved. Materials and joint design must be selected with a knowledge of what each can do. Then the design must be executed with the use of properly trained welders only. Structural welding involves out-of-position work, frequently under awkward conditions, so that a welder must be able to put in good welds under all kinds of conditions.

Standard structural shapes can be used. Frequently pipe makes an excellent structural shape. Scrap materials often can be put to good use. In using scrap, however, it is best to weld with a low-hydrogen E7018-type of electrode, since the analysis of the steel may be unknown and some high-carbon steel may be encountered. The low-hydrogen electrodes will minimize the tendency to crack. This structural scrap frequently comes from old structures, such as elevated railroads being dismantled, which used riveted-quality steel that takes little or no account of the carbon content. Where the quality of the steel is known, an E6010 electrode is used for erection welding. An E7024 or E7014 electrode can be used for fabricating in the shop, if the welding can be done in the flat position. Typical joints that are used in welded structures are shown in Figs. 5-6 to 5-9.

WELDING PROCESSES

Electric-arc welding Electric-arc welding employs the heat of an electric arc to bring metals to be welded to a molten state. In electric-arc welding, the work to be welded is made part of an electric circuit, known as the welding circuit, which has its power source

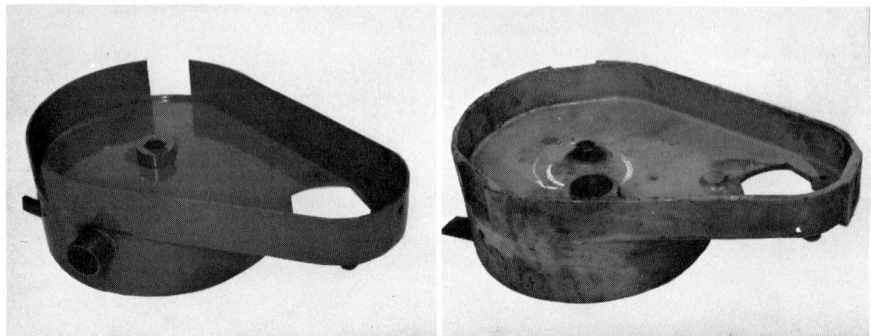

Fig. 5-1 Steel replacement and the cast-iron cover it replaced.

in a welding generator or transformer. One cable carrying current from the power source is attached to the work, and another cable is attached to an electrode holder. An arc is established between the electrode and the work. The arc is moved along the work, melting and fusing the metal as it progresses. Since the arc is one of the hottest commercial sources of heat, this melting takes place almost instantaneously as the arc is applied to the metal.

A variety of welding processes are in common use, employing the electric arc to obtain the welding heat. Each has its particular advantage. All, however, have one problem in

Fig. 5-2 Long delivery prompted welding of this cast-iron punch-press frame.

Arc Welding in Maintenance   1-29

common—that of shielding the arc. Molten steel has a strong affinity for oxygen and nitrogen. If the arc and molten-metal pool are exposed to the atmosphere during welding, the metal will pick up oxygen and nitrogen, forming oxides and nitrides in the weld as it solidifies. These are impurities which will embrittle the weld and thus weaken it.

All the arc-welding processes familiar to the maintenance welder use some method of

Fig. 5-3 Plant-made racks for holding steel.

Fig. 5-4 Typical welding jig and positioner that can be readily fabricated.

shielding the arc and molten pool from the atmosphere, obtaining welds, when correctly made, that are as strong as, or stronger than, the metal being welded. These processes are variations of shielded-metal-arc welding.

Manual and automatic welding   Manual welding, also called hand welding, is welding in which the entire welding operation is performed and controlled by hand. Automatic welding differs from hand welding in that welding equipment mechanically performs the welding operation. The terms *semi* and *fully* are also used to identify automatic welding further in respect to the degree of automation. With semiautomatic welding, the welding

Fig. 5-5   Maintenance-department fabricating trusses. Trusses made from channels and angles. A jig was laid out on plate in the yard.

1-30   Basic Maintenance Technology

equipment is traveled manually along the joint. With fully automatic welding, the welding equipment is traveled mechanically along the joint.

SHIELDED-METAL-ARC WELDING[1]

Shielded-metal-arc welding is by far the most widely used method of arc welding. With this welding method, an electric arc is formed between a consumable metal electrode and the work. The intense heat of the arc, which has been measured at temperatures as high as 13,000°F, melts the electrode and the surface of the work adjacent to the arc. Tiny

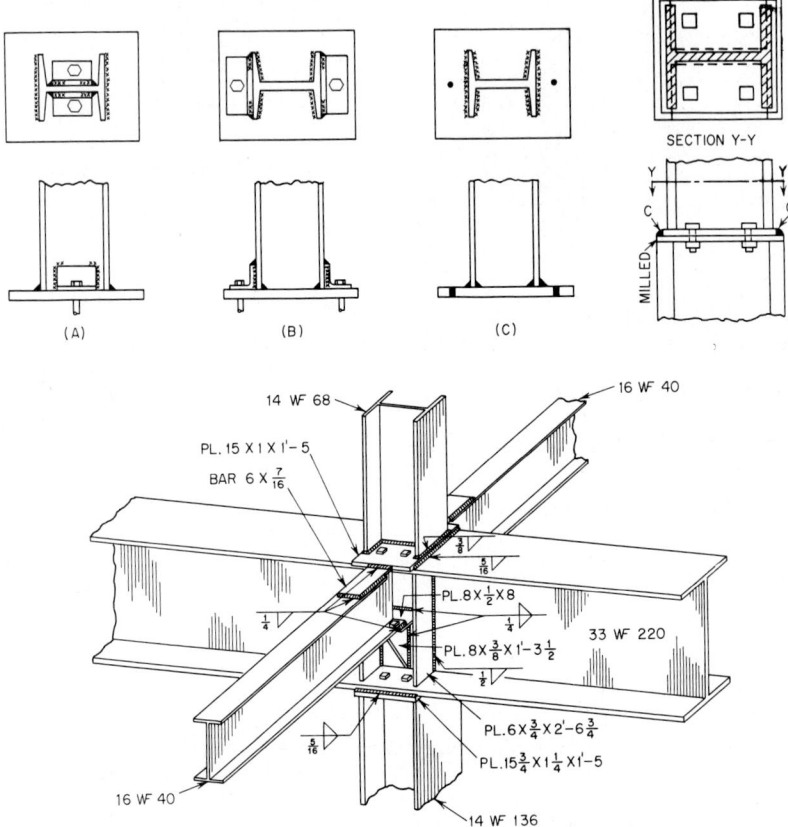

Fig. 5-6 Typical column bases, column splices, and beam-to-column connections that can be used in structural welding.

globules of molten metal rapidly form on the tip of the electrode and transfer through the arc in the *arc stream* into the molten *weld pool* or *weld puddle*, on the work's surface. The actual transfer is induced by the force of gravity, molecular attraction, and surface tension, if the welds are flat or horizontal. Molecular attraction and surface tension are the forces that induce metal transfer from the electrode to the work when the weld is being made in the vertical or overhead position.

[1] Recent welding-process developments and modifications in existing processes are tending to confuse the process classifications established by the American Welding Society. For the most part, the "family grouping" and process name are the same as present American Welding Society designations. In a few instances, however, minor modifications in family-group identification and process name have been made in an attempt to improve clarity and continuity.

In addition to supplying filler metal for the weld deposit, other materials are usually introduced into and/or around the arc; these perform one or all three of the following functions, depending upon the material being welded and the process being used: (1) shielding the arc and preventing atmospheric contamination of the molten metal in the arc stream and the weld puddle; (2) providing scavengers and deoxidizers to protect the molten crater; (3) producing a slag blanket over the very hot but solidified weld. All these functions are necessary to assure the strength and quality of the weld being made.

Self-shielded metal arc welding Electrodes for the shielded metal-arc-welding-process are manufactured by extruding, dipping, or fabricating. The extruded and dipped electrodes, more often referred to as coated, or covered, electrodes, contain the shielding, scavenging, and deoxidizing materials in the covering that surrounds a solid metal core. The fabricated, or cored, electrodes contain the shielding, scavenging, and deoxidizing materials compacted in the electrode core surrounded by a metal sheath. Since both the covered and the fabricated electrodes contain all the materials to accomplish complete arc shielding, they are called self-shieldng electrodes.

The arc-shielding action is essentially the same for both the covered electrodes, as illustrated in Fig. 5-10, and the fabricated electrodes. But the actual method of arc shielding and volume of slag produced will vary with different electrode types.

The bulk of the core or covering materials in some electrodes is converted to a gas by the heat of the arc, and only a small amount of slag is produced. This type of self-shielding electrode, depending largely on a gaseous shield to prevent atmospheric contamination, can be identified by the incomplete or light slag covering of the completed weld.

The other extreme in self-shielding electrode design is the type where the bulk of the covering material is converted into slag in the arc heat with only a small volume of gas being produced. With this type, the tiny globules of metal being transferred in the arc stream are entirely coated with a

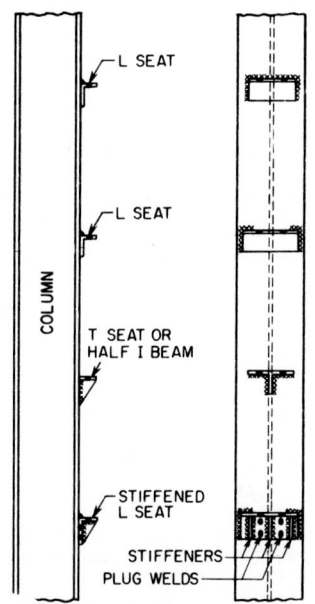

Fig. 5-7 Beam-to-beam framing and methods for seating beams on columns.

thin film of molten slag. This slag floats to the surface of the molten weld puddle before solidifying. The electrodes are identified by the heavy slag deposit that completely covers the surface of the finished weld.

Between these extremes there is a wide variety of electrode types with the ability to produce various combinations of gas and slag shielding. These variations in slag action and arc shielding also influence the performance characteristics of the many different types of self-shielding electrodes available for use in maintenance and manufacturing. For example, an electrode that has a heavy slag action is also one which has a high deposition rate and is suited for making large welds in flat positions. An electrode that develops a gaseous arc shield is one which also has a low deposition rate and smaller molten weld puddle and therefore is suited for making welds in the vertical and overhead positions.

1-32 Basic Maintenance Technology

These and many other performance characteristics are the reasons why one type of self-shielded electrode is preferred over all others for a specific weld in a specific position.

Manual shielded-metal-arc welding Extruded, dipped, and fabricated electrodes are used for manual shielded-metal-arc welding. These electrodes range in length from 9 to 18 in. The consumable welding electrode is placed in a hand-held clamping device called the electrode holder. Welding begins by touching the tip of the electrode to the work to complete the electric welding circuit, then withdrawing the tip, establishing the arc. As the heat of the arc melts the tip of the electrode, the welding operator, called the welder, manually lowers the tip of the electrode, maintaining a uniform distance between it and the work, thereby maintaining a steady arc. Simultaneously, the welder manually moves

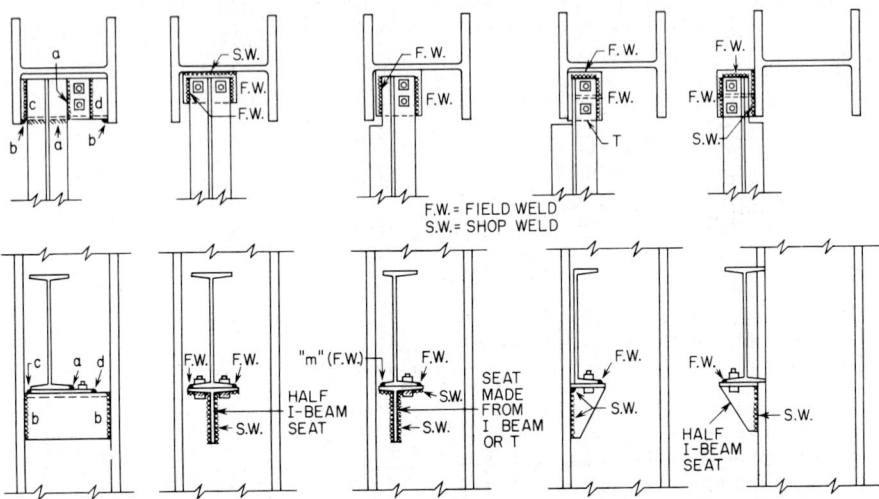

Fig. 5-8 Different ways of connecting beams to columns when an offset is required.

the electrode along the work at a rate of speed that deposits sufficient filler metal to create the needed weld size.

Semiautomatic and fully automatic flux-cored arc welding The electrode used for semiautomatic and fully automatic flux-cored arc welding is mechanically fed through a welding gun or welding jaws into the arc from a continuously wound coil that weighs approximately 50 lb. Only the fabricated flux-cored electrodes are suited for this method of welding, since coiling the extruded flux-cored electrodes would damage the coating. In addition, metal-to-metal contact at the electrode's surface is necessary to transfer the welding current from the welding gun or welding jaws into the welding electrode. This is impossible if the electrode is covered.

Typical applications of semiautomatic and fully automatic equipment for flux-cored arc welding are shown in Figs. 5-11 and 5-12. For a given cross-sectional area of electrode wire, much higher welding amperage can be applied with semiautomatic and fully automatic welding. This is because the current travels only a very short distance along the bare metal electrode, since contact between the current-carrying jaws and the bare metal electrode occurs close to the arc. In hand welding, the welding current must travel the entire length of the electrode, and the amount of current is limited to the current-carrying capacity of the wire. The higher currents used with automatic welding result in a high weld-metal deposition rate. This increases welding speed, reduces welding time, and lowers welding costs.

Submerged-arc welding With submerged-arc welding the arc is completely hidden under a small mound of granular inorganic flux which is automatically deposited around the electrode wire as it is fed to the work (Fig. 5-13). The arc and molten pool are completely blanketed with flux at all times, and there are no visible arc rays or weld spatter.

Under usual welding conditions, the quantity of flux melted weighs approximately the same as the electrode consumed. The unfused flux may be collected and reused. Precautions should be taken to keep the flux and the work clean in order to prevent weld contamination and to maintain weld quality.

The high currents used with submerged-arc welding also develop a deep-penetrating-arc characteristic. Consequently, no groove or a small groove may be used, depending on the thickness of the base metal, with correspondingly smaller additions of filler metal. For example, no chamfering is necessary for two-pass butt joints in steel up to ⅝ in. thick. Complete penetration can also be obtained in fillet welds for material up to ¾ in. thick without chamfering. For joints in thicker material, a double V-groove weld is used. The graph of Fig. 5-14 shows typical relations between penetration and applied current.

With submerged-arc welding, distortion is minimized because of high welding speeds, minimum number of passes, and efficient application of heat. This means that less heat is applied to the weld area and, furthermore, that the heat is applied more uniformly than with hand welding. Distortion caused by an unbalanced heat condition, as in single-groove multiple-pass welded joints, can be corrected by presetting the base-metal parts to offset angular movement. The other methods of controlling distortion, discussed later in this chapter, can also be applied.

Although the submerged-arc-welding process is used primarily for production welding, it also has potential maintenance use which even to this day has been only partially exploited. The process is particularly suited to rebuilding worn surfaces and developing abrasive-resistant surfaces for manufacturing operations encountering severe metal-erosion problems.

Gas-shielded arc welding In gas-shielded arc welding, the arc and weld region is shielded from the air by a protective gas. This gas may or may not be inert. The gases experiencing greatest industrial use are argon, helium, and CO_2. Two variations of the gas-shielded arc-welding process are gas tungsten-arc welding and gas metal-arc welding.

Gas-shielded tungsten-arc welding Gas-shielded tungsten-arc welding with an inert gas was originally developed to weld the corrosion-resistant and other difficult-to-weld metals such as aluminum and copper. Over a period of years, however, its application has expanded to include welding and surfacing operations on practically all commerical metals.

The gas tungsten-arc-welding process obtains the necessary heat for welding by a very intense electric arc which is struck between a virtually nonconsumable tungsten electrode and the metal workpiece (Fig. 5-15). On joints where filler metal is required, a welding rod is fed into the weld zone and melted with the base metal in the

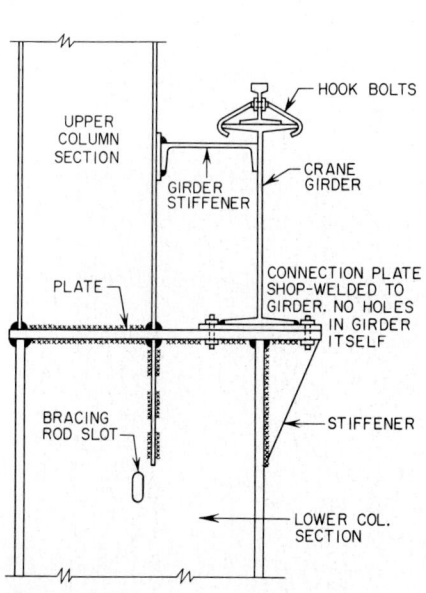

Fig. 5-9 A beam-and-girder connection and a column detail showing craneway.

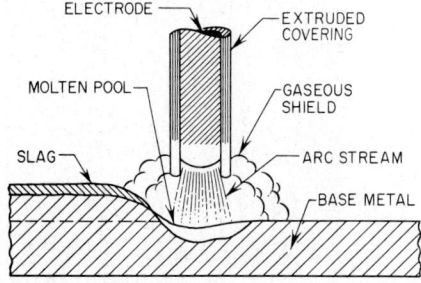

Fig. 5-10 Shielded-metal-arc-welding process.

manner used with oxyacetylene welding. The weld zone is shielded from the atmosphere by an inert gas fed through the welding torch. Either argon or helium may be used.

Inert-gas-shielded tungsten-arc welds, because of 100 percent protection from the atmosphere, are stronger, more ductile, and more corrosion-resistant than welds made with ordinary arc-welding processes. Corrosion due to flux entrapment does not occur,

Fig. 5-11 Fully automatic flux-cored arc welding.

and postwelding cleaning operations are reduced to a minimum. The entire welding action takes place practically without spatter, sparks, or fumes. Fusion welds can be made in nearly all metals used industrially. These include aluminum alloys, stainless steel, magnesium alloys, nickel and nickel-base alloys, copper, silicon copper, copper nickel, brasses, silver, phosphor bronze, plain-carbon and low-alloy steels, cast iron, and others.

Fig. 5-12 Semiautomatic flux-cored arc welding.

Arc Welding in Maintenance 1-35

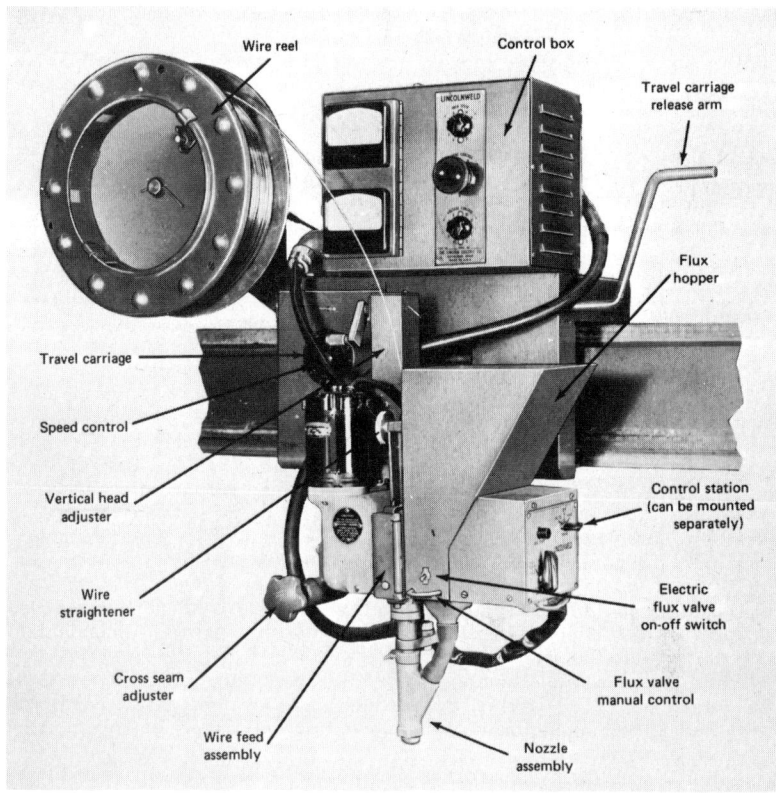

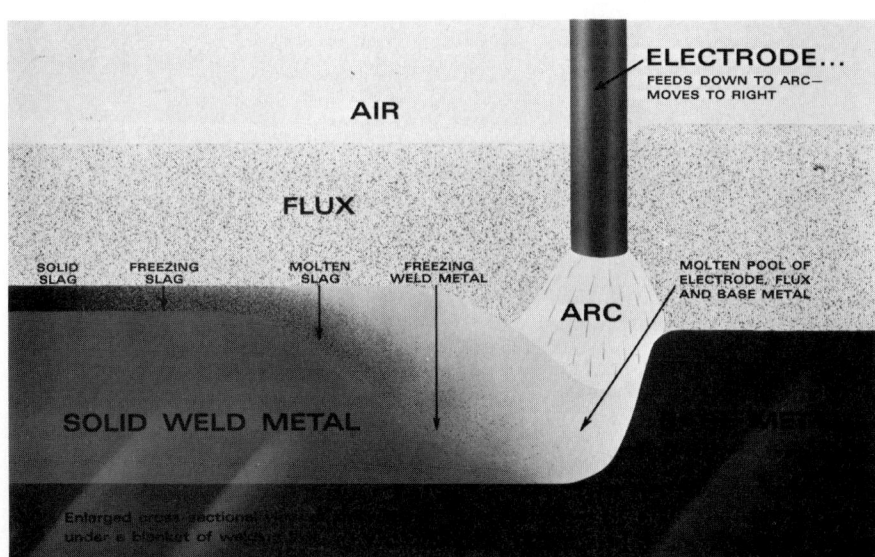

Fig. 5-13 Elements of the submerged-arc-welding process.

The process is also widely used for welding various combinations of dissimilar metals and for applying hard-facing and surfacing materials to steel.

The power supply for inert-gas-shielded tungesten-arc welding may be either alternating or direct current. However, certain distinctive weld characteristics obtained with each type often make one or the other better suited for a specific application.

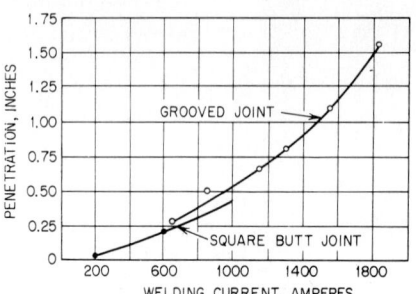

Fig. 5-14 Penetration versus applied current for submerged-arc-welding.

In dc welding, the welding-current circuit may be hooked up as either straight polarity or reverse polarity. The connection for dc *straight-polarity* (DCSP) welding is electrode negative and work positive. In other words, the electrons flow from the electrode to the plate or workpiece, as shown in Fig. 5-16. For dc *reverse-polarity* welding (DCRP), the connections are just the opposite; electrons flow from the plate to the electrode, as shown in Fig. 5-17.

In straight-polarity welding, the electrons hitting the plate at high velocity exert a considerable heating effect on the plate. In reverse-polarity welding, just the opposite occurs; the electrode acquires this extra heat, which then tends to melt the end of the electrode. Thus, for any given welding current, DCRP requires a larger-diameter electrode than does DCSP. For example, a $\frac{1}{16}$-in.-diameter pure-tungsten electrode can handle 125 amp of welding current under straight-polarity conditions. If the polarity were reversed, this amount of current would melt off the electrode and contaminate the weld metal. Hence a $\frac{1}{4}$-in.-diameter pure-tungsten electrode is required to handle 125 amp DCRP satisfactorily and safely.

These opposite heating effects influence not only the welding action but also the shape of the weld obtained. DCSP welding will produce a narrow, deep weld; DCRP welding, because of the larger electrode diameter and lower currents generally employed, gives a wide relatively shallow weld.

One other effect of DCRP, the so-called plate-cleaning effect which seems to occur, is

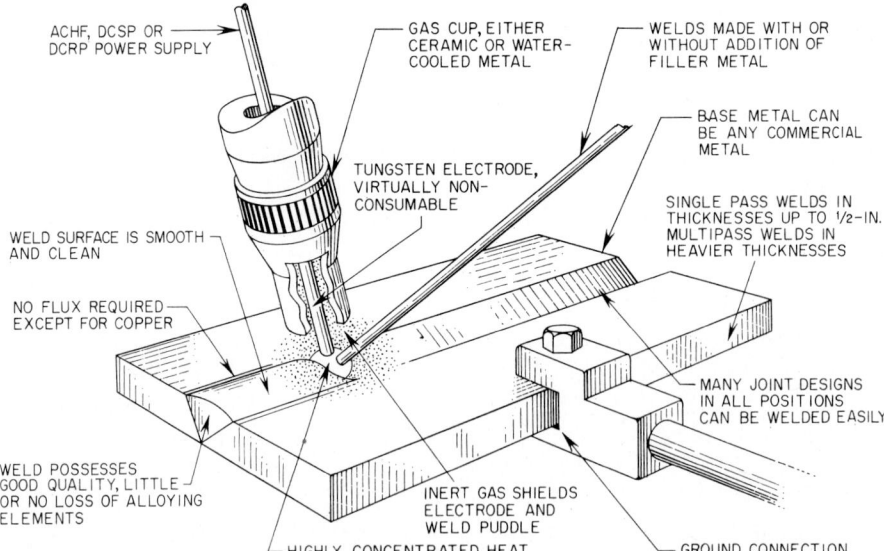

Fig. 5-15 Inert-gas-shielded arc welding with a nonconsumable electrode.

worth mentioning. Although the exact reason for the surface-cleaning action is not known, it seems probable that either the electrons leaving the plate or the gas ions striking the plate tend to break up the surface oxides, scale, and dirt usually present.

Welding with an alternating current is theoretically a combination of DCSP and DCRP welding, since the current flows in one direction and then in the other, or reverse,

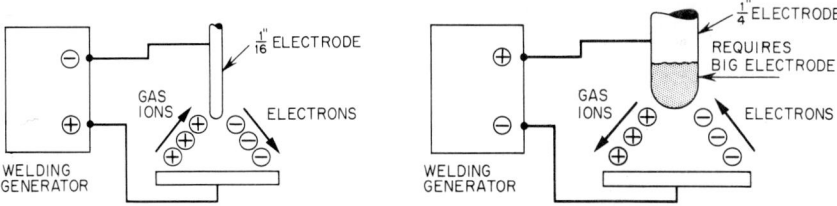

Fig. 5-16 Direct-current straight polarity.   Fig. 5-17 Direct-current reverse polarity.

direction. However, moisture, oxides, and scale on the surface of the plate tend to prevent (partially or completely) the flow of current in the reverse-polarity direction. To ensure proper current flow in the reverse direction when welding with alternating current, it is common practice to introduce into the welding current a high-voltage high-frequency low-power current. This high-frequency current jumps the gap between the electrode and the workpiece and pierces the oxide film, thereby forming a path for the welding current to follow. Superimposing this high-voltage high-frequency current on the welding current gives the following advantages:

1. The arc may be started without touching the electrode to the workpiece.
2. Better arc stability is obtained.
3. A longer arc is possible; this is particularly useful in surfacing and hard-facing operations.
4. Welding electrodes have longer life.
5. The use of wider current ranges for a specific-diameter electrode is possible.

A typical weld contour produced with high-frequency stabilized alternating current is shown in Fig. 5-18, together with DCSP and DCRP welds for comparison.

Tungsten-arc-welding equipment The basic equipment requirement for manual inert-gas tungsten-arc welding is a welding torch plus additional apparatus to supply (1) electric power, (2) argon, and (3) water. Also, certain protective equipment should be employed to protect the operator from the arc rays during welding operations.

The welding current is supplied either by a variable-voltage welding generator or rectifier for dc welding or by a variable-voltage welding transformer for ac welding. When selecting a generator or rectifier, it is important to obtain one which has good current control at the lower end of its current range. This ensures the arc stability required for efficient operation. If you plan to use an older dc welding machine which operates inefficiently in the lower current range, a resistor should be used in the ground line between the generator and workpiece. These resistors are marketed

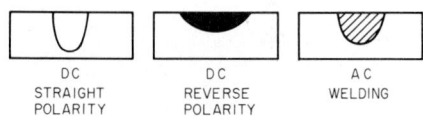

Fig. 5-18 Comparison of weld penetration for the three types of welding current used with inert-gas tungsten-arc welding.

by most manufacturers of dc generating equipment. Several firms manufacture transformers which are special for tungsten-arc welding, some with built-in high-frequency stabilization. Bear in mind that some transformers are designed to produce a balanced wave and can be used at the full rated capacity. Others are not, and should not be used at over 70 percent maximum capacity to avoid overloading the primary. Be certain you know which type you are using. A high-frequency generator used with ac welding can also be obtained from any reputable dealer.

High-purity argon is supplied in steel cylinders, each containing approximately 240 cu ft of argon at a pressure of 2000 psi. A regulator is needed to reduce this pressure to that required for welding, generally about 20 psi. In addition, a flowmeter is required at every

1-38 Basic Maintenance Technology

welding station, since different materials need different flows or amounts of argon for adequate protection. Where a large amount of welding is being done continually, it is advisable to connect a manifold to a bank of cylinders and pipe the argon to each individual work station. Again, a flowmeter is required for each station.

When currents above 130 amp are used, water cooling of the torch and power cable is required. The cooling water for water-cooled torches must be clean; otherwise, restricted or blocked passages may cause excessive overheating and damage to the equipment. Most shops have an adequate supply of cooling water available. However, where welding is done in large shops or outdoor locations, completely self-contained units are available. A typical portable installation is shown in Fig. 5-19.

An inert-gas tungsten-arc-welding torch feeds both the welding current and the inert gas to the weld zone, as shown in Fig. 5-20. The current is fed to the weld zone through the tungsten electrode, which is held firmly in place by the electrode holder. The argon (or helium) is fed to the weld zone through a gas cup at the head of the torch.

The electrode should extend about $1/8$ to $3/16$ in. beyond the end of the gas cup for butt welding and about $1/4$ to $3/8$ in. for fillet welding.

Recommended gas-cup sizes for the various torches and electrode diameters are specified by the manufacturer. Ceramic cups are generally acceptable when the welding current is less than 250 amp. With higher currents or where welding conditions are unusually severe, water-cooled metal gas cups must be used to prevent overheating. Water-cooled cups should never come into contact with the workpiece when the welding current is ON. Conductivity of the hot gases may cause the arc to jump from the electrode to the cup rather than to the workpiece, thus damaging the cup.

As with all industrial equipment, certain common-sense precautions should be

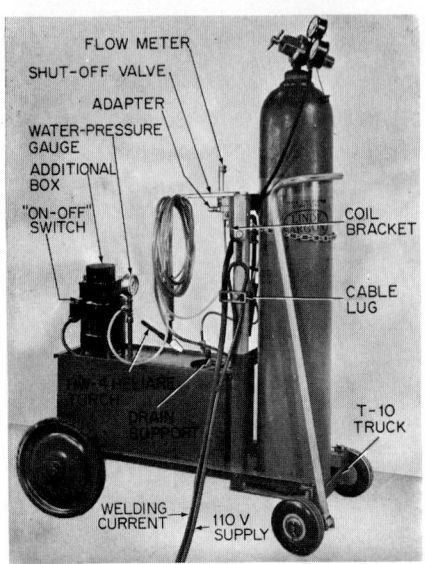

Fig. 5-19 Typical portable installation for inert-gas tungsten-arc welding.

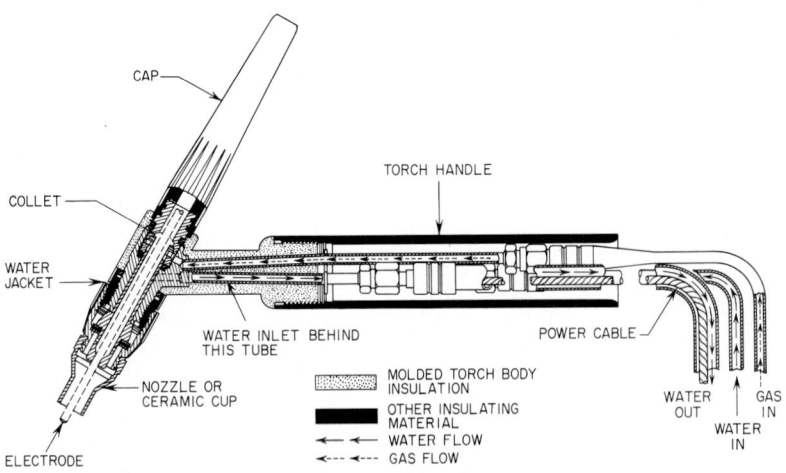

Fig. 5-20 A water-cooled torch for inert-gas tungsten-arc welding.

observed. In the case of tungsten-arc welding, the operator should be properly protected from arc rays. This requires suitable clothing to cover all exposed skin surfaces and a welder's helmet with the proper shade of glass to protect the eyes and face. The shade of the glass lens will depend upon the intensity of the arc. The recommended shades for various current ranges are listed in Table 5-1.

Gas-shielded metal-arc welding   Gas-shielded metal-arc welding is commercially called MIG welding when an inert gas is used. An arc between the consumable wire electrode and the workpiece (Fig. 5-21) is maintained in an atmosphere of inert gases, principally argon. The gases shield the weld zone from possible contamination by the atmosphere and eliminate the need for flux. Quality welds can be produced by either manual or machine welding. Welds made by this process are relatively clean and require little or no postweld finishing.

TABLE 5-1   Lens Shades for Current Ranges

Glass No.	Welding current, amp
6	Up to 30
8	30 to 75
10	75 to 200
12	200 to 400
14	Above 400

With inert-gas-shielded arc welding, you can weld such metals as aluminum, magnesium, copper, nickel, silicon bronze, aluminum bronze, stainless steel, low-alloy steel, and carbon steel. A consumable electrode similar to the metal being welded is used.

The average current density used is about twenty times that recommended for carbon-arc welding and about six times that recommended for covered-metal-arc welding. This high current density results in concentrated heat that produces narrow welds with deep penetration, a small heat-affected zone, and reduced distortion. Conventional dc welding or constant-potential power supplies may be used.

A constant-potential power source is preferred for inert-gas-shielded metal-arc welding with a continuously fed bare-wire electrode. As shown in the accompanying graph (Fig. 5-22) constant potential has a flat volt-ampere characteristic rather than the drooping characteristic of conventional dc power. Since the welding voltage remains essentially constant, the speed of wire feed controls the welding current.

A manual welder for inert-gas-shielded arc welding with a consumable electrode is shown in Fig. 5-23. This particular unit, a portable welder, uses welding currents as high as 500 amp. The electrical control box contains the various control circuits for wire feed, gas flow, and application of welding current. The wire drive unit feeds the consumable wire electrode at the required speed. Once the welding conditions have been set up, the trigger switch on the water-cooled torch stops and starts welding. The remote-control box permits the operator to adjust arc length and to inch out wire electrode for arc striking

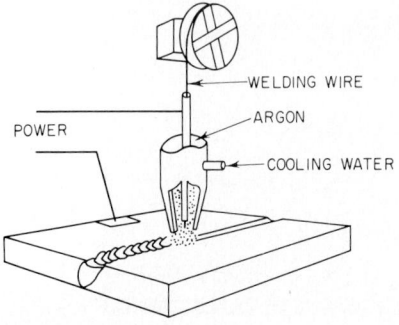

Fig. 5-21   Inert-gas-shielded arc welding with a consumable electrode.

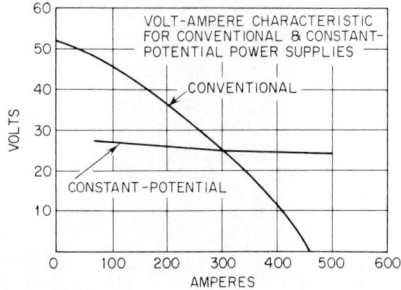

Fig. 5-22   Ampere-volt characteristic of constant-potential and conventional power supplies.

without leaving his welding position. Source of welding current, supply of inert shielding gas (argon, helium, or a mixture of approximately 95 percent argon and 5 percent oxygen), and a supply of cooling water also are required.

The gas-shielded arc-welding processes can successfully weld plain carbon steel, but, in most instances, when compared with other arc-welding processes, cost has prohibited their use.

Gas-shielded metal-arc welding-CO_2 Another version of gas-shielded metal-arc welding uses CO_2 (carbon dioxide) rather than an inert gas to blanket the arc and the surrounding weld area. A typical production welding installation is illustrated by Fig. 5-24. There are two variations of this process. The first uses a solid electrode, the second a fabricated flux-cored electrode. In addition to providing filler metal, these electrodes or the flux contain elements which perform a scavenging and deoxidizing action in the crater to improve weld quality.

The flux-cored-electrode process has the flux within an outer steel sheath.

The gas-shielded metal-arc-welding process with CO_2 is used for production welding of carbon steels and for fabricating industrial piping and sheet metal.

Gas-shielded spot welding This method of welding combines either gas-shielded tungsten-arc-welding or gas-shielded metal-arc-welding equipment with an electrical timing-control system that automatically starts and maintains the arc for a controlled time period. Two lapped pieces of metal are spot-welded together by applying heat from an electric arc to the top surface of the joint. Welding action is controlled by the current input to the arc and the time the arc dwells on the material being welded.

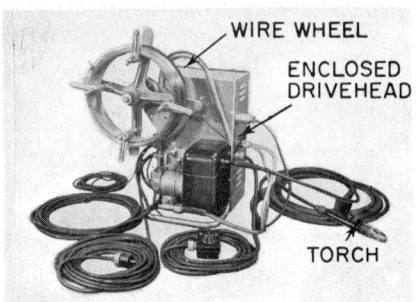

Fig. 5-23 A manual welder for inert-gas-shielded arc welding with a consumable electrode. *(Linde Division, Union Carbide Corporation, New York.)*

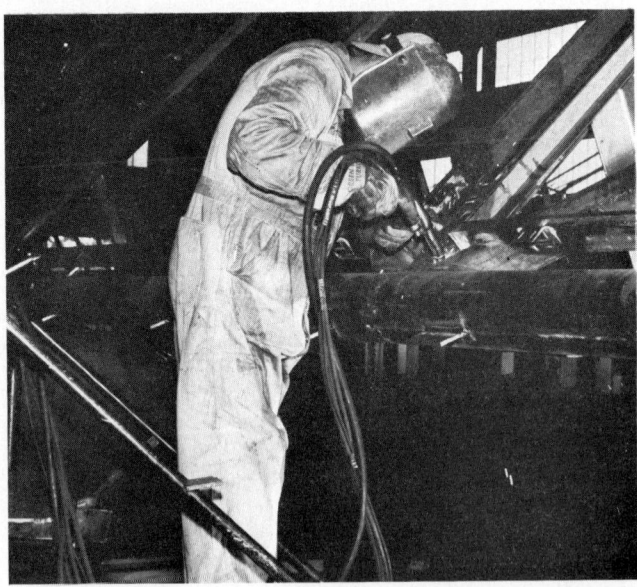

Fig. 5-24 Gas-shielded metal-arc welding.

Shielding of the arc, electrode (consumable metal or nonconsumable tungsten), and fluid weld puddle are similar to those of conventional gas-shielded arc welding.

The resulting spot weld parallels that produced by resistance-welding techniques; however, no electrode pressure is required, and the welding is done from one side of the plate without requiring any weld backup. Both inert-gas and CO_2 spot welding are experiencing expanding industrial use.

RESISTANCE WELDING

Resistance-welding processes are primarily designed for production-welding usage. Nevertheless, a few of the processes can be used effectively by the maintenance department. With this method of welding, the joining of the parts being welded is accomplished by the heat obtained from resistance of the work to the flow of electric current in a circuit of which the work is a part and by the application of pressure.

Fig. 5-25 This resistance spot welder is being used to fabricate sheet metal.

Spot welding Spot welding is the most common resistance-welding process. It is usually employed in the welding of thin metal sheets and is accomplished by placing the sheets between movable copper-alloy electrodes. The electrodes carry the welding current and can be actuated to apply the proper pressure during the welding cycle. A typical production installation in a sheet-metal shop is illustrated by Fig. 5-25. Aluminum presents a special problem because of its high electrical conductivity. So does copper, which has practically the same conductivity as electrode material.

Although maintenance welding departments occasionally have the larger floor-mounted equipment, more often the small portable hand-held spot-welding guns are used for fabricating sheet metal.

OTHER WELDING PROCESSES

The welding processes and equipment described to this point have potential use in the typical maintenance welding department. Many other welding processes are being used which, admittedly, have limited maintenance use. These will be summarized briefly. Additional information about specific processes can be obtained from the American Welding Society (see the references at the end of this chapter).

Atomic-hydrogen welding Atomic-hydrogen welding differs from the other arc-welding methods in that the arc is formed between two tungsten electrodes and the work is not part of the welding circuit. A stream of hydrogen gas is passed through the arc and, in the heat of the arc, changes from molecular to atomic form, giving off an intense heat. The hydrogen acts as an effective heat-transfer medium and results in high heat being applied close to the work. A filler rod is used to supply additional metal to the joint. The process has some advantages in welding thin sheet where a high finish is needed.

Electroslag welding Electroslag welding is the metal-arc-welding process employing the principles of submerged-arc welding. This process involves fusion of the base metal and continuously fed filler metal under a substantial layer of high-temperature, electrically conductive molten flux. By feeding one or a combination of two or three electrodes simultaneously into the arc, plates ranging from 1 to 14 in. thick can be joined in a single pass. Application is generally limited to very heavy weldments. Welds are usually made with the joint vertical and with welding progressing from bottom to top.

Plasma-arc welding Plasma-arc welding exists in several forms. The basic principle is that of an arc or jet created by heating electrically a plasma-forming gas (such as argon with additions of helium or hydrogen) to such a high temperature that its molecules become ionized atoms possessing extremely high energy. When properly controlled, this process results in very high melting temperatures. Plasma-arc welding holds a potential solution to the easier joining of many hard-to-weld materials. When modified for metal cutting, this process achieves unusually high cutting speeds. Another application is the depositing of materials having high melting temperatures to produce surfaces of high resistance to extreme wear, corrosion, or temperature.

Stud welding Stud welding is the end welding of a stud, ordinarily a machine screw, at a particular spot on the work by fusion. An electric arc, struck between the stud serving as the electrode and the baseplate, brings the tip of the stud and the surface of the work adjacent to the stud to a molten state. A light pressure is applied, forcing the stud into the molten weld puddle. Current flow is discontinued, and the stud fuses to the work surface as it cools. A compact unit, called a stud welder, supplies the welding current. The arc may be shielded or unshielded.

Carbon-arc welding Carbon-arc welding employs a carbon rod as an electrode. The arc is formed between the carbon and the work, creating a molten pool on the work surface. This pool is kept molten by playing the arc across it. If extra filler metal is needed to make the weld, it is supplied by introducing a filler rod into the arc, where it is melted into the molten pool. This is a puddling process, and is not applicable to vertical or overhead welding. Shielding may be obtained if desired by introducing a paste, powder, or fibrous flux into the arc.

Carbon-arc welding is used only for specialized applications. The carbon arc is also used for cutting where a precision cut is not necessary or on alloys that cannot be cut by the gas process.

Flash welding Flash welding is a resistance-welding process in which fusion is produced by a high localized heat obtained from the electrical resistance existing between two touching surfaces. This type of resistance is evidenced by a flashing, or shower, of sparks produced by the arcing of current at the joining surfaces. When the temperature of the metal has increased to where the joining surfaces have plasticized, the parts are forced together under pressure to make the weld. A portion of the metal squeezes out (upsets) to form the flash. This must be trimmed off, and the joint then ground or otherwise finished to the section desired.

Percussion welding Percussion welding is a process in which fusion temperature results from an arc created across a gap between two surfaces to be joined, the arc being caused by rapid discharge of electrical energy. A percussive (impact) force is applied during or immediately following the electrical discharge.

Projection welding Projection welding is another method of resistance welding. It differs from those previously described, since it uses projections, or embossments, to localize the current flow and welding heat at predetermined points. These projections, which serve as points of contact, are a part of one or both of the parts to be joined. The parts are supported and pressed together by special dies during welding.

Seam welding Seam welding is fundamentally a spot-welding process. One or two electrode wheels running along a straight line at a fast rate of travel make a series of closely spaced spot welds. When the welded spots are so close that they actually overlap, they form a gastight or watertight seam, as required for a vessel. In other cases, the series of spots may be so spaced that the process becomes a mere tack-welding operation in the assembly of a unit. This is called roll-spot welding and is used to speed up standard spot welding.

Upset welding Upset welding is process in which fusion is produced by the heat obtained from electric resistance through the area of contact of two surfaces held together under pressure. In this case, the force is applied prior to introduction of the electric current and is continued until heating is complete. The continued force produces an upsetting as in flash welding, but since the surfaces are in solid contact with one another, there is no arcing or flashing effect.

Electron-beam welding Electron-beam welding directs a bombardment of electrons at the workpiece placed in a vacuum. Electrons are admitted from a filament, acting as a type of nonconsumable electrode, and are highly accelerated by high-voltage potential between the electrode and the work. The high-velocity energy of the electrons converts to heat when they strike the work. The electron flow is electrically concentrated into a beam by means of an electron gun. Since the operation is carried on in a vacuum, the process can be used to weld highly reactive metals without contamination.

Explosive welding Explosive welding is a process wherein a surface-to-surface bond is achieved by the compressive force of a controlled explosion.

Flow welding This is a process where fusion is produced by heating with molten filler metal poured over the joint until the welding temperature is attained and the required filler metal has fully penetrated the joint.

Hammer welding Hammer welding was commonly employed by the blacksmith of yesteryear; it sees very little industrial usage today. It is also called forge welding.

Friction welding Friction welding is based on the fact that a rapidly moving part in pressure contact with a stationary part generates heat in contacting surfaces. When the fusion temperature is reached, movement is stopped and pressure maintained or increased until the weld is completed.

Induction welding Induction welding depends upon the resistance of the workpiece to the flow of an induced electric current to create heat for fusion. The pieces to be joined are placed within a radio-frequency field, usually developed to the inside of a radiating coil that has been designed to approximate the shape of the intended assembly. Filler metal having a low melting temperature is prepositioned at the joint and distributes through the heated joint by capillary action.

Pressure welding Pressure welding is a process in which two pieces of ductile metal are butt-welded or lap-welded by the application of pressure only, without any of the metal reaching the melting point. Heat, if applied, is sufficient only to facilitate plastic flow of the metal under pressure. Bonding depends upon the ability to bring a large number of atoms on the two surfaces being joined into immediate contact. This requires perfect cleanliness of the surfaces, good alignment, and application of high pressures. The pressure is a squeezing action rather than impact.

Thermit welding Thermit welding is based on the chemical reaction between aluminum and iron oxide. The members to be welded are aligned in proper relation, and a mold is built around the ends to be joined. A pouring gate in the top of the mold receives the molten metal. The Thermit charge is placed in a crucible which has a pouring hole in its bottom. The charge is a mixture of iron oxide and granulated aluminum together with small quantities of alloying elements in the iron oxide. Ignition of the mixture produces a reaction between the iron oxide and aluminum, liberating a large amount of heat. The aluminum combines with the oxygen in the iron oxide and releases free molten steel, which flows into the mold, thus producing the weld.

WELDABILITY OF METALS

The term *weldability* is a relative one. Practically all metals are weldable. Some, however, require special welding procedures in order to preserve the properties and characteristics of the metal for which it was originally alloyed.

Special welding procedures are variants within a limited range of possibilities. If a metal cannot be welded with the regular mild-steel electrodes, E6010 and E6012, for example, some degree of preheat with these electrodes is the next step. Following this, the next alternative is to use a low-hydrogen electrode and finally a stainless-steel type of electrode.

The first aspect of any maintenance welding job is to consider the metal being welded. The behavior of the metal under the heat cycle of welding may or may not be critical. The economy and qualtiy of welding on various metals may be affected by any one or more of the following factors:

1. Oxidation. (*a*) Oxidation producing a gaseous oxide of some one of the elements causing gas holes in the weld metal; (*b*) oxidation producing solid oxides which have a melting temperature higher than the metal, thus causing slag inclusions; (*c*) oxidation producing oxides which are soluble or which are heavier and sink in the molten metal and which render the weld metal brittle or of low strength.

1-44 Basic Maintenance Technology

2. Vaporization. Vaporization of some element in the metal which vaporizes at a temperature lower than the melting point of the metal.
3. Nonmetallic inclusions. Some metals may contain finely divided nonmetallic inclusions which have a melting point higher than that of the metal and therefore did not coalesce when the metal was refined but do melt and coalesce under the high temperature of the arc and then form visible slag inclusions.
4. Change of structure. Change of structure or arrangement of elements within the metal may take place during arc welding, causing change of physical properties or change of resistance to corrosion.
5. Gas solubility of metal. (*a*) Different elements may affect the solubility of various gases at different temperatures, and a decrease in solubility of a gas with a decrease in temperature at the freezing point may cause porosity in weld metal; (*b*) the fluxing out or eliminating of an element during welding may cause the capacity of the metal for a given gas to decrease and thus cause the gas to be given up, producing porosity in the weld metal; (*c*) gases are absorbed during welding to form stable compounds with elements in the metal and thus alter the composition and physical properties of the weld metal.
6. High coefficient of thermal expansion, or high contraction of weld metal upon cooling.
7. Hot shortness, or low strength of the metal at high temperatures.
8. Thermal conductivity, or rate of transfer of heat from fusion zone.
9. Hardenability. Tendency of metal to become hard and brittle in the weld or fusion zone during heat cycle of welding.

The foregoing list indicates why some metals are more satisfactory than others. A careful study of the factors listed indicates that most of the possible undesirable characteristics can be corrected by one or more of the following methods:

1. Selection of metal within the permissible class most suitable for arc welding
2. Use of proper shielded arc
3. Use of proper fluxing material
4. Use of proper electrode or filler material
5. Proper welding procedure
6. In some cases, preheat and postheat treatment

In considering the weldability of any metal, it should be borne in mind that the weld largely depends upon the characteristics of the weld metal which may come from two sources, namely, base metal and electrode or filler metal.

If little or no electrode or filler metal is used, the proper selection of the base metal becomes of prime importance. If the weld metal comes mostly from the electrode or filler metal, then selection of the proper filler metal or electrode becomes of prime importance. However, both electrode and base metal are subjected to similar requirements during arc welding, and both should be of best arc-welding quality, although in many cases the electrode or filler metal serves as a corrective for the base metal.

THE CARBON STEELS

The carbon steels are widely used in all types of manufacturing. The weldability of the different types (low, medium, and high) varies considerably. The preferred analysis range of the common elements found in the carbon steels is shown in Table 5-2. Welding metals whose elements vary above or below the range usually call for special welding procedures.

Low-carbon steels (0.10 to 0.30 percent carbon) Steels of low-carbon content represent the bulk of the carbon-steel tonnage used by industry. These steels usually are

TABLE 5-2 Preferred Analysis Range of Carbon Steels

	Low, %	Preferred, %	High, %
Carbon	0.06	0.10 to 0.25	0.35
Manganese	0.30	0.35 to 0.80	1.40
Silicon		0.10 or under	0.30 max
Sulfur		0.035 or under	0.05 max
Phosphorus		0.03 or under	0.04 max

more ductile and easier to form than higher-carbon steels, and for this reason are used in most applications requiring considerable cold forming, such as stampings and rolled or bent shapes in bar stock, structural shapes, or sheet. Steels with less than 0.13 percent carbon and 0.30 percent manganese have a slightly greater tendency for internal porosity than steels of higher carbon and manganese content.

Medium-carbon steels (0.30 to 0.45 percent) The increased carbon content in medium-carbon steel usually raises tensile strength of the material and also hardness and wear resistance. These steels experience selective use by manufacturers of railroad equipment, farm machinery, construction machinery, material-handling equipment, and similar products. The medium-carbon steels can be successfully welded with the E60XX electrode if certain simple precautions are taken, and the cooling rate is controlled to prevent excessive hardness.

High-carbon steels (0.45 percent and higher) The high-carbon steels are generally used in a hardened condition. In this group are most of the steels used in tools for forming, shaping, and cutting. Tools used in metalworking, woodworking, mining, and farming, such as lathe tools, drills, dies, knives, scraper blades, and plowshares, are typical examples. The high-carbon steels are often described as being difficult to weld, and are not suited to mild-steel welding procedures. Usually, low-hydrogen or other special electrodes are required, and controlled welding procedures, including preheating and postheating, are needed to provide welds that are crackfree.

The higher the carbon content steel, the harder it becomes when it is quenched from above the critical temperature. Welding raises steel above the critical temperature, and the cold mass of metal surrounding the weld area creates a quench effect. Hardness and absence of ductility result in cracking as the weld cools and contracts. Preheating from 300 to 600°F and slow cooling will usually prevent cracking. Figure 5-26A and B shows a calculator for determining preheat and interpass temperatures.

For steels in the higher-carbon ranges (over 0.30 percent) special electrodes are recommended. The lime-ferritic low-hyrogen electrodes (E7016 or E7018) can be used to good advantage in overcoming the cracking difficulties in high-carbon steels. A 308 stainless-steel electrode can also be used to give good physical properties to a weld in high-carbon steel.

Cast iron Cast iron is a complex alloy in which the most important element in welding is the very high carbon content. Quickly cooled cast iron is harder and more brittle than slowly cooled cast iron. The metal also naturally has a low ductility, which results in considerable strain on parts of a casting when one local area is heated. The brittleness and the uneven contraction and expansion of cast iron are the principal concerns in welding it.

Each job must be analyzed to predetermine the effect of welding heat, and procedures correspondingly adopted. Welds can be deposited in short lengths, allowing each to cool. Peening of the weld metal while red hot may be used to stretch the weld deposit.

Either steel or cast-iron electrodes may be used as well as carbon electrodes and nonferrous rods. All oil, dirt, and foreign matter must be removed from the joint before welding. With steel electrodes, intermittent welds no longer than 3 in. and light peening should be used. To reduce contraction, the work should never be allowed to get too hot in one spot. Preheating will help to soften the deposit to make it more machinable.

For welds of such machinability, a nonferrous-alloy rod should be used. A two-layer deposit will have a softer fusion zone than a single-layer deposit. When it is practical, heating of the entire casting to a dull-red heat is recommended, further to soften the fusion zone and burn out dirt and foreign matter. A lower heat can be used if necessary. When the weld to be made is in a deep groove, it is general practice to use a steel electrode for welding cast iron to fill the joint to within approximately ⅛ in. of the surface and then finish the weld with the more machinable nonferrous deposit.

THE ALLOY STEELS

High-tensile low-alloy steels This group of steels is being used increasingly in metal fabricating because their high physical properties permit the use of thinner sections, thus saving metal and reducing weight. They are made with a number of different alloys and can be readily welded with the proper type of electrode designed especially for these metals. Excellent joints of the same high physical properties as the base metal are

1-46 Basic Maintenance Technology

obtained by the use of these electrodes. It is not necessary, as might be suspected, to have a core wire of the same composition as each of the alloys. In some cases, this may even be undesirable, since the electrode metal, in going through the arc, frequently has its analysis and characteristics changed.

Stainless steels Electrodes are made to match various types of stainless steels so that corrosion-resistance properties are not destroyed in welding. The most commonly used types of stainless steels for welded structures are the 304, 308, 309, and 310 groups. Group 304, with 0.08 percent carbon maximum, is a commonly specified type of stainless steel used for weldments.

The general mild-steel welding procedures are used, taking into account the fact of higher electrical resistance, lower thermal conductivity, and higher thermal expansion of the stainless steels. It is important to fit work carefully and clean all edges of foreign material. Light-gauge work must be clamped firmly to prevent distortion and buckling. Small-diameter and short electrodes should be used to prevent loss of chromium and

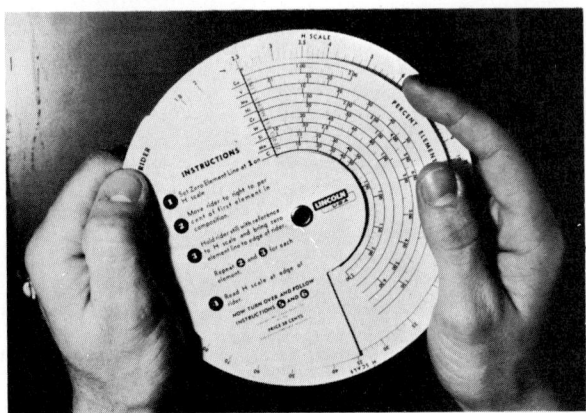

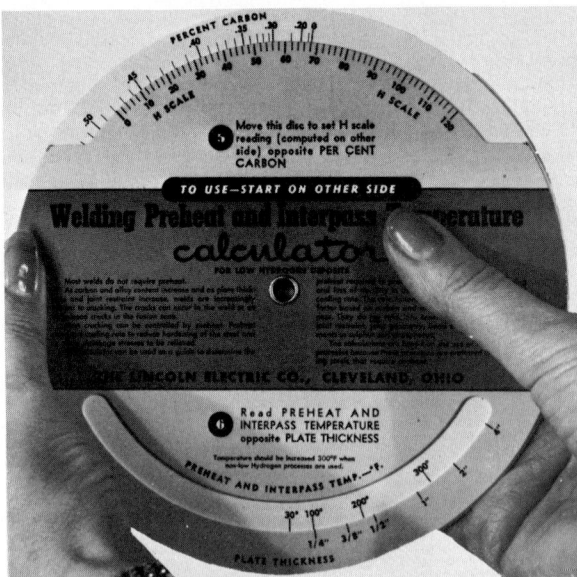

Fig. 5-26 Calculator for determining preheat and interpass temperatures.

undue overheating of the electrode. The weld deposit should be approximately the same analysis as the plate.

Stainless clad steel The significant precautions in welding this material are in joint design, including edge preparation, procedure, and choice of electrode. An electrode should be used of the correct analysis for the cladding being welded. The joint must be prepared and welded to prevent dilution of the clad surface by the steel backing material. The backing material is welded with a mild-steel electrode but in multiple passes to prevent penetration into the cladding. The clad side is also welded in small passes to prevent penetration into the backing material and resulting dilution of the stainless joint. Where in thin-gauge material it is necessary to make the weld in one pass, a 309 stainless electrode should be used for the steel side as well as the stainless side. The design and preparation of the joint can do much to prevent iron pickup as well as reduce labor costs in making the joint.

Chromium steels The intense air-hardening property of these steels, which is proportional to the carbon and chromium content, is the chief consideration in establishing welding procedures. Considerable care must be taken to keep work warm during welding and annealed afterward; otherwise the welds and area adjacent to the welds will be brittle. It is well to consult steel suppliers for specific heat treatment, temperatures, and treatment.

High-manganese steel The tough work-hardening characteristic of this material recommends it for surfaces which must resist abrasion or wear as well as shock. For building up parts of high-manganese steel, an electrode should be used of such type that the physical characteristics of the deposited metal will be approximately the same as those of the base metal.

THE NONFERROUS METALS

Aluminum Most fusion welding of aluminum alloys is done with the inert-gas metal-arc (MIG) process. Weld properties generally are at least equal to those of the base metal at zero temper. Welding speeds are higher than those obtainable with any other arc or gas process. Heat-affected zones are narrower than those with oxyacetylene or covered-electrode arc welding. A dc (reverse-polarity) electric arc, established in an envelope of inert gas between a consumable electrode and the workpiece, is used for welding aluminum by the MIG process.

MIG and another inert-gas shielded-arc process, gas tungsten-arc (TIG), are the principal methods for welding aluminum. The two processes are similar in that an inert gas is used to shield the arc and the weld pool, making flux unnecessary. The chief differences are in the electrodes and the characteristics of the power used.

In MIG welding, the electrode is aluminum filler fed continuously from a reel into the weld pool.

The DCRP action propels the filler metal across the arc to the workpiece in line with the axis of the electrode, regardless of the orientation of the electrode. Because of this and aluminum's density, surface tension, and cooling rate, horizontal, vertical, and overhead welds are made with relative ease. High deposition rates are practical, producing less distortion, greater weld strength, and lower welding costs for a given job than other fusion-welding processes.

TIG welding uses a nonconsumable tungsten electrode, with aluminum-alloy filler material added separately, either from a hand-held rod or from a reel.

Alternating current is preferred by many users of both manual and automatic TIG welding of aluminum. This is because ac TIG achieves an efficient balance between penetration and cleaning.

Copper and copper alloys Copper and its alloys can be welded with shielded-metal-arc, gas-shielded, or carbon-arc welding. Of the three, gas-shielded arc welding with an inert gas is preferred.

Decrease in tensile strength as temperature rises and high coefficient of contraction may make welding of copper complicated. Preheat usually is necessary on thicker sections because of the high heat conductivity of the metal. Keeping the work hot and pointing the electrode at an angle so the flame is directed back over the work will aid in permitting the gas to escape. It is also advisable to put as much metal down per bead as is practical.

CONTROL OF DISTORTION

Distortion in the metal being welded, caused by the heat of welding, may be a problem in welding sheet metal or unrestrained large sections. The following suggestions will help overcome problems of distortion, based on three simple rules applied singly or together:

1. Reduce the effective shrinkage force.
   a. Avoid overwelding. Use as little weld metal as possible by taking full advantage of penetrating effect of arc force.
   b. Use correct edge preparation and fit-up to obtain required fusion at root of weld.
   c. Use few passes.
   d. Place welds near neutral axis.
   e. Use intermittent welds.
   f. Use backstep welding method.
2. Make shrinkage forces work to minimize distortion.
   a. Locate parts out of position so that when weld shrinks they will be in correct position.
   b. Space parts to allow for shrinkage.
   c. Prebend parts so that contraction will pull parts into alignment.
3. Balance shrinkage forces with other forces (where natural ridigity of parts is insufficient to resist contraction).
   a. Balance one force with another by correct welding sequence so that contraction caused by weld counteracts forces of welds previously made.
   b. Peen beads to stretch weld metal. Care must be used not to damage weld metal.
   c. Use jigs and fixtures to hold work in a rigid position with sufficient strength to prevent parts from distorting. Fixtures actually cause weld metal to stretch, thus preventing distortion.

Shielded-metal-arc welding There are two aspects to the problem of selecting the correct electrode for making a good weld under given conditions. The selection must be made according to (1) electrode type as to coating and core-wire analysis and (2) electrode diameter size. In selecting the type of electrode it is necessary to know

1. The position in which the work is to be welded
2. The type and thickness of the metal being used
3. The preparation of the work with regard to fit-up
4. The type of available welding current
5. The class of work (that is, whether deep penetration, surface quality, required physical properties, or code requirements) is the chief essential

The American Welding Society has established specifications for the manufacture of welding electrodes to fulfill the above job requirements. The following specifications have been issued, classifying electrodes as follows:

Mild Steel Covered Arc Welding Electrodes, A5.1-69

Low Alloy Steel Covered Arc Welding Electrodes, A5.5-69

Corrosion-resisting Chromium and Chromium-Nickel Steel Covered Welding Electrodes, A5.4-69

Copper and Copper-alloy Arc Welding Electrodes, A5.6-69

Nickel and Nickel-alloy Covered Welding Electrodes, A5.11-69

In addition to these classifications, electrodes are also manufactured for hard surfacing, welding cast iron, and other miscellaneous applications.

The mid- and low-alloy-steel electrodes are classified with a numbering system for simple identification. E6010 is a typical four-digit classification number. The prefix E designates a metal-arc-welding electrode; the first two digits stand for the minimum allowable tensile strength of deposits in thousands of pounds per square inch. The third digit stands for the welding position or positions in which the electrodes will make a satisfactory deposit, and the last digit indicates various arc characteristics, among them polarity.

Since at least 90 percent of all arc welding is done in mild steel, the following brief descriptions of mild-steel electrode types are included. The significance of the various classification digits as explained for these electrodes is consistent throughout the E70, E80, E90, E100, and E110XX series of steel electrodes. Table 5-3 gives classification characteristics and uses for steel electrodes.

TABLE 5-3 Steel-Electrode Classification, Characteristics, and Uses

Class no.	Work position	Current supply	Basic application
EXX10	All	dc+	Designed to produce good mechanical properties consistent with good radiographic inspection quality. Application is usually structural where multipass welding is employed, such as shipbuilding, bridges, buildings, and piping and pressure vessels
EXX11	All	ac (dc+)	Designed to do the work of XX10, but to employ an ac source. Slightly higher tensile and yield strength
EXX12	All	dc− ac	Especially recommended for single-pass, high-speed, high-current horizontal fillet welds. It is characteristically easy to handle and useful in cases of poor fit-up, both groove and fillet, where a wide range of currents is used. Class 12 has reduced penetration but can meet radiographic standards with single-pass welds
EXX13	All	ac (dc−)	Designed for light-sheet-metal work, but now used widely as an electrode having light penetration. Frequently used in vertical down-welding, even though it produces a flat bead. Particularly well designed for use with low-voltage ac transformers
EXX14	All	ac (dc−)	An iron-powder electrode designed to do the work of 13 with increased deposit rate, although 14 has lower deposition rates than 24 and 27. In the flat position, 13 and 14 have similar welding speeds. Has improved weld appearance and ease of welding in drag technique
EXX15	All	dc+	Offers good physical properties and x-ray quality. A low-hydrogen electrode for difficult-to-weld material such as high-carbon or low-alloy steel. Also, free machining, high-sulfur-bearing steel. Frequently pre- and postheating may be eliminated or reduced by using low-hydrogen rod. Electrode covering cannot perform properly with included moisture. Electrode should be heated before use as recommended by the manufacturer or stored in a moisture-free area
EXX16	All	ac dc+	An electrode similiar to 15 designed to be used with ac and dc + supply
EXX18	All	ac dc−	A 30% iron-powder titania-type electrode. An electrode similar to 15 with a higher deposition rate and an improved weld appearance. Offers better slag removal and higher usable current than the E7016 type
EXX24	HF-F	ac (dc−)	An iron-powder-type electrode ideal for fillet welds. The iron powder in the electrode coating assists in increasing the deposit rate over the 12 class. Electrode can be used in drag technique with ease of handling and good weld appearance. Requires better fit-up than 12, but is of similar application, although limited as to position
EXX27	HF-F	dc− ac	When this high-iron-powder electrode is used in the drag technique, it is faster than the 18 electrode. It is primarily a downhand deep-groove rod, well suited for heavy sections. Second only to 24 in welding speed, but with properties superior to it. Both are equally easy to handle
EXX28	HF-F	dc− (ac)	A 50% iron-powder lime-type electrode. This one yields the highest deposition rates of the low-hydrogen group. The coating also produces an easy-to-maintain arc with a smooth, wide bead. Can be used only in the flat position

HF—horizontal fillet position; F—flat position.

Cellulose-coated electrodes EXX10 and EXX11 The relatively thin coverings of these electrodes contain a high percentage of cellulose. This type of covering produces a small volume of molten slag in the weld crater and light slag coverage of the solidified weld bead. The EXX10 and EXX11 electrodes can be used in all welding positions, as illustrated in Fig. 5-27.

1-50   Basic Maintenance Technology

Types E6010 and E6011   These types may be classified as general-purpose electrodes, since they are used for a wide variety of work and possess high average mechanical characteristics. E6010 is best suited for direct current, electrode positive. In sizes of $\frac{3}{16}$ in. and smaller, in any type of weld, it is suitable in all positions—flat, horizontal, vertical, and overhead (Fig. 5-28). It has deep-penetration qualities and is used very satisfactorily on square-groove butt joints where the electrodes actually scarf or melt the plates. It produces a rather flat bead shape.

Fig. 5-27   EXX10 and EXX11 electrodes are used for all-position welding.

The E6010 electrode has a high cellulose content in the covering. The arc is very penetrating, with a relatively quick solidifying slag and weld-crater action. Protection of the molten metal is obtained principally by gases since only a small amount of slag is produced. The weld metal has excellent physical qualities. Some of the applications are welding pipe, ships, machinery, structures (especially field or erection), and jigs and fixtures.

The E6011 electrode is similar to the E6010 but is designed for ac or either-polarity dc operation. The dc polarity (electrode negative) or reverse (electrode positive) depends upon the type of work being performed. The characterisitics of E6011 electrode design are also high-cellulose covering, penetrating arc, quickly solidifying slag action, similar to E6010, and protection of molten metal obtained principally by gases. As in the case of E6010, this electrode is well suited for making vertical and overhead fillet and butt welds. The applications are the same as for E6010.

These electrodes are generally recommended for use where the weld metal cannot be deposited in the flat, or downhand, position. The deposited metal has good strength and high elongation.

Of the same general characteristics are several electrodes for welding the low-alloy high-tensile steels (E7010, E9010, etc).

Titania-coated electrodes (EXX12, EXX13)   The medium-thick coverings of these electrodes contain a relatively high titania content. This type of covering produces a medium volume of molten slag in the weld crater which simply covers the weld bead when it solidifies. These electrodes can be used in all positions, but are more difficult to control out of position than the cellulose types.

Types E6012 and E6013 The E6012 electrode has a medium-thick covering and is used with direct current with the electrode negative or may be used with alternating current. Sizes of ³⁄₁₆ in. and smaller are suitable for all positions, and larger sizes for welding in flat positions. The electrode may be used for fillet welding, single- or multiple-pass, and can be used for butt welds of the V-groove or U-groove type. Because of its deposition characteristics and ability to build up, it is used to fill gaps in cases of poor fit-up. E6012 coverings are high in titania and low in cellulose. The arc is less penetrating

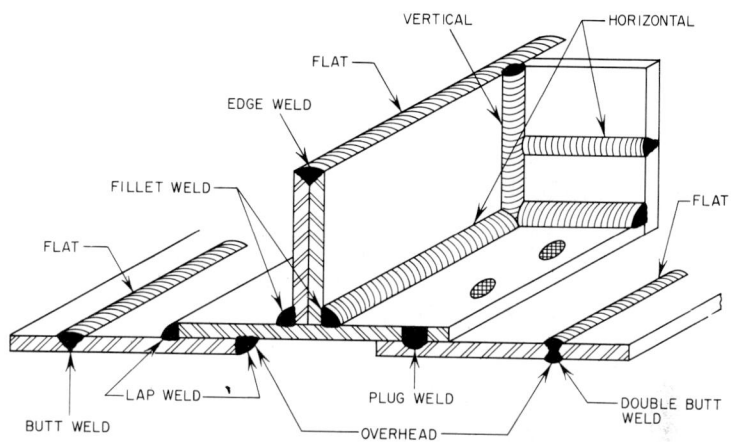

Fig. 5-28 Types of welds encountered in structural and general maintenance welding.

than that of E6010 and E6011, but adequate when correct welding procedures are used. The larger amount of slag gives a better coverage, producing a finer ripple with a more pleasing bead surface.

The E6012 electrode has higher melting rate with lower spatter than E6010 or E6011. It is ideally suited for horizontal and flat fillet welds, for applications where fit-up may be poor, and on steel having characteristics which give poor welding action with electrodes producing greater penetration. It can be used for butt welds of the V-groove or U-groove types. Because it does not penetrate deeply, it is used in cases where dilution of weld and base metal is not desirable. It produces a somewhat convex bead. The weld metal has higher tensile strength and slightly lower elongation than have E6010 and E6011. Some typical applications are welding sheet-metal ducts, tanks, machine guards, and structural work.

E6013 has better ac operation than E6012 and develops a smoother bead appearance. Penetration is similar to that of E6012, so that it works well for poor fit-up. E6013 is more suitable for light-gauge metals than E6012. The bead has a tendency to be convex in making horizontal fillets. The applications are similar to those for E6012.

Iron-powder electrodes The iron-powder-covered electrodes have an exceptionally heavy covering containing a large quantity of iron powder (Fig. 5-29). This type of covering makes welding with these electrodes faster and easier. Welding

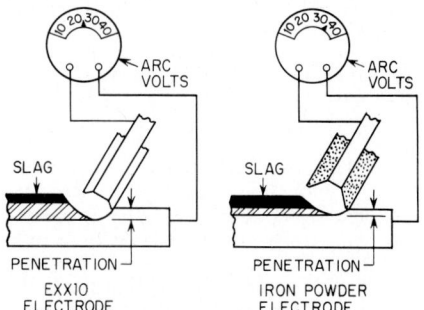

Fig. 5-29 Essential difference between iron-powder and EXX10 electrodes.

speeds are increased as much as 50 percent. Weld appearance is smoother; slag is practically self-cleaning; spatter is eliminated almost completely.

All these advantages result from the nature of the iron-powder coating. The covering more efficiently utilizes the heat of the arc in melting. Welding currents can be increased for a given-diameter electrode, providing greater deposition rates without the difficulties of excessive penetration, gouging, undercutting, and spatter normally encountered when welding with higher currents.

The electrodes operate on either alternating or direct current, but alternating current is preferred. Slightly higher currents than those used with conventional electrodes are required. Also, an electrode one size smaller in diameter is generally used.

They are ideally suited for contact or drag welding techniques, although an arc may be held if desired.

Type E7024 This type of electrode has been designed especially for welding flat and horizontal fillets with either alternating or direct current. It is widely used for production welding in making machinery and structures.

Type E6027 This type of iron-powder electrode has been designed especially for welding flat, deep-groove butt welds with either alternating or direct current. It is also used for flat and horizontal fillets. It is used for code work. The bead has excellent wash-in properties and makes a smooth cover pass. The slag is extremely friable and therefore easily removable under all conditions.

Type E7014 Iron powder has been added to the covering to produce this modified version of an E6013 type of electrode. The result is an excellent electrode having iron-powder characteristics plus the feature of being suited to out-of-position use up to 45° downhill. Although classified as all-position, it is rarely used for vertical and overhead welding.

Buildup and manganese-steel electrodes Several manufacturers are making iron-powder maintenance electrodes for buildup work on worn machinery parts and welding manganese steel. These electrodes carry an official AWS classification, but are usually descriptively named. The high deposition rate of the electrodes results in depositing 35 to 45 percent more metal per minute than is possible with conventional electrodes. This means considerable saving in time when areas being restored to size require the deposition of a large quantity of metal. Properties and characteristics of the electrodes are varied to meet particular service requirements.

Lime-covered low-hydrogen electrode types The low-hydrogen electrode consists essentially of a rimmed-steel core wire upon which a covering of the carbonate of soda and lime type is applied, using other compounds low in hydrogen. This covering is slightly thicker than normal for each diameter, and the electrode is slightly more difficult to use because of the shortness of the arc which must be maintained. A typical analysis of the deposit from this electrode is 0.08 percent carbon, 0.56 percent manganese, and 0.25 percent silicon. The arc is moderately penetrating; the slag heavy, friable, and easily removed; and the deposited metal lies in a flat bead or may be even slightly convex.

The as-welded mechanical and impact properties of deposits made using the low-hydrogen-type electrodes have been found to be superior to those of E6010 and E6011 electrodes depositing weld metal of the same composition. Numerous tests have indicated that the as-welded mechanical and impact properties of deposits from these electrodes approach the properties of stress-relieved deposits of conventional electrodes. Whereas the properties of deposits of conventional electrodes are materially improved when they are stress-relieved, the deposits of low-hydrogen electrodes are changed only slightly. The reduced tendency for underbead cracking and the high quality of as-welded deposits of these electrodes materially reduce the preheat and postheat of weldments, resulting in better welding conditions and reductions in the cost of thermal treatments.

Low-hydrogen electrodes operate best on dc reverse polarity, but most types can also be used with alternating current. They were developed for welding higher-strength high-carbon alloy steels in which the ordinary electrodes are subject to developing *underbead cracking*. These underbead cracks occur along the line of fusion between the parent metal and the weld metal and are caused by the hydrogen present in the conventional electrode covering. Naturally, eliminating the hydrogen tends to help control underbead cracking and permits the welding of the weld steels with little or no preheat, thus making for better welding conditions. Although these cracks do not occur in ordinary steels, they may occur whenever an ordinary electrode is used on high-tensile steels.

Another use for the low-hydrogen-type electrode is the welding of high-sulfur steels. The ordinary electrode deposit on these steels (which contain 0.10 to 0.25 percent sulfur) is badly honeycombed. Low-hydrogen-type electrodes can be used to weld these steels.

Many of the newer high-tensile steels being produced today call for low-hydrogen electrodes ranging up to 110,000 psi tensile strength (E110XX).

The low-hydrogen electrode was developed during World War II for the welding of armor plate, and in addition to its use on alloy steels, high-carbon steels, and high-sulfur steels, it has been found useful on malleable iron, on spring steels, and for welding the mild-steel side of clad plates. Another extensive use has been in the welding of steels which will subsequently be enameled and in all those steels which contain selenium. It is an excellent maintenance electrode, since it can be used with assurance of good welding on steels whose analysis is unknown or may be questioned.

Type E7015 The E7015 electrode was the first of the low-hydrogen types for welding the carbon steels. It was designed exclusively for dc electrode-positive operation. The E7015 electrode can be used in all positions up to and including $5/32$-in. diameter. The larger diameters are useful for fillet and butt welds in the horizontal and flat positions.

Type E7016 The E7016 classification of electrode has all the characteristics of the E7015 classification. The core wire and coating are similar, except for the use of a certain amount of potassium silicate or other potassium salts on the E7016 classification to facilitate its use on alternating current. All that has been said of the E7015 electrode applies equally well to the E7016.

Type E7018 Iron powder has been added in the E7018 type of electrode, thus producing the iron-powdered, low-hydrogen electrode manufactured under this classification. The electrodes have the advantage of low-hydrogen properties plus the excellent operating characteristics associated with iron powder. All that has been said of the E7015 and E7016 applies equally to the E7018. This is an excellent maintenance electrode.

Type E7028 The E7028 classification of electrode combines the advantages of the low-hydrogen types and the heavy-covered powdered-iron types. The electrode manufactured under this classification has a high deposition rate, and is limited in application to horizontal- and flat-position welding.

SUBMERGED-ARC WELDING—EQUIPMENT, ELECTRODES, AND FLUX

Welding equipment The welding heads normally used for fully automatic submerged-arc welding perform the triple function of progressively depositing flux along the joint, feeding the electrode, and transmitting welding current to the electrode. The flux is usually supplied from a hopper either mounted directly on the head or connected to the head by tubing. The bare electrode or wire is fed into the welding head from a coil mounted on a reel. The distance between the end of the electrode and the base metal is maintained constant by special controls which automatically regulate the electrode-feed motor speed or welding current.

Equipment manufactured for semiautomatic submerged-arc welding performs the same functions as that for fully automatic welding. The welding head, however, now consists of a welding gun and wire feeder unit. The flux for semiautomatic welding is supplied by a canister mounted on the welding gun or a continuous-flow flux feed from a pressurized flux tank. With semiautomatic welding equipment, the electrode wire is mechanically fed to the work but the welding gun is manually moved along the joint being welded. This procedure gives added flexibility to this method of welding by permitting its use on irregular shapes and contours, thereby promoting expanded use.

Direct current is used with both semiautomatic and fully automatic submerged-arc welding, whereas alternating current is usually limited to fully automatic submerged-arc welding. The welding voltage for submerged-arc welding will range from 28 to as high as 55 volts. Currents generally used for submerged-arc welding are higher than those used for the other arc-welding processes. They range from a low of 200 amp up to as high as 4000 amp.

Alternating current may be supplied from one or more heavy-duty welding transformers. Direct current may be supplied by one or more motor-generator or rectifier welding machines having capacity suitable for the application. The dc power supplies can be constant-potential or variable-voltage types, depending upon the application and manufacturer's recommendations. Installations of semiautomatic and fully automatic welding equipment are illustrated by Figs. 5-30 and 5-31.

Electrodes and fluxes The ferrous and nonferrous electrodes commonly used for submerged-arc welding are bare rods or wires with clean, bright surfaces to facilitate the introduction of relatively high currents. Electrodes are normally used in the form of coils ranging in weight from a minimum of 25 to 200 lb. On very high-production welding

Fig. 5-30   Semiautomatic submerged-arc welding.

installations, the electrode is frequently fed from a coil in a drum. These drums range up to as high as 1000 lb in weight. Ferrous wire of composition that might readily rust is coppercoated to retard rusting and improve the contact surfaces.

The fluxes used with submerged-arc welding are granulated fusible mineral materials which are essentially free from substances that would create large amounts of gases during welding. These fluxes are made to a variety of chemical specifications which develop particular performance characteristics. The flux has a number of functions to perform, including prevention of atmospheric contamination and performing a scavenging-deoxidizing action on the molten metal in the weld crater. Some special fluxes perform the additional function of contributing alloying elements to the weld deposit, thereby developing specific weld-metal characteristics of higher strength or even abrasion resistance. The choice of flux depends on the welding procedure to be employed, the type of joint, and the composition of the material to be welded.

SPECIAL APPLICATIONS

Sheet-metal welding The welding of sheet metal, as illustrated by Fig. 5-32, has frequent application in maintenance. The principles of good welding practice apply in welding sheet metal as elsewhere, but the nature of the work places special emphasis on several aspects. The problem of distortion requires special consideration in welding thin-gauge metals as well as the problems of burning through the metal. Special attention should therefore be given to all the factors involved in controlling distortion: the speed of welding, the choice of proper joints, good fit-up, position, selection of proper current, use of clamping devices and fixtures, number of phases, and sequence of beads.

Within the limits of good welding appearance, the highest arc speeds and the highest currents should be used. In sheet-metal work, however, there is always the limitation imposed by the threat of burn-through. As the gap in the work increases in size, the current must be decreased to prevent burn-through, which, of course, will reduce welding speeds. A clamping fixture will improve the fit-up of joints and thus make possible the

Fig. 5-31 Fully automatic submerged-arc welding.

Fig. 5-32 Typical sheet-metal welding using the shielded-metal-arc-welding process.

higher speeds. If equipped with a copper backing strip, the clamping fixture will make for easier welding by decreasing the tendency to burn through and will also remove some of the heat which causes warpage. Where possible, sheet-metal joints should be welded downhill at about a 45° angle with the same currents as are used in the flat position, or slightly higher. Tables 5-4 and 5-5 offer a guide to the selection of the proper current, voltage, and electrodes for the various types of joints used with sheet metal ranging from 20 to 8 gauge.

Hard surfacing The building up of a layer of metal or a metal surface by electric welding, commonly known as arc-weld surfacing, has an important and useful application in equipment maintenance. Applications of the process are varied and many, such as restoring worn cutting edges and teeth on excavators, building up worn shafts with low- or medium-carbon deposit, lining a carbon-steel bin or chute with stainless-steel corrosion-resistant alloy deposit, putting a tool-steel cutting edge on a medium-carbon-steel base, and applying wear-resistant surfaces to metal machine parts of all kinds. The dragline of Fig. 5-33 is being returned to new condition by rebuilding and hard surfacing.

Arc-weld surfacing includes, but is not limited to, hard surfacing. There are many building-up applications where hard surfacing is not required.

Wear is the gradual impairment of machinery parts through use. Excluding corrosion, wear results from various combinations of abrasion and impact. Abrasive wear results from one material scratching another and impact wear from one material hitting another.

How to resist abrasive wear Abrasive wear is resisted by materials with a high scratch hardness. Sand wears metals with a low scratch hardness at a high rate, but under the same conditions it will wear a metal of high scratch hardness very slowly. Scratch hardness, however, is not necessarily measured by standard hardness tests. Brinell and Rockwell hardness are not reliable measures for determining the abrasive-wear resistance of a metal. A hard-surfacing material of the chromium-carbide type may have a hardness of 50 Rockwell C. Sand will wear this material at a slower rate than it will a steel hardened to 60 Rockwell C. The sand will scratch all the way across the surface of the steel. On the surfacing alloy the scratch will progress through the matrix material and then stop when the sand grain comes up against one of the microscopic crystals of chromium carbide,

TABLE 5-4 Welding Currents for Sheet Metal

Type of welded joint	20 ga			18 ga			16 ga		
	F*	V*	O*	F	V	O	F	V	O
Plain butt	30†	30†	30†	40†	40†	40†	70†	70†	70†
Lap	40†	40†	40†	60†	60†	60†	100	100	100
Fillet				40†	40†	40†	70†	70†	70†
Corner	40†	40†	40†	60†	60†	60†	90†	90†	90†
Edge	40†	40†	40†	60†	60†	60†	80†	80†	80†

*F—flat position; V—vertical; O—overhead.
†Electrode negative, work positive.

TABLE 5-5 Sizes of Electrodes for Sheet Metal

Type of welded joint	20 ga			18 ga			16 ga		
	F*	V*	O*	F	V	O	F	V	O
Plain butt	$3/32$	$3/32$	$3/32$	$3/32$	$3/32$	$3/32$	$1/8$	$1/8$	$1/8$
Lap	$3/32$	$3/32$	$3/32$	$3/32$	$3/32$	$3/32$	$1/8$	$1/8$	$1/8$
Fillet				$3/32$	$3/32$	$3/32$	$1/8$	$1/8$	$1/8$
Corner	$3/32$	$3/32$	$3/32$	$3/32$	$3/32$	$3/32$	$1/8$	$1/8$	$1/8$
Edge	$3/32$	$3/32$	$3/32$	$3/32$	$3/32$	$3/32$	$1/8$	$1/8$	$1/8$

*F—flat position; V—vertical; O—overhead.

which has a higher scratch hardness than sand. If two metals of the same type have the same kind of microscopic constituents, however, the metal having the high Rockwell hardness will be more resistant to abrasive wear.

How to resist impact wear Whereas abrasive wear is resisted by the surface properties of a metal, impact wear is resisted by the properties of the metal beneath the surface. To resist impact, a tough material is used, one which does not readily bend, break, chip, or crack. It yields so as to distribute or absorb the load created by impact, and the ultimate strength of the metal is not exceeded. Included in impact wear is that caused by bending or compression at low velocity without impact, resulting in loss of metal by cracking, chipping, upsetting, flowing, or crushing.

Types of surfacing electrodes Many different kinds of surfacing electrodes are available. The problem is to find the best one to do a given job. Yet because service conditions vary so widely, no universal standard can be established for determining the ability of surfacing to resist impact or to resist abrasion. Furthermore, there is no ideal surfacing material that resists both impact and abrasion equally well. In manufacturing surfacing electrodes, it is necessary to sacrifice somewhat one quality to gain the other. A material that has a high resistance to abrasion will have a low resistance to impact. High impact resistance is gained by sacrificing abrasion resistance.

Price is no index to quality of electrodes. Simply because an electrode contains an expensive ingredient does not necessarily make it superior for wear resistance. Thus the user of surfacing materials must rely upon the manufacturer's recommendations and his own tests to determine the best surfacing material for his purpose.

How to choose hard-facing material The chart of Fig. 5-34 lists the relative characteristics of manual hard-surfacing materials. It shows in the various columns the ability of each of the materials to resist abrasion, metallic friction, impact, and corrosion. It also gives the relative hardness, ductility, and cost of depositing the material, as well as the physical limitations of weld size in applying each one. This chart is a guide to selecting
   1. The hard-surfacing electrode best suited for a job not hard-surfaced before
   2. A more suitable hard-surfacing electrode for a job where present material has not produced desired results

TABLE 5-4 Welding Currents for Sheet Metal *(Continued)*

Type of welded joint	14 ga			12 ga			10 ga			8 ga		
	F	V	O	F	V	O	F	V	O	F	V	O
Plain butt	85†	80	85†	115	110	110	135	120	115	190	130	120
Lap	130	130	130	135	120	120	155	130	120	165	140	120
Fillet	100	90	85	150	140	120	160	150	130	160	160	130
Corner	90	80	75	125	110	110	140	130	125	175	130	125
Edge	110	80	80	145	110	110	150	120	120	160	120	120

TABLE 5-5 Sizes of Electrodes for Sheet Metal *(Continued)*

Type of welded joint	14 ga			12 ga			10 ga			8 ga		
	F	V	O	F	V	0	F	V	O	F	V	O
Plain butt	1/8	1/8	1/8	5/32	5/32	5/32	5/32	5/32	5/32	3/16	5/32	5/32
Lap	5/32	5/32	5/32	5/32	5/32	5/32	3/16	3/16	5/32	3/16	3/16	5/32
Fillet	1/8	1/8	1/8	5/32	5/32	5/32	3/16	5/32	5/32	3/16	5/32	5/32
Corner	1/8	1/8	1/8	3/16	5/32	5/32	3/16	5/32	5/32	3/16	5/32	5/32
Edge	1/8	1/8	1/8	3/16	5/32	5/32	3/16	5/32	5/32	3/16	5/32	5/32

1-58   Basic Maintenance Technology

Example 1. APPLICATION: Dragline bucket tooth, as illustrated by Fig. 5-35. SERVICE: Sandy gravel with some good-size rocks.

Maximum wear that can be economically obtained is the goal of most hard-surfacing applications. Try to use a material that rates as high as possible in the resistance-to-abrasion column unless some other characteristics shown in the other columns make it unsuited for this particular application.

Fig. 5-33   Shielded-metal-arc welding is used to rebuild and to hard-surface worn areas of a dragline bucket.

First, consider the tungsten-carbide types. Notice that they are composed of very hard particles in a softer and less abrasion-resistant matrix. Although such material is the best for resisting sliding abrasion on hard material, in sand the matrix is apt to scour out slightly, and then the brittle particles are exposed. These particles are rated poor in impact resistance, and they may break and spall off when they encounter the rocks.

Next best in abrasion, as listed in the chart, is the high-chromium carbide type shown in the electrode-size column to be a powder. It can be applied only in a thin layer, and also is not rated high in impact resistance. This makes it doubtful for use in this rocky soil.

The rod-type high-chromium carbides also rate very high in abrasion resistance, but do not rate high in impact resistance. However, the second does show sufficient impact rating to be considered if two or three different materials are to be tested in a field test. Since there is a chance that it has enough impact resistance to do this job, we should not like to pass up its very good wearing properties.

Nevertheless, the semiaustenitic type is balanced in both abrasion and impact resistance. It is much better in resistance to impact than the materials that rate higher in abrasion. Thus semiaustenitic is the first choice on this job, considering that the added impact resistance of the austenitic type is not necessary, since the impact in this application is not extreme.

Example 2. APPLICATION: Same dragline tooth used in Example 1. SERVICE: Soil changed to clay and shale.

The semiaustenitic type selected in the first example stands up well, but the teeth wear only half as long as the bucket lip. With double the wear on the teeth, only half the downtime periods for resurfacing would be needed, and both teeth and bucket could be done together.

Since the impact is now negligible with the new soil conditions, go to a material higher in the abrasion column. Choose a material such as the first high-chromium carbide rod, which could give twice the wear by controlling the size bead applied and still be within reasonable cost.

Example 3. APPLICATION: Same dragline tooth as in Examples 1 and 2. SERVICE: Soil changed to contain large rocks.

If the earth has been changed so that it contains many hard and large rocks and the teeth are failing because of spalling under impact, move down in the abrasion-resisting column to a better impact-resistant material, such as the semiaustenitic type.

From the above, it can be seen that where a dragline operates in all kinds of soils, a material that is good in both abrasion and impact, such as a semiaustenitic type, is the best choice when in doubt as to the conditions that will be met.

When this same type of reasoning is followed in checking the important characteristics, a material can be chosen for any application. And if, for any reason, the first choice does not prove satisfactory, it is usually a simple matter to improve the next application by choosing a material that is rated higher in the characteristic that has caused difficulty.

Where failures occur because of cracking or spalling, it usually indicates that a material higher in impact or ductility rating should be used. Where normal wear alone seems too rapid, a material higher in abrasion rating is indicated.

Check welding procedure Often hard-surfacing failures due to cracking or spalling may be caused by improper welding procedures rather than by improper choice of hard-surfacing material. Before changing to a different hard-surfacing material, serious consideration should be given to the question of whether or not the material has been properly applied.

For almost any hard-surfacing application, very good results can be obtained if the following precautions are observed:

1. Do not apply hard-surfacing material over cracked or porous areas. Remove any defective areas down to sound base material.

2. Preheat. Preheating to 400 to 500°F improves the resistance to cracking and spalling. This minimum temperature should be maintained until welding is completed. The exception to this rule is 11 to 14 percent manganese steel, which should be kept cool.

3. Cool slowly. If possible, allow the finished weldment to cool under an insulating material such as lime, asbestos, or sand.

4. Do not apply more than the recommended number of layers.

When more than normal buildup is required, apply intermediate layers of either medium carbon or stainless steel. This will provide a good bond to the base metal and will eliminate excessively thick layers of hard-surfacing material which might otherwise spall off.

Stainless steel is also an excellent choice for intermediate layers on manganese steels or for hard-to-weld steels where preheating is not practical.

Check before total wear Whenever possible, examine a surfaced part when it is only partly worn. Examination of a part after it is completely worn is unsatisfactory. Did the surface crumble off, or was it scratched off? Is a tougher surface needed, or is additional abrasion resistance required? Should a heavier layer of surfacing be used? Should the surfacing be reduced? All these questions can be answered by examination of a partly worn part and with a knowledge of the surfacing costs and the service requirements.

In case it is impossible to analyze carefully the service conditions, it is always on the safe side to choose a material tougher than is thought to be required. A tough material will not knock or chip off and will offer some resistance to abrasion. A hard abrasion-resistant material is more susceptible to chipping, and surfacing material does not do any good when it is knocked off in large pieces.

After some experience is gained in the use of surfacing materials, various combinations of materials can be tried out to improve product performance. For example, on a part which is normally surfaced with a tough, semiaustenitic electrode, it may be possible to get additional abrasion resistance without sacrificing resistance to cracking. Fuse a little of the powdered chromium-carbide material on critical areas where additional protection is needed.

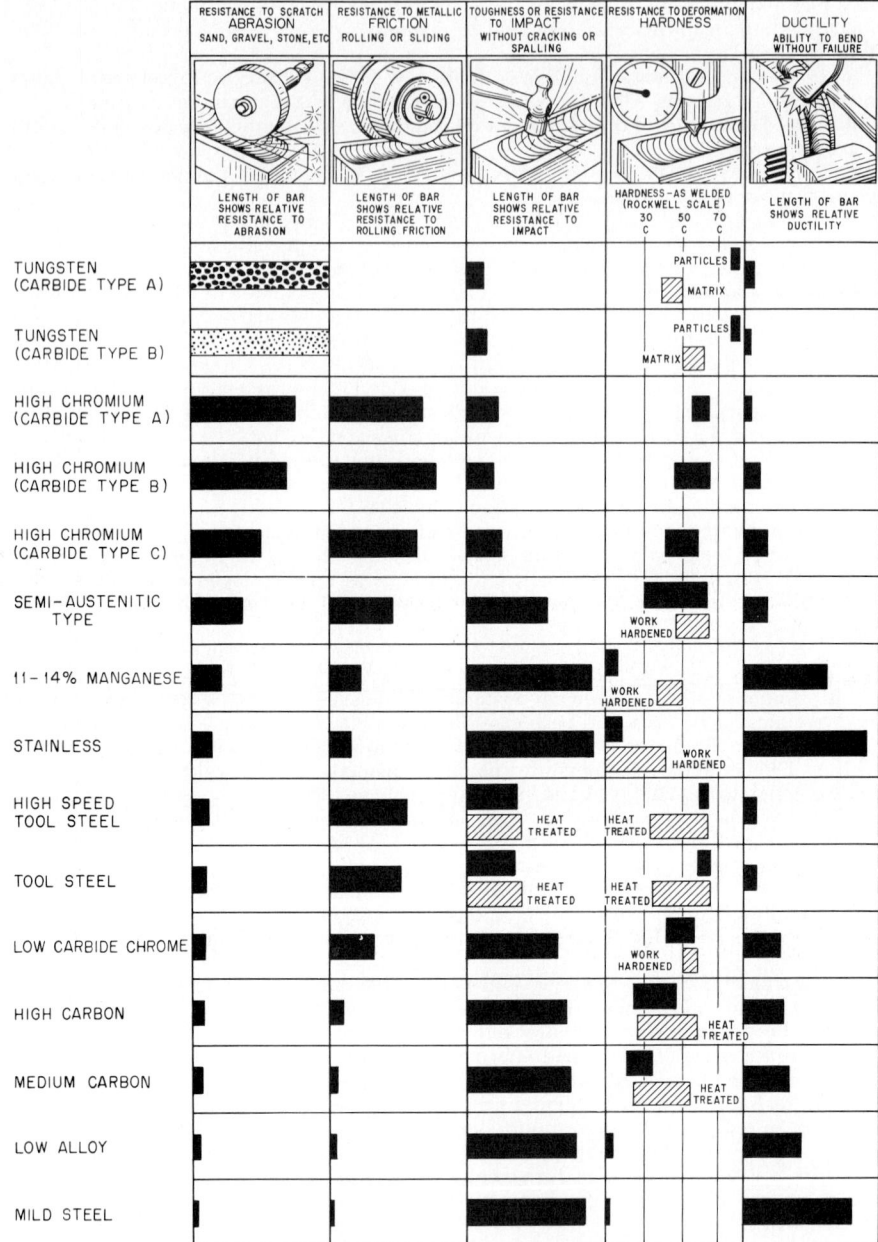

Fig. 5-34 Hard-surfacing guide.

Arc Welding in Maintenance 1-61

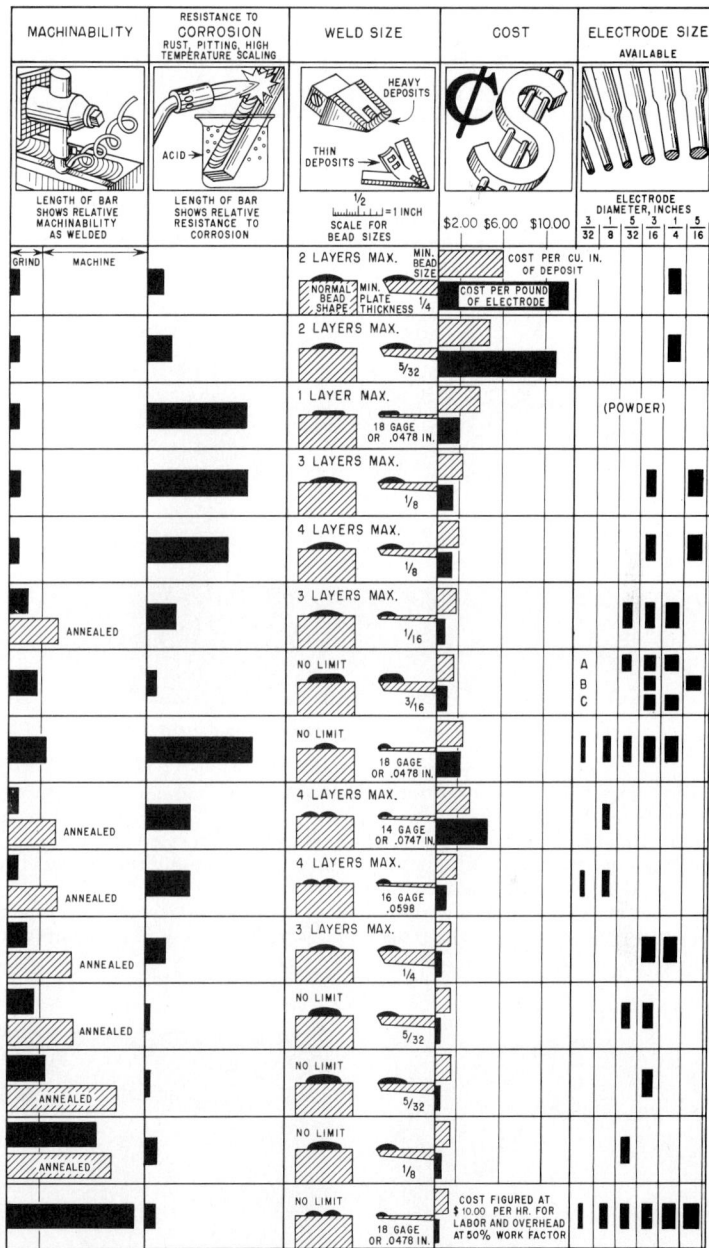

Fig. 5-34 (Continued).

1-62 Basic Maintenance Technology

Many jobs that are badly worn are first built up to almost finished size with a high-carbon electrode. They are then surfaced with an austenitic rod, and finally a few beads of chromium-carbide deposit are placed in spots requiring maximum protection against abrasion.

Regardless of the circumstances, careful analysis of the surfacing problem is well worthwhile. For examples of jobs see Figs. 5-36 to 5-38.

Fig. 5-35 Bucket teeth have been rebuilt and hard-surfaced.

Hard surfacing with submerged-arc process The submerged-arc process offers several advantages for hard surfacing. The greater uniformity of the surface makes for better wearing qualities. The speed of submerged-arc welding creates major economies in hard-surfacing areas which require the deposition of large amounts of metal. These areas may be either flat or curved surfaces. Mixer bottom plates, scraper blades, fan blades, chutes, and refinery vessels are examples of flat plate to be surfaced. Shafts, blooming-mill spindles, skelp rolls, crane wheels, tractor idlers and rollers, and rams are examples of cylindrical surfaces (Figs. 5-39 to 5-42).

Fig. 5-36 Mild-steel die, on the edge of which tool steel has been deposited by means of tool-steel electrode.

Fig. 5-37 Cone used for uncoiling steel. Hard-facing material has been deposited on mild-steel base. Surface is ready for grinding.

The process can be used with either fully automatic equipment or with semiautomatic equipment, the choice depending upon the economics of the application. It is a relatively simple calculation to determine the savings that will result from using the submerged-arc process and thus arrive at a decision as to which type of equipment is warranted. Fully automatic equipment can be quickly fitted with auxiliary accessories which result in more economical metal deposition. An oscillating device can be added to an automatic head to create a bead up to 3 in. wide in a single pass. Another attachment permits the feeding of two electrode wires through a single head and single contact jaw. Both these attachments are useful in hard surfacing.

Hard surfacing with the submerged arc can be done with several different types of materials. The hard-surfacing deposit can be created by using solid alloy wires and a neutral granular flux. It can also be created by using a solid mild-steel wire and an agglomerated alloy flux, the alloys being added to the deposit through the flux rather than through the wire. Also available are tubular wires which contain alloying material in the hollow portion of the mild-steel tube. All the methods have particular advantages. In considering the submerged-arc process, it is well to consult a qualified field engineer who can recommend methods and procedures. With submerged-arc welding, considerable variation in the hard-surfacing deposit can be made by changing the welding procedure to control admixture and the heat-treatment effect of the welding cycle. Procedures should be established with the help of qualified engineers.

Fig. 5-38 Using mild-steel electrode to build up inside diameter and all teeth of 25-year-old cast-steel gear that could not be replaced.

Carbon arc Manual carbon-arc welding can be used to good advantage for welding of copper and its alloys, cast iron, and galvanized sheets and hard surfacing with alloy powder. The hard carbon can also be used for the cutting of steel, cast iron, and the stainless steels, the last two of which cannot be readily cut with the oxyacetylene torch.

The procedures to be used in welding any particular material will vary with the application. In making an edge weld where no filler metal is to be added, the average speeds given in Table 5-6 should be obtained.

Pointing of the Carbons. The diameter of the point should be approximately half the

Fig. 5-39 Steel-mill coke pushers being hard-faced by submerged-arc process using mild-steel wire and alloy flux. Fully automatic equipment in foreground, semiautomatic in background.

diameter of the carbon used. The taper should be gradual back to the point where it is gripped in the holder.

Position of the Carbon in the Holder. The carbon should be gripped as close to the arc as practical because, if a long length of carbon is exposed, the heating causes the carbon to vaporize and burn very rapidly, resulting in excessive wastage.

TABLE 5-6 Average Conditions for Welding

Metal thickness	Arc volts	Arc amp	Carbon size, in.	Welding speed, fph
16 ga. (0.0598 in.)	25	90–100	3/16	135
14 ga. (0.0747 in.)	25	125–135	1/4	125
12 ga. (0.1046 in.)	25	200–250	1/4–5/16	110
10 ga. (0.1345 in.)	25	250–275	1/4–5/16	100

Polarity. Carbon negative should be used in all cases.

Currents. The proper current to be used depends upon the work to be done. Table 5-7 will serve as a guide. The currents given are about the maximum which should ever be used. Smaller currents may be used, depending upon the weight or thickness of the base metal.

TABLE 5-7 Maximum Currents for Hand Carbon Arc

Size of carbon electrode, in.	Maximum current
5/32	50
3/16	100
1/4	200
5/16	350
3/8	450
1/2	700

Arc torch The development of the carbon-arc torch (Fig. 5-43) has further extended the use of the carbon-arc-welding technique to jobs where the application of heat is desired without melting the base metal. With the arc torch, a high-temperature flame is held between the two carbon electrodes clamped in adjustable jaws. The flame is played over the surface of the work, similiar to a gas flame, and as the carbons are consumed, they can be adjusted to maintain a constant distance between them.

The torch is useful for all brazing, soldering of light or heavy copper and galvanized or tinned parts, preheating localized areas prior to welding, and general heating for bending or straightening.

The carbon-arc torch operates at 35 to 40 volts, and since most welder controls are calibrated in amperes at the average metallic arc of 25 to 30 volts, the machine controls should be set 20 percent above recommended current settings. Copper-coated and copper-cored carbons are generally used in 1/4 to 3/8-in. diameter. The current should never be set so high that the copper coating is burned away more than 1/2 in. ahead of the arc. Only enough current should be used to cause the filler material to flow freely on the work. This will avoid consuming carbons too rapidly. The recommended current is between 50 and 75 amps on 5/16-in. carbons. Best results in brazing are obtained when carbons are 1/4 to 3/8 in. away from the work. When possible the joint should be lying horizontally to secure the best flow of molten filler rod.

Cutting Steel can be readily cut with great accuracy by means of the oxyacetylene torch. Not all metals cut as easily as steel. Cast iron, stainless steel, manganese steels, and nonferrous materials cannot be satisfactorily cut and shaped with the oxyacetylene cutting process because of their reluctance to oxidize. In these cases, arc cutting is often used to good advantage.

The cutting of steel is a chemical action. The oxygen combines readily with the iron to form iron oxide. In cast iron, this action is hindered by the presence of carbon in graphite form. Thus cast iron cannot be cut as readily as steel. Higher temperatures are necessary, and cutting is slower. In steel, the action starts at bright-red heat, whereas in cast iron the temperature must be nearer the melting point in order to obtain a sufficient reaction.

Fig. 5-40 Hard-surfacing wire-mill roll by submerged-arc process. Mild-steel wire and alloy flux. Gas torch keeps roll up to temperature.

Because of the very high temperature, the rate of cutting is usually fairly high. However, as the process is essentially one of melting without any great action, tending to force the molten metal out of cut, some provision must be made for permitting the metal to flow readily away from the cut. This is usually done by starting at a point from which the molten metal can flow readily. This method is followed until the desired amount of metal has been melted away.

Fig. 5-41 Automatic head adapted for oscillating and for two electrodes being used to deposit 3-in. beads on a flat mixer bottom plate.

1-66 Basic Maintenance Technology

As an example, the general method is to apply the electric arc on the underside of the work, starting at a lower corner, working toward the center on the lower surface and then up the side, and repeating this action as many times as necessary. This will allow the molten metal to flow out of the cut.

A carbon electrode is generally used. Graphite electrodes are used to some extent

Fig. 5-42 Submerged-arc welding being used to hard-surface a cylindrical surface.

because they permit use of higher currents. Shielded-metal-arc-type electrodes are also effective. In starting a cut, the arc is held at the point selected for the initial cut, as, for example, a lower corner. When the metal begins to flow and run off, the arc is moved along at a rate to permit metal to flow continuously out of the cut.

The width of the cut is dependent upon the ability of the operator to follow a straight line, the size of electrode used, and the thickness of material. The width of the cut is greater on thick sections than on thin.

A process which has come into use for the cutting of materials not readily cut with the oxyacetylene flame is the "oxyarc" cutting process. It cuts by directing a stream of oxygen into a pool of molten metal. The pool is made and kept molten by an arc established between the base metal and the covered tubular cutting electrode, which is consumed during the cutting operation. In addition to providing the arc, the electrode also provides an oxidizing flux and a means of conveying oxygen to the surface being cut.

The tubular cutting electrode is made of mild steel. This is not a detriment when the electrode is used for cutting materials other than mild steel because no contamination of the base metal adjacent to the cut occurs. The possibility of contamination is eliminated by the combination of extremely high heat and oxygen under

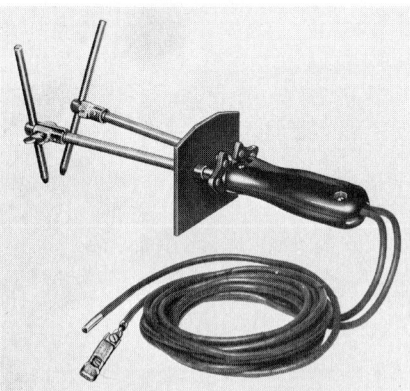

Fig. 5-43 Typical carbon-arc torch.

pressure, which act together to oxidize the electrode and coating at the point of the arc before the electrode metal can fuse with the base metal. The electrode covering helps to maintain arc stability by confining and directing the arc. When in use, it acts as an insulator to prevent the arcing of the rod at undesirable areas.

Quite simple equipment is required for metal cutting by this arc-oxygen process. There is, first of all, a special electrode which is not unlike an arc-welding electrode holder in appearance. This holder serves the double function of conducting current and feeding oxygen for the cutting operation. The tubular cutting electrode can be inserted or removed from the holder with ease. The only other equipment required is an ac or dc arc welder and an oxygen source with usual regulators.

The arc-oxygen process has been used successfully to cut high-chrome and chrome-nickel stainless steels, nickel, cupronickel, Monel, nickel-clad or stainless-clad steel, bronze, copper, brass, aluminum, cast iron, and mild and low-alloy steels. It has not supplanted, however, the oxyacetylene flame for cutting mild steel and other readily oxidized materials, because it is somewhat more expensive.

SELECTION AND MAINTENANCE OF EQUIPMENT

Machines Satisfactory welding can be accomplished with either alternating or direct welding current. Each type of current, however, has a particular advantage which makes it best suited for certain types of welding and welding conditions. The chief advantage of alternating current is its elimination of arc blow, which may be encountered when welding in heavy plate or into a corner. The magnetic fields set up in the plate deflect the path of the arc. Alternating current tends to minimize this deflection and will also sometimes increase the speed of welding with larger electrodes, over $3/16$ in., and with the iron-powder-type electrodes.

The chief advantages of direct current are the stability of the arc and the fact that the current output of the motor-generator-type welder will remain constant in spite of variations of input voltage which affect a transformer-type welder. Direct current, therefore, is a more versatile welding current. Certain electrodes, such as stainless, require a very stable arc, and as yet these electrodes operate much better with direct current. Direct current, because of its stability, is also better for sheet-metal welding where danger of burn-through is present. The dc arc can also be more readily varied to meet different welding conditions. A wider range of control over both voltage and current permits closer adjustment of the arc for difficult welding conditions, such as might be encountered in vertical or overhead welding. Because of its versatility, direct current should be available for maintenance welding.

Direct-current welders (Figs. 5-44 and 5-45) are made either as motor-generator sets or as transformer-rectifier sets. Motor-generator sets are powered by ac or dc motors.

Generators are also powered by small air-cooled gasoline engines (Fig. 5-46). The advantage of this type of set is that, for on-the-spot maintenance welding, it is not necessary to string electric power lines to the welding set, which may have to be used in a location some distance from a power line.

Engine-driven welders powered by gasoline engines are also available and come in larger sizes than the air-cooled engine sets. These are suitable where the size of the plant maintenance operation warrants a larger welder.

For most general maintenance welding, a 250-amp output capacity is ample. Several manufacturers make machines especially for this type of welding, which are compact and readily portable. Higher amperages may be required in particular applications, and, for these, heavy-duty machines should be used. Such a machine is shown in Fig. 5-47.

Another type of welding machine is one which produces both alternating and direct welding current, either of which is available at the flip of a switch (Fig. 5-48). This type of equipment is ideal for maintenance welding, since it makes any kind of welding arc available, giving complete flexibility to the maintenance welding.

Figure 5-49 shows a self-propelled truck with welder driven from power take-off.

Automatic submerged-arc welding is increasingly used as a maintenance process. Both fully automatic and semiautomatic equipment can be used with the process. The chief use of the process in maintenance is for hard surfacing. It permits the rapid deposition of large amounts of uniformly excellent weld metal. Where the maintenance work includes hard

1-68   Basic Maintenance Technology

surfacing, it is well to consider the use of this process. Semiautomatic equipment is relatively inexpensive, and can be adapted to existing welding equipment of larger amperage outputs. Fully automatic equipment is more expensive, and only a large volume of work will justify its installation.

Accessory equipment   The varied and severe service demands made on equipment for maintenance welding require that the best in accessories be used for maximum efficiency. Most maintenance welders make racks for themselves, or other storage conve-

Fig. 5-44   Compact portable welder designed especially for maintenance work. Shows temporary work table for welding angle-iron frame.

niences, which they attach directly to the welding machine for storing and transporting electrodes and accessories. These arrangements will vary to suit individual tastes and needs. The end result of all of them, however, is to have everything immediately available for use.

A part of such accessory equipment mounted on welders should be a fire extinguisher. Maintenance welding may be required in an area containing a fire hazard. At all times, the possibility of fire should be the welder's concern, and in addition to having the proper fire-extinguishing equipment at hand, he should police the area for flammable materials.

Many electrodes holders are available, but only a few combine all the desired features. The operator holds the electrode clamped in the holder, and the current from the welding set passes through the holder to the electrode. The clamping device should be so designed as to hold the electrode securely in position and yet permit quick, easy change of electrodes. It should be light in weight, properly balanced, and easy to handle, yet sturdy enough to withstand rough usage. It should be designed so that it will remain cool enough to be handled comfortably (Fig. 5-50).

Care should be taken in the selection of face or head shields to ensure maximum protection to the operator. These shields are generally constructed of some kind of pressed-fiber insulating material, usually black to reduce reflection. The shield should be light in weight and designed to ensure greatest comfort to the welder. The glass windows in the protective shields should be of such composition as to absorb the infrared rays, the ultraviolet rays, and most visible rays emanating from the arc. The welding lens in the head or face shield should be protected from molten-metal spatter and breakage by a chemically treated clear nonspatter glass covering the exposed side of the lens.

A good protective lens and shield should be used by the operator, and the arc should never be looked at with the naked eye at close quarters. When a new lens is put into the shield, care should be taken that no light leaks in around the glass. If practical, the welding room should be painted a dead black or some other color to prevent reflection. Other workers around an arc can be readily protected by movable or portable screens.

Special goggles are used by welders' helpers, foremen, supervisors, inspectors, and others working close to a welding arc, to protect their eyes from occasional flashes. A good goggle has adjustable elastic headbands and is light, cool, well ventilated, and comfortable. Clear cover glasses and tinted lenses in various shades are available for this type of goggle.

During the arc-welding process, some sparks and globules of molten metal are thrown out from the arc. For protection from possible burns, it is advisable that the operator wear an apron of leather or other protective material. Some operators also wear spats or leggings and sleevelets of leather or other fire-resistant material. Some sort of protection should be provided for the operator's ankles and feet, since a globule of molten metal can cause a painful burn before it can be extracted from the shoe.

A gauntlet type of glove, preferably of leather, is generally used by operators to protect the hands from the arc rays, spatter of molten metal, or sparks. Gloves also provide protection when handling the work.

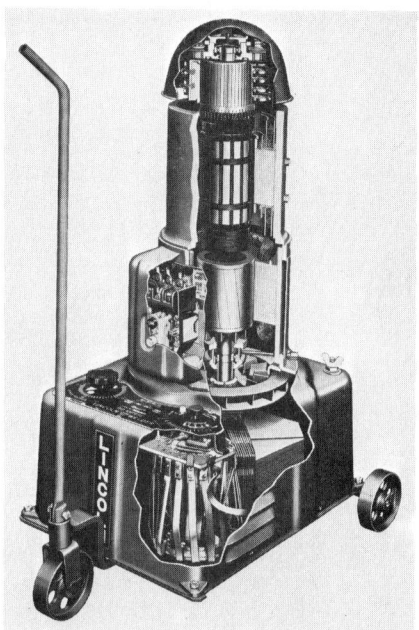

Fig. 5-45 Motor-generator dc maintenance welder.

As a means of protection to other workers from the arc rays, spatter of molten metals, and sparks, the scene of each welding operation should be enclosed by a portable or permanent structure, booth, or screen. Where the welding machine must be taken to the work, it is advisable to surround the scene of the welding operation with portable screens painted dead black to prevent reflection of the arc rays.

Other tools which will prove of value in any shop where welding is done include wire brushes for cleaning the welds, cold chisels for chipping, clamps for holding work in position for welding, wedges, and, where work is large or heavy, a crane or chain block. A drill, air hammer, and grinder are also of value.

INSTALLATION

Good welding begins with proper installation of equipment. Installations should be made in locations that are as clean as possible, and there should be provisions for a continuous supply of clean air for ventilation. It is important to provide separate enclosures if the atmosphere is excessively moist or contains corrosive vapors. If welding must be done where the ambient temperature is high, place the equipment in a different location. Sets operated outdoors should be provided with protection against inclement weather.

1-70 Basic Maintenance Technology

When making an installation, keep the following points in mind:
 1. Consult the local power company to ensure adequate supply of the right type of power.
 2. Provide adequate and even support for the set.
 3. See that there is adequate protection against mechanical abuse and atmospheric conditions.
 4. Make proper provisions for large quantities of fresh air for ventilation and cooling.
 5. Ground the frame of the welding set solidly.

Fig. 5-46 Small engine-driven combination welder and power supply promotes speedy on-the-spot repair.

Fig. 5-47 Mobile 300-amp dc motor-generator set for work in power plant. Welding cables are fed through wall to weld pipe several hundred feet away. Cable reel is mounted on rear of welder platform. Dual-voltage switch permits use of 220 or 440.

Arc Welding in Maintenance   1-71

6. Check electrical connections to make sure they are tight.
7. The fuses for a motor-generator welder should be of the high-lag type and be rated two or three times the input-current rating of the welder.
8. Make sure that the line and welding leads are of sufficient capacity to handle the required current and are well insulated (Tables 5-8 to 5-11).
9. Check over the set before operating to make sure that no parts are visibly loose or not in good condition.

TABLE 5-8 Recommended Wire Sizes for Input Power Cable for Typical Motor-Generator-Type Welder
(Based on National Electrical Code)

Welder size	60-Hz input voltage	Ampere rating	3 wires in conduit or 3-conductor cable. Type R	Grounding conductor
200	230	44	8	8
	460	22	12	10
	575	18	12	14
300	230	62	6	8
	460	31	10	10
	575	25	10	12
400	230	78	6	6
	460	39	8	8
	575	31	10	10
600	230	124	2	6
	460	62	6	8
	575	50	8	8
900	230	158	1	3
	460	79	6	6
	575	63	4	8

TABLE 5-9 Input Cable Sizes for ac/dc Welder

Welder	Volts input	Amp input		Wire size (3 in conduit)			Wire size (3 in free air)		
		With condsr.	Without condsr.	With condsr.	Without condsr.	Ground conduct.	With condsr.	Without condsr.	Ground conduct.
300	200	84	104	2	1	1	4	4	4
	440	42	52	6	6	6	8	8	8
	550	38	42	8	6	6	10	8	8
400	220	115	143	0	00	00	3	1	1
	440	57.5	71.5	4	3	3	6	6	6
	550	46	57.2	6	4	4	8	6	6
500	220	148	180	000	0000	0000	1	0	0
	440	74	90	3	2	2	6	4	4
	550	61	72	4	3	3	6	6	6

OPERATION AND MAINTENANCE

Careful observance of the following precautions and principles will do much to ensure the maximum of satisfactory service from arc welders.

Keep machine clean and cool Because of the large volume of air pulled through welders by fans in order to keep the machines cool, the greatest enemies of continuous, efficient performance are airborne dust and abrasive materials. Where machines are

TABLE 5-10 Welding Cable Sizes, Motor-Generator Welder

Machine size, amp	Cable sizes for lengths (electrode plus ground)		
	Up to 50 ft	50–100 ft	100–250 ft
200	2	2	1/0
300	1/0	1/0	3/0
400	2/0	2/0	4/0*
600	3/0	3/0	4/0*
900		Automatic application only	

*Recommended longest length of 4/0 cable for 400-amp welder, 150 ft; for 600-amp welder, 100 ft. For greater distances, cable size should be increased; however, this may be a question of cost—consider ease of handling vs. moving of welder closer to work.

TABLE 5-11 Welding Cable Sizes 11 ac/dc Welder—for Combined Lengths of Electrode and Ground Cable

Machine size, amp	Lengths up to 70 ft	70–150 ft	150–250 ft
300	0	1/0	3/0
400	2/0	2/0	4/0*
500	2/0	3/0	4/0*

*Recommended longest length of 4/0 cable for 400-amp welder is 150 ft and for 500-amp welder, 120 ft. For longer lengths, cable size should be increased; however, it may be a question of cost and flexibility, so that the welder should be moved closer to the work.

subjected to ordinary dust, they should be blown out at least once a week with dry, clean compressed air at a pressure not over 30 psi. Higher pressures may damage windings.

In foundries or machine shops, where cast-iron or steel dust is present, substitute vacuum cleaning for compressed air. Compressed air under high pressure tends to drive the abrasive dust into the windings. If vacuum-cleaning equipment is not available, compressed air may be used at low pressure.

Abrasive material in the atmosphere grooves and pits the commutator and wears out brushes.

Greasy dirt or lint-laden dust quickly clogs air passages between coils and causes them to overheat. Since resistance of the coils is raised and the conductivity lowered by heat, it reduces the efficiency and can result in burned-out coils if the machine is not protected against overload. Overheating makes the insulation between coils dry and brittle.

Do not block the air intake or exhaust vents, because doing so will interrupt the proper flow of air through the machine.

Keep the covers on the welder. Removing them destroys the proper path of ventilation.

Do not abuse it *Never leave the electrode grounded to the work.* This condition creates a so-called dead short circuit. The machine is forced to generate much higher currents than it was designed for, which can result in a burned-out machine.

Fig. 5-48 Unit which produces both alternating and direct welding current.

Do not work the machine over its rated capacity. A 200-amp machine will not do the work of a 400-amp machine. Operating above capacity causes overheating; so that the insulation may be destroyed or the solder in the commutator connections melted.

Arc Welding in Maintenance   1-73

Use extreme care in operating a machine on a steady load other than arc welding, such as thawing water pipes, supplying current for lighting, running motors, charging batteries, or operating heating equipment. For example, a dc machine, NEMA-rated 300 amp at 40

Fig. 5-49 Self-propelled truck with welding generator driven from power take-off. Has cable reel and cutting torches.

volts or 12 kW should not be used for any continuous load greater than 9.6 kW, and not more than 240 amp. This precaution applies to machines with a duty cycle of at least 60 percent. Machines with lower load-factor ratings must be operated at still lower percentages of the rated load.

Do not handle roughly. A welder is a precisely aligned and balanced machine. Mechancial abuse, rough handling, or severe shock may disturb the alignment and balance of the machine, resulting in serious trouble. Misalignment can cause bearing failure, bracket failure, unbalanced air gap, or unbalance in the armature.

Never pry on the ventilating fan or commutator to try to move the armature. To do so will damage the fan or commutator. If the armature is jammed, inspect the unit for the cause of the trouble. Check for dirt or foreign particles between the armature and frames. Inspect the banding wire on the armature. Look for a frozen bearing.

Do not neglect the engine if the welder is an engine-driven unit. It deteriorates rapidly if not properly cared for. Follow the engine manufacturer's recommendations. Change oil regularly. Keep air filters and oil strainers clean.

Do not allow grease and oil from the engine to leak back into the generator. Grease quickly accumulates dirt and dust, clogging the air passages between coils.

Maintain it regularly

Bearings. The ball bearings in modern welders have sufficient grease to last the life of the machine under normal conditions. Under severe conditions—heavy

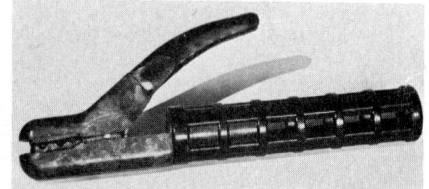

Fig. 5-50 Fully insulated electrode holder designed especially for cool operation.

usage or in a dirty location—the bearings should be greased about once a year. An ounce of grease each year is sufficient for each bearing. A pad of grease, approximately 1 cu in. in volume, weighs close to 1 oz.

Dirt is responsible for more bearing failures than any other cause. This dirt may get into the grease cup when it is removed to refill, or it may get into the grease in its original container. Before the grease cup or pipe plug is removed, it is important to wipe it absolutely clean. A piece of dirt no larger than the period at the end of this sentence may cause a bearing to fail in a short time. Even small particles of grit that float around in factory atmospheres are dangerous.

If too little grease is applied, bearings fail.

1-74　Basic Maintenance Technology

Too light grease will run out. Grease containing solid materials may ruin antifriction bearings; rancid grease will not lubricate.

Dirty grease or dirty fittings or pipes cause bearing failures.

Bearings do not need inspection. They are sealed against dirt, and it is inadvisable to open them unless necessary.

If it is necessary to pull bearings, it should be done with a special puller designed to act against the inner race. These pullers can be bought.

Never clean new bearings before installing. Handle them with care. Put them in place by driving against the inner race. Make sure that they fit squarely against the shoulders.

Brackets or End Bolts. If it becomes necessary to remove a bracket, to replace a bearing, or to disassemble the machine, do so by removing the bolts and tapping lightly and evenly with a babbitt hammer all around the outside diameter of the bracket ring. Do not drive off with a heavy steel hammer.

The bearing housing may become worn and oversized, because of the pounding of the bearing when the armature is out of balance. Bracket bearing housings may be checked for size by trying a new bearing for fit. The bearing should slide into the housing with a light drive fit. Replace the bracket if the housing is oversize.

Brushes and Brush Holders. Set brush holders approximately $\frac{1}{32}$ to $\frac{3}{32}$ in. above the surface of the commutator. If brush holders have been removed, be certain that they are set squarely in the rocker slot when replaced. Do not force the brush holder into the slot by driving on the insulation. Check to ensure that the brush-holder insulation is squarely set.

Tighten brush holders firmly. When properly set, they are parallel to the mica segments between commutator bars.

Use the grade of brushes recommended by the manufacturer of the welding set. Brushes may be too hard or too soft and cause damage to the commutator. Brushes will be damaged by excessive clearance in the brush holder or uneven brush spring pressure. High commutator bars, high mica segments, excessive brush spring pressure, and abrasive dust also will wear out brushes rapidly.

Inspect brushes and holders regularly. A brush may wear down and lose spring tension. It will then start to arc, with damage to the commutator and other brushes.

Keep the brush contact surface of the holder clean and free from pit marks. Brushes must be able to move freely in the holder. Replace them when the pigtails are within $\frac{1}{8}$ in. of the commutator or when the limit of spring travel is reached.

New brushes must be sanded in to conform to the shape of the commutator. This may be done by stoning the commutator with a stone or by using fine sandpaper (not emery cloth or paper). Place the sandpaper under the brush, and move it back and forth while holding the brush down in the normal position under slight pressure with the fingers.

See that brush holders and springs seat squarely and firmly against the brushes and that the pigtails are fastened securely.

Commutators. Commutators normally need little care. They will build up a surface film of brown copper oxide, which is highly conductive, hard, and smooth. This surface helps to protect the commutator. Do not try to keep a commutator bright and shiny by constant stoning. The brown copper oxide film prevents the buildup of a black abrasive oxide film that has high resistance and causes excessive brush and commutator wear.

Wipe clean occasionally with a rag or canvas to remove grease discoloration from fumes or other unnatural film.

If brushes are chattering because of high bars, high mica, or grooves, stone by hand or remove and turn in a lathe if necessary.

Most commutator trouble starts because the wrong grade of brushes is used. Brushes that contain too much abrasive material or have too high a copper content usually scratch the commutator and prevent the desired surface film from building up. A brush that is too soft may smudge the surface with the same results as far as surface film is concerned. In general, brushes that have a low voltage drop will give poor commutation. Conversely, a brush with high voltage drop commutates better but may cause overheating of the commutator surface.

If the commutator becomes burned, it may be dressed down by pressing a commutator stone against the surface with the brushes raised. If the surface is badly pitted or out of round, the armature must be removed from the machine and the commutator turned in a lathe. It is good practice for the commutator to run within a radial tolerance of 0.003 in.

The mica separating the bars of the commutator is undercut to a depth of $\frac{1}{32}$ to $\frac{1}{16}$ in. Mica exposed at the commutator surface causes brush and commutator wear and poor commutation. If the mica is even with the surface, undercut it.

When the commutator is operating properly, there is very little visible sparking. The brush surface is shiny and smooth with no evidence of scratches.

Generator Frame. The generator frame and coils need no attention other than inspection to ensure tight connections and cleanliness. Blow out dust and dirt with compressed air. Grease may be cleaned off with naphtha. Keep air gaps between armature and pole pieces clean and even.

Armature. The armature must be kept clean to ensure proper balance. Unbalance in the set will pound out the bearings and wear the bearing housing oversize. Blow out the armature regularly with clean, dry compressed air. Clean out the inside of the armature thoroughly by attaching a long pipe to the compressed-air line and reaching into the armature coils.

Motor Stator. Keep the stator clean and free from grease. When reconnecting it for use on another voltage, solder all connections. If the set is to be used frequently on different voltages, time may be saved by placing lugs on the ends of all stator leads. This eliminates the necessity for loosening and resoldering to make connections, since the lugs may be safely joined with a screw, nut, and lock washer.

Exciter Generator. If the machine has a separate exciter generator, its armature, coils, brushes, and brush holders will need the same general care recommended for the welder set.

Keep the covers over the exciter armature, since the commutator can be damaged easily.

Controls. Inspect every time the welder is used to ensure that the ground and electrode cables are connected tightly to the output terminals. Loose connections cause arcing that destroys the insulation around the terminals and burns them.

Do not bump or hit the control handles. It damages the controls, resulting in poor electrical contacts. If the handles are jammed, inspect for the cause.

Check the contact fingers of the magnetic starting switch regularly. Keep the fingers free from deep pits or other defects that will interfere with a smooth, sliding contact. Copper fingers may be filed lightly. All fingers should make contact simultaneously.

Keep the switch clean and free from dust. Blow out the entire control box with compressed air.

Connections of the leads from the motor stator to the switch must be tight. Keep the lugs in a vertical position. The line voltage is high enough to jump between the lugs on the stator leads if they are allowed to become loose and cocked to one side or the other.

Keep the cover on the control box at all times.

CONDENSERS

Condensers may be placed in an ac welder to raise the power factor if required. When condensers fail, it is not often apparent from the appearance of the condenser. Consequently, if it is desired to check to see if they are operating correctly, the following should be done: At rated input voltage and with the welder drawing the rated output load current, the input current reading should correspond to the nameplate amperes. If the reading is 10 to 20 percent more, at least one condenser has failed.

Caution: Never touch the condenser terminals without first disconnecting the welder from the input power source; then discharge the condenser by touching the two terminals with an *insulated* screwdriver.

DELAY RELAYS

The delay relay contacts may be cleaned by passing a cloth soaked in naphtha between them. Do not force the contact arms or use any abrasives to clean the points. Do not file the silver contacts. The pilot relay is enclosed in a dustproof box and should need no attention. Relays are usually adjusted at the factor and should not be tampered with unless faulty operation is obvious.

Table 5-12, a troubleshooting chart, may prove to be a great timesaver.

Basic Maintenance Technology

TABLE 5-12 Arc-Welding Troubleshooting Chart

Trouble	Cause	Remedy
Welder will not start (Starter not operating)	Power circuit dead	Check voltage
	Broken power head	Repair
	Wrong supply voltage	Check name plate against supply
	Open power switches	Close
	Blown fuses	Replace
	Overload relay tripped	Let set cool. Remove cause of overloading
	Open circuit to starter button	Repair
	Defective operating coil	Replace
	Mechanical obstruction in contactor	Remove
Welder will not start (Starter operating)	Wrong motor connections	Check connection diagram
	Wrong supply voltage	Check name plate against supply
	Rotor stuck	Try turning by hand
	Power circuit single-phased	Replace fuse; repair open line
	Starter single-phased	Check contact of starter tips
	Poor motor connection	Tighten
	Open circuit in windings	Repair
Starter operates and blows fuse	Fuse too small	Should be two to three times rated motor current
	Short circuit in motor connections	Check starter and motor leads for insulation from ground and from each other
Welder starts but will not deliver welding current	Wrong direction of rotation Brushes worn or missing	Check connection diagram Check that all brushes bear on commutator with sufficient tension
	Brush connections loose	Tighten
	Open field circuit	Check connection to rheostat, resistor, and auxiliary brush studs
	Series field and armature circuit open	Check with test lamp or bell ringer
	Wrong driving speed	Check name plate against speed of motor or belt drive
	Dirt, grounding field coils	Clean and reinsulate
	Welding terminal shorted	Electrode holder or cable grounded

TABLE 5-12 Arc-Welding Troubleshooting Chart (Continued)

Trouble	Cause	Remedy
Welder generating but current falls off when welding	Electrode or ground connection loose	Clean and tighten all connections
	Poor ground	Check ground-return circuit
	Brushes worn off	Replace with recommended grade. Sand to fit. Blow out carbon dust
	Weak brush spring pressure	Replace or readjust brush springs
	Brush not properly fitted	Sand brushes to fit
	Brushes in backward	Reverse
	Wrong brushes used	Renewal part recommendations
	Brush pigtails damaged	Replace brushes
	Rough or dirty commutator	Turn down or clean commuator
	Motor connection single-phased	Check all connections
Welder runs but soon stops	Wrong relay heaters	Renewal part recommendations
	Welder overloaded	Considerable overload can be carried only for a short time
	Duty cycle too high	Do not operate continually at overload currents
	Leads too long or too narrow in cross section	Sould be large enough to carry welding current without excessive voltage drop
	Power circuit single-phased	Check for one dead fuse or line
	Ambient temperature too high	Operate at reduced loads where temperatuer exceeds 100°F
	Ventilation blocked	Check air inlet and exhaust openings
Welding arc is loud and spatters excessively	Current setting too high	Check setting and output with ammeter
	Polarity wrong	Check polarity; try reversing or an electrode of opposite polarity
Welding arc sluggish	Current too low	Check output and current recommended for electrode being used
	Poor connections	Check all electrode-holder, cable, and ground-cable connections. Scrap iron is poor ground return
	Cable too long or too small	Check cable voltage drop and change cable

TABLE 5-12 Arc-Welding Troubleshooting Chart (Continued)

Trouble	Cause	Remedy
Touching set gives shock	Frame not grounded	Ground solidly
Generator control fails to vary current	Any part of field circuit may be short-circuited or open-circuited	Find faulty contact and repair

REFERENCES

"Procedure Handbook of Arc Welding," 12th ed., The Lincoln Electric Company, Cleveland, 1973.
"New Lessons in Arc Welding," The Lincoln Electric Company, Cleveland, 1973.
Jefferson, T. B., and Gorham Woods: "Metals and How to Weld Them," The James F. Lincoln Arc Welding Foundation, Cleveland, 1962.
Rossi, Boniface E.: "Welding Engineering," McGraw-Hill Book Company, New York, 1954.
Austin, John Benjamin: "Electric Arc Welding," American Technical Society, Chicago, 1952.
Morris, Joe Lawrence: "Welding Principles for Engineers," Prentice-Hall, Inc., Englewood Cliffs, N.J., 1951.
"Welding Handbook," American Welding Society, Miami, 1968.
American Welding Society Publications, 2501 Northwest 7th Street, Miami, Fla. 33125.
    Safe Practices for Welding and Cutting Containers That Have Held Combustibles, A6.0–64.
    Recommended Safe Practices for Inert-Gas Metal-Arc Welding, A6.1–66.
    Safety in Electric and Gas Welding and Cutting Operations—ANSI Standard, Z49.1-67.
Code of Minimum Requirements for Instruction of Welding Operators: Part A—Arc Welding of Steel, B2.1-45.
    A Test Program on Welding Iron Castings, D11.1-65.
    Recommended Practices for Repair Welding of Cast-Iron Pipe, Valves, and Fittings, D10.2-54.
Henry, O. H., G. E. Claussen, and G. E. Linnert, "Welding Metallurgy," American Welding Society, New York, 1949.
Sosnin, H. A.: "Arc Welding Instructions for the Beginner," The James F. Lincoln Arc Welding Foundation, Cleveland.
Linnert, G. E.: "Welding Metallurgy," American Welding Society, Miami, 1966.

Chapter **6**

Gas Welding in Maintenance

**Engineers of Union Carbide Corporation, Linde Division,
New York, N.Y.**

AIR-ACETYLENE SOLDERING, HEATING, AND BRAZING

An air-acetylene appliance produces a flame with a temperature of approximately 4000°F by mixing acetylene with atmospheric air in much the same way that air is mixed with city gas in a kitchen range. The correct mixture produces a pale blue flame with a bright, sharp inner cone that is hot enough for light silver soldering (brazing), for most soft soldering, and for hundreds of heating jobs. Air-acetylene appliances are used throughout industry as companion equipment to the oxyacetylene blowpipe for applications requiring clean, ready-to-use heat but not the extremely high temperatures of the oxyacetylene flame.

An air-acetylene outfit consists of torch handle, a torch stem or tip, a pressure-reducing regulator, a cylinder of acetylene, and a hose for connecting the torch to the regulator and tank. Interchangeable stems or tips that give various sizes and types of flames are available (see Fig. 6-1). The acetylene cylinders themselves come in all sizes, including small portable units (see Fig. 6-2).

Soldering The air-acetylene torch is extensively used for all kinds of soldering with both soft and silver (hard) solder. Although soft soldering is more widely used, silver soldering (also referred to as brazing) is sometimes used for soldering sweat-type fittings in addition to the more precise soldering associated with jewelry and instrument manufacturing. With an air-acetylene torch, the silver solder used must have a melting point lower than 1500°F. If sweat-type fittings being silver-soldered are larger than 1½ in. in diameter, or if a great number of joints are being made, an oxyacetylene torch is recommended, since its greater flame temperature speeds the work. Silver soldering commercial metals over $\frac{1}{32}$ in. thick is also best done with an oxyacetylene torch. In contrast to silver soldering, practically all soft soldering can be done with an air-acetylene torch.

Caution: Silver soldering requires a special rod and a special flux, usually in paste form. Care should be taken to follow the manufacturer's directions. The fumes from some silver solders are toxic; therefore, special ventilating precautions are necessary.

When using air-acetylene appliances, you have a choice of two soldering methods:

    1. The open (direct) flame method. The flame heats the workpiece, and the workpiece melts the solder in conjunction with the flame. The advantages of the open (direct) flame method include

        a. Speed (no copper intermediary to be heated).
        b. Greater diversity in the uses to which the flame can be put.
        c. Greater efficiency in the use of fuel (the gas goes further because it is applied directly to the workpiece).
        d. More heat because of direct application of the flame.

1-80 Basic Maintenance Technology

2. The enclosed (indirect) flame method. The flame is applied to the soldering copper. The copper in turn heats the workpiece. The workpiece, in conjunction with the soldering copper, melts the solder where it is needed. The advantages of the enclosed (indirect) flame method include
    a. Heat is better controlled.
    b. Less experience is needed on the operator's part.
    c. More delicate work is possible, especially where damage to the adjacent materials might result from the use of an open flame.

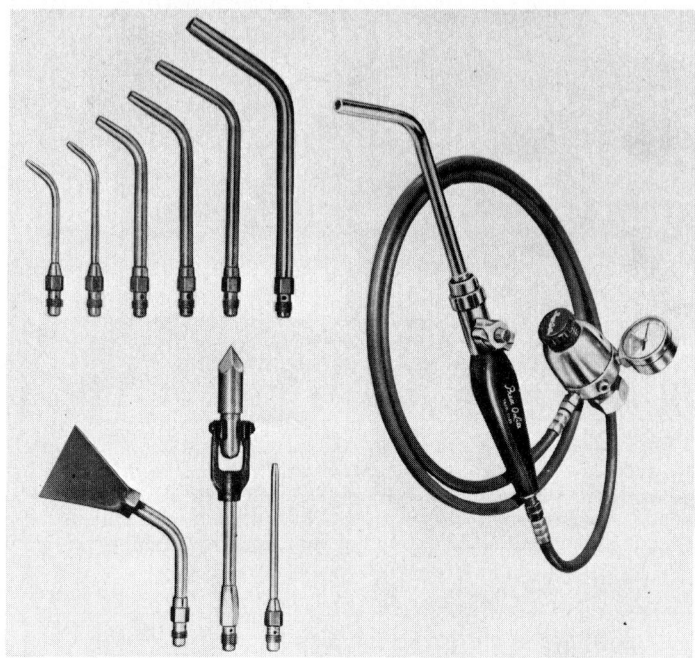

Fig. 6-1 A typical air-acetylene outfit consisting of a regulator, torch handle, and attached torch stem with interconnecting hose. Also shown are some of the typical interchangeable torch stems available. Notice the special-purpose stems: a hatchet-shaped paint-burner stem and a soldering-iron stem.

See Table 6-1 for commonly used soft solders and Table 6-2 for soldering fluxes.

Sheet-metal working Sheet-metal soldering can be done with either the enclosed (indirect) flame method or the open (direct) flame method depending on the choice of the operator. Many types of joints can be made in sheet metal. Joints described on the following pages are most widely used.

    1. The lap joint (see Fig. 6-3):
        a. Thoroughly clean the edges to be joined.
        b. Flux the edges by dipping them in a bath of hydrochloric (muriatic) acid, or using a brush, paint them with it.
        c. If you are using a soldering iron, tin the iron first and then tin the edges. If you are using a soldering torch, tin the edges. The edges should be tinned along their entire length and then placed so that the tinned edges overlap. Use C-clamps to hold them together if you have them.
        d. Next, pressing down on the soldering iron, run it up and down over the seam until a fillet of solder is visible. If you are using a soldering torch, move it back and forth with the flame touching the work until the fillet appears. In both cases, where no fillet appears, add more solder.

e. When making a long seam with a plain lap joint, it is best to tack the seam first. Tacking means applying drops or spots of solder at intervals along a seam to hold it in place. Clean, flux, and tin the entire job. Heat the seam, and apply solder spot by spot. Then do the regular soldering job on the whole seam. If the tacks tend to melt or the seams to pull apart when you near them with the torch, proceed as follows:
  (1) Press the pieces of metal together at the trouble spot with a stick.
  (2) Reheat the tacks and the solder that has been previously applied as tinning. Keep pressing the heated area together with the stick until the solder has cooled and formed a bond. Proceed with the soldering job.

f. When the joint is finished, wipe off all excess solder with a stiff bristle brush and wash off the excess flux with hot water.

TABLE 6-1 Commonly Used Soft Solders

Composition, %	Melting range, °F	Gives best results when used for
Tin, 63; lead, 37	361	Critical electronic work, coatings for printed-circuit boards
Tin, 60; lead, 40	361–374	
Tin, 50; lead, 50	361–420	General purposes
Tin, 40; lead, 60	361–460	Automobile radiators, roofing seams, wiped joints in plumbing, dip coatings
Tin, 35; lead, 65	361–478	
Tin, 30; lead, 70	361–496	Filling dents in automobile bodies
Tin, 20; lead, 80	361–534	Apply by wiping; some dip coating
Tin, 15; lead, 85	438–553	
Tin, 10; lead, 90	514–574	Where higher-melting-point solders are necessary
Tin, 5; lead, 95	574–596	
Tin, 96; silver, 4	430	Food-handling equipment, plumbing, heating, refrigeration tube joints where higher-temperature or higher-strength solders are necessary
Tin, 95; antimony, 5	450–464	Some electrical and copper-tubing joints. Do not use on zinc or galvanized sheet

TABLE 6-2 Soldering Fluxes

Metal	Flux to use*
Aluminum	Aluminum application flux
Block tin	Rosin or zinc chloride
Brass	Rosin, zinc chloride, or muriatic acid†
Cast iron	Zinc chloride or muriatic acid
Chromium	Muriatic acid
Copper	Rosin, zinc chloride, or muriatic acid
Gun metal	Rosin, zinc chloride, or sal ammoniac
Inconel	Strong zinc chloride
Iron (galvanized)	Muriatic acid
Iron (tin-coated)	Rosin or zinc chloride
Lead	Rosin, zinc chloride, or muriatic acid
Monel	Zinc chloride or muriatic acid
Nickel	Zinc chloride or muriatic acid
Pewter	Rosin, pewter application flux
Stainless steel	Strong zinc chloride
Steel (plain)	Zinc chloride
Steel (galvanized)	Muriatic acid
Steel (tin-coated)	Rosin or zinc chloride
Terne plate	Rosin or zinc chloride
Tin	Rosin or zinc chloride
Zinc	Strong zinc chloride or muriatic acid

*Nearly all these fluxes are available commercially in paste form. Pastes are usually preferred because they give excellent results on most jobs and are easy to use.
†Muriatic acid is a mild form of hydrochloric acid.

Basic Maintenance Technology

2. Lock joint (see Fig. 6-4):
   a. Thoroughly clean surfaces that will form the joint.
   b. Form the lock joint between the two sheets.
   c. Pound the joint tight with a composition mallet, or use a block of wood between the sheets and a steel hammer. Try to get the joint as flat and tight as possible.
   d. Apply acid flux along the seam, and heat the seam.
   e. Apply just enough solder to seal the seam. (You have already made the seam mechanically strong by hammering and forming the lock joint.)
   f. If the seam is fairly long, you can run the flame a few inches ahead of the solder instead of heating and soldering a section at a time.
   g. Remove all excess solder with a stiff bristle brush, and wash off excess flux with hot water.
3. Flange joint:
   a. A flange joint is generally used in conjunction with rivets or spot welds. The solder is used to make the seam tight to air, gas, or water.
   b. Before the joint is formed, the area to which the solder will be applied must be thoroughly cleaned and must remain clean until the seam is finished.
   c. A tinning coat of solder can be applied to the seam before it is riveted or spot-welded.
   d. Either use acid core solder, or flux the joint with hydrochloric (muriatic) acid.
   e. Heat the joint with either a torch or soldering iron. Capillary attraction will draw the solder into the seam. Fill the joint with the desired amount of solder.
   f. Remove all excess solder with a stiff bristle brush; wash off excess flux with hot water.

Automobile-body soldering Automobile-body soldering is done to fill in dents that cannot be hammered out completely, rough spots, and welded seams. Either soldering method can be used, direct (open) flame or indirect (closed) flame. Where the deposits of solder to be made are considerable or in places where an open flame would not damage chrome finishes or glass, we recommend the open (direct) flame method because of its speed and the rapidity with which the solder can be deposited. For the places adjacent to glass or chrome finishes, use the enclosed (indirect) flame method. When you have decided which method to use, proceed as follows:

1. Grind away the paint from the dented area, and polish with steel wool or emery cloth.
2. Flux thoroughly and, after heating, apply enough solder to tin the dent.
3. Fill in the dent by adding solder from a bar and smoothing with a maple paddle. Take care not to melt the solder until it runs. Melt it just enough to make it pasty; then smooth with the paddle.
4. When the dent is filled in, heat the solder slightly and smooth it again before letting it cool.
5. Finish the job with rasps, body files, and emery cloth. Clean, prime, and paint.

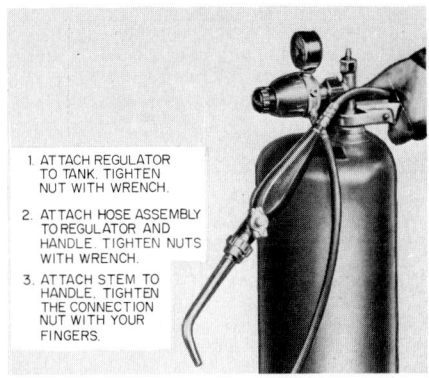

1. ATTACH REGULATOR TO TANK. TIGHTEN NUT WITH WRENCH.
2. ATTACH HOSE ASSEMBLY TO REGULATOR AND HANDLE. TIGHTEN NUTS WITH WRENCH.
3. ATTACH STEM TO HANDLE. TIGHTEN THE CONNECTION NUT WITH YOUR FINGERS.

Fig. 6-2 Connecting a typical air-acetylene outfit. The standard type of portable outfit is shown. Smaller tanks also are available.

Electrical connections For soldering electrical connections (Fig. 6-5) the enclosed (indirect) flame method is preferred. Prepare the electrical connections the way you usually do, and proceed as follows:
1. Thoroughly clean the connections.
2. Apply a noncorrosive flux paste.
3. Tin the soldering iron with a thin coat of solder.
4. Tin the wires, and melt enough solder onto them to be sure you have a good electrical connection.

NOTE: Where very large connections are to be made, an open-flame stem can be used.

Gas Welding in Maintenance 1-83

Installing sweat-type fittings The following is the most efficient method for making sweat-type joints as recommended by two of the leading copper-tube manufacturers. The air-acetylene torch with a direct (open) flame is used by literally thousands of plumbers and is universally recognized as the best means of making these joints. The torch saves time and money, and a relatively inexperienced worker can do a good job with very little training and practice.

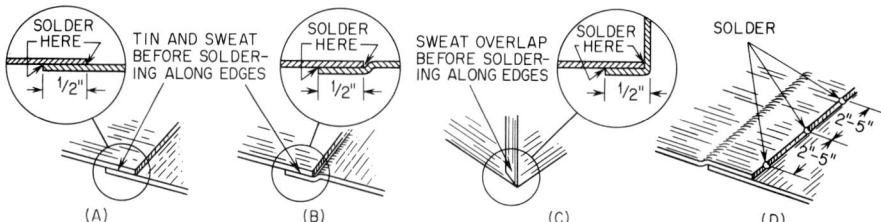

Fig. 6-3 Three variations of a lap joint (A, B, and C); at the right (D), notice how to tack a long seam by applying spots of solder at intervals.

There are two basic types of sweat-type fitting: the plain type and cast type. With the plain type the solder is fed at the point where the fitting and the tube join. With the cast type the solder is fed through precast holes in the fitting itself. The instructions below will work equally well with both types (see Figs. 6-6, and 6-7).

1. Cut the tube to the length required with a hacksaw (32 teeth to the inch), or a disk cutter. Make certain that the tube ends are cut square. Special vises which hold the tube securely and guide the saw blade are furnished by a number of manufacturers.

2. Ream the tube, and remove burrs on the outside. Use a sizing tool if necessary to correct any possible distortion of the tube from handling. The point of a sizing tool is inserted in the end of the tube and is hammered until the tube is again round.

3. Clean the outside surface of the tube and the inside surface of the fitting until the metal is bright. All traces of discoloration must be removed. This must be done even though the tube may appear to be perfectly clean, and it is particularly important when soldering larger-size joints. No. 00 steel wool is very satisfactory for cleaning tubes and fittings. Do not use files or rough sandpaper, as they score the surface and may result in a poor joint.

4. Apply a thin, uniform, and complete coating of a reliable brand of soldering flux or paste to the cleaned portion of both tube and fitting. Do not apply the flux too thickly, as excess flux may form bubbles when heated and prevent the solder from creeping into the joint. After the tube has been inserted into the fitting as far as it will go, revolve the fitting once or twice to spread the flux evenly.

5. Apply the flame evenly all around the circumference of the fitting, and as it

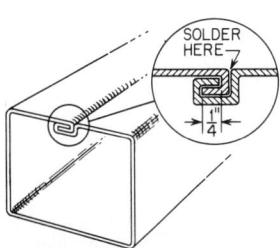

Fig. 6-4 A lock joint where mechanical strength is provided by the joint rather than by the solder bond.

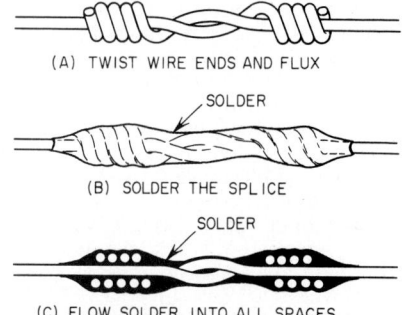

Fig. 6-5 Soldering electrical connections.

becomes heated, move the flame back and forth to prevent overheating. Occasionally test the heat by touching the fitting with solder where the tube and fitting join. Do not let the flame touch the solder while testing the temperature of the joint.

It is important not to overheat the joint. If the connection is heated too much, the flux may be burned out from inside the joint and the solder will not spread properly. An overheated joint causes the solder to seep through the joint and run away.

During the heating operation, adjacent wood surfaces should be protected from the heat by means of sheet asbestos. Because of its narrow, concentrated flame, the air-acetylene torch can be used very close to wood surfaces without scorching them.

6. Remove the flame, and apply solder to the edge of the fitting where it comes in contact with the tube as soon as the fitting has reached the correct temperature to melt the solder. Be sure that enough solder is used.

Enough solder to make an efficient joint will be automatically sucked in by capillary attraction. When a line of solder shows completely around the fitting a fillet of solder appears in the chamfer at the end

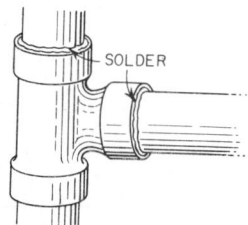

Fig. 6-6 Where to solder plain-type sweat fittings.

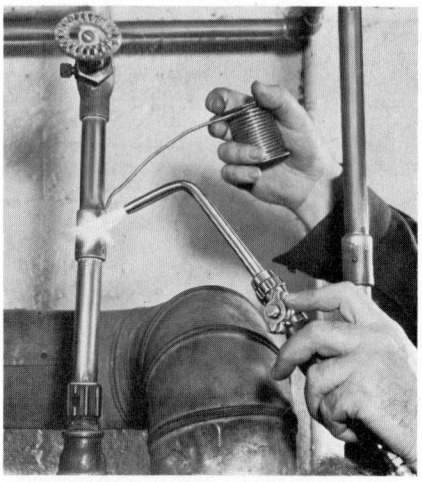

Fig. 6-7 Soldering 1-in. copper tubing and fitting with precast holes.

of the fitting, and the joint has all the solder it will take. Wipe off any excess solder or flux.

7. Slightly reheat the connection in order to help the solder permeate the metal. Remove the flame, and continue to feed solder to make certain the joint is filled.

8. Permit the connection to cool for a fraction of a minute. A rag or wad of waste saturated with water will hasten the cooling. Remove all surplus solder from around the edges with a brush. This operation will show whether or not the solder has filled the joint.

9. When disconnecting a soldered tube from a fitting on which other soldered connections are to be left intact, the application of wet cloths to the parts which are not to be disconnected will prevent melting of the solder at such connections.

10. More than ordinary care should be exercised in soldering fittings 2½ in. in diameter and larger. It is essential that the heat be uniformly distributed around the entire circumference of the fitting and not concentrated in one spot.

When making large-diameter joints, a tip producing a large flame should be used. The flame should be directed on the fitting to avoid any unnecessary annealing of the tube.

For assembling lines 3 in. in diameter or over, it may be advisable to use two or three torches. Solder should then be applied simultaneously at two or more points.

11. In applying solder to a tee, feed solder from both ends of the fitting.

12. Solder when confined between two surfaces will run uphill (by capillary attraction), and joints can be made in almost any position.

13. In sweating male and female adapters, care should be taken to allow more time for the solder to set, as these heavier fittings do not cool so quickly.

Paint burning An air-acetylene torch with a paint-burning stem is a quick, easy, and economical means of removing old, cracked, and checked paint from a surface that can stand a moderate amount of heat. The number of coats of paint is not important; it just takes a little more time to remove them. Paint can be removed from wood, canvas, brick, stone, or metal.

Caution: Avoid inhaling any dust or fumes that may be given off in the paint-burning operation. Such dust and fumes may be toxic, particularly if the paint being removed contains lead or cadmium compounds.

There are two methods of removing paint. They are listed below as Method A and Method B. We suggest you try both methods. You can then use the one that suits your particular type of work. Once the old paint is removed and rough spots smoothed, the surface is ready for a new coat of paint.

Method A (Fig. 6-8):

1. Hold the paint burner in your left hand. Hold the putty knife (with a stiff blade about 3 in. wide) in your right hand.

2. Move the torch backward and forward 1 in. from the painted surface about 6 in. at a stroke. Follow the movements of the torch with a steady forward movement of the putty knife, keeping the putty knife hot with the flame.

NOTE: You will find it advisable to wear asbestos or other heavy flame-resistant gloves when burning paint. The putty knife gets very hot after a while; so you should protect your right hand. Cloth (cotton) gloves are not satisfactory.

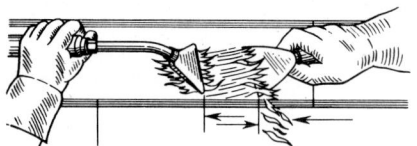

Fig. 6-8 Paint burning (Method A).

Fig. 6-9 Paint burning (Method B).

3. Moving the torch back and forth changes the paint to a plastic state and keeps the putty-knife blade hot. A hot blade reduces the tendency of the paint to stick to it.

Method B (Fig. 6-9):

1. Move the torch more or less steadily from right to left over the painted surface. Bring the paint to a bubbly plastic state. Scrape off the paint as soon as it bubbles. Do not let the flame touch the blade of the putty knife.

2. The putty knife should have a back-and-forth motion which will intermittently expose the scraped area to the flame. This method is recommended for particularly heavy or stubborn paint.

Miscellaneous air-acetylene applications There are many repair and maintenance jobs aside from soldering and brazing that can be done efficiently with an air-acetylene torch. A few of these applications are given below.

Loosening Nuts and Bolts. Frequently you come across a bolt that resists all attempts to loosen it with a wrench. Heat the nut for several minutes and let it cool; then try the wrench again. Generally, you will now find the nut ready to turn.

Freeing Frozen Shafts. A frozen shaft of small diameter can be freed by heating the collar that holds it. Heat the collar, not the shaft. You will find that you can separate the parts quite quickly no matter how tightly they are frozen together.

Lead Working. The air-acetylene torch can be used to build up lead battery terminals. Any of the standard stems can be used depending on the amount of work to be done and the speed with which you want to do it.

It is recommended that you use a form, where possible, to keep the lead in the shape of a battery terminal and to prevent it from running on the battery. Put the form over the old terminal and keep adding melted lead until the desired height and shape of the terminal are attained.

The air-acetylene torch can be used to repair lead-lined vats, wipe joints in lead pipe and lead-covered cable, and solder battery-cable lugs.

Caution: When working with lead in a confined space, be very sure of your ventilation and, if possible, use a suitable air line mask.

Anchoring Bolts in Concrete or Stone. Firmly anchoring a large bolt in concrete or stone can be solved as follows:

1. Drill a hole in the concrete or stone with a star or other type drill. It is best to dish or widen the bottom of the hole slightly to increase the stability of the bolt after the solder sets. Make certain all free moisture or water is removed from the cavity.

2. Heat the solder (bar solder is best) in a ladle with an air-acetylene torch until the solder is molten.

3. Place the bolt in the hole thread-end up, and pour molten solder around it until the solder is level with the floor.

4. This type of mounting will give years of satisfactory service; if it should become loose, just reheat the solder with the torch and it will be as tight as ever.

Cutting Asphalt Tile. The air-acetylene torch has been used with good success by asphalt-tile contractors for heating tiles that have to be bent, formed, or cut. After a few seconds of heating, the tile can be shaped or cut with great ease.

Cutting Safety Glass. Using an air-acetylene torch with a medium-sized stem, the following procedure can be used when cutting safety glass:

1. Score both sides of the glass with a glass cutter and break the glass.
2. Soften the plastic filler by running the torch back and forth along the lines of the cut.
3. Wobble the glass from side to side several times. Then hold the glass to one side while you cut the heat-softened plastic filler with a razor blade.

Precautions and safe practices

1. Do not let acetylene escape near any possible source of ignition. Accumulations of acetylene in certain proportions may explode if ignited.

2. *Never* store acetylene tanks in a closed or confined space, such as a closet.

3. *Never* solder a container that contains or has contained flammable liquids or vapors (including gasoline, benzene, solvents, and other similar or dissimilar materials) unless the container has been thoroughly purged of all traces of flammable material and vapors. Be sure that any container you work on is vented. We urge that before you do work of this kind, you get Booklet A-6.0.40 from the American Welding Society, 2501 Northwest 7th St., Miami, Fla. 33125.

4. *Never* use a tank with a leaking valve.

5. *Do not* make any repairs to an acetylene tank, except to tighten the packing-gland nut on the valve.

6. *Do not* abuse or drop tanks or handle them roughly.

7. *Never* use a tank as a roller. Never use a wrench or pliers on the tank valve. Always use a valve key.

8. *Never* allow full tank pressure to enter a stopped hose. Always use a regulator when there is a needle valve on the torch handle.

9. Examine your hose for leaks frequently. Dipping it in a bucket of clean water, with the pressure in the hose, is the quickest and easiest way.

10. *Do not* use hose that is worn or any equipment that is in need of repair.

OXYACETYLENE WELDING, CUTTING, GOUGING, AND HARD-FACING

Metal production, fabrication, and repair, as they are known today, would be impossible without the oxyacetylene process and its flame of approximately 6000°F. The oxyacetylene process is built on two principles: (1) acetylene burned with an equal amount of oxygen produces an intensely hot flame that will melt and fuse most metals, and (2) a jet of oxygen striking a piece of ferrous metal that has been heated to its kindling temperature will rapidly burn the metal away.

Welding and brazing Welding with an oxyacetylene blowpipe is simple. You put two pieces of metal together, then melt the edges with an oxyacetylene flame. The molten metal flows together and forms a single, solid piece of metal. Welding rod similar to the base metal is usually added to strengthen the joint. This is known as *fusion welding.* If you use a steel rod, the process is sometimes called *steel welding.*

Braze welding is another method. In braze welding, the two pieces being joined are heated to a dull red. They are not melted. A flux is added to clean the metal and protect it from the air. When the pieces are dull red, molten-bronze welding rod is added to form a strong bond. This bronze weld is generally as strong as the base metal.

Building up worn parts with bronze or steel welding rod, heating and forming of metals, gouging, hard facing, and soldering are other jobs done by the oxyacetylene flame.

Cutting Oxygen cutting is similar to the eating away of steel by ordinary rusting, only it is very much faster. In rusting, the oxygen—in the air or in water—affects the metal

Gas Welding in Maintenance 1-87

slowly. Directing a jet of pure oxygen at metal heated almost to the melting point actually speeds up chemical reaction of rusting to such an extent that the metal ignites and burns away. The iron oxide melts and runs off as molten slag to expose more iron to the action of the oxygen jet. This makes is possible to cut iron and steel leaving a smooth, narrow cut. An oxyacetylene outfit is shown in Fig. 6-10.

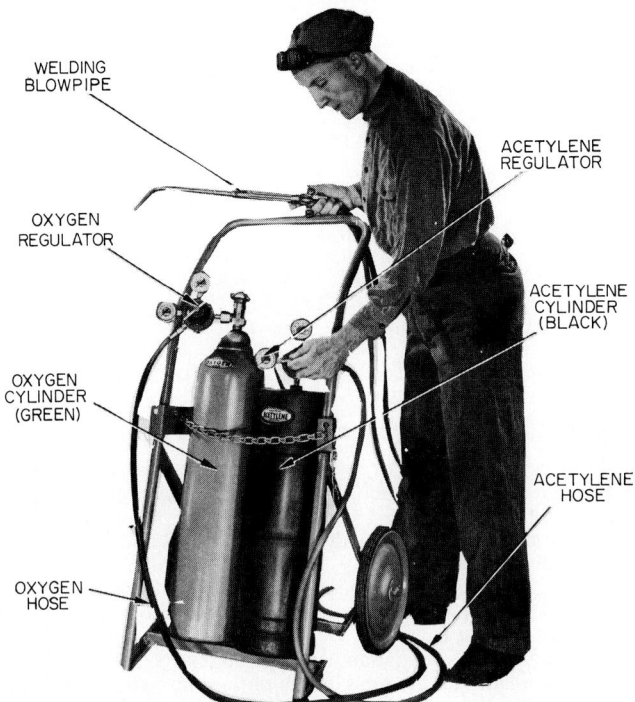

Fig. 6-10 A complete oxyacetylene outfit. The welding blowpipe shown may be adapted for cutting by exchanging the welding tip with a cutting attachment. Blowpipes designed especially for cutting only are also available.

Setting up an outfit Suggestions and recommendations for safe handling of oxyacetylene equipment have been set forth by the International Acetylene Association. They are included here in brief form.
1. Fasten your cylinders.
    a. Use a cylinder truck, or tie them to a post or bench with a chain, wire, or strap iron.
2. "Crack" the cylinder valves.
    a. Stand behind them; open each a fraction of a turn and close immediately. This blows out any dirt that may have collected in the valve.
3. Connect the regulators.
    a. Oxygen connections have right-hand threads.
    b. Acetylene connections have left-hand threads.
4. Loosen the pressure-adjusting screws on the regulators.
5. Open the oxygen cylinder valve slowly, then as far as you can.
6. Open the acetylene cylinder valve only 1½ turns.
    a. Stand to one side of the regulator gauges when opening these valves.
    b. Leave the T wrench in the acetylene cylinder in case you have to shut it off quickly.

1-88   Basic Maintenance Technology

7. Attach the hoses to the regulators, then to the blowpipe.
    a. Use oxygen to blow out new hoses.
8. Attach a welding head to the blowpipe.
    a. Tighten the connection nut.
9. Turn off valve at blowpipe, admit pressure to hose, and test for leaks.
10. Adjust the oxygen pressure.
    a. Open the blowpipe oxygen valve.
    b. Turn in the regulator pressure-adjusting screw until the regulator gauge shows desired pressure.
    c. Close the blowpipe valve.
11. Adjust the acetylene pressure.
    a. Same procedure as for oxygen.

Flame adjustment   The three basic types of flames for an oxyacetylene blowpipe are shown in Fig. 6-11.

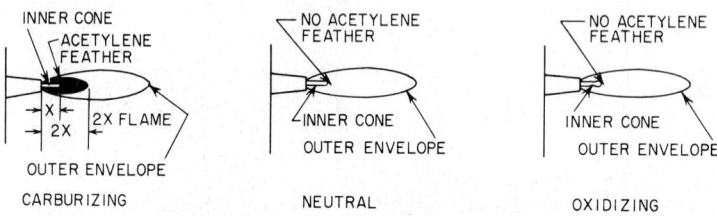

Fig. 6-11   The three basic types of flames for an oxyacetylene blowpipe.

1. To adjust to an excess acetylene (carburizing) flame, start with both blowpipe valves closed. Then
    a. "Crack" the blowpipe oxygen valve; open the acetylene valve about a full turn. Light the blowpipe.
    b. Increase the oxygen supply until you see three distinct parts to the flame: a brilliant inner cone, a whitish acetylene feather, and a bluish outer envelope. This is a carburizing flame.
    c. The amount of excess acetylene in the flame is expressed as a ratio of the total length of the feather to the total length of the inner cone. Thus, in a 2X flame, the acetylene feather is twice as long as the inner cone.
2. To adjust to a neutral flame
    a. Proceed as above, but keep adding oxygen until the acetylene feather just disappears.
    b. This leaves two parts to the flame: a brilliant white inner cone and a bluish outer envelope.
    c. At this point, the blowpipe is burning equal amounts of oxygen and acetylene. This is a neutral flame.
3. To adjust to an oxidizing flame
    a. Proceed as above until a neutral flame is obtained.
    b. Keep adding oxygen beyond the point where the acetylene feather disappears until the inner cone shortens (about 20 percent shorter than a neutral inner cone) and becomes "necked-in."
    c. A harsh sound also characterizes this oxidizing flame unless a very low flow is used.
4. Carburizing, neutral, and oxidizing flames can be harsh or soft. You get a harsh flame when using almost the maximum flow through a tip; you get a soft flame when using less than normal flow.
    a. In a harsh flame, the pressures approach blowoff; that is, a slight increase in pressure causes a gap to appear between the flame and the tip.
    b. In a soft flame, the gas flow is reduced with the blowpipe valves. The inner cone is about half as long as that in a harsh flame.

Braze welding   (see Fig. 6-12) Braze welding is a process which enables you to weld various metals and alloys without melting the base metal. Using a bronze rod (which melts between 1500 and 1650°F) as a filler metal, you can make strong joints in many metals and alloys. The process is similar to soldering, the difference being that solder melts at a much lower temperature than the bronze rod does and is of much lower strength.

In braze welding, a slightly oxidizing flame is generally used, since a carburizing flame gives off certain gases that dissolve in the molten puddle and leave weak, porous spots in the weld.

You can braze-weld cast, malleable, wrought, and galvanized iron; carbon steels and alloy steels, copper, brass and bronze; nickel; Monel; Inconel; and other metals. The following are some of the features and advantages of braze welding:

1. Braze welding can be used for many repair and fabrication jobs.
2. It is faster than fusion welding because less heat is required to melt the filler metal.
    a. This means that you use less gas; so costs are lower.
    b. Less heat means less distortion in the piece being braze-welded.
    c. More work can be done in less time with this fast process.

Fig. 6-12 A braze weld.

Fig. 6-13 A fusion weld.

3. Braze welding produces good strong joints.
    a. Bronze rod, properly deposited, can have a tensile strength up to 56,000 psi.
    b. Tensile strength of plain low-carbon steel is about 52,000 psi.
4. Braze welding can be used for joining dissimilar metals: cast iron to steel, iron to copper.
5. Braze welding can be used to join malleable iron parts and to repair large castings.

Fusion welding   (see Fig. 6-13) Fusion welding is the joining of metal by melting and fusing the edges together. The joint is a thorough mixture of the base metal and the welding or filler rod used to build up the seam. There is no sharp line of demarcation as with a braze weld. The filler rod is used in all cases, except when you are welding sheet metal, and should be about the same composition as the base metal. For example, you use steel rods when welding various plain-carbon or alloy steels while cast-iron rod is used for welding iron castings.

Fusion welding is used mainly where you cannot use braze welding. It has a wide appeal to small users for light-gauge, mostly nonproduction work and for maintenance jobs, although it has largely been displaced for large-scale production work by electric-arc welding. Fusion welding, rather than braze welding, is necessary for parts that will be in use at high temperatures. As a braze-welded joint becomes heated, it loses its strength rapidly, since the filler metal will melt at about 1650°F.

Parts subjected to great tensile stresses, i.e., great pulling loads, should be fusion-welded. For example, some steels have tensile strengths up to 90,000 psi and more. These exceed the tensile strength of a braze weld (up to 56,000 psi) and must be fusion-welded with a special steel rod if the joint strength is to equal or exceed the strength of the base metal.

Fusion welding can also be used where an approximate color match between welded parts is necessary.

Fusion welding uses a neutral or slightly oxidizing flame, since a carburizing flame can cause entrapments in the filler metal.

Weld preparation
1. As a part of your preparation for welding, you should select and prepare the proper joint design for your work (see Fig. 6-14).
    a. Square-edge butt and flange welds are commonly used in sheet-metal work. In the latter, the edges are turned up and melted to form the joint and no filler rod is needed.
    b. Lap joints are rarely used except where one cylindrical section fits inside another.

c. The butt joint with beveled edges (the V) is most widely used. For plate over 3/16 in., bevel the edges by oxygen cutting, grinding, or machining to an included angle of 90°.
d. The corner fillet and double-V joints can also be used, depending on the demands of the job.
e. In the double-V joint, the plates are welded from both sides. It is generally used for work thicker than ½ in.

2. The second step in preparation is to clean the edges of any oil or grease, dirt, scale, or rust with steel wool, a wire brush, or some other means.

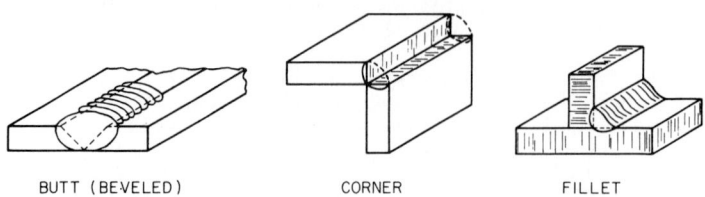

BUTT (BEVELED)   CORNER   FILLET

Fig. 6-14 Butt, corner, and fillet joints.

3. Before welding, the pieces must be properly spaced, since they will tend to expand during welding. Two types of spacing are used to counteract expansion. On material ⅛ in. thick or less, the edges are generally placed parallel to each other about ⅛ in. apart. They should then be tack-welded every 6 in. or so to prevent undue distortion. This space between the plates allows the molten filler metal to flow to the bottom of the joint. Good penetration is thus assured. For material over ⅛ in. thick, a progressive method of spacing is used (see Fig. 6-15). For every foot of weld length, the pieces should be spread apart about ⅛ in. For example, welding pieces 2 ft long, you would leave the pieces approximately ¼ in. apart at the finishing end of the weld. At the starting end, the pieces should be spaced about 1/16 in. Slightly more spacing is required for unbeveled edges.

4. Tack-weld the pieces at start and finish. In sheet metal, tack-weld about every 6 in. This keeps the pieces in alignment and prevents them from drawing too close together as they are heated during welding. There is a great deal of strain on a tack weld as a result of these internal expansion and contraction stresses. So make your tack welds carefully, and make them strong.

Blowpipe motion There is no hard-and-fast rule which will tell anyone exactly how to move a blowpipe when welding. Each welder develops a natural motion after a little practice. Figure 6-16 gives a suggested pattern for moving the blowpipe while welding. The motion is effectively a series of semicircles, wide enough to ensure heating beyond the limits of beveling and moving forward to a slight extent in each blowpipe swing. For fusion welding, it is important to try to move the flame around in front of the rod at the end of each sweep so that complete melting of the edges is obtained. With braze welding, it is not necessary to melt the edges of the joint; so the blowpipe motion is generally faster.

Making a weld The procedures for making a braze or fusion weld are essentially the same except that, for fusion welding, the edges of pieces being joined are actually melted, while for braze welding, the edges of the pieces being joined are heated to a dull red. In both fusion and braze welding, the filler rod is melted to furnish the filler metal for the seam. The following brief discussion applies to both fusion and braze welding. Nevertheless, the difference between the two welding methods should always be remembered.

1. Steel thicker than 5/16 in. can be welded in two or more passes, while material over ½ in. should always be multipass welded (see Fig. 6-17).

2. Lay in the first pass or root weld from 2 to 3 in. long. After making this beginning section of the root weld, go back and build up the finishing weld to the desired reinforcement.

3. There are two reasons for making a root weld for about 2½ in. and then returning for the second pass or finishing weld before continuing with the root weld (see Fig. 6-18):

a. You take advantage of the heat left in the plate when you made the beginning section of the root weld. If you continued all the way across the plate, this heat would be lost when you returned for a second pass.

b. Experience shows that best results in strength, uniformity, and appearance are achieved when this system is used.

4. The following points are important in making a good weld:

   a. Do not add filler rod until you have formed a molten puddle (fusion welding) or heated to a dull red (braze welding).

   b. Keep the rod in the puddle.

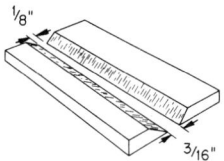
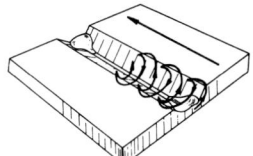
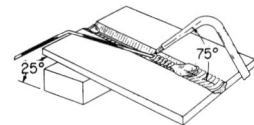

Fig. 6-15 When making a butt weld of pieces over 1/8 in. thick, progressive spacing will counteract distortion caused by expansion.

Fig. 6-16 Blowpipe motion for fusion and braze welding.

Fig. 6-17 Braze or fusion welding of two pieces thicker than 3/16 in. is best done with two passes. Notice that the weld proceeds in stages.

   c. Keep your eye on the leading edge of the puddle to ensure that you always have thorough fusion or heating.

   d. Remember that the rod is deposited evenly by constantly melting it into the molten puddle, not by applying the flame directly to the rod.

5. The blowpipe should be tilted slightly to an angle of 75° with the plate surface, to ensure a certain amount of preheat as the weld proceeds. The plate may be tilted upward to an angle of perhaps 25° to aid in even buildup.

6. The blowpipe should be directed squarely into the V between the plates so that both sides will be heated evenly.

7. Proper weld sequence is shown below. The first root weld is made for about 2½ in.; the first finishing weld is about half this length. Each successive pass, both root and finishing, is about the same length as the first section. The final section of finishing weld will be a bit longer than any other part to make up for the shortness of the first finishing weld.

8. Never make a flush weld if maximum strength is desired. Always provide reinforcement; that is, make sure that the lowest ripple is 1/16 in. above the surface of the plate.

Heavy braze welding The following points apply to such heavy jobs as repairing heavy steel or iron castings:

1. Preparation of the work is important. First, vee out the crack with your gouging nozzle on steel parts or by chipping, grinding, or machining if the piece is cast iron. Be sure to clean thoroughly a generous space on each side of the V to permit the crown of the weld to lap over and give additional strength.

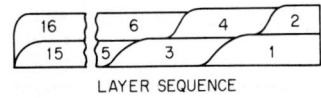

LAYER SEQUENCE

Fig. 6-18 Sequence of root and finishing welds for a weld made in two passes (layers).

2. On cast-iron pieces, it is fairly certain that graphite (pure carbon) flakes are embedded in the surface and have been smeared by machining or grinding. In the presence of an oxidizing flame, this carbon will unite with oxygen and burn off as a gas. Use steel wool or a wire brush on the surface to complete the cleaning job.

3. Choose a location where it will be possible to set up a temporary preheat furnace. The reasons for preheating are:

   a. It is easier to braze-weld if heat is stored up in heavy pieces. A fairly small and convenient welding flame can be used if the pieces are at about 500°F.

   b. If the pieces are cold, heat from the welding flame would be rapidly drawn away.

  c. Preheating will help prevent excessive internal stresses from occurring as the piece cools.

 4. Depending on the shape of the piece, it may be possible to make a double V and have a welder work on each side of the joint.

 5. Since it is easier to build up a weld in successive horizontal layers, position the work if possible so that your weld line is flat.

  a. If possible, it is desirable to support your starting weld on a carbon block or piece of firebrick. If this cannot be done, use a piece of 10-gauge sheet or carbon plate fitted to the bottom of the abutting pieces. When the weld is finished and the casting is cool, you can remove the sheet if necessary by chipping.

  b. Use plenty of flux or flux-coated rod so the tinning action will take place automatically and stay well ahead of the weld itself.

 6. When completed, the weld can be cleaned by starting at the top and working downward with a large oxidizing flame, melting the runovers.

 7. In cases where the castings are spread out, be sure that they are well supported, since cast iron is weakened when heated to a high temperature.

Fusion-welding cast iron

 1. Cast iron does not have the strength and ductility of steel. Without careful cleaning before and proper cooling after welding, a casting may become hard and brittle and possibly crack.

  a. Clean off any dirt, scale, and grease that might weaken the final weld with a wire brush, a grinder, or a file.

  b. In order to equalize internal expansion and contraction stresses introduced during welding, preheat small castings locally with your blowpipe. Large castings should be placed entirely in a preheat furnace and raised to a temperature of approximately 500°F. The stresses of concentrated welding heat might crack the casting without this preheating.

 2. Molten cast iron is very fluid and may tend to fall through. It is also a good idea to weld "in the flat" with some sort of backup where possible. Carbon blocks may be removed after the weld has cooled.

 3. Bevel the edges, by chipping or grinding, to an included angle of about 90°.

 4. To help further in cleaning the edges so that a clean, sound weld will be obtained, use a flux that will chemically float out dirt, slag, and oxide inclusions.

 5. Add just enough flux so that all the impurities are cleaned and fluxed out of the weld zone.

 6. Use only one pass. It is not necessary to fill in a root weld and then a finishing weld as was the case with steel.

 7. Cast iron must be cooled slowly after welding. Sudden chilling of a recently welded cast-iron part can cause it to crack. Fast cooling also makes a casting hard, brittle, and subject to being cracked easily; slow cooling imparts softness and ease of machinability. Small parts can be placed into a can of lime or cement or some similar material so that they will cool properly. Larger castings can be left in the preheat furnace for slow, even cooling.

 For recommended welding methods see Tables 6-3 and 6-4.

 Oxygen cutting Iron burns (all burning is an oxidation process) as readily as wood or paper if it is heated to the right temperature and is exposed to a large amount of pure oxygen. Metals like aluminum, stainless steel, and magnesium also oxidize, but it takes even more heat to melt their oxides than it does for iron oxide. Other means must be used to cut them. Oxygen cutting is primarily intended for cutting ferrous metals (iron or steel).

 The first step in oxygen cutting ferrous metals is to preheat the metal until it is red hot. At this point, the metal is said to be at its kindling or ignition temperature—it is ready to burn away. The actual cut in the metal is started by directing the pure oxygen stream from a cutting blowpipe at the preheated metal. The hot iron and the oxygen react instantly, producing so much heat that the oxide formed melts and flows or is blown away. As the oxide flows away, the cut progresses through the metal as the next layer of metal is exposed to the oxygen. When the blowpipe is moved along the line of cut, the heat of the reaction between the iron and oxygen raises the temperature of these successive layers of metal.

TABLE 6-3 Recommended Welding Methods (Ferrous)

Metal	Welding method	Flame adjustment	Recommended welding rod	Flux
Steel, cast.........	Fusion weld	Neutral	High-test steel	None
Steel pipe	Fusion weld	Neutral	High-test steel	None
	Steel welding	Carburizing	CMS steel	
Steel plate	Fusion weld	Neutral	Drawn iron	None
	Steel welding	Carburizing	High-test steel	
			CMS steel	
Steel sheet	Fusion weld	Neutral	Drawn iron	None
	Bronze weld	Slightly oxidizing	High-test steel Bronze	Brazing None
High-carbon steel	Fusion weld	Carburizing	High-test steel CMS steel	None
Manganese steel...	Fusion weld	Slightly oxidizing	Same composition as base metal	None
Cromansil steel....	Fusion weld	Neutral	High-test steel CMS steel	None
Wrought iron	Fusion weld	Neutral	High-test steel	None
Galvanized iron ...	Fusion weld	Neutral	Drawn iron	None
	Fusion weld	Neutral	High-test steel	None
	Bronze weld	Slightly oxidizing	Bronze	Brazing
Cast iron, gray.....	Fusion weld	Neutral	Cast iron	Ferrous
	Bronze weld	Slightly oxidizing	Bronze	Brazing
Cast iron, malleable	Bronze weld	Slightly oxidizing	Bronze	Brazing
Cast-iron pipe, gray	Fusion weld	Neutral	Cast iron	Oxweld ferrous
	Bronze weld	Slightly oxidizing	Bronze	Brazing
Cast-iron pipe	Fusion weld	Neutral	Cast iron Same composition as base metal	Ferrous
Chromium-nickel ..	Bronze weld	Slightly oxidizing	Bronze	Brazing
Chromium-nickel steel castings	Fusion weld	Neutral	Same composition as base metal 25-12 chromium-nickel steel Columbium-bearing 18-8 Stainless steel	Stainless steel
Chromium-nickel steel (18-8)	Fusion weld	Neutral	Columbium-bearing 18-8 Stainless steel	Stainless steel
Chromium-nickel steel (25-12)	Fusion weld	Neutral	Same composition as base metal	Stainless steel
Chromium steel ...	Fusion weld	Neutral	25-12 chromium-nickel steel Columbium-bearing 18-8 Stainless steel	Stainless steel
Chromium steel (4–6 percent)	Fusion weld	Neutral	Columbium-bearing 18-8 Stainless steel	Stainless steel
Chromium iron	Fusion weld	Neutral	25-12 chromium-nickel steel Columbium-bearing 18-8 Stainless steel Same composition as base metal	Stainless steel

1-94 Basic Maintenance Technology

Oxygen cutting is used almost everywhere—for cutting straight lines and circles in plate, for cutting shapes to accurate dimensions in single pieces and in stacks, for trimming plate to size and beveling it for welding, for piercing holes, for cutting I beams and other structural members to size, and for many other uses. The oxygen-cutting blowpipe is also a prime fabricating tool in industry for preparing plates and cutting

TABLE 6-4 Recommended Welding Methods (Nonferrous)

Metal	Welding method	Flame adjustment	Recommended welding rod	Flux
Aluminum	Fusion weld	Slightly carburizing	Aluminum	Aluminum
Brass	Fusion weld	Oxidizing	Bronze	Brazing
	Bronze weld	Slightly oxidizing	Bronze	
Bronze	Fusion weld	Neutral	Bronze	Brazing
	Bronze weld	Slightly oxidizing	Bronze	
Copper (deoxidized)	Fusion weld	Neutral	Deoxidized copper	None
	Bronze weld	Slightly oxidizing	Bronze	Brazing
Copper (electrolytic)	Fusion weld	Neutral	Cupro	None
	Bronze weld	Slightly oxidizing	Bronze	Brazing
Everdur bronze	Fusion weld	Slightly oxidizing	Everdur bronze	Silicon bronze
Nickel	Fusion weld	Slightly carburizing	Same composition as base metal	None
Monel metal	Fusion weld	Slightly carburizing	Same composition as base metal	Monel
Inconel	Fusion weld	Slightly carburizing	Same composition as base metal	Inconel
Lead	Fusion weld	Slightly carburizing	Same composition as base metal	None

structural members in the shipbuilding, heavy-machinery, and building-construction industries. Oxygen cutting is also extensively used for demolishing and scrapping of machinery, obsolete equipment, unsafe or unwanted structures; for cutting heavy scrap to smaller size; for removing bolts and rivets; and for similar work (see Fig. 6-19).

Oxygen cutting is very versatile in that steel, wrought iron, and cast iron can be cut in almost any form, of almost any thickness. Hand cutting is restricted to thicknesses of about 1 in. Machine cuts have been made, however, in material of about 6 in. in thickness.

The process is inexpensive. Initial equipment cost and subsequent upkeep costs are very low compared with other means of doing the same job. The gas costs are almost negligible when you consider the variety and quality of the work done. The equipment needed for oxygen cutting is easily portable and can be taken almost anywhere for on-the-job use. The process is very fast. Depending on the thickness of the material, speeds up to 500 fph can be attained. The process is easily learned. The correct techniques can be studied and picked up in a few minutes.

Oxygen-cutting equipment   (See Fig. 6-20) Oxygen cutting requires the same equipment needed for welding, including a welding blowpipe fitted with a cutting attachment and a special nozzle. Where you are going to do oxygen cutting for long periods, a cutting blowpipe is more desirable than a cutting attachment. The cutting nozzles come in various sizes. The thickness of the metal and its surface condition determine the size of the nozzle needed. For example, five different-sized nozzles handle all thicknesses up to 12 in.

Various accessories, which supplement basic equipment, are available for making special types of cuts. In freehand guided cutting, the blowpipe head can be drawn along a bar or straightedge. This will assure an accurate square or beveled straight-line cut. Circles or disks with 2 in. or greater diameters can be accurately made with circle-cutting

Gas Welding in Maintenance 1-95

attachments. Where high accuracy is required in cutting straight lines, circles, or shapes, special machines are available which mechanically hold, guide, and advance the blowpipe over the work. Little or no finishing is required on these high-quality machine cuts.

Preparation for cutting

1. First, select a suitable place for working—make sure there is no combustible material at hand. Use asbestos or sheet-metal shields to protect floors of wood or other materials, where necessary. Protect your legs and feet from sparks and slag.

2. A clean metal surface means lower gas consumption and a good-quality cut. So remove all the dirt and paint you can by scraping or wire brushing.

3. Look at the instruction sheet for your cutting attachment or blowpipe to find out what size nozzle to use for the thickness of metal you are cutting.

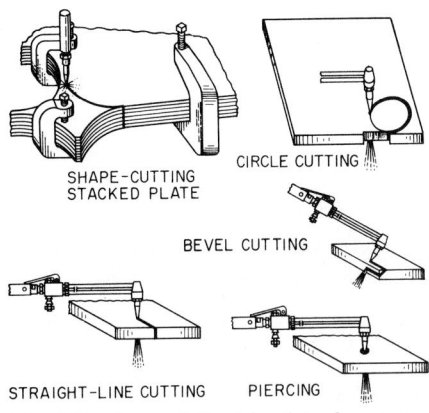

Fig. 6-19 Some of the jobs done by oxygen cutting.

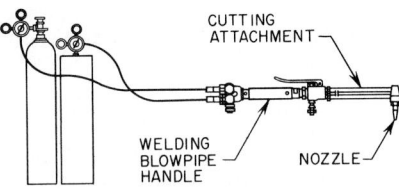

Fig. 6-20 Oxygen-cutting equipment.

4. The adjustment of the flame for a cutting attachment or blowpipe is different from that for a welding blowpipe because the latter has no cutting oxygen stream.
   a. If the cutting oxygen valve is opened after the preheat flames are adjusted to neutral, the preheat flames will lack oxygen. This is because both preheat and cutting oxygen come from the same source and part of the preheat supply has been diverted to form the cutting-oxygen stream.
   b. To correct this, the preheat flames should be adjusted with the cutting oxygen level down.
   c. Also, oxygen-flow adjustments must always be made with the needle valve on the cutting attachment. Open the blowpipe oxygen valve wide, leave it that way while cutting, and adjust the flame with the other valve.

Making the cut

1. During cutting, hold the blowpipe in one hand and guide the blowpipe by resting it on your other hand.
   a. A piece of firebrick on the plate will provide a rest for your hand as well as indicate the proper spacing of the blowpipe from the work.
   b. Make sure nothing will prevent you from finishing the cut without interruption.
2. Hold the blowpipe so that the preheat cones just lick the work surface. Preheat the starting point on the edge to a bright red (see Fig. 6-21).
3. Start the cut by slowly pressing down the cutting-oxygen lever (see Fig. 6-22).
   a. Keep the tip vertical and always the same height above the work.
   b. Do not advance the blowpipe until the cut is completely through the metal.
   c. Continue the cutting action by moving the blowpipe along the line of cut at a uniform rate (see Fig. 6-23).
4. If you move the blowpipe too slowly, you will melt over the edges of the cut and give it a ragged appearance.
5. If you move the blowpipe too fast, the cutting jet will not penetrate the metal completely and you will lose the cut. In this case, release the cutting-oxygen lever, go back to where you lost the cut, and start over again.

    a. Experience is the only way to learn exactly how fast to move the blowpipe.
    b. When the cut is finished, release the cutting-oxygen lever and turn off the preheat flames.

Gouging   Gouging or grooving is merely a special type of oxygen cutting. It is a means of removing a narrow strip of metal from the surface of a plate. You use the same equipment for gouging that you use for cutting, except that you must have a large-bore,

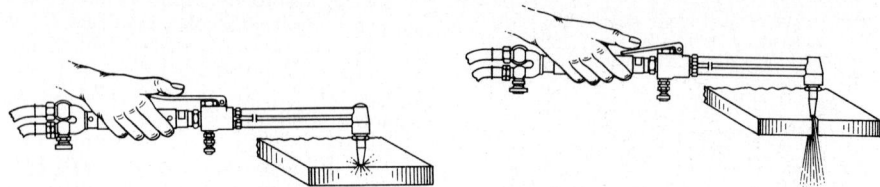

Fig. 6-21   Preheating.   **Fig. 6-22**   Beginning a cut.

low-velocity nozzle. As in cutting, the operation centers around three main steps: preheating, starting the groove, and progressing. Other things to be watched during gouging include
    1. Pulling the nozzle back along the plate surface after preheating, then opening the cutting-oxygen lever. This ensures that the stream will fall on hot metal, not on relatively cold metal ahead of the preheated spot.
    2. Keeping the flames low. If the inner cones of the preheat flames on the lower side of the nozzle are just barely touching the work, you will get maximum efficiency from the preheat flames.
    3. Keeping the blowpipe moving in a straight line. When making a long groove, there is a tendency to move the blowpipe toward you as the groove proceeds and describe a long arc instead of a straight line in the plate.

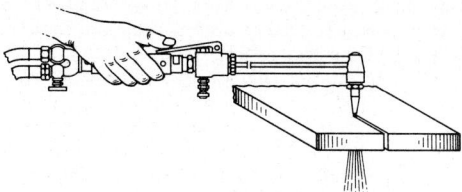

Fig. 6-23   Cutting.

    With the step-back method of gouging, you will have less tendency to lose the cut or swing out of line than if you gouge in one continuous pass.
    1. The groove is carried progressively across the plate in a series of short gouges.
    2. Start the groove, then continue it for about 3 in. Lift up the blowpipe, bring it back about ½ in., and restart the groove.
    3. As each short pass is completed, the nozzle is drawn back slightly to restart the groove.
    4. Repeat these steps until you have reached the full length of the desired groove.
    Gouging is used in three main applications:
    1. Removing defective welds. When a weld does not have a good appearance or is not as strong as it should be, it can be removed by gouging and replaced. You can also remove the old weld and have the piece ready to be rewelded all in one operation by gouging.
    2. Opening up cracks in castings so that sound repairs can be made by welding.
    3. Dismantling welded structures to permit reuse of most of the parts, thus obtaining maximum salvage.

Using a special gouging nozzle, you can cut grooves from ⅜ to ½ in. wide by ⅛ to ⁷⁄₁₆ in. deep. These variations in groove dimensions are controlled by three factors:

1. By the angle of the nozzle with respect to the work (see Fig. 6-24). A flat angle gives a shallow groove, and a steeper angle a deep groove.

2. By the speed of travel of the blowpipe. The faster you move, the shallower the gouge becomes.

3. By the oxygen pressure. High pressures wash a bit more metal out of the groove than lower pressures.

Hard-facing   Hard-facing is the process of applying a layer of special alloy on a metal part or surface to protect it from wear. The big difference between hard-facing and the fusion welding is that the hard-facing alloy does not mix with the base metal to any extent. In fusion welding, complete penetration is necessary, but in hard-facing, it should be avoided. This is important because mixing of the base metal with the hard-facing alloy would dilute and soften the deposit. In hard-facing, the surface of the steel picks up carbon from an excess acetylene flame. The carbon lowers the melting point of the steel and causes it to melt quickly to a depth of only a few thousandths of an inch. This very thin film of melted steel fuses with the hard-face deposit to make a strong bond between the deposit and the steel.

Metals that can be hard-faced include carbon and low-alloy steels (covering 95 percent of the wear problems you will normally encounter), all forms of cast iron (except chilled), and many other special alloys.

With the longer life of hard-faced parts (2 to 25 times longer), the reduction of maintenance labor and of replacement parts used is dollars saved. Here are a few typical examples of how hard-facing increases the life of parts:

Part Hard-Faced	Times Longer Life
Pump shaft	3
Clutch plate	7
Valves, valve seats	7
Valve-seat inserts	15
Hand shovels	3
Spray nozzle disks	12
Cams	6
Shear blades	10
Mill hammers	5
Punches	13

Hard-facing rods   There are a number of hard-facing rods available to help you solve particular wear problems resulting from such factors as abrasion, impact, corrosion, and heat. Very often more than one cause of wear is present. Your problem then is to choose the hard-facing alloy best suited to combat the combination of factors. You should consider every job as a special problem. The same rod used for one job will not necessarily work on the same or similar part in another instance. If you are in doubt about which rod to select, test several under actual conditions. Manufacturer's data will usually help you select the proper rod, but often you must make the final decision in the light of what you can find out about the wear conditions involved. Tables 6-5 and 6-6 show the particular characteristics of some of the hard-facing rods available.

How to hard-face steel

1. Clean the surface to be hard-faced by filing, wire brushing, or grinding. Edges or corners that might become overheated during hard-facing should be grooved out as shown and filled with hard-facing deposit. Use your cutting blowpipe or attachment and grooving nozzle (see Fig. 6-25). If an edge or corner of the part takes a lot of pounding or impact in use, machine the corner or edge as shown in Fig. 6-26. The dotted lines in the illustration show how the hard-face deposit should be built up to the original contour of the part.

2. Parts more than 2 in. in thickness should be preheated throughout to prevent the deposit or the part itself from cracking when it cools. You can preheat medium-sized parts with your blowpipe. Use a neutral flame. Move the flame in a wide circle over the part. Gradually make the circles smaller and smaller until the part turns a dull red color. Large surfaces or bulky parts should be preheated in a furnace. Heat the part until it turns a dull red color.

3. Deposits up to ⅛ in. in thickness can be made in one pass. Best impact resistance is obtained from deposits ¹⁄₁₆ in. in thickness, never over ⅛ in. If you want to build up a

Codes, Specifications, and Welding Standards*

Title	Published by	Field of application	Source
	General		
AWS Terms and Definitions, A3.0	AWS,‡ 1976, 80 pp.	Welding, cutting, brazing	AWS‡
Resistance Welding Equipment, ANSI C88.2	ANSI,¶ 1969	General welding	ANSI¶
AWS Welding Symbols, A2.0, ANSI Y32.3	AWS,‡ 1968, 90 pp.	Engineering-shop drawings	AWS‡
Master Chart of Welding and Allied Processes	AWS,‡ 1976	Wall size 22 by 28 in. Desk size 8½ by 11 in.	AWS‡ AWS‡
Electric Arc-Welding Apparatus, ANSI C87.1	National Electrical Manufacturers Association, 1970	General welding	ANSI¶
Safety in Welding and Cutting, ANSI Z49.1	ANSI,¶ 1973	General welding	ANSI¶
Safety Standard for Transformer-Type Arc-Welding Machines, ANSI C33.2	Underwriters' Laboratories, 1972	General Welding	ANSI¶
Welding Symbols Chart	AWS,‡ 1976	Wall size 22 by 28 in. Desk size 11 by 17 in.	AWS‡ AWS‡

Boilers and pressure vessels		
ASME Boiler and Pressure Vessel Code		
Sec. I, Power Boilers	ASME,§ 1977	Power boilers in stationary service	ASME§
Sec. III, Div. 2, Code for Concrete Reactor Vessels and Containments	ASME,§ 1977	Nuclear power plants	ASME§
Sec. IV, Heating Boilers	ASME,§ 1977	Boilers in operation at less than 15 psig and for hot-water heating and supply	ASME§
Sec. VIII, Pressure Vessels	ASME,§ 1977	Pressure vessels	ASME§
Inspection and testing		
Appendix, Inspection of Welding, 3d ed.	API,‡ 1978	Refinery equipment	API†
Standard Methods for Mechanical Testing of Welds	AWS,‡ 1974	Welding shops and fabricators of welded structures	AWS‡

*All are available from sponsoring organization. For convenience, AWS is given as source when possible.
†American Petroleum Institute, 2101 L St., N.W., Washington, D.C. 20037.
‡American Welding Society, 2501 N.W. 7th St., Miami, Fla. 33125.
§The American Society of Mechanical Engineers, Order Department, United Engineering Center, 345 East 47th St., New York, N.Y. 10017.
¶American National Standards Institute, 1430 Broadway, New York, N.Y. 10018.

1-99

Codes, Specifications, and Welding Standards* *(Continued)*

Title	Published by	Field of application	Source
Piping			
Power Piping, ANSI B31.1	ANSI,¶ 1977	Pressure piping systems	ANSI¶
Standard for Welding Pipelines and Related Facilities (Std 1104)	API,‡ 1977	Cross-country petroleum and natural-gas pipelines	API†
Standard for Qualification of Welding Procedures and Welders for Piping and Tubing, D10.9	AWS,‡ 1969, 72 pp.	All piping systems	AWS‡
Structural (building)			
Rules for Arc and Gas Welding and Oxygen Cutting of Steel Covering the Specifications for Design, Fabrication, and Inspection of Arc and Gas Welded Steel Structures and Qualifications of Welders and Supervisors. Cal. 1-38-SR.	Board of Standards and Appeals, New York, 1968, V. 53, Bull. 51, p. 1338; Amendments: 1974, V. 59, Bull. 4, p. 34; V. 59, Bull. 13, p. 259; 1977, V. 62, Bull. 22, p. 500	Buildings in New York	Board of Standards and Appeals, City of New York, 80 Lafayette St., New York, N.Y. 10013
Structural Welding Code, D1.1	AWS,‡ 1975, 166 pp.	Highway, railway, bridges, buildings, and tubular structures	AWS‡
Safe Practices for Welding and Cutting Containers That Have Held Combustibles, A6.0	AWS,‡ 1965, 16 pp.	Shops engaged in welding or cutting operations on containers of combustible solids, liquids, or gases	AWS‡

1-100

Specifications for Field-Welded Tanks for Storage of Production Liquids, 8th ed. (Spec. 12D)	API,‡ 1977	Oil-field service—capacities over 500 bbl	API†
Specifications for Shop-Welded Tanks for Storage of Production Liquids, 7th ed. (Spec. 12F)	API,‡ 1977	Oil-field service—capacities to 440 bbl	API†
Recommended Rules for Design and Construction of Large, Welded, Low-Pressure Storage Tanks, 6th ed. (Std. 620)	API,‡ 1977	Petroleum products storage—for internal pressures of 15 psig or less	API†
Welded Steel Tanks for Oil Storage, 6th ed. (Std. 650)	API,‡ 1977	Oil storage at atmospheric pressure	API†
Standard for Welded Steel Elevated Tanks, Standpipes, and Reservoirs for Water Storage, D5.3	American Water Works Association and AWS,‡ 1973	Elevated steel water tanks, standpipes, and reservoirs	AWS‡
Water Tanks for Private Fire Protection, NFPA Std. No. 22	National Fire Protection Association, 1976	Field-welded tanks, gravity and pressure towers, etc.	National Fire Protection Association, 470 Atlantic Ave., Boston, Mass. 02210

*All are available from sponsoring organization. For convenience, AWS is given as source when possible.
†American Petroleum Institute, 2101 L St., N.W., Washington, D.C. 20037.
‡American Welding Society, 2501 N.W. 7th St., Miami, Fla. 33125.
§The American Society of Mechanical Engineers, Order Department, United Engineering Center, 345 East 47th St., New York, N.Y. 10017.
¶American National Standards Institute, 1430 Broadway, New York, N.Y. 10018.

Codes, Specifications, and Welding Standards* *(Continued)*

Title	Published by	Field of application	Source
Qualifications			
Qualifications Test for Gas Welders (General Specifications for Inspection of Material, Appendix VII, Welding, Part E, Sec. E-2)	Bureau of Supplies and Accounts, U.S. Dept. of the Navy	All gas welding done for the Navy Dept.	Bureau of Supplies and Accounts, U.S. Dept. of the Navy, Washington, D.C. 20350
Welding Procedure and Performance Qualification, B3.0	AWS,‡ 1977, 97 pp.	Industry, welding instructors, and code-writing bodies wishing to prescribe methods	AWS‡
Welding and Brazing Qualifications, Boiler and Pressure Vessel Code, Sec. IX	ASME,§ 1977	Boilers and pressure vessels	ASME§

*All are available from sponsoring organization. For convenience, AWS is given as source when possible.
†American Petroleum Institute, 2101 L St., N.W., Washington, D.C. 20037.
‡American Welding Society, 2501 N.W. 7th St., Miami, Fla. 33125.
§The American Society of Mechanical Engineers, Order Department, United Engineering Center, 345 East 47th St., New York, N.Y. 10017.
¶American National Standards Institute, 1430 Broadway, New York, N.Y. 10018.

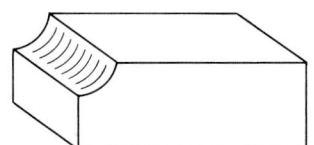
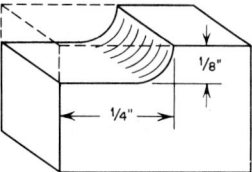

Fig. 6-24 The angle of the nozzle with respect to the work controls the depth of the groove.

Fig. 6-25 A grooved edge for hard-facing.

Fig. 6-26 A machined edge for hard-facing.

TABLE 6-5 Characteristics of Hard-Facing Rods

Hard-facing rod	Tensile strength, psi	Hardness on Rockwell C scale	Melting point, °F	Contains
Haynes:				
90	63,000	45–55	2390	Iron, chromium, carbon
92	25,000	64	2012	Iron, molybdenum, carbon
93	43,000	57–62	2225	Iron, chromium, molybdenum, cobalt, vanadium, carbon
94	60,000	50–61	....	Iron, chromium, boron, cobalt, carbon
Haynes Stellite:				
1	47,000	46–54	2828	Cobalt
6	105,000	33–44	2327	Chromium
12	76,000	37–47	2306	Tungsten
1016	....	58	....	Cobalt, chromium, tungsten, carbon
Hascrome	40,000	28–43	2500	Iron, chromium, manganese

TABLE 6-6 Characteristics of Hard-Facing Rods

Hard-facing rod	Resistance to			
	Abrasion	Impact	Corrosion	Hot abrasion
Haynes:				
90	A	C	C	NR
92	A	NR	NR	NR
93	A	D	C	NR
94	A	D	D	NR
Haynes Stellite:				
1	A	NR	A	A
6	B	A	A	B
12	A	B	A	A
1016	A	NR	C	A
Hascrome	C	A	D	NR

A, excellent; B, high; C, good; D, fair; NR, not recommended.

badly worn surface with hard facing to a depth greater than ⅛ in., you should use more than one deposit.

4. Hard-facing rods are applied with a carburizing flame—a flame using more acetylene than oxygen. The extra acetylene shows up as a whitish feather around the inner cone. Use the amount of excess acetylene specified by the rod manufacturer.

5. Low- and medium-carbon steels are the most widely used metals and are the easiest to hard-face. The following instructions are for the hard-facing of these steels:

6. If you have selected a rod, prepared the part, and set up your welding outfit, you are ready to start hard facing. Begin by heating the part (see Fig. 6-27).

7. Now adjust to a carburizing flame. Reduce the amount of oxygen until you have the proper flame, depending on the rod you are using.

8. Hold the carburizing flame over the heated area. The tip of the inner cone should be just off the steel surface—about ⅛ in. as shown. Hold the flame there until the metal under the flame starts to sweat.

9. Next lift the welding blowpipe a little and put the rod into the flame so that it just touches the sweating surface. Lower the blowpipe until the inner cone of the flame just

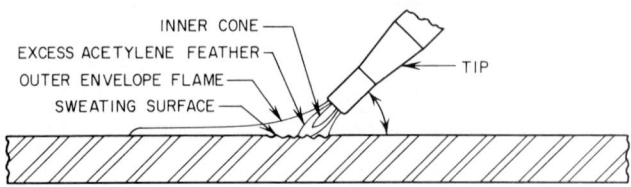

Fig. 6-27 Heating the surface before depositing the hard-facing rod.

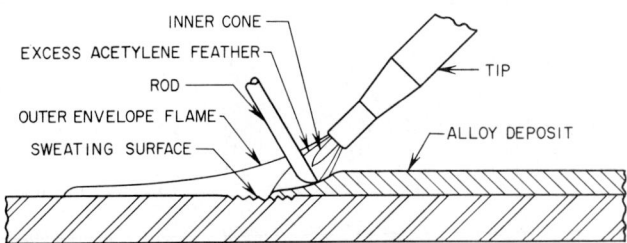

Fig. 6-28 Depositing the hard-facing rod.

touches the rod and is about ⅛ in. from the steel surface, as shown in Fig. 6-28. A small puddle of melted rod will form on the sweating surface. If the first few drips of the melted rod foam or bubble or do not spread evenly, the surface is too cold. Take the rod away and start over again.

10. Next take the rod out of the puddle. Spread the puddle over the sweating surface by pointing the flame into it—do not use the rod to spread it. If there is not enough hard-facing deposit to cover the wearing surface, continue the process.

11. Point the flame so that it touches the forward end of the puddle and the steel surface.

12. When the surface sweats, add more metal to the puddle from the rod. Then, as you did before, remove the rod and spread the puddle with the flame. Repeat until the entire surface is covered.

13. Allow the part to cool slowly to prevent cracks and stresses in the hard face. Small- and medium-sized parts can be cooled in air. Large or bulky parts should be wrapped in asbestos paper or buried in asbestos, slaked lime, wood ashes, or another insulating material until they cool. Parts that are liable to crack should be put in the preheating furnace while they are still hot from hard-facing. Then they should be brought to an even red heat and, with the heat turned off, allowed to cool overnight in the furnace with the door closed.

Hard-facing cast iron

1. Cast iron does not sweat like steel, and it melts at about the same temperature as the rod. *So be careful*—do not melt the base metal too deeply.

2. Use a little less acetylene in the flame than you would for steel.

3. Use cast-iron brazing flux when you apply the rod.

4. A crust will form over the surface of the cast iron when it is heated. To get a good bond, you will have to break the crust with the end of your rod.

5. Very thin cast-iron parts should be backed up with wet asbestos or carbon paste to keep them from melting.

Finishing the hard-facing deposit

1. *Heat treating* of the hard-faced parts is *usually not necessary.* The only time you will heat-treat a part after hard facing is when you want to toughen the base metal. To do that, heat the whole part to a dull red heat. Then dip it in oil. Do not use water for the quench because it may crack the deposit and base metal.

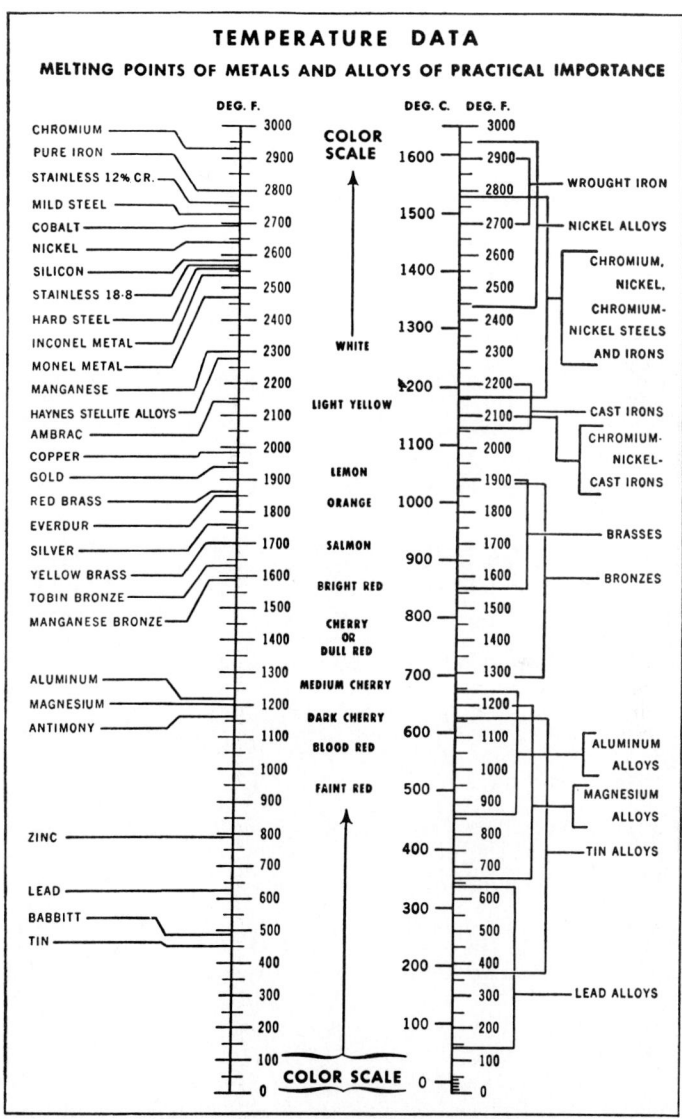

Fig. 6-29 Melting points of metals and alloys.

2. *Surface cracks* are usually caused by insufficient preheat or by cooling the part too quickly. You will find, however, that a surface crack will not harm the properties of the hard face or the strength of the part. If you want to repair a cracked surface
    a. Preheat the piece as for hard facing.
    b. Heat the metal around the crack to a dull red.
    c. Then melt the edges down into the crack.
    d. Add a little metal from the rod.
    e. Now slowly move the flame away from the hot spot to prevent quick cooling.

3. You can grind a hard-faced part to exact size or remove high spots on the surface. Use a grinding wheel not coarser than 46 or finer than 60 in Grade I or J of the Norton system. The speed of the wheel should be between 2800 and 4200 sfpm. Higher speeds might crack the hard-face surface.

See Fig. 6-29 for melting points of metals and alloys.

REFERENCES

"The Oxy-Acetylene Handbook," Union Carbide Corporation, Linde Division, 3d ed., 1976.
Linnert, G. E.: "Welding Metallurgy," American Welding Society, 3d ed., vol. 1, *Fundamentals*, 1965; vol. 2, *Technology*, 1967.
"Brazing Alcoa Aluminum," Aluminum Company of America, 1967.
"Welding Alcoa Aluminum," Aluminum Company of America, 1972.
"Stellite Hard-Facing Products," Stellite Division, Cabot Corporation, 1975.
Jefferson, T. B.: "The Welding Encyclopedia," Jefferson Publications, 17th ed., 1974.
"Welding Handbook," American Welding Society, 7th ed., vol. 1, *Fundamentals of Welding*, 373 pp.; vol. 2, *Welding Processes: Arc and Gas Welding and Cutting, Brazing, and Soldering*, 570 pp.

Chapter **7**

Metal Resurfacing

K. N. KELLEY[1]
Manager, Marketing Communications, Stoody Company,
Industry, Calif.

Construction equipment is subjected to constant battering from wear-inducing forces to a degree perhaps unequaled in any other industry. Minimizing or circumventing these forces through wear-avoidance practices ranging from design factors through preventive maintenance is of primary concern to both the design and the maintenance engineer.

Yet it is not possible to insulate many working parts from the destructive influences which cause metal wear. Recourse in these cases is to increase the inherent wear resistance of the part as much as possible. Manufacturers have a number of options at their disposal, including a variety of hardening and coating techniques. At the maintenance level the most widely used and effective procedure is surfacing by welding processes.

INTRODUCTION

These introductory paragraphs will briefly review several techniques for increasing wear resistance that are commonly employed in the manufacture of metal parts. Note that these procedures are usually not feasible in the typical maintenance shop nor are they suitable for restoring parts where significant amounts of wear have already occurred; the remainder of this chapter will deal more fully with the subject of rebuilding and hard-facing as practiced generally in construction machinery maintenance shops.

Manufacturing techniques include through-hardening by conventional heat-treating methods and the following surface-modification procedures:

Flame hardening   The flame-hardening process hardens the surface or specific portions of a hardenable steel part. It is a selective heat treatment accomplished by the direct impingement of a flame on the surface, followed by an air or water quench. The chemistry of the surface is not changed; wear resistance is enhanced simply by increased hardness resulting from a metallurgical transformation.

Hardness can be increased to ¼ in. in depth. Control factors in flame hardening are of critical importance since unequal stress leading to fracture or warpage can easily be generated; however, some maintenance shops have mastered flame-hardening techniques for specialized applications.

Diffusion alloying   Carburizing and nitriding are well-known examples of a procedure for surface chemistry modification called *diffusion alloying*. Alloying of the surface with carbon or nitrogen is accomplished in the presence of a reactive gas or salt at relatively high temperature. Increased wear resistance is a result of a carbide or nitride constituent at the surface—usually about 1 percent—in combination with elements such as chromium or molybdenum which may be present in the base metal.

Carburizing and nitriding are highly suitable processes for the mass production of small

[1]The author is now retired.

parts, and they are reasonably inexpensive. However, the treatment is essentially superficial, seldom exceeding 0.020 to 0.025 in. in depth, and its benefits are limited in service under normal loads to resisting the effects of sliding or rolling friction.

Other examples of diffusion alloying include the addition in a similar manner of chromium or aluminum to steel surfaces. The presence of chromium in the first few thousandths of an inch of the surface will increase wear resistance, provided sufficient carbon is present in the base metal.

Hard chrome plating Hard chrome plating is an effective means of protecting steel parts from both abrasive wear and corrosion, provided the allowable limits of wear do not exceed the depth of the relatively thin layer deposited. The process is also frequently used to add stock to parts which have been mismachined undersize.

Hard chrome plating is an electrolytic procedure generally available from specialty shops. It differs from decorative plating chiefly in that the deposited surface layer is essentially pure chromium metal as opposed to the copper-nickel-chrome combinations normally used for decorative work. Substantially greater amounts of buildup are feasible in hard chrome plating, up to as much as 0.040 in., or even more. However, because the deposit is relatively brittle and subject to fracture or peeling under certain types of stress loading, thickness is normally limited to 0.020 in. or less.

Unlike decorative plating, hard-chroming requires finishing by grinding and therefore the deposit is made oversize to provide grinding stock. The thickness of the chromium layer is quite uniform except at sharp corners such as the crest of threads where excessive buildup can occur; platers use special techniques to equalize thickness at such points when this is necessary.

WELDED OVERLAYS

The process of depositing a layer of metal on a steel part by welding is commonly called *surfacing*. When the weld metal is added simply for the purpose of restoring a worn part to its original configuration, the term correctly applied is *rebuilding*. Hard-facing (or hard-surfacing) is defined as the addition of an overlay of wear-resistant alloy to worn, partially rebuilt or new parts.

Rebuilding and hard-facing is a distinct branch of the welding art. It had its origins in the early 1920s and from the beginning has been closely associated with the construction industry. It is probable that the total usage of welding alloys to protect and restore equipment used in earth-moving, digging, and crushing operations exceeds that of all other industries combined.

While the basic equipment, processes, and skills required for weld-surfacing are essentially the same as those used in metal-joining, the materials and many of the techniques are quite specialized. A competent welder will have no difficulty in adapting his skills to the requirements of rebuilding and hard-surfacing; however, in the choice of alloy and procedures, a thorough understanding of the basic principles involved is essential (Fig. 7-1).

FUNCTIONS OF REBUILDING AND HARD-FACING

While rebuilding and hard-facing are typically classified as maintenance functions, the benefits of the latter process often clearly surpass those that might be expected of a simple maintenance procedure, although this is often not fully understood. By extending part life or by modifying or improving part performance, hard-facing can result in economic rewards considerably greater than those related merely to the salvage of worn parts.

Obviously, the reduction of replacement part costs through the restoration of used parts is the principal goal of rebuilding applications where the deposited metal has a wear resistance approximately equal to that of the original part. Here the economic criteria for the application simply involve balancing the cost of rebuilding versus the replacement cost. Frequently the cost saving is significant, especially where large parts are no longer usable after relatively small amounts of metal loss have occurred or where the rebuilding procedure can be automated for minimum application cost.

Where the use of hard-facing, which typically may extend the wear life of parts from 3 to 10 times, is feasible, the added economic advantages become very significant. First, hard-faced parts can generally be refaced and put back in service, compounding the initial cost benefit. Further, by extending part life hard-facing can eliminate one or more downtime periods. In some applications the cost of unscheduled equipment shutdowns for part replacement or repair can be extremely costly.

Because hard-faced parts maintain their original size and shape longer, equipment operates with greater efficiency. The economic benefit which can be gained with increased production from crusher hammers or rolls that stay close to size may sometimes totally overshadow any other cost considerations which might be associated with a decision concerning hard-facing as opposed to replacement.

The performance of parts with a cutting edge, such as shovel teeth, blades, or tillage-type tools, can be enhanced by a technique which takes advantage of the bimetallic nature of hard-faced surfaces. By placing the alloy on only one side of the cutting edge, selective

Fig. 7-1 Typical hard-facing activity shows group 3 alloy weld-cast into lengths of pipe driven into worn sheepsfoot tamps.

wear is encouraged; as the unprotected side wears away, a fresh sharp edge of hard-facing material is continuously exposed.

Finally, hard-facing can be used to modify the surface of parts in other ways which may be desirable, and the nature of the alloys is such that the surface will tend to retain the modified contours in service. For example, weld beads running across the surface of a roll can improve its gripping action; beads running parallel to material flow can help a part slide more freely. Tungsten carbide alloys, which produce a heterogeneous deposit, can be used to provide a nonslip surface or a serrated edge to improve the cutting action of various parts.

REBUILDING AND HARD-FACING PROCESSES

Almost all the welding methods commonly used for joining metal parts are also used in hard-facing and rebuilding operations. These include manual gas and electric arc, semiautomatic and fully automatic processes. Thermal spraying (flame spraying and plasma arc) is primarily a surfacing process.

The selection of the most appropriate method is limited to some extent by the equipment available but is primarily dictated by the requirements of the application. Labor costs, the type of alloy to be used and its cost, and part configuration must all be considered in making this decision.

Oxyacetylene welding Oxyacetylene or gas welding is the slowest and therefore often the most costly method of depositing overlay material. It is, however, a preferred method for certain types of parts and for one important group of hard-facing alloys (tungsten carbide).

Use of the torch permits welding on thin-edge tools without burning through and is sometimes the only practical method for parts of this type. Gas welding also is used when dilution of the overlay alloy with the base metal must be held to a minimum. Finally, it is a desirable method for depositing tungsten-carbide-type rods since the low heat input and fast-freezing puddle minimize the tendency of the carbide particles to dissolve in the molten steel.

Flame spraying, which utilizes alloy wires or powders instead of welding rods, is essentially a variation of oxyacetylene welding and offers the same advantages but is limited in both deposit thickness and the variety of alloys available.

Manual arc welding Manual electric-arc welding is still the most widely used method for rebuilding and hard-facing applications (although the semiautomatic process is increasingly popular, especially in the construction industry). Almost all types of the commonly specified alloys are available as coated electrodes, including tungsten carbide and nonferrous cast rods.

Manual arc welding is relatively fast and many of the electrodes available are quite inexpensive. It is an excellent procedure for general-purpose use on a variety of parts.

Semiautomatic welding Semiautomatic welding is a superior choice for a wide variety of applications where part size and shape will permit its use. Deposition rate is high and the process utilizes continuous wires which eliminate the waste of stub ends and lost time for electrode changes.

Semiautomatic welding equipment is commonly available. Both open-arc and gas-shielded arc (GMAW) methods are used but open-arc is common for hard-facing because most of the materials do not require external shielding.

Automatic welding Automatic welding equipment, which eliminates the need for manual control of the weld bead, is available in considerable variety; several types are especially designed for construction equipment—specifically for tractor undercarriage components and for various types of crushers.

Automatic systems facilitate the use of the submerged-arc welding method wherein the molten weld puddle is shielded from the atmosphere by a layer of granular flux. This makes possible the use of relatively low alloy and inexpensive materials which are not suitable for open-arc welding. The equipment will also handle open-arc wires and may be adapted to gas shielding where required.

Conventional machines are most commonly used for cylindrical parts which are rotated in a positioner to advance the weld bead. Equipment is also available for flatwork where the welding head travels while the work remains stationary. General-purpose equipment, which can manipulate the arc in two directions to provide a variety of overlay patterns on flat surfaces, has been introduced in recent years and is especially suited to maintenance shop applications.

The advantages of automatic welding are high deposition rates and precise control of deposit appearance and chemistry. Its use is limited, however, to applications where part size and shape are appropriate, where the deposit can be programmed in advance and, except for several applications using specialized equipment, to instances where the work can be brought to the equipment.

SELECTION OF REBUILDING AND HARD-FACING ALLOYS

A large variety of rebuilding and hard-facing alloys are commonly available. To select an alloy for a specific application it is necessary to recognize the wear factors present and the way in which they operate under the particular service conditions involved.

Types of wear Materials engineers recognize many types of wear. For a practical understanding of wear as it relates to the selection of surfacing materials the following five broad categories will be useful:

1. *Abrasion.* Abrasion is defined as the grinding or scratching action of hard particles, such as sand or rocks, rubbing or sliding against a surface. When combined with heavy loading the condition is often called gouging.

2. *Impact.* Surface deterioration which results from a blow or series of blows is classified as impact-type wear. It can take the form of deformation, fracture, or spalling (spalling means that the surface peels or breaks off in pieces).

3. *Corrosion.* Corrosion is a type of wear associated with chemical or electrochemical attack; rust is a common example. Corrosion problems can often be anticipated and avoided by the use of appropriate materials or by painting, plating, or rubber-coating steel

parts, but where corrosive attack is combined with other wear factors such as abrasion the use of special wear and corrosion-resistant alloys is indicated.

4. *Heat.* Heat affects wear resistance by softening metal surfaces, making them more susceptible to abrasive and corrosive attack. Where heat is the only problem it can be dealt with, as in the case of corrosion, by more direct means than hard-facing. When combined with abrasion, the use of specialized hard-facing alloys is required.

5. *Stress-related wear.* Stress related wear is a result of metal fatigue. Adhesion, or galling, between two metallic surfaces is an example. Fretting, caused by vibration between parts, is another type of wear generated by stress.

Most wear problems encountered under actual service conditions involve more than one of these factors. A careful analysis should be made of the degree to which each type of wear is present in a particular application, and the selection of material should be predicated on that determination.

Wear typically starts at a specific location on a part and its progressive effects may be observed throughout the part's service life. A study of this process will provide useful guidance in establishing hard-facing procedures and material choices. Sometimes a small amount of alloy applied at the point where wear begins will effectively deter its further progress. In many instances a combination of two or more surfacing materials is indicated, depending on the severity of the wear attack on various areas of the part.

Types of rebuilding and hard-facing alloys Despite occasional claims to the contrary there is no universal hard-facing alloy. The many products on the market are each designed to satisfy certain specific and varied requirements of the many applications for which they are used. Cost, weldability, soundness of deposit—in addition to the basic requirement of resisting one or more of the various wear types—are all of concern to the user. The enhancement of one feature is generally a trade-off accomplished at the expense of another.

Therefore, the choice of the correct material for a particular job is a matter of considerable importance and is as vital to the success of the application as the skill of the welder. The recommendations provided by the various manufacturers are of assistance in making the most advantageous choice, but they are sometimes quite contradictory. It will be helpful if the basic principles of the metallurgy of wear-resistant materials and the categories into which they can be grouped are understood.

Almost any hard material, if it is compatible with the base metal, can be used for surfacing and will provide some added wear resistance; cast iron and certain tool steels are examples. But most alloys classified as hard-facing materials have one characteristic in common: Their microstructure consists of a network of metal carbides evenly distributed throughout a matrix of softer material. These carbides are precipitated as the molten metal in the weld deposit cools. They are much harder than steel and harder than the apparent hardness of the deposit itself. They carry the wear load and, as the matrix wears away between them, fresh surfaces are constantly exposed. Thus a hard-facing deposit with an apparent hardness of less than 50 Rockwell C will outwear hardened steel with a hardness over 60 Rockwell C; the hardness reading reflects the matrix hardness primarily and the carbides to only a small extent.

Hard-facing alloys typically consist of iron—or in some cases cobalt or nickel—with varying amounts of carbon plus chromium, tungsten, molybdenum, or other carbide-forming elements added. As a rule, more carbides are formed and the abrasion resistance of the weld deposit is increased as the percentage of these added elements is increased.

Alloys used for rebuilding and hard-facing can be conveniently classified into five major groups. The appropriate category into which a particular alloy should be placed can usually be determined from the manufacturer's literature, and the material's classification provides guidelines for its use.

Group 1. Build-Up Materials. The alloys in this category are not hard-facing materials but are important in the rebuilding of worn parts. They are designed to provide a tough underbase for hard-facing deposits in areas where severe wear has occurred.

Materials used for rebuilding carbon-steel parts contain up to 6 percent alloys (other than iron). They combine strength and toughness, can be deposited crack-free in multiple layers, and are usually readily machinable. The value of these alloys as an underbase is that they will not readily deform under the more brittle hard-facing layers. This can occur if ordinary steel welding rods are used for buildup, with the result that the hard-facing cracks and spalls.

These materials are often used as the final overlay on parts subject to impact, metal-to-metal sliding or rolling—shafts, gears, and tractor rails are typical examples—or where a machined finish is required. Also included in this category are materials designed for buildup on austenitic manganese steel parts (also called Hadfield's manganese or, simply, manganese). These alloys typically contain about 14 percent manganese with smaller amounts of nickel or molybdenum and sometimes chromium.

Like the base metal itself, manganese alloys are tough and work-harden in service under impact. They are used both for buildup prior to hard-facing and as an overlay on parts subject to impact such as shovel bucket and rock-crusher components.

Group 2. Low-Alloy Ferrous Materials. Most authorities classify iron-base materials up to approximately 20 percent alloy content as low alloy. They have greater shock resistance but less abrasion resistance than materials with alloy contents above 20 percent.

Alloys in this group form relatively few, if any, carbides but owe their wear resistance to the special characteristics of their various chemistries. Their properties are inherent, as deposited, and not developed as a result of heat treatment as in the case of machinery-grade steels. Group 2 alloys are appropriate for hard-facing nonlubricated metal-to-metal rolling or sliding parts and parts subject to considerable impact and low-to-moderate abrasion. Examples include sprockets and drive tumblers plus a variety of machinery parts.

Low-alloy ferrous materials are generally the least expensive hard-facing materials and are commonly available as manual (stick) electrodes, often with iron-powder coatings for high deposition rates. They are also widely used in wire form in submerged-arc automatic applications on parts such as tractor rolls but are seldom applied by the oxyacetylene method.

Group 3. High-Alloy Ferrous Materials. These are the most common hard-facing alloys, containing over 20 percent alloying elements. They form carbides and, as a rule, the higher the alloy content within this group, the greater the abrasion resistance and the lower the impact strength.

High-alloy deposits typically cross-check upon cooling; cross-checks are normal and generally desirable. Deposits are usually not machinable but can be ground if necessary—though sometimes with difficulty. In most applications, deposits should be limited to two layers. Where more buildup is required to reach finish size, it is advisable to rebuild to within two layers with a group 1 material.

Applications for group 3 alloys include those involving wear from abrasion by rocks, sand, ore, and cement and bulk materials which contain dirt or other gritty particles. Typical parts hard-faced with these materials are ripper teeth, dozer end bits, shovel teeth and buckets, draglines, conveyor parts, buckets and mixer paddles. Group 3 alloys are available as bare rods, coated electrodes, and wires for semiautomatic and fully automatic applications (Figs 7-2 and 7-3).

Group 4. Tungsten-Carbide Materials. This special group includes the most wear-resistant hard-facing alloys. These are steel tubes which contain crushed particles of tungsten carbide in various mesh-size ranges. In depositing these materials, only the matrix material is melted; the carbide particles are not (they will, however, *dissolve* in the molten weld puddle and care must be taken to minimize this effect). The method by which the deposit resists wear is the same as that of the group 3 alloys, but in this case the carbides are of visible size (ranging from a fine powder to almost ⅛ in. in diameter) and are pure tungsten carbide, substantially harder than the microscopic carbides formed as a group 3 alloy deposit cools. Tungsten-carbide deposits are unique in that the deposit is itself abrasive as well as abrasion-resistant. Deposits of all but the finest mesh sizes are sandpaperlike on the surface. They can provide a nonslip surface that won't wear out and can improve the cutting action of some parts such as posthole augers. Group 4 alloys provide maximum abrasion resistance for parts in contact with the earth, including digging tools and most tillage implements.

Mesh-size ranges (these are expressed as 5–8, 8–10, 10–20 and so on—the higher the number, the smaller the particle size) are determined by the application. Deposits made with the larger particles provide better cutting action and more impact strength; deposits made with the smaller particles are more wear-resistant. If the application involves abrasive materials which are very fine, tungsten-carbide materials may be unsuitable since it is possible to cut between the carbide particles and undermine them. Group 4 alloys should *never* be used on metal-to-metal applications.

Tungsten-carbide materials are available as bare rods, coated electrodes, wires, and powders. They are expensive, so the economics of the application must be carefully analyzed, but for jobs where extreme abrasion is present they are often the best possible choice.

Group 5. Nonferrous Materials. The nonferrous alloys are those in which iron is not a principal ingredient. They are most often cobalt base or nickel base and, like the group 3 alloys, precipitate carbides in the weld deposit which are responsible for their wear resistance.

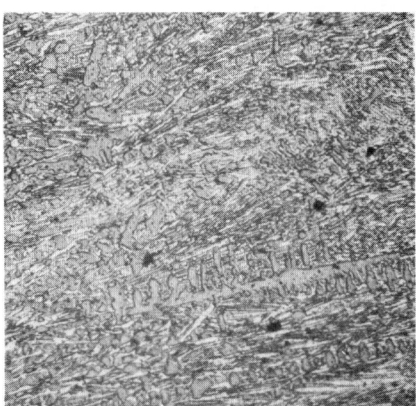

Fig. 7-2 Micrograph of medium-alloy group 3 material shows small eutectic carbides (white areas).

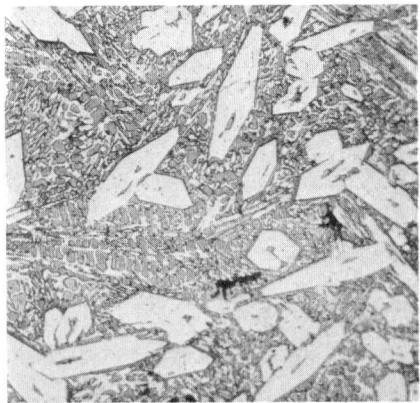

Fig. 7-3 High-alloy group 3 material reveals both primary (large white areas) and eutectic carbides.

Cobalt-base alloys are extremely versatile hard-facing materials. Depending on the amount of carbide-forming elements contained (principally, chromium and tungsten, with silicon and sometimes nickel, boron, or molybdenum), the abrasion resistance of the various grades is comparable with the group 3 alloys from the low to the high end of the scale. Deposits are resistant to corrosion and retain useful hardness at temperatures up to 1500°F. Cobalt-base alloys are resistant to galling and are excellent for metal-to-metal applications such as bearings, seals, and valves.

The nickel-base alloys are similar to the cobalt base and are useful in certain corrosion situations; no single alloy or class of alloys is resistant to all corrosive media. Nickel-base alloys are the predominant type used for thermal spraying.

Nonferrous hard-facing alloys are relatively expensive (though not as expensive as the group 4 alloys) and principally available as bare or coated rods or powders. Several types can be machined with carbide tools; the harder ones must be ground if a finish is required.

Not all materials sold for wear resistance will fit precisely into the groups above. There are copper aluminum and copper nickel brazing alloys with good wear resistance which are used for bearings and similar parts. Certain tool steel rods are quite similar to group 2 alloys. High-carbon 420 stainless steel is a useful hard-facing material in some applications. The best guidelines for selecting the right material are those outlined in the preceding paragraphs, plus manufacturers' recommendations and experience gained by observing the performance of parts that have been rebuilt.

Forms of rebuilding and hard-facing alloys Rebuilding and hard-facing alloys are available as bare and coated rods, continuous wires, and powders. They often differ in form from conventional welding materials. Because they are inherently hard they cannot be pulled through drawing dies to form wire, as is ordinary carbon steel; therefore, manufacturers use several other methods to produce them. One is to melt the alloy in a furnace and cast it into rod form. Cobalt- and nickel-base alloys are commonly made in this manner, as are several high-alloy iron-base products.

Low- to medium-alloy iron-base electrodes are often produced by adding the carbon and alloy requirements to the coating applied to a carbon steel core. The coating thus serves a dual purpose, shielding the arc and contributing to the analysis of the weld deposit.

Many hard-facing electrodes and almost all hard-facing wires for semiautomatic and automatic application are produced by a method called *fabrication*. Here the alloying ingredients are contained in a tube formed of carbon steel, often with shielding or fluxing materials included. As with the alloy-coated electrodes, the welder actually creates the alloy during the welding process.

REBUILDING AND HARD-FACING METHODS

Base metals In all types of welding, it is important to be aware of the chemical composition of the base metal. This is essential in hard-facing because parts intended for severe service are frequently made of high-carbon or -alloy steels, cast iron, or Hadfield's manganese; these can present welding problems.

Cast iron is extremely crack-sensitive and not all cast-iron parts can be hard-faced. High preheat temperatures and slow cooling are necessary and the deposit should be peened to help relieve stresses. Many hard-facing alloys are not recommended for cast iron; manufacturers' literature will usually specify those which are.

Since austenitic manganese steel is widely used for construction machinery parts, it is most important to identify this material which is significantly affected by welding temperatures. Hadfield's manganese steel derives its unique toughness and work-hardening characteristics not only from its composition—12 to 14 percent manganese and 1.2 to 1.4 percent carbon—but also from a rapid water quench from 1800 to 2000°F which makes it soft and ductile. The manganese-alloy electrodes used for rebuilding this material contain additional alloying elements which compensate for the fact that the weld deposit will not have the benefit of this special processing, but the properties of the base metal itself can be damaged or destroyed if sufficient heat input occurs during welding to alter the effects of the original heat treatment.

In addition, intermediate alloys formed from a mixture of Hadfield's manganese and ordinary carbon- or low-alloy steels generally have extremely poor physical properties. These will occur in the fusion zone between the deposit and the base metal. In this case and in the case of excessive heat input, spalling of the welded overlay will occur in service.

Austenitic manganese steel can be identified with a pocket magnet. Parts are nonmagnetic when new, weakly magnetic after work hardening. Heat should always be held to a minimum when welding this base metal, and only an alloy recommended for manganese welding should be used.

An ideal base metal for hard-facing is a medium-carbon steel, 0.30 to 0.50 percent. Manufacturers who design parts with hard-facing in mind frequently use medium-carbon steel and equipment owners ordering replacement parts often specify this base metal to facilitate rebuilding.

Preparation for welding Surface preparation is especially important in rebuilding and hard-facing because worn parts may be contaminated with a variety of foreign substances. Dirt, oil, and rust should be removed, as failure to clean the surface to be welded can result in porosity in the deposit and spalling.

Deep cracks should be gouged out to their full depth with a torch by grinding or machining and the metal replaced with a compatible welding material. Fatigued or rolled-over metal should also be removed. Heat-checks (alligator hide) are not necessarily harmful. Stress cracks, however, which run in fairly straight lines at right angles to the direction of the load, should be repaired.

When rebuilding parts which have been previously hard-faced, the alloy applied should be compatible with the previous deposit. If some spalling of the old deposit has occurred, make sure that portions of the remainder have not loosened. Tap the part with a hammer to locate these areas, and remove them. If only small amounts of the old hard-facing deposit remain and if they are sound, it's fairly safe to proceed with reapplication. If more than 20 percent of the original thickness remains, a good rule is to remove it or return the part to further service until more has worn away.

Preheat Preheat is an important element in many rebuilding and hard-facing jobs. It is necessary in some instances to avoid damage to the base metal or distortion of the part. Inadequate preheat can also result in deposits that will crack or spall.

As the carbon content of carbon-steel parts increases, the preheat requirements increase. Alloy steels also generally need preheat in proportion to the amount of alloy present; chrome nickel stainless steel alloys are an exception. Cast iron requires a very high preheat, preferably to a dull red color. Preheating of austenitic manganese steel should generally be avoided, although in very cold weather parts may be warmed to no more than 200°F.

Temperature-indicating crayons may be used to determine when proper preheat temperatures are reached. The manufacturers of these materials provide charts which show the precise temperature ranges recommended for welding various base metals. When parts require preheat, they should be held at the specified temperature a sufficient length of time for it to reach the core. The part should be held at or near this temperature throughout the welding process. On large parts the use of a supplementary heating torch may be necessary. All parts that require preheat should be slow-cooled.

The use of some hard-facing materials makes it necessary to preheat parts even when the base-metal composition does not. Metals expand when heated and contract as they cool. When a weld bead is first deposited it is fully expanded; the base metal, if not adequately preheated, is not. As the weld cools it will contract and must either stretch, crack, or distort the part. Most hard-facing alloys, especially those in groups 3 and 5, have almost no ductility (ability to stretch); so preheat is especially

Fig. 7-4 Deposit cross sections (magnified) compare penetration of straight versus reversed polarity.

important. When adequate preheat cannot be accomplished—because the part is austenitic manganese, because it is too large, or because time or equipment is not available—special attention must be given to the hard-facing alloy selected and to welding procedures.

Welding procedures Rebuilding and hard-facing are fusion welding processes. Fusion between the deposit and the base metal is of concern regardless of the method of welding used because it affects both the soundness of the bond and the chemistry of the deposit. Insufficient penetration can result in spalling; too much will diminish the wear-resistant properties of the overlay because of dilution. Hard-facing alloys are designed to tolerate some dilution with the base metal, but penetration should be held to the minimum amount consistent with a strong bond (Fig. 7-4).

Oxyacetylene Hard-Facing. Oxyacetylene hard-facing is most suitable for thin-edge tools and for applications where dissolving tungsten-carbide particles or dilution must be held to a minimum. It is not recommended for cast iron or large parts where a high preheat is required.

Surface impurities will adversely affect oxyacetylene deposits, as will oxides formed during welding. Therefore it is necessary to clean worn parts thoroughly by grinding, and it is advisable to lightly grind new parts prior to surfacing. The area ground should be slightly larger than the area to be surfaced. The part should be positioned for downhand welding and a jig should be used if it will have to be turned during the surfacing operation. If preheating is required, it is helpful to group small parts to simplify this operation.

The diameter of the rod used is determined by the size and thickness of the part; small-diameter rods are required for delicate work but larger diameters will reduce both time and material costs. A tip size about three times larger than normally used for mild-steel welding with the same-diameter rod is recommended.

A carburizing flame should be used to reduce oxides; it will also add carbon to the surface of the part which lowers its melting point and aids fusion. A 3X flame (the feather is three times as long as the inner cone) is recommended for cobalt-base alloys and fine-mesh tungsten-carbide rods. A 4X flame is used for larger-mesh tungsten-carbide rods and most iron-base hard-facing alloys (Fig. 7-5). Nickel-base alloys are an exception to the rule; most should be applied with a neutral flame.

The entire part should be preheated with the torch and then the flame should be concentrated on the area to be hard-faced until it reaches red heat. The flame should then be directed on the point where the deposit will start, until the surface appears shiny and

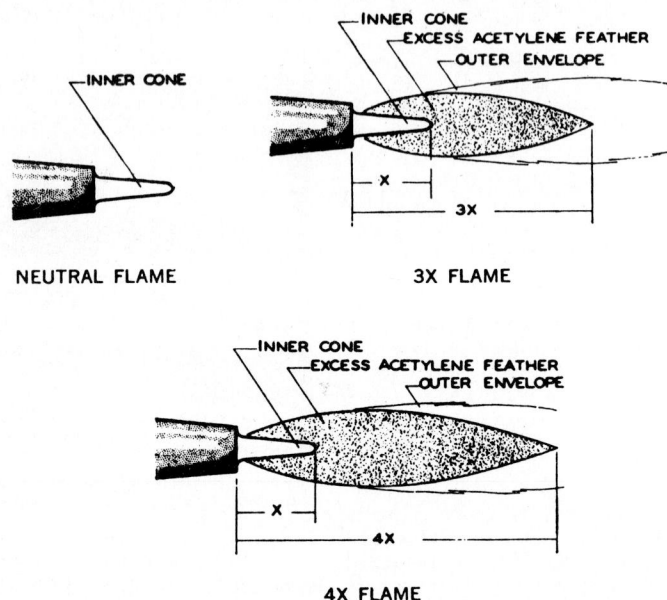

Fig. 7-5 Comparison of 3X and 4X flames with neutral flame.

watery. This is the sweating temperature—it indicates that the surface is beginning to melt. The tip of the rod should be moved under the edge of the flame so that a drop will melt and fall into the sweating zone. As the torch is moved forward, the drop will spread and follow the heat of the flame. As more alloy is added in this manner, the heat should be directed more at the tip of the rod than on the work, but enough heat should be kept on the base metal that it remains at sweating temperature ahead of the alloy puddle.

Different alloys vary substantially in their ability to flow or to stack, and either characteristic may be desirable in a particular application. The skillful oxyacetylene welder, however, can exert considerable control overpenetration, deposition rate, and the shape of the deposit.

Manual Arc Hard-Facing. Penetration is of primary concern in arc welding because it's easy to overdo it. It is generally desirable to hold penetration to the minimum amount that will ensure a sound bond. The use of straight polarity will result in less penetration than reverse polarity; however, reverse polarity must be used when recommended for a particular electrode. The lowest possible amperage will minimize penetration, but amperage should not be less than the lowest recommended for the electrode. The use of the largest-diameter electrode compatible with the part will also minimize penetration.

The arc should be directed toward the pool of molten alloy rather than the base metal to limit penetration. A tight overlap of adjacent stringer beads will significantly reduce penetration, but don't exceed 50 percent—too much overlap can reduce penetration to a point where the bond is inadequate and spalling will occur.

The use of two layers is the most effective way to reduce the effect of dilution and is recommended for all alloys except those in group 4 (tungsten carbide) which should be limited to a single pass. However, no *more* than two passes are recommended for most group 3 and group 5 alloys.

Stringer beads, weave beads (Fig. 7-6) and horseshoe-shaped beads are all used in manual arc surfacing. In general, an oscillated bead is less likely to spall than stringer beads, especially when depositing hard-facing alloys over areas rebuilt with group 1 alloys. The manufacturer's literature for a particular rod will usually suggest the best technique.

Semiautomatic Hard-Facing. The semiautomatic method is increasingly popular for rebuilding and hard-facing, particularly in the construction field where large amounts of metal must be deposited on large parts. Most of the rules for manual electric surfacing apply to semiautomatic operation.

Wires for semiautomatic hard-facing are of the tubular (fabricated) type. Because they will crush under pressure, special care must be given to the feed mechanism of the semiautomatic machine. Feed rolls must be carefully aligned and, if the unit is not already so equipped, should be replaced with a special geared type which grips without excessive pressure. Most semiautomatic hard-facing wires run open arc. Welding should be done in the flat (downhand) position whenever possible. Some wires can be welded vertically or semivertically if necessary.

The manufacturer's recommendations for voltage and amperage settings should be followed; most wires operate dc reverse polarity within a range of 200 to 400 amp. The shortest possible arc is usually recommended to minimize spatter.

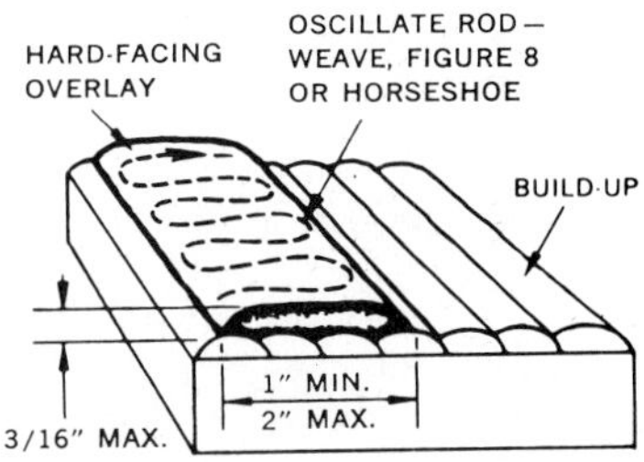

- Use weaving bead instead of stringer bead when applying hard-facing.
- Limit single pass bead thickness to 3/16".
- Use same technique for second layer.
- Avoid severe quench.

Fig. 7-6 Proper process for applying hard-face overlay.

Automatic Rebuilding and Hard-Facing. A great deal of construction equipment rebuilding and hard-facing is done on automatic welding machines, in dealer shops, job shops, some large contractor-owner maintenance shops and, with specialized equipment designed for specific applications, in the field. Where a considerable amount of welding is to be done on a regular basis on a particular part, the investment in automatic equipment is worthy of consideration.

In addition to the usual welding considerations of voltage, amperage, and polarity, the operator of automatic equipment is concerned with such variables as travel speed, step-over, wire extension, and, when welding round parts, the "lead." Each of these factors is important to the success of the application.

Wire manufacturers' recommendations should be heeded with regard to the welding parameters. In general, amperage requirements are in the range from 300 to 400 amp, though some wires can be deposited at 500 amp or more; best results will usually be obtained at the low end of the recommended range. Voltage influences bead shape because the width will increase as the voltage increases; the normal range for most wires is 28 to 32 volts. Straight polarity results in less dilution, higher hardness, and faster deposition; reverse polarity yields a sounder and smoother deposit.

Once the welding parameters have been set, the speed at which the arc travels is a primary determinant of bead shape. About 20 to 30 in./min is normal for most wires. When welding rotating round parts, the lead—the distance between top center and the point at which the arc contacts the work—also influences bead shape. It is normally set at 1½ to 2 in., depending on part diameter.

As in manual arc welding, step-over affects penetration. A tighter step-over will reduce penetration, but step-over should not exceed 50 percent.

Wire extension—the distance between the contact tip and the work—is normally set at 1½ to 2 in. Deposition rate can sometimes be increased by increasing the extension but this can also result in some loss of precision in the placement of the weld bead.

Automatic welding machines designed for such parts as tractor rolls and rails, which are rebuilt with low-alloy materials, are equipped with flux hoppers which supply a constant

1-118 Basic Maintenance Technology

layer of neutral submerging flux to the surface of the part for arc shielding. Machines designed for mounting directly onto roll or impact crushers use small-diameter high-alloy wires and will operate open arc (Figs. 7-7 and 7-8).

Thermal Spraying. Thermal spraying is used not only for hard-facing but for a variety of surfacing operations including the buildup and repair of flawed or mismachined parts. Several distinct processes are included in this classification. Many of the alloys are

Fig. 7-7 Automatic impactor-bar rebuilding machine rebuilds parts in place. Similar equipment is designed for roll crushers.

actually brazing materials; the procedures used produce an overlay with a bond that ranges from surface adhesion to true fusion (Fig. 7-9).

In flame spraying, the alloy in powder form is injected into the gas stream of an oxyacetylene torch from a hopper connected to the torch. The flame is directed on the work and the particles pass through it and adhere to the surface. This type of spraying is often done with a mounted gun on round parts in a lathe. The as-sprayed deposit is

Fig. 7-8 Submerged-arc automatic welding system is designed specifically for rebuilding tractor undercarriage rollers.

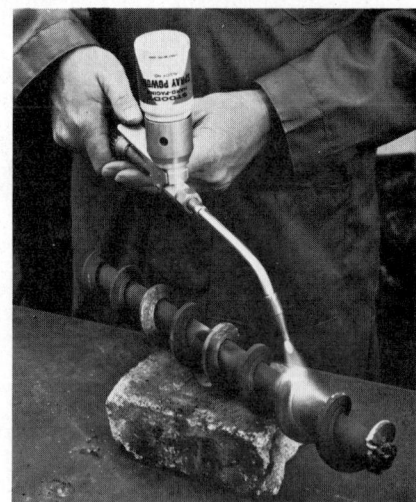

Fig. 7-9 Hand-held spray torch is effective for thin deposits on small parts.

essentially held by a mechanical bond, and it is then fused with the torch or in a furnace. Hand-held spray torches, with a powder hopper mounted on the torch, are commonly used for small or thin parts. Here the procedure is very similar to oxyacetylene welding although fusion with the base metal is minimal.

Most spray powders used for manual torch application are nickel base and should be applied with a neutral flame (manufacturer's instructions should be followed for other types). The torch should be adjusted for a slightly reducing flame; it will change toward neutral when the trigger is pulled to inject the powder. Preheat the part lightly with the torch 1½ to 3 in. away before starting to apply the powder. When it reaches sweating temperature and appears shiny, the trigger should be pulled. The powder should flow on and fuse. It may be necessary, from time to time, to release the trigger and fuse the overlay with the torch.

Fig. 7-10 A one-bead-high, 1 in. sq. waffle pattern on a crusher roll produces 7/16-in. chips.

In torch spraying and all thermal spraying, parts must be very clean—free of oxides, grease, moisture, or other contaminants. Spray parts as soon as possible after cleaning; even an 8-hr delay can allow an undesirable surface film to build up.

Plasma-arc spraying makes possible the use of a wider variety of powders as it results in true fusion; penetration can be easily controlled and overlays from as little as 0.010 in. to as much as ¼-in. thick can be deposited.

Metallizing, on the other hand, produces a deposit which is essentially mechanically bonded. It is used to apply both powders and wires. The wire melts in the flame of the special gun and is carried in droplets onto the work by compressed air. Any metal that can be drawn into wire form can be applied with this process.

SURFACE-CHECKS IN HARD-FACING

Surface-checks, also called cross-checks or relief-checks, are not only normal for most medium- and high-alloy materials but must occur to relieve stresses which would otherwise be locked in and result in major cracks or spalling, either immediately or in service (Fig. 7-10). Manufacturers' literature should identify those materials which are supposed to surface check. A good pattern will exhibit fine tight cracks across the weld beads at intervals of ½ to ¾ in. They should not open up and must not extend into the base metal.

If an adequate pattern does not occur naturally, it should be induced by sponging the deposit with a wet cloth or spraying it with a fine water mist. It can also be accelerated during the cooling period by occasionally striking the deposit with a hammer.

HARD-FACING PATTERNS

While a smooth surface is sometimes required for hard-facing deposits and some parts are machined or ground after surfacing, it may not be necessary or even desirable. Placement of the weld beads in the proper locations and in the proper pattern can increase wear resistance beyond that provided by the alloy itself; correct placement can save material and even improve performance of the part (Fig. 7-11).

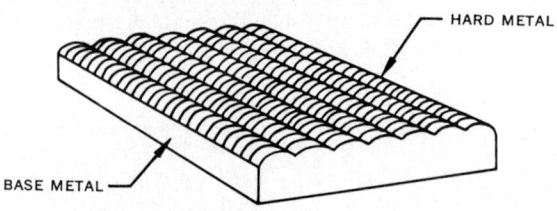

CROSS-CHECKS ARE NUMEROUS, TIGHT, CLOSELY SPACED AND DO NOT EXTEND INTO THE BASE METAL.

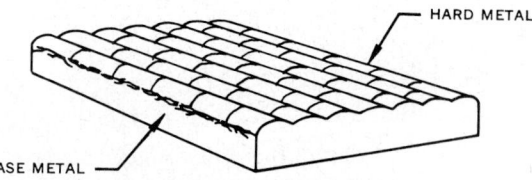

IF CROSS-CHECKING DOES NOT OCCUR IN HIGH ALLOY DEPOSITS, CRACKS WILL PROBABLY APPEAR. HAIRLINE CRACKS AT THE INTERFACE CAN FOLLOW AND CAUSE SPALLING.

Fig. 7-11 Proper and improper cross-checking.

Often it is desirable to put less hard-facing material on a part. In rocky soil, parallel stringer beads, spaced some distance apart, act as rails to carry the material if they run in the same direction as the load. Parts working in fine sand, however, can be improved by a pattern of spaced beads at right angles to the flow of the material. Here the material will build up between the beads and thereafter wear against itself as long as the beads remain. A cross-hatched or waffle pattern is often effective for the same reason and is especially suitable for working in slightly damp soil. A pattern of regularly spaced dots is useful for protection against solid surfaces and saves substantially in both material and application time (Figs. 7-12 to 7-14).

Applying hard-facing to both sides of a tooth or cutting edge will result in the part becoming more and more blunt as it wears. Applying it to one side only will produce a part which self-sharpens as the unfaced side wears away. Experience will show which side is best for surfacing.

The best guide for establishing the pattern for hard-facing any part is to observe, at intervals during its operation, the wear pattern that develops. Thicker deposits can be applied at the points of first and severest wear and lesser amounts at subsequent points. Some parts can be adequately protected by as little as a single drop of hard-facing applied to the precise point where wear begins.

HARD-FACING APPLICATION EXAMPLES

There are hundreds of construction machinery parts that can be salvaged, protected, or improved by rebuilding and hard-facing. The typical examples that follow illustrate basic techniques that can be applied to a variety of parts.

Metal Resurfacing   1-121

Cable Sheaves. Most cable sheaves should be preheated and slow cooled. Sheaves should be placed on a jig that can be turned or on a rotating positioner for downhand welding (Fig. 7-15). Badly worn areas should be built up with group 1 alloys. They should then be finished with circumferential beads (around the part) using a slight weave and either a group 1 or tough group 2 alloy. Cable sheaves are good parts for semiautomatic application and best for full automatic.

Fig. 7-12 Waffle pattern on shovel bucket traps soil; compacted soil provides added protection from wear.

Fig. 7-13 Dozer-blade resurfacing shows intricate pattern of checkerboard beads.

Chutes (Baffle Plates). Stringer beads should be applied ½ to 1½ in. apart in the direction of material flow; spacing will depend on size of the material being processed. An intermediate group 3 material should be used and the application repeated before the deposit is entirely worn away.

Shafts. Shafts are generally made of medium- or low-alloy steels and can be rebuilt manually with group 1 alloys. Thorough preheat and slow cooling is required; the need for preheat increases with larger parts and care should be taken to maintain temperature during welding. Grinding or machining of worn areas prior to welding is advisable. Weave beads about 1 in. wide running longitudinally should be used and the part should be built oversize about ⅛ in. to allow for finish machining.

Large shafts can often be rebuilt most economically by the automatic submerged-arc process. Shafts subject to severe wear, as in a packing gland, are often rebuilt manually with a machinable group 5 (cobalt-base or nickel-base) alloy. Where minimum buildup is required, flame spraying is a simple and effective procedure.

Swing Hammers. Swing hammers come in many shapes and sizes. Almost all are subject to severe wear and lose efficiency in the process. Many small hammers may often be hard-faced when new, before being put into service. A small amount of group 4 alloy on the striking edge will keep them out to size for full working efficiency even though unprotected areas may wear away behind the deposit (Fig. 7-16).

Large hammers have often lost a considerable portion of their original size before

Fig. 7-14 Dot pattern saves alloy and welding time, limits heat input on manganese shovel.

1-122   Basic Maintenance Technology

being removed from service. Copper plates clamped around the part to form a cavity the shape of the required finish size can be used to simplify the rebuilding process. The process of filling the cavity with semiautomatic wire is sometimes called *weld casting*.

Where substantial buildup is required, a group 1 alloy that is compatible with the base metal should be used. Choice of overlay alloy will generally be from groups 3 or 4.

Engine Valves. Almost all internal combustion engine valves can be hard-faced; the procedure is economically most feasible with the large diesel types (many are hard-faced by manufacturers in original equipment).

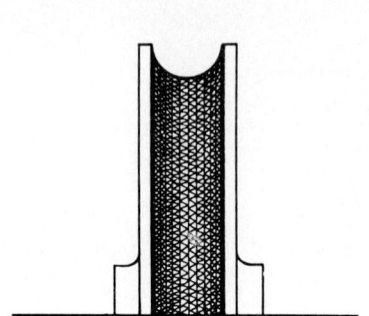

Fig. 7-15 Cable sheave.

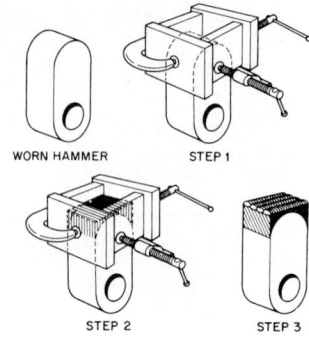

Fig. 7-16 Swing crusher hammers.

The area to be hard-faced should be undercut $\frac{1}{8}$ in. by machining or grinding. Use a jig so that the part can be positioned for downhand welding and rotated during the process. Apply a nonchecking grade of cobalt-base (group 5) alloy by the oxyacetylene method. The seating area can be built up sufficiently oversize to allow for finishing by grinding or machining.

Teeth. Teeth at the leading edges of shovel buckets, ditch diggers, dredge cutters, and similar equipment are generally hard-faced, often even though they may be of a throwaway, replaceable type or made of special wear-resistant alloys. One reason is that hard-facing can be applied in a manner that will improve their operating characteristics while increasing service life. The hard-facing is placed chiefly on one side of the part. The unprotected side will wear away more rapidly, producing a continuously self-sharpening point (Fig. 7-17). Care should be taken to place the deposit on the leading edge of the tooth so that it will be supported as wear occurs.

The type of material to be handled by the equipment should be considered. Beads running the length of the tooth, parallel to material flow, will act as runners for large particles such as rock or slag, but fine materials like sand or dirt can work

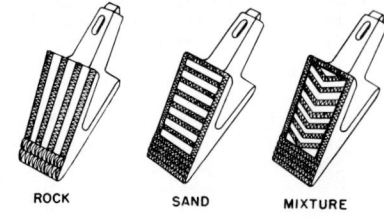

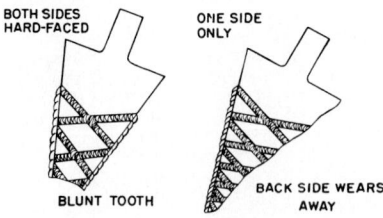

Fig. 7-17 Hard-facing for self-sharpening.

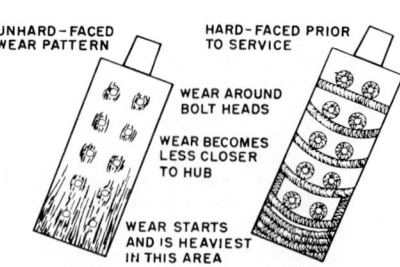

Fig. 7-18 Applying hard-facing to match wear pattern. Note (top) how material flow is controlled.

between them and erode the tooth itself. Spaced beads across the tooth or a waffle (cross-hatch) pattern will result in fine materials becoming packed between the beads, giving additional protection to the tooth (Fig. 7-18).

Groups 3 and 4 alloys, singly or in combination, are generally used for teeth; tungsten-carbide materials are often applied just at the leading edge with the remainder of the surface faced with a group 3 material. Many large teeth are made of austenitic manganese steel-base metal and proper welding precautions should be observed.

Augers and Conveyor Screws. Augers and conveyor screws should be positioned vertically in a rotating fixture for hard-facing the flight faces, horizontally for hard-facing

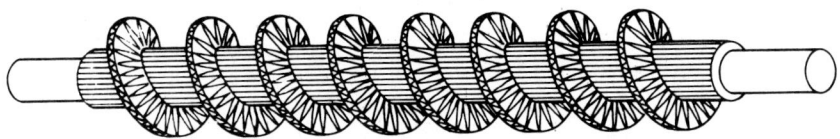

Fig. 7-19 Curbing-machine augers. Hard-face flight faces and peripheries and shaft with coated rod. Reapply hard-facing as necessary.

the periphery. The wear pattern on worn parts should be observed as a guide to material placement. A fine-mesh (40-Down) group 4 rod is a good choice for severe wear applications. A group 3 material may be used when the edge must be ground or in lesser wear situations.

Hard-face posthole auger cutter teeth on the top side only for self-sharpening effect (Fig. 7-19).

Crusher Roll Shells. Semiautomatic or automatic welding is recommended for rebuilding or hard-facing rock-crusher roll shells whenever possible. If the crusher is used regularly and is subject to fairly severe wear, an automatic unit which may be temporarily mounted to rebuild the rolls during normal downtime periods is a good investment. These machines use semiautomatic-type open-arc wires and are capable of both circumferential welding and a variety of transverse or cross-hatch patterns.

Most crusher roll's shells are made of austenitic manganese steel. Normal precautions for rebuilding this material should be observed and special care taken that excessive heat doesn't build up in the roll during the welding process. The temperature of the roll should never exceed 500°F nor should even moderate temperatures be maintained for extended periods of time. High temperatures in localized areas must also be avoided.

Size is a factor to be considered in relation to heat input; most crusher roll shells are large enough to provide sufficient mass for uniform heat distribution and dissipation. However, water cooling may be necessary on some smaller rolls. High-arc travel speeds achieved with automatic equipment or skip welding techniques with manual or semiautomatic wires are the best ways to avoid localized heating.

New shells should be put into service for a short time—about one shift—to remove residual stresses before welding. They may then be hard-faced with minimum buildup required and maintained on a regular schedule.

In rebuilding roll shells, the surface must be thoroughly cleaned prior to welding, and cracks repaired. Wedge bolts should be loosened if extensive rebuilding is required and retightened after the roll has cooled.

Where substantial buildup is required, a compatible buildup alloy (manganese alloy for Hadfield's manganese shells) should be used to bring the roll to within two layers of finish size. Transverse crescent-shaped weave beads, ½ to ¾ in. wide, are generally deposited when using manual or semiautomatic welding. Three or four beads are deposited in the area of deepest wear, then the roll is rotated to a new position. This process is repeated

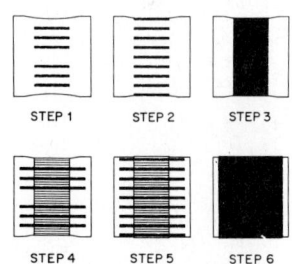

Fig. 7-20 Skip weld process used in restoring worn crusher roll shell to size and contour. Skip welding provides proper heat distribution, minimizing thermal stress.

until the pattern covers the entire circumference, then continued until the spaces between the beads are filled in. Successive layers of transverse beads are added until the roll is built up to the required diameter for hard-facing (Fig. 7-20). Circumferential beads are usually applied when building up worn rolls automatically.

Two layers of a group 3 alloy are applied for hard-facing. In both rebuilding and hard-facing the deposit should not come closer than 1 in. to the extreme edge as cracking can start at this point. A variety of patterns may be applied for the final overlay and will affect the size and shape of the crusher's output.

Chapter **8**

Maintenance of Plain Bearings

W. N. GOLDENBOGEN
Supervisor, Material Quality, Gould, Inc., Clevite Engine Parts Division, Cleveland, Ohio

DESIGN

Materials Plain bearings, or sleeve bearings, are designed to support rotating or oscillating shafts and at the same time protect them from damage. The ideal bearing offers low friction, low journal wear, conformability, and embeddability. During abnormal operating conditions the bearing metal should yield rather than damage or distort the shaft. Observation of this yielding should indicate to the operator or mechanic that unusual conditions exist. This should be a warning sign to investigate thoroughly and make minor repairs or adjustments before costly maintenance becomes necessary.

Designers, machinists, and metallurgists have combined their efforts to improve bearing performance. Refinements in materials, manufacturing methods, tolerances, and lubricants have increased the life and capacity of bearings.

Chief among design factors are materials, loads, size tolerances, temperature, and lubrication. Except for load-carrying capacity, the well-known white metals, tin-base and lead-base babbitt, are still the best all-around bearing metals. Where size is no limitation and unit loads can be kept low (2000 psi maximum), these metals have operated on shafts with hardnesses as low as 170 Brinell. With increased power ratings from smaller machines, the increased loading has demanded bearings with greater load capacity. Bronzes, copper-leads, cadmium base, aluminum base, and silver have been developed to meet this need. Some surface qualities have been sacrificed to boost the fatigue resistance (load capacity), and a corresponding increase in shaft hardness has been required. Plated overlays of lead- or tin-base material have been used in the range of 0.001 in. thick to improve the surface action of the bronzes and silver. The fatigue life of babbitt increases as the thickness drops below 0.008 in. (see Fig. 8-1). Automotive-bearing thicknesses are 0.002 to 0.004 in. today. The use of these thin MICRO[1] layers is not practical where conditions of dirt or wear may expose the steel backing.

Loads Modern bearing design demands that bearing loads be accurately determined. After due consideration of inertia, deflection, distortion, shaft whip, radial loads, and shaft speeds, it is possible to select a bearing material and a suitable bearing area that can be expected to perform satisfactorily. Figure 8-2 gives some idea of the relative load-carrying capacity of several typical bearing materials.

Tolerances After selection of a suitable bearing material, attention is given to establishing dimensional tolerances. Permissible variations are listed in Tables 8-1 to 8-5.

[1] Registered trademark of Clevite Corp.

TABLE 8-1 Case Tolerances

Bore tolerance:
  0.001 in. up to 10-in. bore
  0.002 in. over 10-in. bore

Taper tolerance:	Normal service	Heavy-duty service
1-in. length	0.0002 in.	0.0001 in.
1- to 2-in. length	0.0004 in.	0.0002 in.
Over 2-in. length	0.0005 in.	0.0003 in.

Out-of-round tolerance:
  0.001 in. max allowed if horizontal is larger than vertical
Alignment:
  Alignment bar with diameter 0.0005 to 0.00075 in. under low limit of case bore should turn freely with the use of small lever when cap bolts are properly torqued

TABLE 8-2 Shaft Tolerances

	Automotive	Heavy duty
Diameter tolerance:		
Up to 1½-in. journal	0.0005 in.	0.0005 in.
1-up to 10-in. journal	0.001 in.	0.001 in.
Over 10-in. journal	0.002 in.	0.002 in.
Diametral-taper tolerance:		
Up to 1 in. of length	0.0002 in.	0.0001 in.
1 up to 2 in. of length	0.0004 in.	0.0002 in.
Over 2 in. of length	0.0005 in.	0.0003 in
Out-of-round condition:		
Up to 3-in. diameter	0.0003 in	0.0002 in.
3- to 5-in. diameter	0.0005 in.	0.0003 in.
Over 5-in. diameter	0.001 in	0.0004 in.
Maximum misalignment:		
Adjacent main journals	0.001 in.	0.0005 in.
Crankpin parallel with main journals	0.001 in	0.0005 in.
End clearances:		
Shaft diameter		
2–2¾ in.	0.003–0.007 in.	0.003–0.007 in.
2¾–3½ in.	0.005–0.009 in.	0.005–0.009 in.
3½–5 in.	0.007–0.011 in.	0.007–0.011 in.
Over 5 in.	0.009–0.013 in.	0.009–0.013 in.
Shaft hardness:		
Brinell	200	300
Shaft-journal finish (all applications):		
Microinches	15 max	
Waviness	0.0001 in. max	
Lobing	0.0001 in. max	
Chatter	0.00005 in. max	

TABLE 8-3 Connecting-Rod Tolerances

	Automotive	Heavy duty
Diameter:		
Up to 3¼-in. diameter	0.0005 in.	0.0005 in.
3¼ to 10-in. diameter	0.001 in.	0.001 in.
Over 10-in. diameter	0.002 in.	0.002 in.
Taper, hourglass, or barrel shape:		
1-in. length	0.0002 in.	0.0001 in.
1- to 2-in. length	0.0004 in.	0.0002 in.
Over 2-in. length	0.0005 in.	0.0003 in.

Out-of-round:
  0.001 in. max if rod is larger horizontally than vertically
Parallelism and twist between rod bore and wrist-pin bore when measured 6 in. from the end of wrist-pin bushing 0.001 in. max

Maintenance of Plain Bearings 1-127

TABLE 8-4 Spread

Free spread (width across the open ends):
  Main bearings—crankcase bore plus 0.005–0.020 in. depending on the thickness and structural stiffness of the bearing

TABLE 8-5 Recommended Oil Clearances

Bearing-oil clearances:
  The general rule for the size of the oil clearance for pressure-lubricated bearings is to allow 0.001 in. for each inch of journal diameter, subject to modification according to the bearing metal alloy used

Bearing alloy	Shaft diameters	
	2–2¾ in.	2¹³⁄₁₆–3½ in.
Lead- and tin-base babbitts	0.0015–0.0025 in.	0.0025–0.0035 in.
Cadmium	0.002–0.003 in.	0.003–0.004 in.
Copper-lead	0.0025–0.0035 in.	0.0035–0.0045 in.

Lubrication To make any bearing perform with maximum life requires adequate lubrication. Not only should the supply of oil be maintained, but also an oil of proper viscosity should be used. A weight of oil that will provide a liquid film and consume the least amount of power is normally selected. Usually the equipment manufacturer makes a recommendation for the grade of lubricant to be used. The ideal design will provide for a large enough volume of oil to keep the bearing reasonably cool. The volume of oil to be pumped is a function of the temperature rise and will determine the size of the grooves and clearances. A temperature rise of 60°F is considered a safe figure for satisfactory operation, with maximum of 290°F.

Grooving Grooving design is important to the operation of plain bearings because it distributes the lubricant from the point of entry to the places where it is needed. The edges of the axial grooves must be well blended to avoid shearing of the oil film on the sharp corners. Properly designed grooves prevent lubricant leakage from the ends of the bearing. Introduction of the oil ahead of the highly loaded area is essential to the development and maintenance of a hydraulic film to separate moving surfaces. Figures 8-3 to 8-7 show examples of grooving design.

CARE AND MAINTENANCE

Lubricant Selection of lubricants for modern complex machinery should not be done entirely on a dollars-and-cents basis. The least expensive oils may eventually cost the most. Nor will high-priced oils provide the greatest protection against breakdown. Usually the equipment manufacturer has decided which of the many additives are required for any particular condition. For maintenance purposes it is necessary only to see that the recommended lubricants are properly applied.

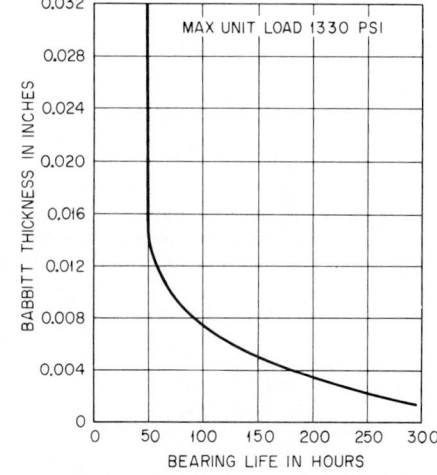

Fig. 8-1 Bearing life versus babbitt thickness.

By this we mean that drain and replacement schedules are observed, filtration is provided if necessary, contamination is minimized, and operating temperatures are kept as low as possible.

The statement that a lubricant never wears out may be true. However, in many cases, before the lubricant has a chance to wear out, it has been contaminated with foreign material (water, metal chips, abrasive particles, or acidic organic compounds) which will cause wear or loss of bearing material. Changing the lubricant becomes necessary to remove these potential sources of bearing wear and destruction.

Cleanliness The life of a lubricant can be extended by continuous removal of contaminants (filtration) and good housekeeping to exclude dirt and water. Such items as air filters, clean lubricant containers or transfer equipment, and tightly fitting covers on oil reservoirs are important.

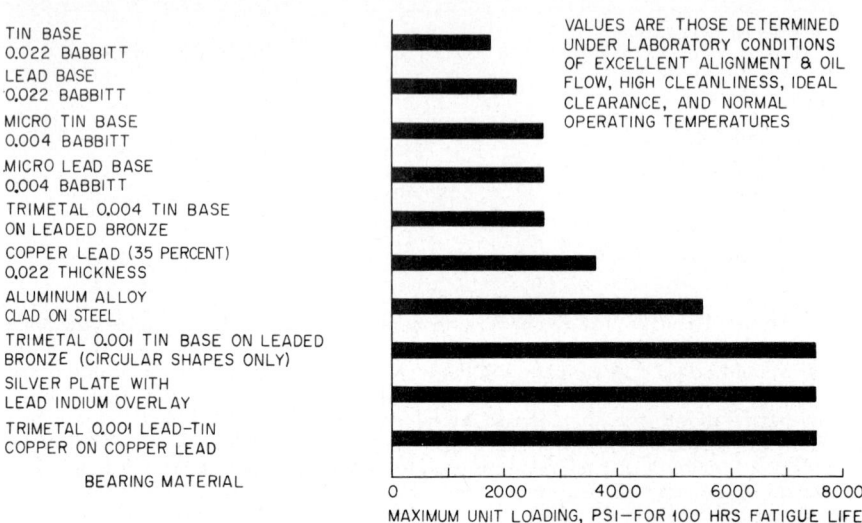

Fig. 8-2 Relative load-carrying capacity.

Temperature In addition to the maintenance of the lubricant, there are at least two other precautions which can be observed in bearing operations. They are temperature and pressure. High temperatures rob bearing materials of their strength (Fig. 8-8). High temperatures (above 275°F) also promote rapid breakdown of lubricants to form sludges or corrosive compounds. Temperature control of bearings therefore becomes an important factor in their life. By suitable means, either a temperature gauge at the bearing or in the oil or by feeling, the temperature of a bearing should be determined at frequent intervals. If an abnormally hot bearing is discovered, some means must be found to cool it immediately. If necessary, shut down the machinery.

Pressure When pressurized lubrication systems are used, a normal operating pressure is established. A frequent look at the pressure gauge will reveal any great variation from this normal value. Either an abnormally high or low pressure can be a danger signal.

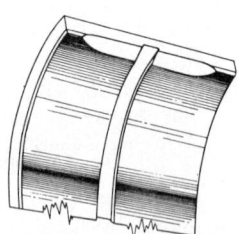

Fig. 8-3 Spreader groove and chamfer at parting line.

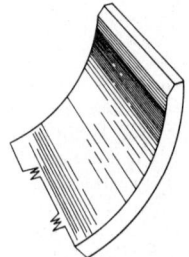

Fig. 8-4 Chamfer at parting line.

Maintenance of Plain Bearings 1-129

If a pump or line fails, lubrication of a part will be impaired and dangerous conditions may develop. On the other hand, the oil flow may be restricted. Although the gauge reading may be high, the flow of lubricant may be inadequate to prevent trouble. Investigate these conditions, and take corrective measures immediately.

INSPECTION AND RECONDITIONING

When it becomes necessary to dismantle a bearing, certain precautions should be observed. It is extremely important to mark or identify each part so that when the

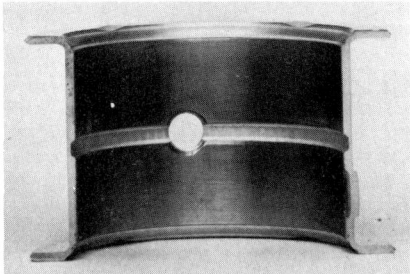

Fig. 8-5 Main-bearing oil groove and hole.

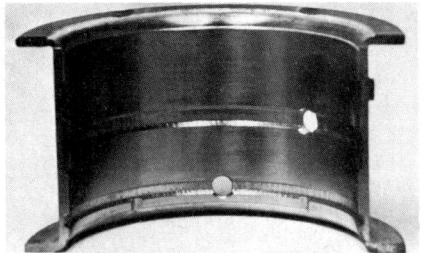

Fig. 8-6 Grooving in main bearing.

machinery is reassembled, it can be installed in its original position. The matching of parts during the original (break-in) period makes them unfit for operation in other positions.

Journal Examination of the parts of a bearing can be a valuable source of information for the immediate trouble and for averting possible future difficulty. If the journal surfaces are excessively worn, ridged, or scored, the shaft must be reground for further service. For minimum specifications of a shaft to continue in service see Table 8-2.

If the shaft is to be used without reconditioning, it should be thoroughly cleaned, including oil passages. Also, carefully measure the diameters of the journals, because these values will be used to select the size of the replacement bearings that will provide the proper oil clearance.

Surface areas of a reground shaft are made up of a series of tiny sharp ridges. These are created by the cutting action of the abrasive grains on the face of the grinding wheel. Although these ridges are scarcely detectable, they present an unsatisfactory surface which will cause excessive bearing wear unless the roughness is reduced by a finishing operation. For final polishing, set the shaft in V blocks and polish off the ridges with a fine emery cloth and light machine oil, using a reciprocating motion. Some prefer to place the shaft in a lathe and polish while the shaft rotates in the *same direction* as it would *in an engine*.

All fillets should be checked to ensure against interference with the ends of the bearings. The conditioned shaft should be thoroughly washed, and all oil passages cleaned.

Connecting rods If connecting rods are involved, two conditions require checking: parallelism and twist (see Table 8-3). The rod bore and piston pin should be parallel within 0.001 in. in 6 in. A bent rod will cause an uneven distribution of load

Fig. 8-7 Thrust-face tapered lands.

on the bearing area, forcing the piston skirt out of parallel with the cylinder bore. This will result in uneven and excessive bearing wear, piston-ring wear, and out-of-round cylinder wear.

Twist in a rod should also be limited to 0.001 in. in 6 in. Out-of-roundness beyond 0.001 in. for rod and main bores causes variation in oil clearance. A maximum out-of-round rod should never be matched with a maximum out-of-round shaft. Bore finish should be coarse enough to ensure proper bearing contact, yet smooth enough to promote good heat transfer; 60 to 90 microinches is recommended.

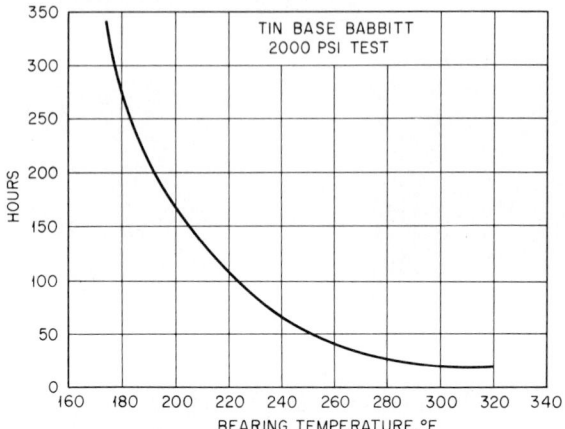

Fig. 8-8 Temperature versus fatigue life.

Bearing replacement The usual practice is to replace or renew the bearing surfaces at the time of overhaul. However, these may be times and conditions under which it is not necessary to do so. The condition of the bearings must be evaluated and compared with the cost of a subsequent overhaul. If the bearings are not worn, fatigued, or damaged in any way and all the applicable tolerances are within established limits (consult manufacturer's service manual), it is not necessary to replace the parts.

Many teardowns reveal the bearings to be ready for renewal or replacement. Such evidence as wear, edge loading (Fig. 8-9), fatigue (Fig. 8-10), embedment (Fig. 8-11), scoring (Fig. 8-12), lack of clearance (Fig. 8-13), and hourglass journal damage (Fig. 8-14) without a doubt calls for replacement or renewal of the bearings. After the journals have been examined or repolished, the job of selecting or fitting the bearings to the shaft is begun. It is advisable to follow the manufacturer's original equipment specifications for bearing materials and running clearances.

Fig. 8-9 Edge loading has caused fatigue at upper edge of bearing.

Fig. 8-10 Condition of bearing surface resulting from fatigue.

REASSEMBLY

Crush It is of utmost importance that the bearing inserts have good contact with the housing or seat. To assure this, the diameter of the two inserts at right angles to the parting line when placed together is slightly greater than the diameter across the parting surface when the bearing is in place, thus requiring this amount to be compressed when the bearing is drawn up tight. For example, each half shell is made slightly in excess of an exact half circle. The excess is called crush (see Fig. 8-15), and its purpose is to permit the shell to be firmly clamped in the bearing seat. If the bearing does not have the proper amount of crush, it will not be held securely and will have a slight degree of movement during operation.

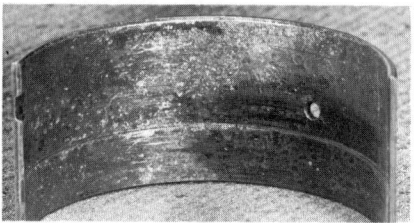

Fig. 8-11 Result of high temperatures caused by dirt.

Fig. 8-12 Scoring caused by circulating hard particles.

Loose inserts also will allow oil to work in between the back of the bearing and the housing. This cuts down the heat conductivity and tends to raise the bearing temperature. Also, a certain amount of flexing of the insert will take place, which adds to the normal friction heat, and under a retarded rate of heat transfer, will lead to a premature bearing failure due to overheating.

Insufficient crush can be due either to filing of the parting surfaces of the shells or to the presence of dirt and foreign matter between the parting faces of the bearings and bearing caps. The dirt will act as shims to prevent the faces from coming together as they should. Under no circumstances are the parting surfaces of the bearing insert, the cap or shank, or the saddle to be filed. In assembling the bearing, be absolutely sure that no dirt, nicks, or burrs remain upon the parting faces of either the cap or saddle.

Do not attempt any operation on the bearing insert other than correcting the spread, and this only when necessary. The spread (see Table 8-4) is built into the bearings so that the inserts have to be lightly pressed into place. If the parts have excessive spread, they can be tapped gently on the end to cause close-in. If they are too loose, the insert can be opened by placing it on a wooden block with convex side up and tapping with a mallet.

Bolt torque On all service installations it is an absolute necessity to use recommended bolt torque values and a torque wrench (Fig. 8-16) when tightening the bearing nuts or cap screws. Almost all the engine builders perform their boring operations with

Fig. 8-13 Wiping caused by insufficient clearance.

the bolts torqued to the same specifications as recommended in their service manuals. It is well to remember that any variation in bolt torque may seriously affect the crankcase or rod-bore sizes, bearing crush, clearances, and resulting bearing performance.

Oil clearance The various bearing metals have individual requirements for oil clearances. A general rule for the amount of oil clearance for pressure-lubricated bearings is to allow 0.001 in. for each inch of shaft-journal diameter. Table 8-5 lists recommended oil clearances for bearing alloys and shaft sizes.

Measure oil clearance The inside diameter of the bearing, in assembly, can be measured with inside micrometers if dial indicator bore gauges are not available. The journal diameter is best measured with micrometers that read in ten-thousandths of an inch.

There are various practical methods for determining oil clearance. A material which will deform can be squeezed between the journal and the bearing with the cap bolts properly torqued (see Fig. 8-17). After removal of the cap, the flattened material is compared with a prepared chart and the tolerance can be read in thousandths directly from the chart. Several commercial items of this nature are available from automotive-parts manufacturers.

Fig. 8-14 Fatigue caused by hourglass journal.

An alternative method is the use of lead or brass shims whose thickness ranges are standardized. A shim of suitable thickness, shorter than the bearing and about ¼ in. wide, should, when clamped between the shaft and bearing, allow the shaft to turn easily. A shim 0.001 in. heavier than the required clearance should lock the shaft from rotation. This check requires experience and care to avoid damaging the bearing inserts, and it is made with all bearing caps loose except for the position under consideration. It is necessary to apply the correct torque to the clamping bolts. Extreme care must be used to eliminate false readings that can be caused by housing bore or journal misalignment. If out-of-roundness is found to exist, use the largest journal diameter, because minimum clearance is the critical condition.

End clearance Table 8-2 lists the recommended values of end clearances for various sizes of shafts. Checking this dimension is absolutely necessary, since the lack of clearance can easily cause thrust-bearing failure. End movement can be measured by forcing the shaft in each direction and checking either with shims at the thrust faces or with an indicator on the flywheel face.

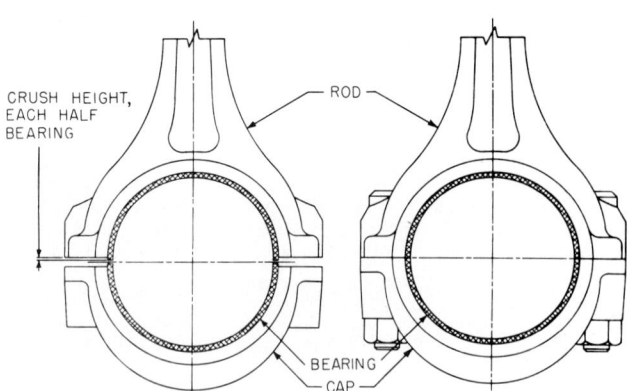

Fig. 8-15 Diagrammatic illustration of bearing crush.

Maintenance of Plain Bearings 1-133

Final checking In all final assembly operations and after finished surfaces have been prepared, use the utmost care to exclude dirt, chips, and all foreign matter from bearing surfaces. If included in the assembly, these particles will damage the bearing and the journal by scoring and wearing during the initial revolutions of the engine. Excessive conditions of dirt have been known to cause almost immediate failure through wear and high temperatures due to friction.

Preliminary lubrication Another precaution is to flood the unit whenever possible with clean oil just prior to initial start-up after the final assembly. This will provide temporary lubrication until the normal supply of oil is available through the lubrication system.

Free rotation If there is any doubt about the clearance and contact area between journals and bearings, it can be checked by manual rotation of the

Fig. 8-16 Torque wrench.

Fig. 8-17 Gauging oil clearance by means of plastic material.

unloaded shaft. It must rotate freely. Any indication of binding must be traced, and its cause removed. The best way to make this check is with the block mounted on a stand. Assemble the shaft with a light, uniform coating of bluing on the journals, and after rotating the shaft a couple of revolutions, turn the block over and rotate the shaft again. The transfer of blue to the bearing shells will indicate alignment condition and causes for binding. Lack of contact on a bearing also means trouble. It could be excessive clearance, or it could be shaft and/or case bore misalignment. A suitable blue pattern covering 45 to 90° at the center of all bearings will predict good bearing performance. This check is also a good place to pick up evidence of hourglass journals, taper, shaft burrs, and fillet ride.

RENEWING CAST-BABBITT LINERS

Rebabbitting large cast-in-place bearings may be not only desirable but practical in maintaining bearings used in heavy stationary machinery. During the teardown, one should look for the usual signs of misalignment, uneven wear, lack of lubrication, excessive shaft wear, excessive dirt, and high temperatures so that an attempt can be made to correct these conditions during reassembly or in the method of operation after overhaul.

Reclamation of housing Removal of the old babbitt liner can be done by heating the inside surface with a "buffalo" torch. Use as low a temperature as possible to avoid distortion of the steel back. Wiping the molten surface with a dry cloth will effectively remove all but a thin layer of the babbitt. If the removal of the old babbitt can be accomplished without severe oxidation, the residual layer will serve as the bonding layer for the new babbitt. The surface should be light golden or straw colored; otherwise, do not attempt to flux and rebabbitt without further preparation. If the remaining metal is brown or black colored or if there appear to be cracks in the surface, a light machine cut should be taken to expose sound metal of the steel shell.

For proper tinning, the steel surface must be chemically clean and slightly etched. Dip the whole shell into a hot (180°F) alkaline metal cleaner until no water breaks appear, which will indicate a clean surface. After rinsing in clean running water, pickle the shell in 1 part water and 1 part muriatic acid at 160°F for 2 to 4 min or until the surface has a gray matted finish. Remove from the acid, and rinse in clean water. Immerse the shell in soldering flux (commercial brands available) kept above 150°F. After removal from the flux, dip the shell into molten pure tin at approximately 550°F. Use caution when dipping the wet shell in the molten tin to avoid spattering of hot metal because of steam generation. Wear safety clothing to protect from possible burns. Allow the shell to remain in the tin pot long enough to approach the temperature of the tin. The length of time required for this will depend on the mass of metal present.

To pour the babbitt, remove the shell from the tin bath and attach the heated core and end plates. Immediately pour the heated babbitt (700 to 800°F) into the annular space between the core and shell. Pour sufficient metal at one time to fill the entire space. This will prevent lamination and segregation. If at all possible, it is desirable to cool the shell quickly by quenching with water applied on the bottom side. As soon as the babbitt has set, the assembly can be thoroughly cooled and knocked apart.

Determine the dimensions to which the bearing is to be machined by checking the journal size and making sure that its condition is satisfactory or by reconditioning it. Apply the manufacturer's recommendations for tolerances and grooving. Use sound machining, locating, and measuring techniques. The final cut should be made in such a manner that the best possible surface finish is obtained.

Precautions in checking clearances, alignment, lubrication, and cleanliness as described earlier are recommended.

Chapter **9**

Maintenance of Rolling Bearings

W. RALPH GOOD
SKF Industries, Inc., King of Prussia, Pa.

GENERAL

Reliable bearing performance is a key factor in optimizing maintenance costs. If this goal is to be attained, it is necessary to follow the equipment builder's operating and maintenance recommendations. For special cases, most bearing manufacturers maintain service departments to render technical assistance. Also, most areas of the world are serviced by distributors, who usually represent more than one bearing manufacturer. As with all mechanical equipment, bearings should be handled carefully and sensibly to prevent damage from mechanical abuse and contamination.

BEARING DESIGNS AND NOMENCLATURE

Nine basic types of rolling bearings and their standard nomenclature are shown in Fig. 9-1. Some of these basic types are available in many variations; for instance, single-row deep-groove ball bearings are generally available in nine different configurations as shown in Fig. 9-2. Tapered roller bearings can come in more than 20 different configurations, some of which are shown in Fig. 9-3. Cylindrical roller bearings may be obtained with one, two, or four rows of rollers. The other basic types do not come in large numbers of configurations, but it should be noted that all types of rolling bearings may vary greatly in internal design, depending on the manufacturer. It is not within the scope of this book to describe all the various designs of rolling bearings used in machinery. Rather it alerts maintenance personnel to their variety. Details are given in manufacturers' catalogs.

BOUNDARY DIMENSIONS

In general, ball, spherical roller, and some cylindrical roller bearings are made to metric boundary dimensions and tolerances which have been standardized by ISO (International Standards Organization). Therefore, bearings from all manufacturers throughout the world are physically interchangeable. Most tapered roller bearings are made to inch dimensions. However, the AFBMA (the U.S. Anti-Friction Bearing Manufacturers Association) has recently proposed to ISO several new series of metric-dimensioned bearings which in all probability will become available in the near future. Also, some metric tapered roller bearings to the present ISO boundary plans are made in Europe and Asia. Interchangeable units are thus available from several manufacturers. In most cases identical numbers are used.

1-136   Basic Maintenance Technology

Inch-dimensioned cylindrical roller bearings do not conform to an ISO standard and will vary depending on the manufacturer.

BEARING SERIES

For any given bore size all types of rolling bearings are manufactured in several series for different severity of service. For instance, most ball bearings are made in three series for light, medium, and heavy duty. These are designated as the 2-, 3-, and 4-diameter series. Spherical roller bearings are normally available in eight different series as shown in Fig. 9-4. Tapered roller bearings have a larger number of series or duty classifications, but all series are not necessarily available in every bore size.

BEARING NOMENCLATURE

The illustrations below identify the bearing parts of the nine SKF basic bearing types. The terms used conform with the terminology section of the AFBMA* Standards—and are mutually accepted by the anti-friction bearing manufacturers.
*Anti-Friction Bearing Manufacturers Association, Inc.

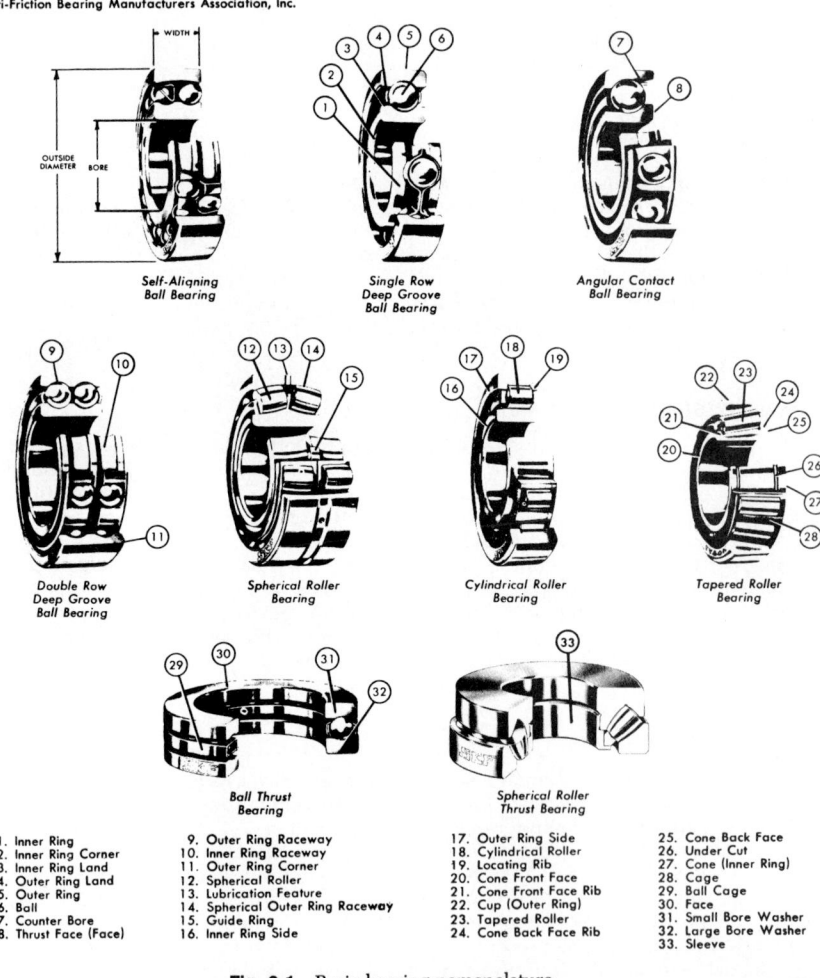

1. Inner Ring
2. Inner Ring Corner
3. Inner Ring Land
4. Outer Ring Land
5. Outer Ring
6. Ball
7. Counter Bore
8. Thrust Face (Face)
9. Outer Ring Raceway
10. Inner Ring Raceway
11. Outer Ring Corner
12. Spherical Roller
13. Lubrication Feature
14. Spherical Outer Ring Raceway
15. Guide Ring
16. Inner Ring Side
17. Outer Ring Side
18. Cylindrical Roller
19. Locating Rib
20. Cone Front Face
21. Cone Front Face Rib
22. Cup (Outer Ring)
23. Tapered Roller
24. Cone Back Face Rib
25. Cone Back Face
26. Under Cut
27. Cone (Inner Ring)
28. Cage
29. Ball Cage
30. Face
31. Small Bore Washer
32. Large Bore Washer
33. Sleeve

Fig. 9-1   Basic bearing nomenclature.

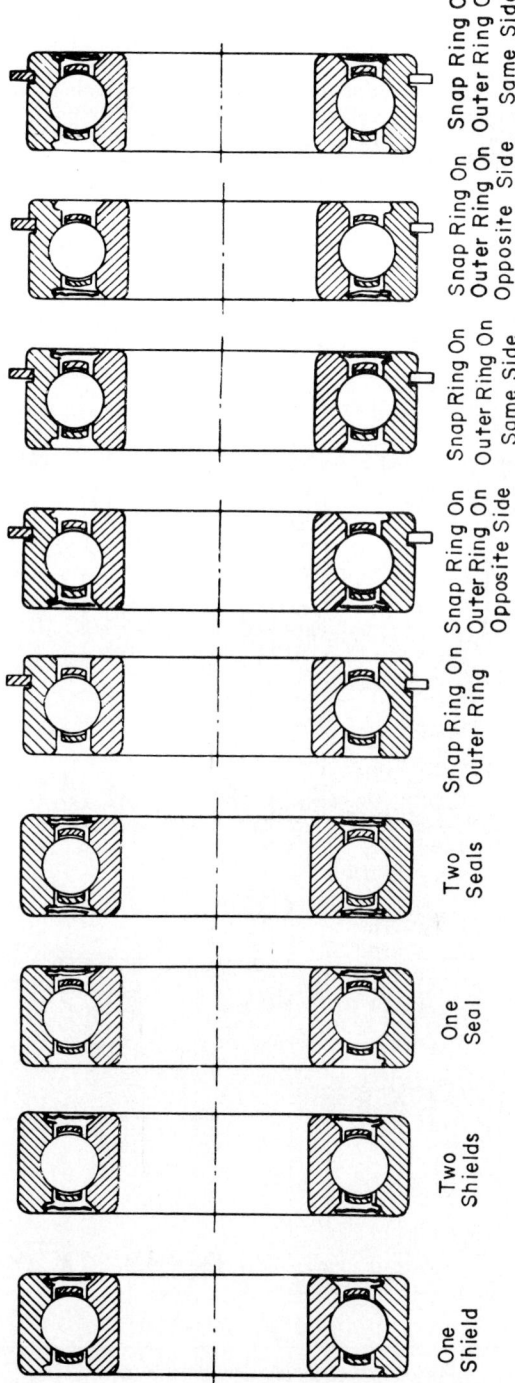

Fig. 9-2 Shields, seals, and snap rings.

1-138   Basic Maintenance Technology

LOAD RATINGS

All manufacturers of rolling bearings establish a load rating for each bearing produced. An industry-approved method for calculating this rating exists, but not all manufacturers use the method. The unfortunate situation therefore exists that two almost identical bearings produced by different manufacturers can have vastly different ratings.

Ratings are expressed as the load which will give a rating life of a certain number of revolutions. Rating life is defined as the number of millions of revolutions at a given constant speed that 90 percent of the bearings will complete or exceed before first

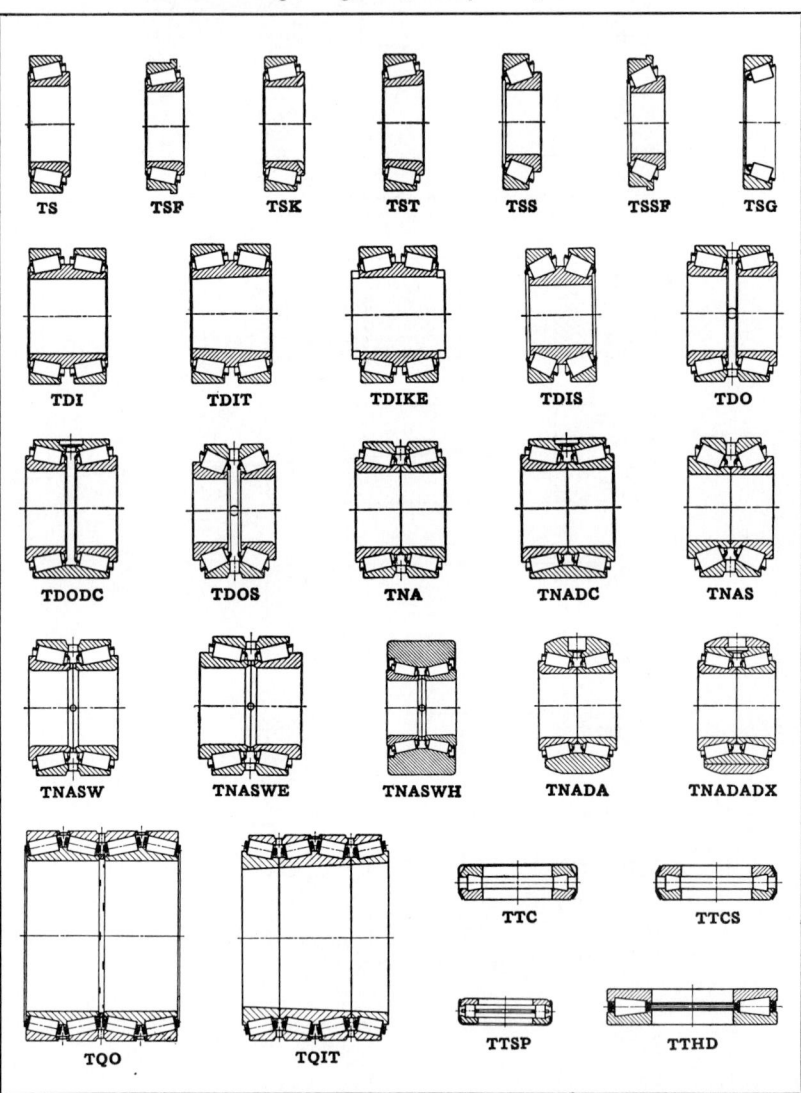

Fig. 9-3   Types of tapered roller bearings.

evidence of fatigue develops. In other words, this is a reliability or statistical rating, the only mechanical component so rated. The most common basis of rating is 1 million revolutions, but tapered roller bearings are rated on the basis of 90 million revolutions, usually expressed as 500 rpm and 3000 hr. Hence it can easily be seen that comparing manufacturers' ratings as published in their catalogs can be misleading if appropriate adjustments are not made to published values.

SHAFT AND HOUSING FITS

It is a basic rule of design that one ring of a rolling bearing must be assembled with its shaft or housing with an interference fit, since it is virtually impossible to prevent creep by clamping the rotating ring axially. Generally it is the rotating ring that is tight, but more

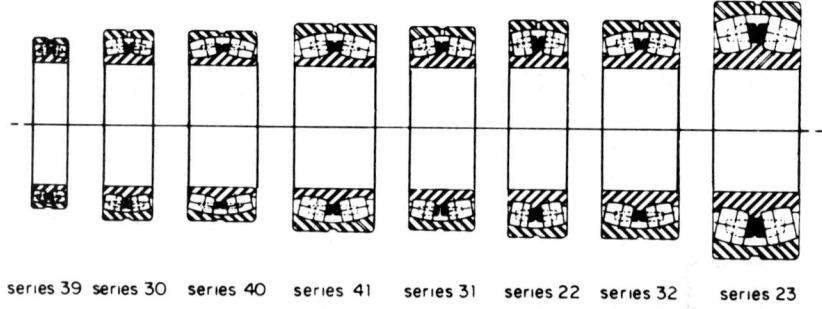

Fig. 9-4 Roller bearing series.

correctly stated it is the ring which rotates relative to the load. In some special cases this is not the rotating ring; for instance, in a vibrating unit where vibration is produced by eccentric weights, the load rotates with the rotating ring and it is best to have the stationary ring have the tight fit.

Except for special cases as illustrated above, the stationary ring can normally be assembled with shaft or housing with a slip fit.

The magnitude of the interference fit will vary with the severity of duty and type of bearing. Ball bearings under normal-load conditions will have approximately 0.00025 in. interference per inch of shaft when the inner ring is the tight fit. Roller bearings will have fits of approximately 0.0005 in. per inch of shaft. Fits will be increased for heavy-duty service and decreased for light duty. In general, when the outer ring is the tight fit, the interference is less than a corresponding shaft fit.

All bearing manufacturers show recommended fitting practices for their bearings in their general catalogs. With the exception of tapered roller bearings, the recommendations are normally expressed in ISO standards. ISO standards designate the fit between the bearing outside diameter and the housing by a capital letter and a number such as H7, J6, or P6. Fits between the shaft and bore of the bearing are designated by lowercase letter and number such as g6, m5, or r7. In the ISO system the letter indicates the class or type of fit and the number the tolerance range.

BEARING MOUNTINGS

When a shaft is mounted on rolling bearings, some provision must be made for thermal expansion and/or contraction of the shaft. Also, the shaft must be located and held axially so that all machine parts remain in the proper relationship dimensionally. When the inner ring has the tight fit, it is usually locked axially relative to the shaft by locating it between a shaft shoulder and some type of removable locking device. A specially designed nut as shown in Fig. 9-5 is normal for a through shaft. A clamp plate as shown in Fig. 9-6 is normally used when the bearing is mounted on the end of the shaft. For the locating or held bearing of the shaft, the outer ring is clamped axially, usually between housing shoulders or end-cap pilots. This type of mounting restricts axial movement in the shaft to the end movement resulting from the internal clearance of the bearing. If required, this

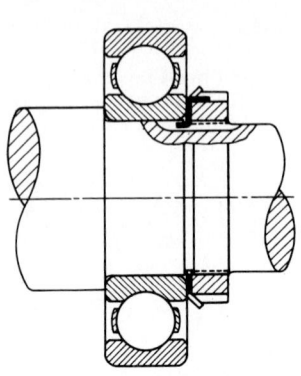
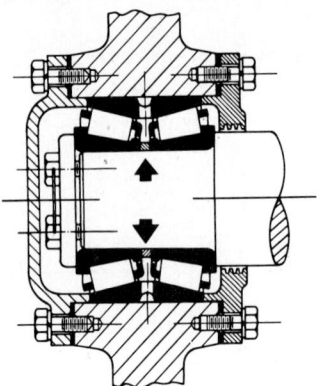

Fig. 9-5 Special nut for through shaft.

Fig. 9-6 Cone-spacer adjusting device.

can be zero if the appropriate bearing type is used. The outer rings on all other bearings on the shaft should not be secured axially, and enough clearance should be provided between the side face of the stationary ring and the nearest housing shoulders to allow for anticipated expansion or contraction. A typical held-free mounting is shown in Fig. 9-7.

Certain types of cylindrical roller bearings are capable of absorbing shaft expansion internally simply by moving one ring relative to the other as shown in Fig. 9-8. The advantage to this type of mounting is that both inner and outer rings may have a tight fit. This may be desirable or even mandatory if significant vibration and/or unbalance exists in addition to the applied load.

Where bearing centers are short and minimum thermal expansion is expected, an opposed mounting as shown in Fig. 9-9 may be used. In addition to its simplicity, this mounting has the advantage that thrust in one direction will be taken on one bearing and thrust in the other direction on the other bearing. Obviously, the clearance between the side face of the bearing and the housing shoulder must be carefully controlled or the shaft will shift excessively in an axial direction.

Single-row tapered roller bearings and angular-contact ball bearings require special consideration. For example, if a radial load is applied to a single-row tapered roller

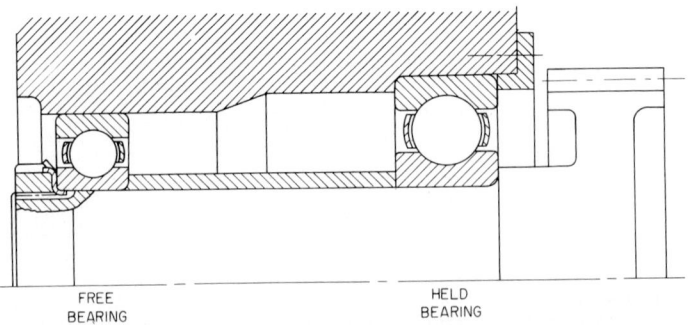

Fig. 9-7 Axial shaft locations, one free and one held bearing.

bearing, an axial component of the load is generated by the angle of the roller set which tends to separate the bearing unless this induced thrust is resisted by another bearing properly mounted to resist the movement. The other bearing is normally another single-row tapered roller bearing. A mounting of this type may have the bearings arranged in one of two ways as shown in Fig. 9-10. The upper portion of Fig. 9-10 shows the included angles of the conical portions of the bearing, or the cup roller track, open away from each other. This is known as an *indirect* mounting. The lower half shows the included angles of the conical portions of the bearings open toward each other. This is known as a *direct*

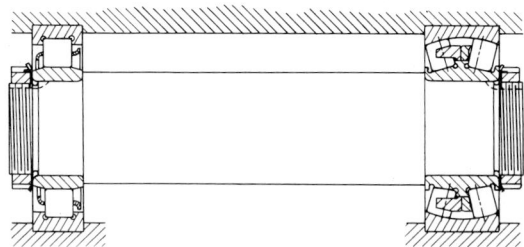

Fig. 9-8 Axial shaft locations, roller bearing used for floating location.

mounting. It should be noted before progressing further with a description of this type of mounting that the point of reaction of the load on the centerline of the shaft, or the effective center, is not at the geometric center of a single-row tapered roller bearing but at some point "O" as determined by the angle of the roller relative to the centerline of the shaft.

Therefore, if the bearings of two different mountings are physically located the same distance apart with an indirect mounting, the effective centers of the bearings are farther apart than with the direct mounting, a more desirable arrangement when an overturning load exists.

With either a direct or an indirect tapered roller bearing mounting it is necessary to set the running clearance of the bearings when they are assembled. This is done by adjusting the cones in an indirect mounting and the cups for a direct mounting. Figures 9-11 and 9-12 show two ways of adjusting cones by nuts, and Fig. 9-13 shows a method of shimming for cone adjustment. Figures 9-14 to 9-16 show three ways of shimming cups in a direct mounting. Proper running clearance is controlled by measuring the end movement, or end lateral, of the shaft. The machine builder's recommendation for proper end lateral should be strictly followed. It will usually be indicated on the drawing of the particular part or given in the maintenance manual for the equipment.

Obviously, the only provision for thermal expansion in either of these mountings is the end lateral of the assembly. For that reason they should be used only where bearing centers are relatively short or where little temperature variation is anticipated.

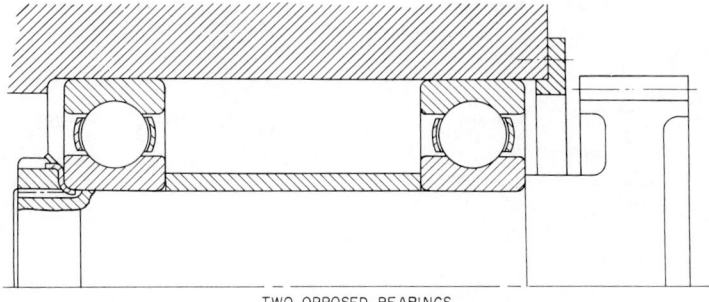

Fig. 9-9 Axial shaft locations, opposed mounting of ball bearings.

1-142 Basic Maintenance Technology

Two-row tapered roller bearings are mounted the same as other types of bearings. Proper end lateral is established in the factory.

Angular-contact ball bearings are rarely used singly. However, if they are, they must be mounted in a similar manner to single-row tapered roller bearings. Thus much smaller running clearances used in ball bearings make a mounting of single angular-contact ball

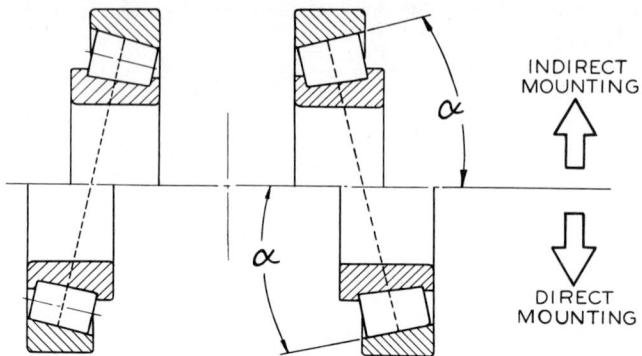

Fig. 9-10 Indirect and direct mounting.

bearings very difficult to adjust properly. Angular-contact bearings could be substituted for the tapered roller bearings of Fig. 9-10, and the same comments and nomenclature would apply for single-bearing mountings.

However, angular-contact ball bearings are normally used in pairs as shown in Fig. 9-17. The side faces of these bearings are especially ground in the factory to permit them to be mounted side by side as shown in Fig. 9-17. Face-to-face (Fig. 9-17A), back-to-back (Fig. 9-17B), and tandem (Fig. 9-17C) is the common terminology for these mountings. When two or more bearings are stacked in tandem for high thrust loads, usually another bearing in the assembly is mounted face-to-face or back-to-back with the tandem stack. When mounted in any of these arrangements, they may be considered as one multiple-row bearing. Because methods of face grinding may differ from one manufacturer to

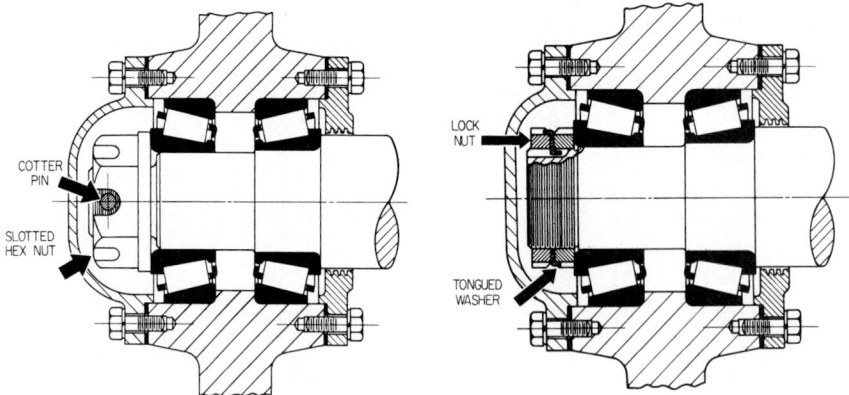

Fig. 9-11 Slotted-nut adjusting device.

Fig. 9-12 Double-nut and lock-washer adjusting device.

Maintenance of Rolling Bearings   1-143

another, it is advisable not to mix brands in a pair of tandem-group bearings. The bearing number should indicate in some way that the bearings have been properly ground for mounting in pairs. Bearings for single mounting are available and should not be used as part of a pair.

A large percentage of spherical roller bearings are made with tapered bores. Some ball,

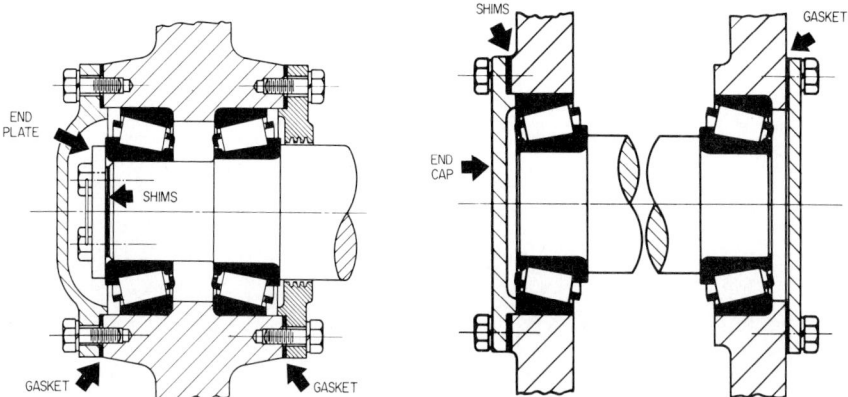

Fig. 9-13 End-plate and shims adjusting device.   **Fig. 9-14** End-cup and shims adjusting device.

tapered roller, and cylindrical roller bearings are also available with tapered bores. These bearings may be mounted directly on the shaft as shown in Fig. 9-18. However, many tapered-bore bearings are mounted on one of two types of sleeves as shown in Figs. 9-19 and 9-20. European machinery builders are particularly partial to use of sleeve mountings.

The adapter sleeve may be mounted as shown in Fig. 9-19 or with a shaft shoulder ring as shown in Fig. 9-21. With a removable type of sleeve as shown in Fig. 9-20, the bearing must always be against a shaft shoulder.

The taper is 1 to 12 on diameter in all but the widest series of spherical roller bearings when a flatter 1 to 30 taper is used. Some four-row cylindrical-roller rolling-mill bearings will also use a 1 to 30 taper in the bore of the inner ring.

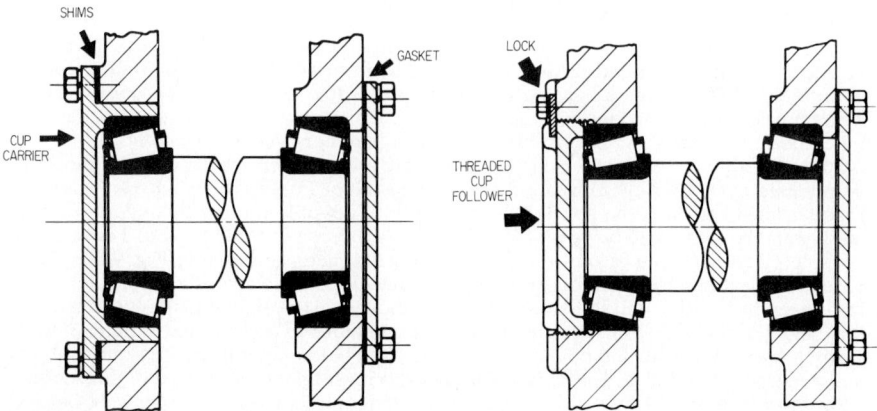

Fig. 9-15 Cup-carrier and shims adjusting device.   **Fig. 9-16** Threaded cup-follower adjusting device.

MOUNTING AND DISMOUNTING OF ROLLING BEARINGS

General The most important thing to remember when mounting or dismounting a rolling bearing of any type is to apply the mounting or dismounting force to the side face of the ring with the interference fit. Keep this force from passing from one ring to the other through the ball or roller set. This is particularly important during mounting. Cleanliness is, of course, extremely important. Not only the bearing but also the shaft and housing must be free from chips, burrs, and dirt.

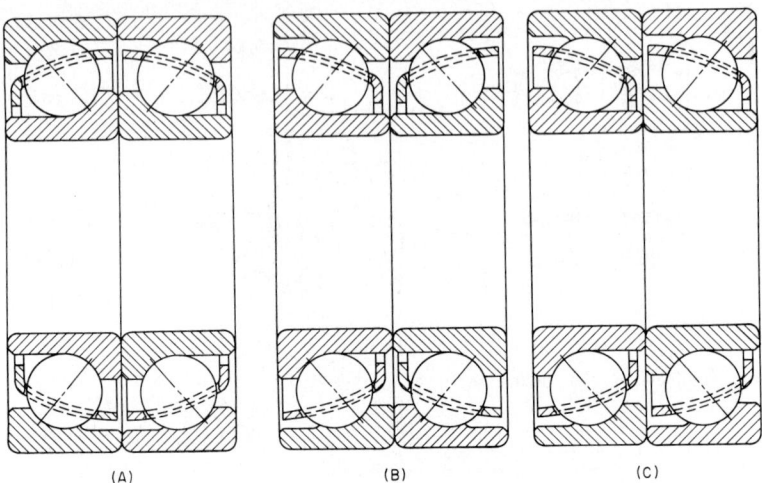

Fig. 9-17 Methods of mounting two flush-ground angular-contact bearings. (A) Face-to-face. (B) Back-to-back. (C) Tandem.

Bearings should be kept wrapped until the last possible moment. Since most modern rust preventives used by bearing manufacturers are compatible with petroleum-base lubricants, the slushing compound is normally not removed. However, there are exceptions to this rule. If oil-mist lubrication is to be used and the slushing compound has hardened in storage or is blocking lubrication holes, it is best to clean the bearing with kerosene or other appropriate solvent. Obviously, the other exception would be if the slushing compound has been contaminated with dirt or foreign matter before mounting. It is also permissible and sometimes desirable to wipe the rust preventive from the bore or outside diameter of the bearing, depending on which surface will have the tight fit. Before mounting or dismounting a bearing, always take the time to collect the proper tools and accessories. The use of inappropriate tools is a major cause of bearing damage. Also, remember, never strike a bearing directly with a hammer, sledge, or mallet.

Cold mountings All small bearings (4-in. bore and smaller) may and sometimes must be mounted cold by simply forcing them on the shaft or into the housing. However, it is important that this force be applied as uniformly as possible around the side face of the bearing and to the ring to be press-fitted. Mounting fixtures should be used. These can be a simple piece of tubing of appropriate size and a flat plate as shown in Fig. 9-22. Do not try to use a drift and hammer, because the bearing will become cocked on the shaft. Force may be applied to the simple fixture described above by striking the plate with a hammer or by an arbor press as shown in Fig. 9-23. It is a good idea to apply a coat of light oil to the bearing seat on the shaft and bore of the bearing itself before forcing on the shaft. It should be noted that all sealed and shielded ball bearings must be mounted cold in this manner.

Temperature mountings The simplest way to mount any open straight-bore bearing, no matter what size, is to heat the entire bearing and simply push it on its seat and hold in place until it cools enough to start gripping the shaft. For tight outside-diameter fits the housing may be heated if practical; if not, the bearing may be cooled by dry ice. However,

Maintenance of Rolling Bearings   1-145

if the ambient conditions are humid, cooling the bearing introduces the possibility of condensation on the bearing which will induce corrosion later.

There are several acceptable ways of heating bearings. Some of these are as follows:

1. Hot plate: a bearing is simply laid on an ordinary hot plate until it reaches the approved temperature. The disadvantage of this method is that the temperature is difficult to control. A Tempilstik or pyrometer should be used to make certain the bearing is not overheated.

2. The temperature-controlled oven: This method needs little comment. The bearings should be left in the oven long enough to heat thoroughly. However, never leave bearings in a hot oven overnight or over a holiday or weekend.

3. Induction heaters are available which can be used to heat bearings for mounting. One of these is shown in Fig. 9-24. It must be remembered that this is a very quick method of heating and that some method of sensing the ring temperature must be used or the bearing may be damaged. A Tempilstik or pyrometer can serve this purpose.

4. A hot-oil bath may also be used to heat the bearing and, in fact, is the most practical means to heat larger bearings. This method has some drawbacks, as the temperature of the oil is difficult to control and may overheat the bearing or even become a fire hazard. A mixture of soluble oil and water can eliminate both these disadvantages. Make the mixture 10 to 15 percent soluble oil. This solution will boil at approximately 210°F, which is hot enough for most bearing fits.

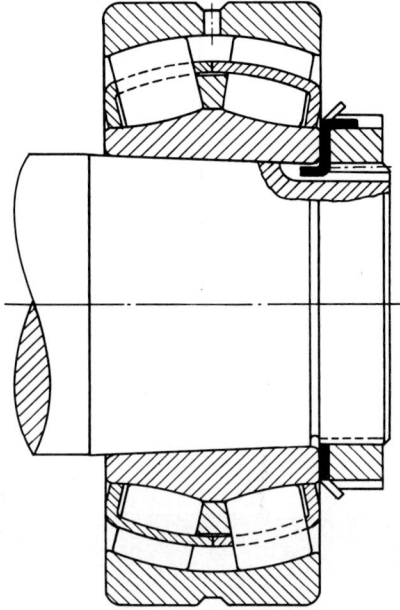

Fig. 9-18 Direct mounting.

The heating solution should be placed in a tank or container which has a grate several inches off the bottom, as shown in Fig. 9-25. This will allow any contaminants to sink to the bottom and keeps the bearings off the bottom of the container.

As mentioned above, 210°F is hot enough to mount most bearings. If using one of the other methods of heating or another solution, 250°F will do the bearing no harm. However, this temperature should not be exceeded for small ball bearings (2-in. bore and smaller). Larger bearings can be heated somewhat higher than this without harm, but metallurgical damage will occur at over 300°F.

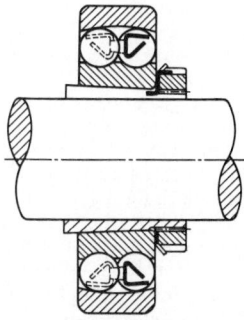

Fig. 9-19 Mounting with an adapter sleeve.

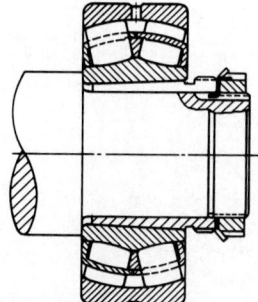

Fig. 9-20 Mounting with removable type of sleeve.

1-146 Basic Maintenance Technology

Mounting tapered-bore bearings Tapered-bore bearings can be mounted simply by tightening the locknut or clamping plate, which will locate it on the shaft until the bearing has been forced up the taper the proper distance. However, especially for large bearings, this technique will require a good amount of brute force. There are special techniques that may be used to reduce the amount of force required.

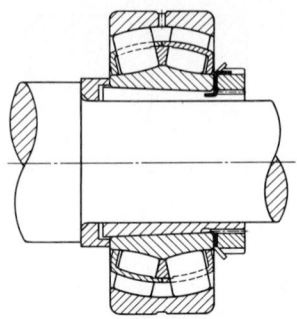

Fig. 9-21 Mounting with shaft shouldering.

Fig. 9-22 Mounting using flat plate.

Before reviewing the mounting techniques for tapered-bore roller bearings, we will discuss the special case of self-aligning ball bearings. The bearing should be put on its tapered seat and the locknut hand tightened until all looseness is removed between adjacent parts. Then using a spanner wrench, not a drift and a hammer, tighten the nut one-eighth turn farther. Bend a lockwasher tab into the nut slot nearest to a washer tab in a tightened direction. At this point, the outer ring should rotate as well as swivel freely.

Tapered-bore spherical roller bearings can be mounted a bit more scientifically. Since the internal clearance in a roller bearing is significantly larger than in a ball bearing, this clearance can be measured with a thickness gauge. As the bearing inner ring is pushed up the tapered seat, the inner ring expands, thereby reducing the internal clearance. Hence the amount of this reduction is a direct function of the interference fit between the bore of the bearing and the shaft. Therefore, if we measure the internal clearance of the bearing unmounted and control the amount the clearance is reduced during mounting, we control the shaft fit within very close limits.

The internal clearance of a spherical roller bearing is measured as follows: The bearing is unwrapped and placed on a table so that it can be easily handled. With one hand grasping the lower portion of the inner ring, oscillate the inner ring and roller set in a circumferential direction to seat the lower rollers properly in the sphere of the outer ring, on the roller paths of the inner ring and against the separate guide ring between the two rows of rollers. Select a gauge blade of perhaps 0.003- or

Fig. 9-23 Mounting using an arbor press.

0.004-in. thickness or less for small bearings. The usable length of the blade should be somewhat longer than the length of a roller. It should not be equal to or greater than the width of the bearing. While pushing the top roller against its guiding surface, insert the blade between two rollers and the outer ring and slide the blade circumferentially toward the roller at the top of the bearing, as shown in Fig. 9-26. The blade should pass between

Fig. 9-24 Induction heater, as used to heat bearing prior to mounting. *(Reed Electric Sales & Supply Inc., Portland, Ore.)*

the uppermost roller and the inside of the outer ring. Do this with successively thicker feeler blades until a blade will not pass. Move it so that it approaches the bite between a roller and the outer ring sphere; then with one hand grasping the inner ring as described earlier, slowly roll the uppermost roller under the feeler blade. With the blade between the uppermost roller and the sphere, attempt to swivel the blade and withdraw it axially. The swiveling motion helps to center the roller in its proper operating position, and withdrawing it with the characteristic wiping feel of a line-to-line contact will show that thickness to be the looseness over that roller. If the blade becomes much looser during the swiveling and withdrawing process, attempt the same procedure with a blade 0.001 in. thicker and continue until a blade cannot be swiveled or withdrawn. The internal clearance over that roller will be the blade that can be swiveled and withdrawn after a thicker one has jammed.

Repeat this procedure in two or three other locations by resting the bearing on a different spot on its outside diameter and measuring over different rollers in one row. Either repeat the above procedure for the other row of rollers or measure each row alternately in the procedure described above. Make a note of this unmounted internal clearance.

After the unmounted radial clearance is measured, the bearing is placed on its tapered seat. If the shaft provides for a locknut, it is then assembled, but the lock washer is left off the shaft at this point. The locknut should then be tightened against the bearing, pushing it up the taper until the internal clearance is reduced by the specified amount as shown in Table 9-1. An impact-type spanner wrench as shown in Fig. 9-27 is ideal for tightening the nut.

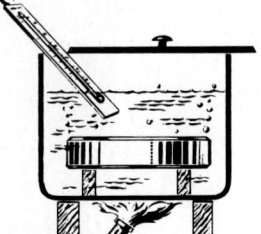

Fig. 9-25 Hot-oil bath for bearing.

The amount of force required to drive a tapered-bore bearing can be greatly reduced if the shaft is drilled and grooved as shown in Fig. 9-28. If these fittings are available, attach a hydraulic pump to the connection at the end of the shaft. Drive the bearing on the taper just enough so there is some interference; then build up hydraulic pressure under the bore of the bearing. A

TABLE 9-1 Recommendation for Driving a Spherical Roller Bearing on a Tapered Seat
(Values in inches)

Bearing bore, mm		Reduction in radial internal clearance		Min permissible final clearance after mounting bearings with clearance		
Over	Incl.	Min	Max	Normal	C3	C4
40	50	0.0010	0.0012	0.0008	0.0012	0.0020
50	65	0.0012	0.0015	0.0010	0.0014	0.0025
65	80	0.0015	0.0020	0.0010	0.0016	0.0030
80	100	0.0018	0.0025	0.0014	0.0020	0.0030
100	120	0.0020	0.0028	0.0020	0.0025	0.0040
120	140	0.0025	0.0035	0.0022	0.0030	0.0045
140	160	0.0030	0.0040	0.0022	0.0035	0.0050
160	180	0.0030	0.0045	0.0024	0.0040	0.0060
180	200	0.0035	0.0050	0.0028	0.0040	0.0065
200	225	0.0040	0.0055	0.0030	0.0045	0.0070
225	250	0.0045	0.0060	0.0035	0.0050	0.0080
250	280	0.0045	0.0065	0.0040	0.0055	0.0085
280	315	0.0050	0.0075	0.0043	0.0060	0.0095
315	355	0.0060	0.0085	0.0047	0.0065	0.0100
355	400	0.0065	0.0090	0.0050	0.0075	0.0115
400	450	0.0080	0.0105	0.0050	0.0080	0.0120
450	500	0.0085	0.0110	0.0065	0.0090	0.0135

NOTE: The axial displacement of the bearing is approximately 16 times the clearance reduction.

pressure of 3000 to 6000 psi will be needed, but with this pressure between the bore of the bearing and the shaft it is possible to float the bearing up the taper with much less torque applied to the locknut or clamp plate than in a dry mounting.

Another convenient way to mount a tapered-bore bearing is to use a hydraulic nut or mounting tool as shown in Fig. 9-29. This technique can also be adapted to sleeve mountings that are large enough to be drilled and grooved.

Cylindrical and tapered roller bearings with tapered bores are not as common as their spherical counterparts, and the manufacturer will have specific mounting instructions for each application.

Fig. 9-26 Determining internal bearing clearance.

Maintenance of Rolling Bearings   1-149

Dismounting of bearings   A wide variety of tools are available commercially which are designed to remove a rolling bearing from its seat without damage. Typical bearing pullers are shown in Fig. 9-30. In removal we should again keep in mind the basic rule to apply force to the ring with the tight fit. Pullers can normally be applied to bearings so that

Fig. 9-27   Use of impact-type spanner wrench.

this rule is observed. However, sometimes supplementary plates or fixtures may be required.

For smaller bearings, an arbor press is equally effective at removing as well as mounting bearings. Also techniques such as the one shown in Fig. 9-31 may be used where size permits.

Hydraulic removal   Where shafts have been designed to apply hydraulic pressure to the fit between shaft and bearing, removal is quite simple. First the locking device,

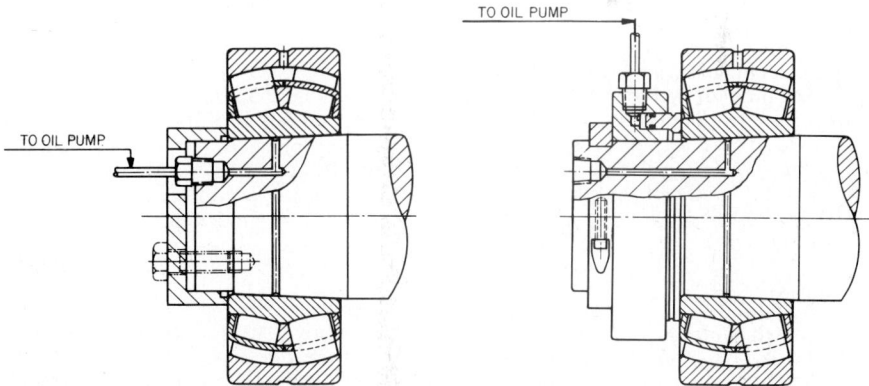

Fig. 9-28   Drilling and grooving to reduce driving force.

Fig. 9-29   Use of hydraulic nut.

1-150　Basic Maintenance Technology

whatever it is, should be backed off a distance greater than the axial movement of the mounting; ¼ in. will be sufficient in virtually every case. Then connect a hydraulic pump to the fitting provided at the end of the shaft as shown in Fig. 9-32A and start building up pressure. When the pressure becomes great enough to break the fit, usually about 3000 to 6000 psi, the bearing will literally jump off the taper with a sharp bang. The retaining

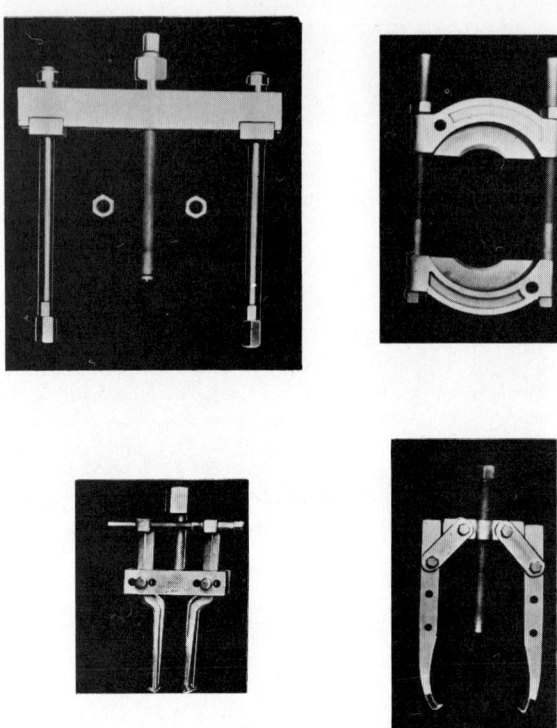

Fig. 9-30　Typical bearing pullers.

device, still being loosely connected, will prevent the bearing from coming off the end of the shaft.

Hydraulic pressure may be used with straight-bore bearings, but a puller must be used in conjunction with the hydraulic pump, as there will be no axial component of the hydraulic pressure to blow the bearing off its seat. See Fig. 9-32B.

Larger sleeve mountings may also be designed to utilize hydraulic pressure for dismounting. If this feature is available, follow the same procedure as outlined above. However, if the sleeve mounting does not have this feature, other techniques such as shown in Fig. 9-33 must be used. For withdrawal sleeves a special nut must be used as shown in Fig. 9-34. For larger withdrawal sleeves a hydraulic nut is desirable for dismounting.

LUBRICATION

The primary purpose of lubrication in a rolling bearing is to separate the contacting surfaces, both rolling and sliding. This purpose is rarely achieved 100 percent, and boundary lubrication or partial metal-to-metal contact frequently occurs. By far the most common lubricants are petroleum products in the form of grease or liquid oil.

Generally the machine builder decides whether a bearing will be grease- or oil-lubricated and normally will recommend the basic specifications of the required lubri-

Maintenance of Rolling Bearings   1-151

cant. However, because the machine designer cannot foresee all the variable conditions under which his equipment will operate, some judgment is required on the part of maintenance personnel. Some knowledge of the lubricants' specifications is therefore useful.

Oil lubrication For oil lubrication, the Annular Bearing Engineers Committee (ABEC) has issued the following recommendations:

The friction torque in a ball bearing lubricated with oil consists essentially of two components. One of these is a function of the bearing design and the load imposed on the bearing, and the other is a function of the viscosity and quantity of the oil and the speed of the bearing.

It has been found that the friction torque in a bearing is lowest with a very small quantity of oil, just sufficient to form a thin film over the contacting surfaces, and that the friction will increase with greater quantity and with higher viscosity of the oil. With more oil than just enough to make a film, the friction torque will also increase with the speed.

The energy loss in a bearing is proportional to the product of torque and speed, and this energy loss will be dissipated as heat, and cause a rise in the temperature of the bearing and its housing. This temperature rise will be checked by radiation, convection and conduction of the heat generated to an extent depending upon the construction of the housing and the influence of the surrounding atmosphere. The rise in temperature, due to operation of the bearing, will result in a decrease in viscosity of the oil, and therefore, a decrease in friction torque compared with the friction of starting, but soon a balanced condition will be reached.

With so many factors influencing the friction torque, energy loss, and temperature rise in a bearing lubricated with oil, it is evidently not possible to give definite recommendations for selection of oil for all bearing applications, but two general considerations are dominant:

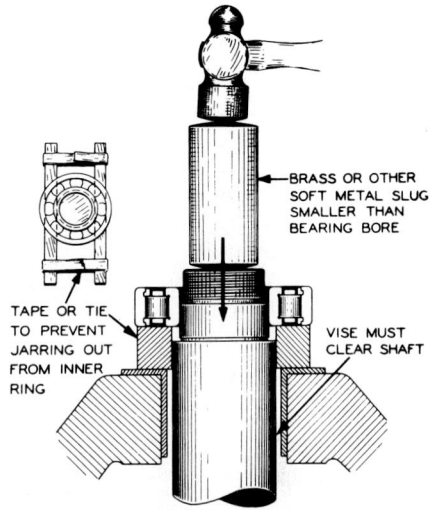

Fig. 9-31 Use of an arbor press to remove small bearings.

1. The desire to reduce friction to a minimum, which requires a small quantity of oil of low viscosity.

2. The desire to maintain lubrication safely without much regard for friction losses, which results in using larger quantities of oil and usually of somewhat greater viscosity in order to reduce losses from evaporation or leakage.

This second condition is most frequently met when bearings have to operate in a wide range of temperatures. An oil that has the least changes with changes in temperature, i.e., an oil with high viscosity index, should be selected.

In the great majority of applications pure mineral oils are most satisfactory, but they should, of course, be free from contamination that may cause wear in the bearing, and they should show high resistance to oxidation, gumming, and to deterioration by evaporation of light distillates and they must not cause corrosion of any parts of the bearing during standing or operation.

It is self-evident that for very low starting temperatures an oil must be selected that has sufficiently low pour-point, so that the bearing will not be locked by oil frozen solid.

In special applications, various compounded oils may be preferred, and in such cases the recommendation of the lubricant manufacturer should be obtained.

For cases where the bearing load is not known, it is a good rule to select an oil that will have at least the following viscosities at the operating temperature:

For ball bearings and cylindrical roller bearings .... 70 SSU
For spherical roller bearings ....100 SSU
For spherical roller thrust bearings ....150 SSU

It should be kept in mind that the temperature of the oil is usually 5 to 10°F higher than the bearing housing. For example, assuming that the temperature of the bearing housing is 170°F, the temperature of the oil will usually be 175 to 180°F. The viscosity of the oil at this temperature should therefore be 70, 100, or 150 Saybolt Seconds Universal (SSU), depending on the bearing type as indicated above. To obtain the approximate viscosity at any other temperature, the diagram in Fig. 9-35 may be used as a guide. For example, refer to the diagram: If the oil viscosity at 180°F should be 70 SSU, read up from 180 on the temperature coordinate to the intersection of 70 on the viscosity coordinate. From this point, follow a line parallel to one of the oblique lines to point 100 on the temperature coordinate. Then, reading over to the viscosity coordinate, find that an oil having a viscosity of about 360 SSU at 100°F is required. This is the customary way of specifying

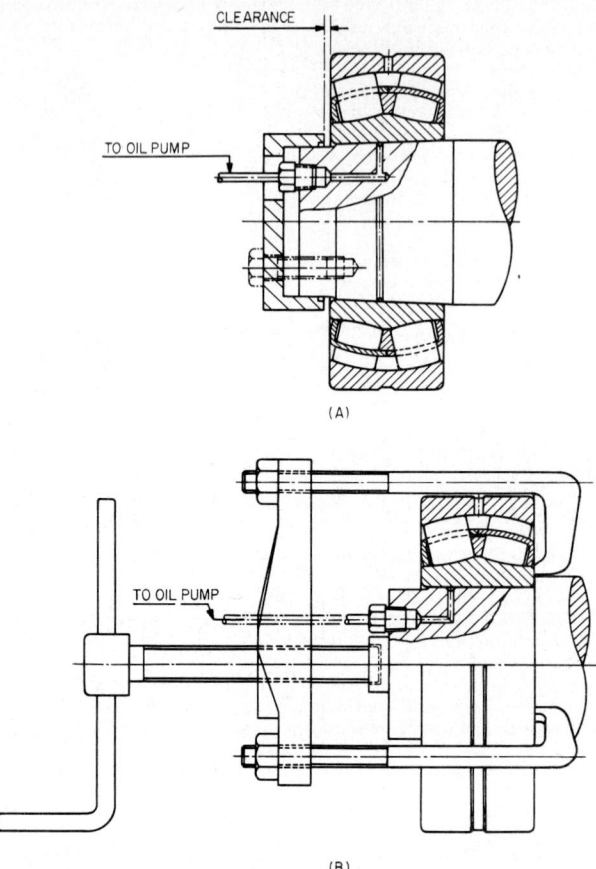

Fig. 9-32 Hydraulic removal. (A) By connection to pump. (B) In conjunction with a puller.

oil viscosities, since the viscosity ratings are usually given at temperatures of 100 and 210°F, the latter being for very heavy oils. Information regarding change in viscosity with change in temperature for a particular oil should, however, be obtained from the oil supplier.

For bearings operating at extremely slow speed, that is, 10 rpm or less, heavy oils which

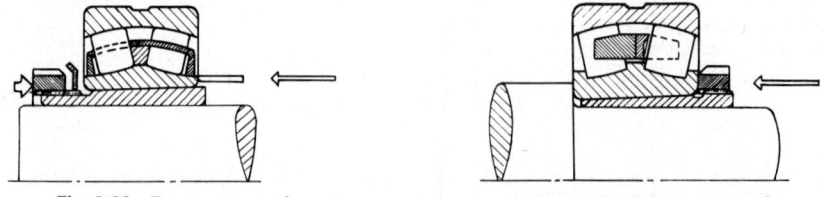

Fig. 9-33 Bearing removal.   **Fig. 9-34** Bearing removal.

have high viscosities at operating temperatures are required. For these applications, it is best to consult the bearing manufacturer.

Grease lubrication Where grease lubrication is used, we need to consider a few of the basic physical and chemical characteristics of the lubricant. Greases are a mixture of lubricating oil and usually a soap base. The base merely acts to keep the oil in suspension.

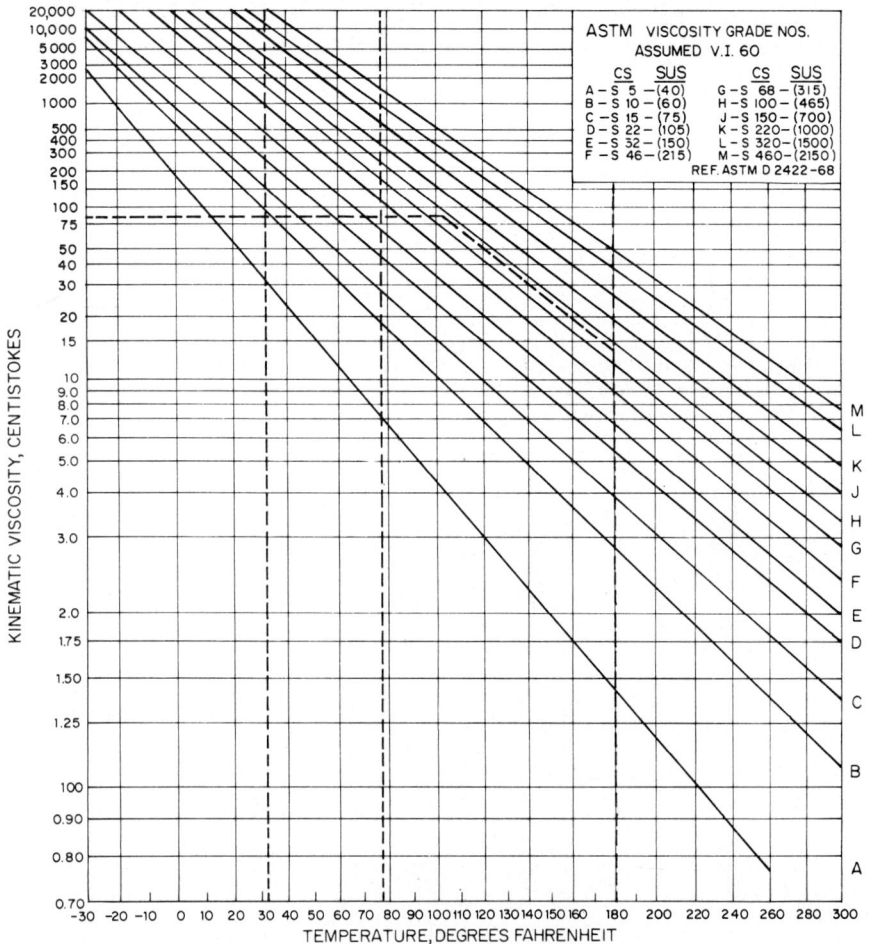

Fig. 9-35 Temperature-viscosity diagram.

When moving parts of a bearing come in contact with the grease, a small quantity of oil will adhere to the bearing surfaces. Oil is therefore removed from the grease near the rotating parts. Bleeding of the grease obviously cannot go on indefinitely; so new grease must come in contact with the moving part or a lubrication failure will result.

Many maintenance departments want to use one grease to lubricate all bearings in the plant. Some lubricant suppliers even advocate this technique. However, it is a risky procedure at best, since there is no true universal ball- and roller-bearing grease. A ball bearing is best lubricated with a fairly stiff grease which will channel. On the National Lubricating Grease Institute (NLGI) code, greases of the number 2 consistency, or 265 to 295 worked penetration, are normally recommended. For roller bearings a grease stiff enough to channel is not desirable, since the full width of the roller track would soon be starved for lubricant if the grease is not soft enough to slump back into the bearing when it is pushed aside. This generally means greases in the number 0 or 1 consistency class with worked-penetration numbers of 355 to 380 for grade 0 and 310 to 340 for a number 1 grease.

Whatever the consistency of the grease, it is still the properties of the oil compounded in the grease that determine if the bearing will be satisfactorily lubricated. All statements and guidelines outlined above in the discussion of oil lubrication also apply to grease-lubricated bearings.

Another characteristic of a grease that must be considered is its drop point. This is the temperature at which the grease passes from a semisolid to a liquid. Typical dropping points are as follows:

Calcium	160–210°F
Sodium	275–350°F
Lithium	350–400°F
Bentone	500°F plus
Silicone	500°F plus
Calcium complex	500°F plus
Aluminum complex	450°F plus

The drop point is the characteristic referred to when a grease is advertised as being good up to 400°F. Whether it will lubricate a bearing or not is still a function of the viscosity of the lubricating oil, not of the drop point of the base. In fact, common industrial bearings made of standard through-hardened or case-hardened materials have temperature limitations of 200 to 300°F depending on the material and how it was heat-treated. The bearing manufacturer should be consulted for specific information.

Chapter **10**

Maintenance of Mechanical Power Transmission Equipment

A. BISHOP
Product Manager, Mounted Bearings, Dodge Division, Reliance Electric Company, Mishawaka, Ind.

R. ELSON
Product Specialist, Mounted Bearings, Dodge Division, Reliance Electric Company, Mishawaka, Ind.

M. D'HOORE
Product Specialist, Babbitted and Bronze Sleeve Type Bearings, Dodge Division, Reliance Electric Company, Mishawaka, Ind.

C. HUEMMER
Product Manager, V-Belt Drives, Tapered Bushings, Dodge Division, Reliance Electric Company, Mishawaka, Ind.

G. BOEYER
Product Manager, Shaft Mounted Speed Reducers, Flexible Couplings, Dodge Division, Reliance Electric Company, Mishawaka, Ind.

W. STREJC
Product Specialist, Chain Drives, Dodge Division, Reliance Electric Company, Mishawaka, Ind.

H. BURTON
Product Manager, Dry Fluid Drives and Couplings, Dodge Division, Reliance Electric Company, Mishawaka, Ind.

A. SAREEN
Product Specialist, Dry Fluid Drives and Couplings, Dodge Division, Reliance Electric Company, Mishawaka, Ind.

MAINTENANCE: TAKE CARE OF IT NOW—OR PAY LATER!

Too often, maintenance is done only *after* the damage is done. That's exactly when it's too late. An expensive and probably critical piece of equipment is down and you're stuck with costly repairs, to say nothing of even costlier downtime losses.

Basic Maintenance Technology

The right time to consider maintenance is *before* equipment failure. There is absolutely no substitute for a program of planned preventive maintenance, especially for mechanical power transmission components, to prevent the hair-raising problems of catastrophic equipment failure and the ensuing cost-raising problems of expensive replacement, installation, and lost production time.

As a guide to proper maintenance of the many mechanical power transmission components used in construction equipment, this chapter concentrates on various facets of installation—as a function of maintenance—and on maintenance procedures themselves. Its objective is to help you get all the life and performance capabilities out of your equipment the manufacturers have built into it.

MOUNTED BEARINGS

Bearings are used to support rotating shafts carrying loads of various types. Mounted bearings (Fig. 10-1) such as pillow blocks, flange bearings, and take-ups are frequently used since they offer additional advantages over the use of bare bearings:
1. Their integral housings make them easier and quicker to install.
2. Bearings are protected by seals to give them longer service life.
3. They are available in a wide variety of housing configurations, types of seals, mounting devices, rolling elements, and a range of sizes to match most application requirements.
4. Manufacturers such as Dodge® supply most of their mounted bearings completely assembled, adjusted, and lubricated, saving installation time and eliminating the possibility of contamination during installation.

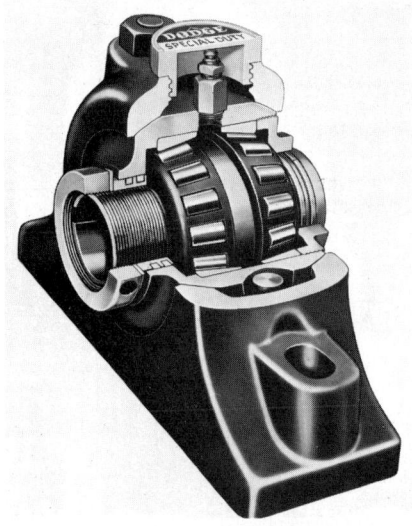

Fig. 10-1 Special-duty pillow-block mounted bearing.

Because these bearings are often used on equipment subject to shock loads, vibration, and operation in dirty, wet, or abrasive environments, proper selection, installation, and maintenance are critical to obtaining maximum performance and service life.

Selection of the correct mounted bearings is primarily a function of the equipment manufacturer. However, the user should also be concerned since he will be operating the equipment. Proper selection, installation, and maintenance are cheaper than repeated bearing failures and will result in reduced operating costs.

All manufacturers of mounted bearings supply installation manuals, which should be followed for correct installation and maintenance procedures. The following recommendations will help you know what to look for and what to avoid in order to assure long life and trouble-free performance.

Mounting There are three basic types of arrangements used to mount bearings to a shaft: setscrew, eccentric collar, and adapter. The setscrew and eccentric collar types are used for small to medium sizes of light to moderately loaded bearings. Bearings having these types of mountings are positioned on the shaft and then secured to the shaft.

Setscrew Types. Setscrews are mounted either in the inner race or in overcollars. They should be properly tightened to keep the inner race of the bearing secured to the shaft.

Eccentric Collar Types. These consist of cam locking of collar and inner race of bearing to shaft. Eccentric collar mounts must be tightened in direction of rotation only.

Adapter Types. These result in the equivalent of a press fit yet can be used with economical commercial shafting. Adapter type mountings are recommended for applications involving heavy loads with shock and/or vibrations.

Maintenance of Mechanical Power Transmission Equipment

Shafting Proper shafting is a critical factor of mounted bearing installations. Industry standards for shafting tolerances, such as those listed in Table 10-1, should be consulted to make sure that the shafting is within recommended values on critical applications. Shaft fits for other than adapter-type mount bearings should be as snug as practical.

TABLE 10-1 Recommended Shaft Tolerances

Nominal shaft size, in.	Commercial shaft tolerance*	Recommended Shaft Tolerances		
		Setscrews in inner race or collar	Eccentric lock collar	Adapter mounting
Up to 1½	+.000−.002	+.0000−.0005	+.0000−.0005	+000−.002
Over 1½ to 2½	+.000−.003	+.0000−.0010	+.0000−.0010	+.000−.003
Over 2½ to 4	+.000−.004	+.0000−.0010	. . . .	+.000−.004
Over 4 to 6	+.000−.005	+.0000−.0015	. . . .	+.000−.005
Over 6 to 8	+.000−.006	+.0000−.0020	. . . .	+.000−.006

*Cold-finished low-carbon bars (Ref. A.I.S.I. Tables 5-1 and 5-2).

Alignment Bearing alignment is another installation factor to be considered. Bearings should be carefully aligned at installation. Careful alignment procedures optimize bearing and sealing capabilities.

Load direction Preferred direction of load is from the shaft into the base. If this is not practical, housing, cap and base bolt strengths, and bolt-tightening procedure, must be considered.

Seals Mounted bearings probably fail more from contamination and lack of proper lubrication than from fatigue of the bearing elements. Since the bearing seals keep contamination out and lubricant in, selection of the correct seal will affect overall bearing performance. Clearance seals operate with small clearance between seal and shaft or inner race. Misalignment will cause a seal to close up on one side and open up on the opposite side.

Labyrinth Seals. These are a type of clearance seal in that they have no rubbing parts. Labyrinth seals have radial passages generally requiring at least reversal in directions through which the grease passes to keep contaminants out of the bearing.

Lip Seals. These seals offer more initial resistance to entrance of foreign material; however, the seal and the surface against which it rubs are subject to damage from abrasive contaminants. The seal can also be damaged by heat conducted through the shaft or generated by the friction between the seal and the rubbing surface. Replacement should be made when loss of seal flexibility impairs its capability.

Combination Seals. These incorporate both labyrinth and rubbing-seal action. The rubbing-seal portion is generally protected by a metallic labyrinth seal and is within the confines of the seal.

Flingers Locking collars and adapter nuts used to lock a bearing to a shaft also act as flingers, adding to sealing efficiency and protecting the seal. While some bearings have no device other than a seal to act as flingers, many types of bearings are furnished with collars at either one or both ends of the bearing.

Lubrication Bearing manufacturers' installation and operating manuals provide specific information regarding lubrication schedules for applications at various operating speeds. Relubrication schedules established by experience are often required because of the wide range of operating conditions involving varying amounts of dirt, moisture, and heat.

The type of grease is important since grease not only acts as a lubricant but also provides a barrier against contamination of rolling elements. Lithium-base NLGI #2 greases are commonly used because of their water-resistant qualities and wide operating-temperature range. Other greases may be better suited for special application requirements, such as extreme temperatures, extreme loads, or chemical compatibility. Static oil, circulating oil, or oil mist are used for higher speed applications.

Too much grease can cause a bearing to run hot, particularly at higher operating speeds. Removal of a grease fitting until excess grease purges from a bearing will generally relieve this situation.

Troubleshooting Discovering and correcting operational problems in time will result in continued operation instead of unexpected failure. Careful observation generally pinpoints problems. Tips on what to look for as part of a regular maintenance program include:

    1. Check seals. Look for signs of wear and heat. Fresh grease showing at seals is desired.

    2. Check lubricants. There should be no pronounced changes in consistency or color and a minimum of contaminants should be present.

    3. Check for heat, noise, or vibration. All of these can be signs of lack of lubrication. If the addition of a small amount of grease does not eliminate these symptoms, check with vibration analysis equipment.

    4. Check mountings. Examine all fastening devices periodically for proper tightness.

After an initial short run-in period, also check base bolts, cap bolts, setscrews or collars, and adapter nut for proper tightness. Check bearings for freedom of operation and operating temperatures. It is not uncommon for a bearing operating temperature to be 70° to 80°F above the ambient temperature at higher operating speeds.

BABBITTED AND BRONZE SLEEVE-TYPE BEARINGS

Sleeve-type bearings have stood the test of time in literally thousands of construction applications where they have been proved for stock product economy, long life, and minimum maintenance characteristics.

The sleeve bearing can be simply described as a bearing housing which holds a sleeve of soft material—normally, babbitt or bronze—with a lubricant between the sleeve and the shaft. Theoretically, the bearing can deliver infinite life if the lubricant film is properly maintained.

The following guidelines are suggested for proper installation and maintenance of sleeve bearings to get dependable performance and long service life.

Maintain adequate grease lubrication High starting friction is a fact of life with sleeve bearings because of their large shaft and sleeve contact area. Yet, it is this large contact area that gives the bearing its heavy-load-carrying capabilities, quiet operation, and vibration- and shock-dampening characteristics.

In repetitive start and stop operation, lubrication tends to break down, resulting in metal-to-metal contact and destructive wear of the soft metal sleeve. Care should be taken to see that adequate lubrication is maintained under all operating conditions and especially when a bearing is operating under 10 rpm.

Maintain proper bearing alignment With non-self-aligning bearings, proper alignment is critical for uniform distribution of load under all operating conditions. If misalignment occurs, the shaft will tend to wear the cap (top) and base (bottom) on the opposite sides of the same sleeve bearing. Misalignment can be corrected by shimming the bearing housing.

Do not exceed prescribed rating loads Any good power transmission engineering catalog contains load rating tables for sleeve bearings. These ratings are the industrial standards established by the Mechanical Power Transmission Association (Standard No. 401). Starting and occasional peak loads may be allowed to exceed these ratings but never by more than 100 percent.

A general rule of thumb is that babbitted sleeve bearings will take approximately 250 psi. To find an approximate load rating at slow speed, use the formula

$$\text{psi} = \text{load} \div \text{shaft diameter} \times \text{length through bore}$$

Load ratings are slightly reduced with an increase in speed.

Maintain correct load direction Direction of load should not be closer than 30 deg to the grease groove. If a bearing has a cap which is not gibbed or doweled to its base, the load should be on the base and not closer than 30 deg to the joint between cap and base. When the direction of load is in the joint area between cap and base, angle babbitted and bronze pillow blocks are recommended. Typical applications are conveyor head shafts or shafts carrying heavy gear or chain drives.

Maintain shaft journal surface The finish on a shaft journal must be equal to that of commercial steel shafting (about 32 microinches) and the diameter must be within the tolerance of commercial steel shafting. A shaft must also be clean and free from rust, nicks, or burrs which would hamper operation and reduce bearing life. The better the shaft finish, the longer the sleeve bearing will operate with normal maintenance.

Inspect for wear periodically Soft bushing material is used in sleeve bearings so that wear is on the bearing and not on the shaft. By visually inspecting the ends of sleeve bearings periodically, replacement can be scheduled to eliminate downtime and catastrophic failure.

Do not operate at temperatures higher than recommended Ambient operating temperatures should not exceed 130°F for babbitted sleeve bearings and 300°F for bronze sleeve bearings. These temperature extremes are industry standards. If a shaft transmits heat from any source, shaft temperature at the bearing should not exceed these extremes.

Protect against adverse operating conditions The soft bushing material used in sleeve bearings does have self-healing characteristics. However, it is best to keep contaminants out or to remove them as soon as they get in. This is especially important in the dirt and dust of construction service. Greasing bearings to purge contaminants and to maintain adequate lubrication cannot be done too often in the effort to ensure long bearing life and dependable performance.

TAPERED BUSHINGS

Tapered bushings, such as the Taper-Lock® manufactured by Dodge, offer the most practical alternative to the restrictions of shrink-fit or bored-to-size bushings (Fig. 10-2). The advantages of tapered bushings over these other types are numerous. They allow you to meet a maximum of application requirements with a minimum inventory of bushings. They provide easy-on, easy-off installation, good gripping strength, and excellent concentricity, and they are readily available in a wide range of sizes without special machining.

Selection of the tapered bushing is based on the torque capacity of the component—sheave, sprocket, clutch, etc.—and the standard commercial shafting being used. Incorrect selection or installation can result in the components flying off while in use, causing costly equipment damage and personal injury. The following suggestions will help you get dependable performance and long life from tapered bushing installations.

Stick to manufacturer's size recommendations Generally, the component manufacturer will indicate the correct bushing size to meet torque and shaft requirements. For best results, stick to those recommendations.

Fig. 10-2 Taper-Lock bushing.

Consider wall thickness Avoid bushings whose walls are relatively too thick or too thin. A thick wall bushing has poor flexibility and will resist good gripping action on the shaft. On the other hand, a thin wall bushing is too fragile and could easily be broken, especially in the removal process.

Clean shaft bushing and component To take maximum advantage of a bushing's capabilities, be sure that the shaft is free of nicks and burrs and that no dirt or chips are on the bushing and component taper bore. Such contamination will impair the precision-machined fit of the tapered bore, the bushing, and the bushing shaft. The result will be lack of concentricity, an imbalanced component, or gripping strength on the shaft that is less than optimal.

Tighten mounting screws Follow the bushing manufacturer's recommended torque values in tightening mounting screws. To assure proper seating, tighten screws alternately, using a torque wrench and cheater bar to get sufficient torque values.

Instructions for Dodge Taper-Lock bushings recommend that a bushing be rapped in with a hammer and drift as part of the installation sequence. This procedure supplements the mounting screws in assuring proper seating of the bushing in the component. Flanged bushings should *not* be rapped in.

When removing the component from a shaft, the opposite technique is often helpful. In other words, after the mounting screws are loosened, the hub is rapped rather than the housing. This tends to back the bushing out of the component and makes disassembly easier than with flanged bushings.

V-BELT DRIVES

V-belt drive performance depends on four basic factors: drive design, component selection, installation, and maintenance. If any of these are neglected, the system cannot function properly and premature failure is likely.

Generally, the user of construction equipment has little control over drive design and component selection. These are functions of the equipment manufacturer and the equipment in which the drives are installed. However, installation as it relates to component or belt replacement and maintenance as it relates to prevention of trouble are factors which the user can control for optimum return on the investment in V-belt drive equipment.

The following is a quick guide to proper installation and maintenance procedures.

Proper Installation Minimizes Future Maintenance

Inspecting sheaves Sheaves that are worn or in less than perfect condition can seriously shorten belt life. Using a sheave groove gauge, check for groove wear or distortion. If grooves are worn, the belt will bottom out, resulting in slippage and belt charring or burning. If sidewalls are dished, the bottom shoulder of the sheave will wear the bottom corner of the belt, causing premature failure. Also check sheave grooves for rust and wipe clean of oil and grease. Inspect grooves for cracks, chips, or burrs which could damage the belts. Don't use a sheave that has been cracked or damaged in transit; it could fly apart and cause serious damage to equipment and injury to operating personnel.

Mounting new sheaves Three types of sheave bushings are in general use today: the split-taper-keyed bushing; the Q-D, or quick-disconnect; and the Taper-Lock bushing manufactured by Dodge. For best results, bushed sheaves should be mounted and tightened according to specific instructions supplied by the manufacturer. Refer to the instruction/installation manual. Always be sure that the mating surfaces of the sheave and hub are free of foreign material—dirt, grease, paint, or burrs. Sheave wobble could indicate the presence of foreign material and should be corrected.

Checking alignment V-belt drives do not require as tight alignment tolerances as most other drives. However, unless the belts run through the sheave grooves in a relatively straight line, wear will be accelerated. Using a straightedge, steel tape, or even a piece of string, see if both the driver and driven sheaves are properly aligned to ensure maximum operating efficiency. Another common cause for misalignment is incorrectly mounted sheaves. Rotate drive slowly to check for wobble, which indicates improper installation.

Selecting belts Always select belts to match sheave grooves. The instruction/installation manual will usually include selection tables and conversion charts. A sheave-groove gauge will also be useful in this procedure to determine the proper cross section of belt required. Belt brands should never be mixed on a multibelt drive. Always use a matched set from a single manufacturer; otherwise belt and drive life will be substantially shortened because of differing belt-performance characteristics.

Installing a belt Prior to installing belts, be sure that power is turned and locked off so there is no chance of accidental or inadvertent start-up while work is being done. Never pry or roll belts into the sheaves. It is not only dangerous to an installer's hands, but it also tends to shorten belt life even if there is no visible damage. Instead, use the drive take-up and drop belts into the grooves. Use a bar if necessary but only to move the motor, never to pry the belts.

Tensioning belts properly Proper tensioning is probably the most critical factor contributing to long V-belt drive life with minimum maintenance. Too much tension shortens belt and bearing life, while too little can cause a belt to slip under load. Generally, the best tension is the least tension at which belts will not slip under full load.

Maintenance of Mechanical Power Transmission Equipment

Refer to manufacturer's specifications for proper tensioning and use a tension tester to meet recommended values.

Proper Maintenance Minimizes Costly Failures

Prevent oil and grease buildup Standard service belts exposed to excessive oil and grease contamination are likely to fail prematurely. Check and replace bearings and/or bearing seals if necessary to remove the most common source of oil and grease leakage. If the operating environment is naturally oily or greasy, special oil-resistant belts should be used. Drip shields should be provided where necessary to prevent oil or grease leakage onto the belts from equipment above.

Prevent dirt accumulation Dirt buildup on belts or in sheave grooves accelerates wear and impairs traction. Check and clean when necessary for optimum performance and long belt life.

Never use belt dressing Belt dressing of any kind is unnecessary and actually harmful to V-belts which operate from the wedging action of the belt in the sheave groove. Belt dressing causes belt slippage and attracts dirt and grit which accelerate wear of both the sheave and the belt.

Troubleshooting V-belt drives are an extremely reliable and efficient means of power transmission that require few special tools or procedures for good maintenance. The main ingredients of a successful preventive maintenance program are visually inspecting, listening, and then correcting apparent drive problems. An easy-to-use table of common problems, symptoms, and required corrective action is included for convenience in troubleshooting V-belt drives on a regular basis (see Table 10-2).

SHAFT-MOUNTED SPEED REDUCERS

A shaft-mounted speed reducer is an expensive piece of machinery usually performing a critical function in construction equipment (Fig. 10-3). Install and maintain it properly and it will give you years of dependable service. Abuse or ignore it and you will pay a high price in lost time and emergency repairs.

Maintenance Begins with Proper Installation

The first step in maintaining a shaft-mounted reducer is to install it properly. Begin by obtaining the correct reducer. This may sound elementary, but, on a job site where several reducers are being installed, it can save embarrassing switching later.

Check the driven shaft The driven shaft should be examined for burrs and to make sure it is straight. Bent shafts can cause excessive reducer movement and will eventually loosen the torque arm bolts.

Position and tighten the reducer Position the reducer on the driven shaft as close as possible to its supporting bearing. Then, tighten it to the shaft, using the procedure recommended by the manufacturer. Most manufacturers use either setscrew or tapered bushings to secure the reducer against axial movement on the driven shaft. Rotate the reducer to its running position and secure the torque arm to a rigid support.

Remove shipping tape With the reducer installed in its running position, locate the magnetic drain plug in the bottom hole and the vent plug in the top hole. It is extremely important that shipping tape covering the vent hole be removed. Failure to do this will result in excessive internal pressure which can blow the reducer's oil seals or cause oil to leak past the seals.

Fig. 10-3 Shaft-mounted speed reducer.

TABLE 10-2 V-Belt Drive Failures and Remedies

V-belt drive failures and remedies

Maintenance of Mechanical Power Transmission Equipment 1-163

Add proper lubrication Use only the recommended quantity and grade of oil specified on the name plate and installation instructions. As a safety check, most manufacturers provide an oil-level plug which should be used during filling and subsequent level checks. For reducers equipped with backstops, do not use oils that contain graphite or molybdenum disulfide because these oils can cause sprags to malfunction.

Mount input sheave Proper mounting is important because it affects the life of the input shaft bearings. The sheave should be as large or larger than the minimum recommended by the manufacturer. It should be mounted as close to the reducer bearing retainer as possible to minimize the overhung load.

Check alignment and adjust V-belts Both the motor and reducer sheave should be checked for alignment with a straightedge. Realign, if necessary, and adjust V-belt tensions. The use of a tension tester will prevent overtightening the belts, which could shorten the life of the motor and reducer bearings.

Continued Maintenance Prolongs Life

If all the reducers that failed prematurely were broken down by categories, the number one culprit would be poor maintenance. This covers a variety of sins, but first and foremost among them is failure to change oil as recommended.

Change oil frequently Most manufacturers state the maximum operating period between oil changes should be 6 months or 2500 hr of operation, whichever occurs first. Such figures are fine for normal installations, but, under the extreme operating conditions of the construction industry where dirt, dust, and occasional high temperatures are common, the oil should be changed every 1 to 3 months.

Consider seasonal changes Where seasonal changes are extreme, grades of oil should be changed to suit warm or cold weather conditions. When changing oil, a careful inspection of the magnetic drain plug can provide valuable information about the condition of the gears and bearings and thereby possibly prevent costly unscheduled downtime.

Operating inspection In addition to regular oil changes, proper maintenance should also include periodic inspections of the reducer while it is operating. In general, such inspection will provide numerous clues that will prevent premature failure and expensive equipment shutdowns.

Troubleshooting The following list enumerates symptoms of trouble to look for and what each may indicate:

1. Excessively high running temperatures can indicate overload or too high an oil level.
2. Excessively loud noise levels can indicate impending bearing or gear failure.
3. Oil on the reducer can indicate a damaged seal or loose oil plugs.
4. Loose or turned V-belts can indicate that the torque arm may not be properly fastened to a rigid support.

If any of these conditions exists, stop the machinery and provide whatever service is necessary to correct the malfunction before it does excessive or permanent damage.

CHAIN DRIVES

Since the early 1890s when chain drives were developed to power bicycles and the first automobiles, they have emerged as one of the most economical methods of power transmission available to industry. In the face of rising costs, chain drives offer one of the best horsepower-per-dollar packages for meeting numerous application demands in construction equipment. However, to take full advantage of the economy and savings inherent in chain drives, correct lubrication and maintenance are vital. Without these, an investment in chain-driven equipment can become a burdensome expense because of early failure and performance problems.

The following suggestions are recommended to ensure that a chain drive delivers the life and performance capability built into it by the component manufacturer.

Installation Recommendations

Machinery Preparation Prior to mounting drive components, be sure that rigid support members are present to oppose vibration and drive movement during operation. Shafting and shaft-support bearings must also be able to withstand the bending movement imposed by drive operation. Nonrigid support, excessive shaft displacement, or improper bearings can all lead to premature drive failure.

Alignment Both shafts must be aligned parallel to each other (Fig. 10-4), so that they rotate in the same plane of operation. It is also necessary to make sure that shafts are level. These alignment factors can be verified by using a machinist's feeler bar and a spirit level.

On portable machinery, where misalignment could be a frequent problem, the use of a wide-clearance- or offset-sidebar-type chain might be desirable. Such chains can take 4 in. of lateral displacement or 8 deg of twist per 4-ft length.

Cleaning components All shafts, sprockets, bushings, and keys, must be thoroughly cleaned to remove chips, dirt, and burrs that could interfere with drive operation and wear-drive components.

Mounting sprockets Two mounting methods are generally in use: setscrew with key, and tapered bushing. Both offer the user certain advantages and restrictions within the capabilities of their design. But industry experience tends to prove that the taper bushed sprocket—because of its excellent concentricity—leads to maximum drive life.

Installing chain Any length of chain with an even number of pitches requires only one connecting link for assembly. Chain with an odd number of pitches requires one connecting link plus an offset link. Because odd-pitch chain and offset links tend to weaken a chain assembly, their use should be avoided whenever possible.

Proper tensioning Do not install chain too tightly or it will impose bearing loads greater than normal. On the other hand, loose installation creates drive noise and chain-speed pulsations. Either way, the result is abnormal chain and sprocket wear. To tension

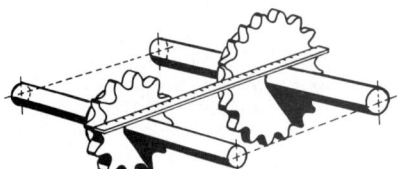

Fig. 10-4 Sprocket alignment.

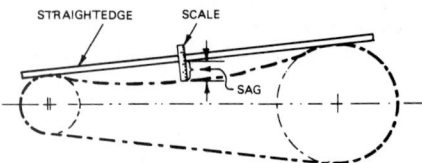

Fig. 10-5 Chain-sag measurement—horizontal drive.

chain properly, turn the free sprocket until lower strand is tight. This creates a sag in the upper strand. With the use of a straightedge and scale (Fig. 10-5), adjust the sag so that it does not exceed 2 or 3 percent of the horizontal length between sprockets or the center distance of the total chain.

Maintenance Recommendations

Inspection As with all forms of power transmission equipment, there is no substitute for a systematic, periodic inspection of chain drives to prolong service life. Because the greatest amount of adjustment is usually necessary during the first few hours of run-in after drive installation, inspection should occur more frequently during the first 8 hr of operation but can become less frequent after that period.

Lubrication The importance of maintaining adequate lubrication of chain-drive components can't be emphasized enough. Most chain-drive failures result from a lack of lubrication that can cause destructive metal-to-metal contact of moving parts. Most chain-drive-component suppliers such as Dodge provide factory prelubrication. However, it is critical that some method of regular lubrication maintenance be provided by the chain-drive user—either manual, drip, bath, or oil stream. Consult the manufacturers' instructions for their recommendation for a particular drive speed and application.

Cleaning Chain should be cleaned at least once a year or immediately if the presence of corrosion or foreign material is detected. Proper procedure for cleaning roller chain is as follows:
1. Completely remove the chain strand from the sprockets.
2. Thoroughly brush-clean chain in kerosene.
3. Inspect the chain strand for signs of wear or corrosion.
4. Soak the chain in an oil bath so lubricant thoroughly penetrates components.
5. Allow chain to hang freely to drain off excess lubricant.

Maintenance of Mechanical Power Transmission Equipment   1-165

6. Clean and inspect sprockets and shafts.
7. Reassemble chain strand around the sprockets.

Replacement Should excessive wear of either the chain or the sprockets be detected, they should be replaced immediately and simultaneously. Placing good chain on badly worn sprockets or worn chain on good sprockets tends to shorten drive life. When replacing chain, length should be determined by the number of pitches rather than by the measured length, which might actually be longer than necessary because of some chain elongation.

Troubleshooting Periodic inspection can reveal several drive problems which, if caught and corrected soon enough, can save the drive, your temper, and catastrophic failure with its ensuing downtime losses. For convenience and easy reference, these problems, their causes, and the corrective actions necessary are listed in Table 10-3.

FLEXIBLE COUPLINGS

Flexible couplings such as the Para-Flex® by Dodge are turning out to be the most maintenance-free available to users of construction equipment (Fig. 10-6). Properly installed, Para-Flex will provide years of trouble-free service and will even give an early warning that element failure is approaching.

Proper alignment is vital As in all elastomeric shear-type couplings, service life is a function of alignment. The closer the coupling is aligned, the longer its service life will be. Check carefully during installation.

Tightening the clamping bolts One factor that more than any other can cause premature failure and operating problems is insufficient tightening torque on clamping bolts. Bolts supplied are grade 8, with a minimum tensile strength of 150,000 psi, so there is little likelihood of stripping threads or snapping off the bolt head in tightening a bolt properly. Use of a torque wrench is recommended.

Periodic visual inspection The coupling element should be checked periodically for signs of cracking. These cracks indicate that the element is reaching the end of its useful life and should be replaced in the next 3 to 6 months.

Fig. 10-6 Para-Flex coupling.

The element will not fail catastrophically except under extreme overloads, in which case it acts like a mechanical fuse to protect the driven machinery. Under normal operating conditions, it will give adequate warning before failure and replacement can be scheduled during routine maintenance shutdown.

DRY FLUID DRIVES AND COUPLINGS

The Flexidyne® dry fluid drive and coupling manufactured by Dodge is extensively used in construction equipment to provide smoother starts, prevent breakage, and reduce maintenance on motors, gears, bearings, and driven machinery (Fig. 10-7).

The fluid is heat-treated steel shot, a measured amount of which—called the *flow charge*—is contained in a housing keyed to the motor shaft. When the motor is started centrifugal force throws the flow charge out to the perimeter of the housing, packing it between the housing and rotor. In this operating mode, power is transmitted to the load. After the starting period of slippage between the housing and rotor, the two become locked together and full-load speed is achieved with operation continuing without slip and at 100 percent efficiency.

TABLE 10-3 Solutions for Chain and Sprocket Failures

Common symptoms	Probable cause	Corrective action
Nonsymmetrical wear on sprockets or rollers	Shafts out of parallel or not in same plane	Realign shafts
Wear on inside of roller plates or side tooth form of sprocket teeth	Sprockets offset on shafts (misaligned), or out of parallel	Realign sprockets
Wear on tips of sprocket teeth	Chain elongated excessively	Replace chain
	Improperly cut sprockets	Replace with correct sprocket
Worn or hooked sprocket teeth	Unhardened sprockets	Replace with hardened sprockets
Wear on edges or sides of link plates	Chain contacting case or fixed object	Increase case clearance or move fixed object
Excessive vibration	Excessive eccentricity or face run out in sprocket	Replace with properly machined sprocket
	Broken or missing roller	Repair or replace chain
Premature elongation	Inadequate or contaminated lubrication, or underchaining	Increase oil flow or redesign
Brown-red oxide in chain joints and oil	Inadequate lubrication	Improve lubrication

Chain jumps sprocket teeth	Wear to vertical limit or excess initial slack	Adjust centers or idler
	Wear to mismesh on large sprocket	Replace chain
Broken chain parts	Drive overloaded	Redesign or avoid
	Excessive slack causing chain to jump teeth	Periodically adjust center distance
	Foreign object	Prevent entry
	Excess chain speed	Redesign or avoid
	Poorly fitting sprockets	Replace
	Inadequate lubrication	Proper lubrication
	Corrosion	Prevent or use noncorrosive chain
Excessive noise	Chain contacting fixed objects	Remove objects
	Inadequate lubrication	Improve lubrication
	Broken or missing rollers	Repair or replace chain
	Misalignment	Check shaft and sprocket and realign
	Chain jumping sprocket teeth	Adjust center distance

1-168 Basic Maintenance Technology

Properly installed, the Flexidyne requires only minimum maintenance to deliver years of economical service. An instruction/maintenance manual is supplied with each unit and should be consulted for installation and maintenance procedures. However, the following guidelines are suggested to provide dependable, long-term performance.

Placing the unit If the output shaft or motor shaft extends too far into a driven hub, the Flexidyne will not slip—no matter how much flow charge is used. Excessive vibration may also be caused by this type of improper installation. A quick inspection through the output hub will determine whether this is the reason for no slippage.

Mounting and aligning Such factors as flexible mounting structure and misalignment of other components are major causes of vibration and shortened service life. Dodge

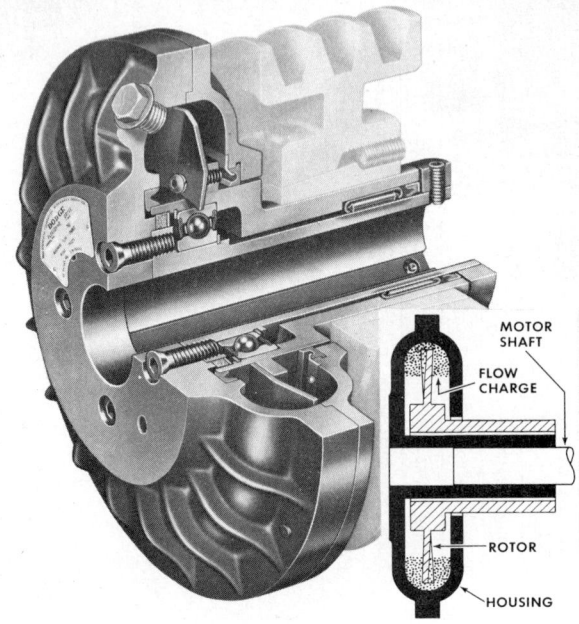

Fig. 10-7 Flexidyne drive and coupling.

recommends that units be aligned with a dial indicator and that motor bases be doweled to assure positive alignment.

Adverse operating conditions In applications where a Flexidyne is subjected to unusually corrosive elements or a very humid condition, stainless steel flow charge will provide better service than the standard cast-steel charge. Another use of stainless steel flow charge is in installations subjected to long periods of idleness where the drive is exposed to wide ranges of temperature or humidity.

High-speed operation Generally, considerably higher-than-normal motor speeds should be avoided when using the Flexidyne. However, where high speeds are unavoidable, it is important that the Flexidyne be connected to the driven machinery or that some resistance be offered to the rotation of the rotor. The rotor must slip for a minimum of 1 to 1½ sec in the flow charge so that the charge will be evenly distributed in the housing. Otherwise, an undesirable out-of-balance condition is likely to result.

Changing operating characteristics Acceleration time and starting torque provided by the Flexidyne can be changed with the addition or removal of flow charge. Consult the instruction/maintenance manual for amounts of flow charge to be used. The charge can be poured in or out of the housing through an easily accessible filler plug.

Overload protection A Dodge speed-drop cutout is included routinely with large Flexidyne units and is available as an option on smaller units. It must be installed to protect against heat damage in applications where overload conditions are anticipated.

The cutout can be connected to interrupt current or to activate a bell, light, or other type of warning device. For hazardous atmospheres, a special explosion-proof cutout is available.

Frequent starting For normal service, involving three or four starts a day of not over 6-sec acceleration time each, the flow charge should be changed after 10,000 hr of operation. Under adverse conditions, such as more frequent starting or extended acceleration time, flow charge should be changed more often. For very extreme conditions, it may be necessary to change the flow charge every 3 to 4 months. If in doubt, consult the instruction/maintenance manual included with each unit.

Lubrication The Flexidyne is lubricated at the factory by Dodge and requires no further lubrication. Never apply oil or grease to the flow charge.

Erratic acceleration If a Flexidyne unit is experiencing erratic or nonrepetitive acceleration, the reason is usually a breakdown of the flow charge into a powdery form. This deterioration is caused by prolonged use without adding new shot. The old flow charge needs to be removed and replaced with new shot.

Slippage while running If the Flexidyne slips while running, it is an indication that the flow charge has worn out and needs replacement. Also, under long service, the rotor will eventually wear and replacement will be necessary. The need for new shot or a new rotor will usually be signaled by a loss in the power transmitting capacity of the Flexidyne.

Chapter **11**

Maintenance of Scaffolds and Ladders

LEONARD SAFIER
President, Patent Scaffolding Company, A Division of Harsco Corporation, Fort Lee, N.J.

Major strides have been made in recent years in the availability of different types and wider choices of scaffolds and ladders for above-the-ground work. Today's manager has more varied and versatile equipment available than ever before; consequently, without realistic guidance, it is easier to make the wrong choice of equipment.

Additionally, more consideration has to be given to employee safety, since current OSHA regulations require safety and protective components which in preceding years were never a major issue. In pre-OSHA days, it was important only to supply a guardrail; under today's working conditions, this guardrail must be accompanied by a midrail and toeboard plus, under certain circumstances, 19-gauge wire mesh installed between guardrail and toeboard.

In general, the wise manager should insist upon having a variety of access equipment at his disposal; the days have passed when you could make do with one old weatherbeaten stepladder and one extension ladder in poor condition. Certain types of ladders and scaffolds are especially suited to certain specific tasks but can often be most unsuitable for other work of another nature requiring a scaffold more specifically suited for the work. Consideration must always be given to the type of equipment which will produce the highest specific work output and at the same time accomplish this in a safe manner with safe equipment. Sturdy work platforms of high stability and ample size result in freely moving, confident workers; the solidity of the work platform is a very important psychological factor and one which is commonly overlooked. Many people are unable to climb more than a few feet above the ground without feeling the necessity of having to hang onto something tightly with at least one hand; consequently, a man assigned to work on a ladder or at a precarious perch could produce only a small fraction of his work output doing the same work at ground level.

In considering the type of equipment for specific tasks, the following typical questions should be asked:

1. How high is the work?
2. Is it spot work or is it continuous in a horizontal- or vertical-pass direction?
3. Is it work which can be most easily done from top to bottom over a length of 10, 20, or 30 ft?
4. Is it a large area which will require a combination of frequent horizontal and vertical passes?
5. Is the area below the work suitable for support from the ground?
6. Is there a wall or an unobstructed space to support a ladder or scaffold?

Basic Maintenance Technology

7. Must people or conveyances pass unobstructed beneath the work?
8. Depending on question 2, how frequently is it necessary to move the work support?
9. How many men and how much equipment must be supported on the scaffold?

The following is a list of scaffolds and ladders, followed by basic descriptions to enable you to determine the best equipment to use for individual circumstances:

    Stepladders and extension ladders—wood, aluminum, and fiberglass
    Welded aluminum folding or sectional scaffolding (rolling towers)
    Welded sectional steel scaffolds and rolling towers
    Safety swinging scaffolds, one- and two-point suspension
    Tube and coupler scaffolds (steel and aluminum)
    Special-design scaffolds

STEPLADDERS AND EXTENSION LADDERS

Stepladders and extension ladders are preferably used for spot work at relatively low heights with no obstructions. The size and weight of a ladder are important; there is room for only one man on a ladder, but if two men are required to carry and lift the ladder into position, then one man is working only a small part of the time.

Choice of materials for ladders—wood, metal, fiberglass.

1. Wood is preferred for regular use in relatively low-height operations and is the workhorse of the industry. However, a wood ladder (or any other type for that matter) can be abused through overfamiliarity. Frequent inspection, properly documented, is essential. Wood is often abused because of its ability to withstand high excessive loads for short periods of time, such as walking or working on it while it is in a *flat* position. Wood is organic and therefore biodegradable; therefore, wood-ladder life can be extended by water-repellent and antifungicide treatments such as pentachlorophenol.

2. Aluminum ladders are preferred by many for their lighter weight. However, Underwriters' Laboratories, Inc., issue the following CAUTION printed on their listing label:

    ELECTRICAL SHOCK HAZARD—METAL LADDERS SHOULD NOT BE USED WHERE CONTACT MAY BE MADE WITH ELECTRICAL CIRCUITS. REFER TO INSTRUCTION LABEL.

3. Fiber-glass ladders are preferred for use in workplaces involving electrical hazards, are about the same weight as wood ladders, and have the additional advantage of insulation properties. Fiber glass is less subject to the effects of abuse (resulting from being dropped or otherwise mishandled) than either wood or metal ladders, since fiberglass side rails have superb recovery from bending or distortion. For proper maintenance, fiber-glass ladders should be coated with a hard floor wax or car wax to reduce the tendency of glass fibers to bloom at areas of friction and scraping.

Some things are so simple, both in the way they are made and in the ways they are used, that the scientific principles upon which they are based are barely visible. With all the simplicity it has constantly retained, the ladder as we know it in its best form today represents a high degree of engineering skill, scientific accuracy, and, most important, dependable safety.

Vital differences in ladders may not be detected except by an expert; hence industry finds it profitable to seek expert ladder advice. Proper weight, exact balance, scientific proportions, dependable quality of materials, character of workmanship, and, of utmost importance, adaptability of a certain type of ladder to the particular kind of service for which it is intended—these factors are essential in the modern ladder even though they may not always be visible to the uninitiated.

Most of the states and principal industrial groups take the ladder question seriously and have acknowledged the importance of ladder efficiency and safety by establishing rigid codes designed to ensure the use of ladders built for the special kind of service required.

A program of ladder upkeep and care should be as much a part of any company's safety program as its maintenance of equipment or any other efficiency or safety devices. The correct use of ladders must be considered no less than the matter of choice when purchases are made.

Three ladder groups Generally, ladders can be classified in three groups: extension ladders, single (straight) ladders, and stepladders. Also, there are special-purpose ladders, which may constitute a fourth classification.

Extension ladders are used in building construction, painting, plastering, maintenance, and almost everywhere that an adjustable ladder is needed.

Single ladders, which of course are not adjustable for height, are generally used when one type of work is to be done at a more or less standard height.

Both extension and single ladders should be fitted with ladder feet to prevent slipping.

Stepladders are what the name implies. They also are self-supporting. Selection for both efficiency and safety is of prime importance. The use of the proper type of stepladder, whether standard classification or special purpose, is receiving more and more attention from industrial executives responsible for their selection.

For design and construction details, the "Safety Code for Portable Wood Ladders" (A14.1-1968), published by the American National Standards Institute (ANSI) is a reliable guide. Since not all ladders meet the standards of safety necessary in industrial plants, the following condensed information on the most important types of ladders should be helpful to those responsible for procurement and factory maintenance. In addition, Underwriters' Laboratories, Inc., approval of a ladder indicates it meets basic standards of design and construction for safety for various categories of use.

Extension ladders Extension ladders (Fig. 11-1) consist of two sections with three-section ladders also available for longer lengths. The strength and safety of such ladders come from the type and quality of wood used in the side rails, as well as their size. Rung size also is important. The required thickness of the side rails depends on the length of the ladder and the type of wood used. The distance between the side rails of the bottom section of a parallel-side ladder should be at least 14½ in. inside to inside for ladders of extended lengths up to 28 ft. Between 28 and 40 ft it should be a minimum of 16 in., and for ladders over 40 ft it should be 18 in. Two-section extension ladders longer than 60 ft are not allowed under the ANSI code.

Good-grade rope (5/16 in.-diameter minimum) and pulley (1¼-in. diameter minimum) for raising the upper section are essential features of well-built extension ladders, because weakness at this point may result in serious injuries. All holes to hold the wood rungs must either extend through the side rails or be bored to give at least ⅞-in. length of bearing to the rung tenon.

Extension ladders should be equipped with automatic spring locks, which enable the worker who is on the ground or floor to raise and lower the upper section. No manual adjustment is necessary. This automatic spring lock, like all the other ladder hardware, is better if plated to resist rust.

Single ladders Single ladders longer than 30 ft are not permitted under the ANSI code. Rungs should be not more than 12 in. apart, and all holes for rungs should be drilled in the same manner as for extension ladders. Rungs must be tight in the hole and secured in place with nails to prevent turning. Pressed-steel rung braces under several rungs is one earmark of a good single ladder.

Fig. 11-1 Extension ladder meeting ANSI Safety Standard A14.1 and listed by UL, Inc.

The width between side rails at the base, inside to inside, must be at least 11½ in. for all ladders up to and including 10 ft. This minimum width increases ¼ in. for each additional 2 ft of length.

Because ladders are subjected to rough usage in such trades as masonry and building construction, the extra-heavy-duty single ladder is preferred. Both the side rails and the rungs are heavier than in the standard single ladder, although the length is limited to 30 ft. Mason's ladders should measure at least 12 in. between side rails (inside to inside) up to 10 ft, with ¼ in. for each additional 2 ft of height. Rungs must be between 8 and 12 in. apart.

Stepladders Stepladders should be made so that the treads will be level in the open position. Good-quality stepladders are designed so that, when open, the slope of the front section is at least 3½ in./ft and the slope of the back section at least 2 in./ft for each 12 in. of side rail. Stepladders, in accordance with ANSI and UL code requirements, should never be furnished in lengths greater than 20 ft, and steps should be uniformly spaced not more than 12 in. apart. Good-quality ladders are equipped with steel safety spreaders, so designed that they will not injure hands when opening and closing. The spreaders also act as braces between the front and rear side rails.

The inside-to-inside measurement of side rails at the top should be at least 11½ in., with an increase of at least 1 in. for each foot of ladder length. This assures a safe, wide base.

Fig. 11-2 Platform stepladder. Preferred because it permits worker to have both hands free.

Safe footing is assured in high-quality ladders by reinforcing steps by means of trussing and bracing, substantially attached by rivets, bolts, or screws. Ladders should be checked frequently to be sure that steps are securely fastened.

Stepladders for heavy-duty use are usually identified by a rung-type back, this construction being more rigid than the slatted back. It permits working from either side.

The minimum dimensions of parts of the heavy-duty ladder (or equivalent cross section developing an actual working stress per square inch as required by the ANSI Code) are shown in the table on the following pages.

The platform stepladder (Fig. 11-2) is by far the most popular, having proved itself by reducing the number of accidents resulting from falls and dropped objects. The 14- by 19-in. platform gives the worker a a firm footing and a guard on three sides, at the same time permitting him to work with both hands.

In the safety platform ladder, steps are truss-rodded and also knee-braced. The steel spreader is of the safety type, a shield over the joint preventing injury to the worker's hands. Good spreaders do not permit the ladder to fold up accidentally. Holes in the top are used as a tool rack, thus reducing danger from falling tools.

Special-purpose ladders There are many types of special-purpose ladders, such as shelf, fruit pickers', trolley, decorators, and paperhangers'. One of the more familiar types is the sectional which, as the name implies, is made in interlocking sections, either continuous taper or interchangeable. It is used widely by window cleaners. A big advantage is portability, since it can be knocked down into small units. Sectional ladders should not be longer than 31 ft and should have an overlap of at least 1 ft.

All special-purpose ladders should conform to the ANSI standards.

Metal ladders In recent years, single, extension, and stepladders made of aluminum or magnesium alloys have come on the market. They are light in weight and resist climatic conditions. However, because they are conductors of electricity, they should not be used around electrical equipment. It is well to tag or paint instructions to this effect on the ladder.

Upon receipt, metal ladders should be examined for sharp edges and burrs on the side rails, tops, and bottoms; such defects can cause painful cuts. The bottoms should be protected to prevent the marring of floors. The best method is to use safety shoes, which also help to prevent slipping.

Metal ladders are now covered by ANSI code A14.2, effective Feb. 1, 1972.

Precautionary measures Where special groups are using ladders, such as plumbers, electricians, and millwrights, the ladders should be properly identified, with the members of each craft held responsible for their particular ladders. The use of just any ladder the worker comes across is likely to lead to costly disaster. Instructing workers as to ladder usage is extremely important.

LADDER SAFETY RULES

LADDER SAFETY RULES #402W FOR WOOD SINGLE & EXTENSION LADDERS

FOLLOW THESE INSTRUCTIONS FOR YOUR SAFETY AND THAT OF OTHERS

1. Inspect ladder carefully on receipt and before EACH use. Test all working parts for proper attachment and operation. Ladders found to be damaged, defective or with missing parts should be withdrawn from use and marked "DO NOT USE." Never use a ladder known to have been dropped until it has been carefully re-inspected for damage of any nature.

2. Install and use this ladder in compliance with the Regulations of the Occupational Safety and Health Act – 1970, and with all other applicable governmental regulations, codes and ordinances. Ladder usage must be restricted to the purpose for which the ladder is designed.

3. Keep nuts, bolts, and other fastenings tight. Oil moving metal parts frequently. Obtain replacement parts from the manufacturer. Do not allow makeshift repairs. Replace frayed or badly worn rope promptly. Keep rungs free of grease, oil, paint, snow, ice or other slippery substances.

4. Ladders must stand on a firm level surface. Always use safety feet and other suitable precautions if ladder is to be used on a slippery surface. Never use an unstable ladder. Ladders should not be placed on temporary supports to increase the working length or adjust for uneven surfaces.

5. Face ladder when ascending or descending. Always place ladder close enough to work to avoid dangerous over-reaching. Keep work centered between side rails. Side loading should be avoided.

6. Sectional (Window Cleaners) Ladders must be assembled in proper sequence with a base section at the bottom, and equipped with safety feet where slippery conditions exist. Maximum assembled length must not exceed 21 ft. for Standard Sectional Ladders or 31 ft. for Heavy Duty. Do not intermingle Sectional Ladders of different types or strength. All Safety Rules printed herein apply to Sectional Ladders except nos. 10 & 11.

7. Never place ladders in front of doors or openings unless appropriate precautions are taken.

8. Before installing an extension or single ladder always insure that working length of ladder will reach support height required. It should be lashed or otherwise secured at top to prevent slipping and should extend at least 3 feet above a roof or other elevated platform. Never stand on top three rungs of an extension or single ladder.

9. Install a single or extension ladder so that the horizontal distance of that ladder foot from the top support is $1/4$ of the effective extended length of the ladder ($75\frac{1}{2}°$ angle). Always insure that both side rails are fully supported top and bottom. Never support ladder by top rung.

10. Overlap extension ladder sections by at least:
    3' each overlap for total nominal lengths up to & including 36'; 4' each overlap for total nominal lengths over 36', up to and including 48'; 5' each overlap for total nominal lengths over 48', up to & including 60'.

At overlaps, fly or upper sections must always be outermost so as to rest on lower section(s).

11. Be sure all locks on extension ladders are securely hooked over rungs before climbing. Make adjustments of extension ladder heights only when standing at the base of the ladder. Never

1-176 Basic Maintenance Technology

extend a ladder while standing on it. For 3-section ladders, always fully extend top section first. Ladders must not be tied or fastened together to provide longer sections other than manufactured for.

12. Water conducts electricity. Do not use wet ladders where direct contact with a live power source is possible. Use extreme caution around electrical wires, services and equipment. Provide for temporary insulation of any exposed electrical conductors near place of work.

13. A ladder is intended to carry only one person at a time. Do not overload. For support of 2 persons special ladders are available. NEVER use a ladder in a horizontal position, never sit on a ladder when it is on edge and never use a ladder in a flat position as a scaffold plank.

14. Store ladders on edge in such a manner to provide easy access for inspection. Provide sufficient supports to prevent sagging. Never use ladders after prolonged immersion in water, or exposure to fire, chemicals, fumes or other conditions that could affect their strength.

15. Only premium grade extension ladders should be used in conjunction with ladder jacks and stages or planks.

16. For further instruction on the care of Wood Single and Extension Ladders refer to the American National Standard, Safety Code for Portable Wood Ladders, ANSI A14.1 – 1968.

LADDER SAFETY RULES #402A FOR ALUMINUM SINGLE & EXTENSION LADDERS
FOLLOW THESE INSTRUCTIONS FOR YOUR SAFETY AND THAT OF OTHERS

1. Inspect ladder carefully on receipt and before EACH use. Test all working parts for proper attachment and operation. Ladders found to be damaged, defective or with missing parts should be withdrawn from use and marked "DO NOT USE." Never use a ladder which has been dropped until it has been carefully re-inspected for damage of any nature.

2. Install and use this ladder in compliance with the Regulations of the Occupational Safety and Health Act – 1970, and with all other applicable governmental regulations, codes and ordinances. Ladder usage must be restricted to the purpose for which the ladder is designed.

3. Keep nuts, bolts and other fastenings tight. Oil moving metal parts frequently. Obtain replacement parts from manufacturer. Do not allow makeshift repairs. Never straighten or use a bent ladder. Replace frayed or badly worn rope promptly.

4. Ladders must stand on a firm, level surface. Always use safety feet. If ladder is to be used on a slippery surface take additional precautions. Never use an unstable ladder. Ladders should not be placed on temporary supports to increase the working length or adjust for uneven surfaces.

5. Face ladder when ascending or descending. Always place ladder close enough to work to avoid dangerous over-reaching. Keep work centered between side rails. Side loading should be avoided.

6. Keep rungs free of grease, oil, paint, snow, ice, or other slippery substances.

7. Never place ladders in front of doors or openings unless appropriate precautions are taken.

8. Before installing an extension or single ladder always insure that working length of ladder will reach support height required. It should be lashed or otherwise secured at top to prevent slipping and should extend at least three feet above a roof or other elevated platform. Never stand on top 3 rungs of an extension or single ladder.

9. Install a single or extension ladder so that the horizontal distance of that ladder foot from the top support is $1/4$ of the effective working length of the ladder ($75\frac{1}{2}°$ angle). Always insure that both side rails are fully supported top and bottom. Never support ladder by top rung.

10. Overlap extension ladder sections by at least:
    3' each overlap for total nominal lengths up to & including 36'.
    4' each overlap for total nominal lengths over 36' up to & including 48'.
    5' each overlap for total nominal lengths over 48', up to and including 60'.
At overlaps, fly or upper sections must always be outermost so as to rest on lower section(s).

11. Be sure all locks on extension ladders are securely hooked over rungs before climbing. Make adjustments of extension ladder heights only when standing at the base of the ladder. Never extend a ladder while standing on it. For 3-section ladders, always fully extend top section first. Ladders must not be tied or fastened together to provide longer sections than manufactured for.

12. Metal and water conduct electricity. Do not use metal, metal reinforced or wet ladders where direct contact with a live power source is possible. Use extreme caution around electrical wires, services and equipment. Provide for temporary insulation of any exposed electrical conductors near place of work.

13. A ladder is intended to carry only one person at a time. Do not overload. For support of 2 persons special ladders are available. NEVER use a ladder in a horizontal position, never sit on a ladder when it is on edge and never use a ladder in a flat position as a scaffold plank.

14. Store ladders on edge in such a manner to provide easy access for inspection. Provide sufficient supports to prevent sagging. Never use ladders after exposure to fire, chemicals, fumes or other conditions which could affect their strength.

15. Portable ladders are designed as one-man working ladders, including any material supported by the ladder. There are three classifications:

Maintenance of Scaffolds and Ladders   1-177

Type I – Heavy Duty for users requiring not more than a 250 pound load capacity for maintenance, construction or heavy duty work.

Type II – Medium Duty for users requiring not more than a 225 pound load capacity for painting, or other medium duty work.

Type III – Light Duty for users requiring not more than a 200 pound load capacity for service requirements such as general household use. Not for use with stages or planks.

16. Only Type I and Type II extension ladders should be used in conjunction with ladder jacks and stages or planks.

17. For further instructions on the care of Aluminum Single and Extension Ladders refer to the American National Standard, Safety Code for Portable Metal Ladders, ANSI A14.2 – 1972.

LADDER SAFETY RULES #403W
FOR WOOD STEPLADDERS

FOLLOW THESE INSTRUCTIONS FOR YOUR SAFETY
AND THAT OF OTHERS

1. Inspect ladder carefully on receipt and before EACH use. Test all working parts for proper attachment and operation. Ladders found to be damaged, defective or with missing parts should be withdrawn from use and marked "DO NOT USE." Never use a ladder that has been dropped or tipped over until it has been re-inspected for damage of any nature.

2. Install and use this ladder in compliance with the Regulations of the Occupational Safety and Health Act – 1970, and with all other applicable governmental regulations, codes and ordinances. Ladder usage must be restricted to the purpose for which the ladder is designed.

3. Keep nuts, bolts, and other fastenings tight. Oil moving metal parts frequently. Obtain replacement parts from the manufacturer. Do not allow makeshift repairs.

4. Ladders must stand on a firm level surface. Always use safety feet and other suitable precautions if ladder is to be used on a slippery surface. Never use an unstable ladder. Never "walk" a stepladder while on it. Ladders should not be placed on temporary supports to increase the working length or to adjust for uneven surfaces.

5. Face ladder when ascending or descending. Always place ladder close enough to work to avoid dangerous over-reaching. Keep work centered between side rails. Side loading should be avoided.

6. Keep steps free of grease, oil, paint, snow, ice, or other slippery substances.

7. Insure that stepladders are fully opened with spreaders locked. Do not stand on top, pail rest or rear rungs of stepladders.

8. Never place ladders in front of doors or openings unless appropriate precautions are taken.

9. Water conducts electricity. Do not use wet ladders where direct contact with a live power source is possible. Use extreme caution around electrical wires, services and equipment. Provide for temporary insulation of any exposed electrical conductors near place of work.

10. A ladder is intended to carry only one person at a time. Do not overload. For support of 2 persons special ladders are available. NEVER use a stepladder in a closed or horizontal position, never sit on a ladder when it is on edge and never use a ladder in a flat position as a scaffold plank.

11. Store ladders on edge in such a manner as to provide easy access for inspection. Provide sufficient supports to prevent sagging. Never use ladders after exposure to fire, chemicals, fumes or other conditions which could affect their strength.

12. Portable ladders are designed as one-man working ladders, including any material supported by the ladder. There are three classifications:

Type I – Industrial – for Heavy Duty and Industrial use.

Type II – Commercial – for Medium Duty and Light Industrial use.

Type III – Household – for Light Duty such as light household use.

13. For further instructions on the use and care of Wood Stepladders refer to the American National Standard, Safety Code for Portable Wood Ladders, ANSI A14.1 – 1968.

LADDER SAFETY RULES #403A
FOR ALUMINUM STEPLADDERS

FOLLOW THESE INSTRUCTIONS FOR YOUR SAFETY
AND THAT OF OTHERS

1. Inspect ladder carefully on receipt and before EACH use. Test all working parts for proper attachment and operation. Ladders found to be damaged, deformed, defective or with missing parts should be withdrawn from use and marked "DO NOT USE." Never use a ladder that has been dropped until it has been carefully re-inspected for damage of any nature.

2. Install and use this ladder in compliance with the Regulations of the Occupational Safety and Health Act – 1970, and with all other applicable governmental regulations, codes and ordinances. Ladder usage must be restricted to the purpose for which the ladder is designed.

3. Keep nuts, bolts, and other fastenings tight. Oil moving metal parts frequently. Obtain re-

placement parts from manufacturer. Do not allow makeshift repairs. Never straighten or use a bent ladder.

4. Ladders must stand on a firm, level surface. Always use safety feet. If ladder is to be used on a slippery surface, take additional precautions. Never use an unstable ladder. Never "walk" a stepladder while on it. Ladders should not be placed on temporary supports to increase the working length or to adjust for uneven surfaces.

5. Face ladder when ascending or descending. Always place ladder close enough to work to avoid dangerous overreaching. Keep work centered between side rails. Side loading should be avoided.

6. Keep steps free of grease, oil, paint, snow, ice, or other slippery substances.

7. Insure that stepladders are fully opened with spreaders locked. Do not stand on top, pail rest or rear rungs of stepladders.

8. Never place ladders in front of doors or openings unless appropriate precautions are taken.

9. Metal and water conduct electricity. Do not use metal, metal reinforced or wet ladders where direct contact with a live power source is possible. Use extreme caution around electrical wires, services and equipment. Provide for temporary insulation of any exposed electrical conductors near place of work.

10. A ladder is intended to carry only one person at a time. Do Not overload. For support of 2 persons special ladders are available. NEVER use a stepladder in a closed or horizontal position, never sit on a ladder when it is on edge and never use a ladder in a flat position as a scaffold plank.

11. Store ladders on edge in such a manner as to provide easy access for inspection. Provide sufficient supports to prevent sagging. Never use ladders after exposure to fire, chemicals, fumes or other conditions which could affect their strength.

12. Portable ladders are designed as one-man-working ladders, including any material supported by the ladder. There are 3 duty classifications:

Type I – Heavy Duty – for users requiring not more than 250 pound load capacity for maintenance, construction or heavy duty work.

Type II – Medium Duty – for users requiring not more than a 225 pound capacity for painting, or other medium duty work.

Type III – Light Duty – for users requiring not more than a 200 pound load capacity or service requirements such as general household use. Light duty ladders should not be used with scaffold planks.

13. For further instructions on the care of Aluminum Stepladders refer to the American National Standard, Safety Code for Portable Metal Ladders, ANSI A14.2 – 1972.

WELDED ALUMINUM SCAFFOLDS

This type of scaffold affords firm, solid work platforms for use by one or more men. Because of their lightness, they are fast and easy to erect and are therefore preferred where a number of off-the-floor jobs are required to be done in a large number of positions. Their lightness, mobility, and ease of erection make them most suitable for light-duty work, especially where the equipment requires frequent erection and dismantling.

These scaffolds are prefabricated from high-strength aluminum-alloy tubing and are equipped with casters as necessary for easy mobility of the erected scaffold. The types usually most practical for maintenance work are aluminum rolling scaffolds with internal stairways, and aluminum ladder scaffolds.

Folding ladder scaffolds (Fig. 11-3) are built in one-piece base sections which speed the erection and dismantling process. The ladder-type base sections are 29 in. or 4 ft 6 in. wide, with spans of 6, 8, or 10 ft between frames. In this type of unit, the two diagonal braces and one horizontal brace are integral parts of the folding unit. Intermediate, extension, and guardrail sections can be placed atop the folding unit, using individual end frames and braces.

A larger folding-type scaffold has base dimensions of 4 ft 6 in. by 6 ft. This unit has an internal stairway, and the upper sections as well as the base section are one-piece folding units. When the scaffold must be erected higher than recommended for a base of this size, outriggers can be used. They clamp to the legs of the base section. Means for leveling, to compensate for uneven ground, are part of the leg equipment. The casters on the legs are locked at both wheel and swivel. Folding scaffold sections are, of necessity, heavier than individual components of demountable sectional scaffolding.

Sectional aluminum stairway scaffolds are designed with end frames of various heights to provide different working levels, adjustable bottom sections with casters but without the folding feature, intermediate sections, half sections, and guardrail sections. All components are demountable so as to be light and easy to handle for erection and dismantling. Internal stairways are used. Outriggers may be used to increase the base area (see Fig. 11-4).

The folding, sectional-stairway, and ladder scaffolds are used for outdoor cleaning and maintenance work—ladder scaffolds for low to medium height and one-man jobs, and

Fig. 11-3 Folding aluminum ladder scaffolds are designed so that the end frames will not fall over at any point during the erection or dismantling process. It is a completely freestanding unit at all times.

folding or sectional-stairway types for higher or heavier work. They are especially suitable when the work is horizontal. Indoors, they simplify work on walls and ceilings, and often are suitable for group lamp replacement.

WELDED SECTIONAL STEEL SCAFFOLDS

Used in situations similar to those of aluminum scaffolds, welded sectional steel scaffolds are heavier and therefore more suitable for heavy-duty work requiring relatively infrequent erection and dismantling. They can be assembled as a rolling scaffold and have mobility similar to aluminum scaffolds but are heavier and more cumbersome to handle. Some end frames have integral exterior ladders. Adjustable extension legs may be used for leveling. Casters lock at the wheel and swivel.

Steel ladder scaffolds, similar to the aluminum ones, are used for heavier-duty work in restricted spaces. Steel-pivoted diagonal cross bracing is used as with larger steel frames. An often useful accessory for the steel ladder scaffold is the bridging trestle. This replaces the diagonal cross braces at the bottom level and permits the scaffold to clear obstructions or permits the passage of traffic beneath the scaffold without interference.

For access to high work areas where relatively heavy work has to be done, such as the replacement of a crane motor or large heating unit, the steel scaffold is unsurpassed in

Fig. 11-4 Outrigger supports should be used when working at platform heights greater than four times the smallest dimension. Shown is an aluminum sectional scaffold 4 ft 6 in. wide.

Fig. 11-5 Sectional rolling scaffold. Steel frames are 5 ft wide and joined by pivoted diagonal braces of lengths from 4 to 10 ft. A hook-on type of access ladder is used with a 3-ft-high grabrail at top, guarded by steel-wire rope with snap hooks. The complete lift of the ladder is set back 7½ in. from the scaffold frames.

Fig. 11-6 Safety swinging scaffold (two-point suspension). Utilizes steel-wire rope with ratchet-action raising and crank-handle lowering. Note use of guardrails, midrails, and toeboards. Similar scaffolds also available with power machines and aluminum platforms.

strength and versatility. It is now available with OSHA-complying external stand-off access ladders (Fig. 11-5), or even *internal* stair-type systems called step units.

SAFETY SWINGING SCAFFOLDS

These are generally two-men work platforms, either wood or aluminum, suspended from roof supports either inside or outside buildings (see Fig. 11-6). They are most suited for successive work operations vertically above each other. Swing-stage platforms are generally available in lengths from 8 to 32 ft, and the industry norm is 28 in. in width. They can be used in conjunction with hoisting apparatus consisting of rope blocks and falls, manually operated steel-wire-rope hoisting mechanisms, and air or electrically operated machines at the platform level. All hanging equipment requires most extreme care in safe rigging procedures by experienced personnel. In general, they are seldom economical for work at heights of less than 30 ft, *unless* access to such work is impractical by other types of ladders or scaffolds.

Swinging scaffolds are particularly suitable for cleaning and painting, tuck pointing, window washing, and similar jobs on exterior walls or tanks where large vertical range and quick up-down mobility are required. They are used also where the surface below the work is crowded, or where conditions are unsuitable for support of ground-based scaffolding. They are recommended for light and medium loads.

Safety belts and separately attached lifelines are essential for proper worker safety as well as insistence on the installation and use of guardrails, midrails, and toeboards. Wire mesh between guardrail and toeboard is required by OSHA *only* when employees are required to work under the scaffold.

As well as the conventional two-point suspension swing-stage platforms, additional items for specialized work are available in the form of one-man work cages and bosun's chairs. The cages are often available with extensions attached to each side of the cage for use by two men. Generally, the cages are used with power-operated winches and the bosun's chairs with powered winches or blocks and falls.

TUBE AND COUPLER SCAFFOLDS (STEEL AND ALUMINUM)

These scaffolds provide the greatest versatility in scaffolding odd shapes such as processing works and refineries and in erecting to extreme heights. They are erected from four basic components: baseplates, interlocking tubing or pipe, bolt-activated couplers for making right-angle connections, and adjustable couplers for making connections at other than right angles.

Horizontal runners can be placed at any point on the vertical posts and, in turn, bearers at any point on the runner, thereby obtaining maximum versatility. This type of scaffolding is unsurpassed in providing work platforms for spheres, cylinders, and other odd-shaped vessels such as those in refineries. It also can be used to build storage racks of virtually any size and capacity, and is even the most suitable type of equipment to scaffold certain buildings having uneven exteriors and projections (Fig. 11-7).

This scaffold also is available in all-aluminum components (Fig. 11-8), making it particularly useful in corrosive atmospheres. Both types can be made into rolling scaffolds with the addition of special casters at the base.

SPECIAL-DESIGN SCAFFOLDS

Special-purpose scaffolds are available specifically designed for access purposes where standard scaffolding components cannot be used easily or are even impossible to use. Such scaffolds, stationary and mobile, are frequently used in the aircraft and aerospace industries, as well as for special requirements on many projects.

Safety requirements Almost all present-day steel and aluminum scaffolds are listed under the Reexamination Service of Underwriters' Laboratories, Inc. The UL seal on maintenance scaffolds means not only that the product is properly designed to sustain the loads for which it was intended but also that certain manufacturing standards are included in the design to assure maximum strength. One of the first things a maintenance department should look for is a UL listing sticker or label. After purchase of the equipment, it becomes the maintenance department's responsibility to make sure not only that the

1-182   Basic Maintenance Technology

Fig. 11-7 This shows Tubelox scaffolding used to scaffold building surfaces with a difficult roof-overhang condition. The versatility of Tubelox is similarly utilized for unusually shaped structures.

Fig. 11-8 Aluminum ladder scaffold. Basic unit consists of 6-ft ladder frames, diagonal braces, platform, and adjustable casters. Note use of guardrails, midrails, and toeboards.

scaffold is used properly but also that it is maintained properly. Of course, all equipment must comply with the appropriate OSHA regulations concerning its manufacture, installation, and use. Near the end of this section an OSHA scaffolding checklist for scaffolds of various types is reproduced which can be used to assist in assuring that OSHA compliance is being maintained.

Safe use and safety rules Any reputable manufacturer of ladders and scaffolds will furnish (with or attached to their equipment) information on specific items and include a set of safety rules which, if followed, can drastically reduce many industrial accidents associated with the use of this equipment. Copies of safety rules should be freely available from the manufacturer of the product—not only at the time of first delivery, but cheerfully and freely in later years when requested. Printed safety rules are *not* intended to be retained with the delivery slip (to which they are frequently attached) and sent into the office for billing purposes and thence forgotten. Neither are they for the purpose of padding out a foreman's hip pocket while the men involved in doing the work have never seen or read them. Safety rules must be read and clearly understood by the men *doing* the work; it is up to the judgment of the workman's immediate superior as to whether best results are obtained by having the workman read the safety rules or by having them read to him along with an explanation of the reasons why certain safety precautions are vital to freedom from injury.

Recently, in the promulgations of OSHA, two sets of regulations for scaffolding were made. One set is for use in construction (1926.451) and the other set is for use in general industry (1910.28), although both can apply in certain circumstances. The need for maximum employee safety in the erection and use of scaffolds is strongly emphasized. The OSHA regulations should be thoroughly read and observed, especially where the use of certain safety components has traditionally *not* been customary. Briefly, the OSHA regulations merely describe in detail the safe use of scaffolds, and therefore it is to the benefit of employer and employee to be completely familiar with such requirements pertinent to the work to be done. There is no shortcut to OSHA compliance; however, it need not be dwelled on at length in this chapter, since the general requirements and principles are familiar.

OSHA SCAFFOLDING CHECKLIST

This highlights certain basic safety precautions concerning scaffolds and can be used as a base for expansion to cover your own particular maintenance requirements. This list is basic and does not purport to be all-inclusive or to encompass all circumstances. Such determinations must be made by the employer in his own individual circumstances.

OSHA SCAFFOLDING CHECKLIST

Project _____
Inspection area _____ Area supervisor _____
Inspected by _____ Date _____

Ladders and scaffolding	Yes	No	Action/comments
Manually propelled mobile scaffolds 1. When using freestanding mobile scaffold towers, do you restrict tower heights to four times the minimum base dimension? (3 and $3\frac{1}{2}$ in some states) 2. Do casters have a positive locking device that will hold the scaffold? 3. Are mobile scaffolds properly braced by cross and horizontal bracing? 4. Is the cross bracing of such length as will automatically square and align vertical members so that the erected scaffold is plumb, square, and rigid? 5. Are all brace connections secured? 6. Are platforms tightly planked for the full width of such scaffolds, except for the necessary entrance opening? 7. Do you provide a ladder or stairway for proper access and exit?			

OSHA SCAFFOLDING CHECKLIST (Continued)

Project _____
Inspection area _____ Area supervisor _____
Inspected by _____ Date _____

Ladders and scaffolding	Yes	No	Action/comments
8. Is such a ladder or stairway affixed to or built into the scaffold?			
9. Is it so located that when in use it will tend to tip the scaffold?			
10. Are landing platforms provided at intervals not greater than 35 ft? 30 to 35 ft?			
11. Is the force necessary to move your mobile scaffold applied near or as close to base as practicable?			
12. Do you make adequate provision to stabilize the tower during movement from one location to another?			
13. Do you permit scaffolds to be moved only on level floors, free of obstructions and openings?			
General scaffolds			
14. When scaffolds are in use, do they rest upon a suitable footing?			
15. Are they plumb?			
16. Have you installed guardrails at all open sides and ends of the scaffolds?			
17. Are such guardrails made of not less than 2- by 4-in. lumber (or other material providing equivalent protection)?			
18. Are they approximately 42 in. high, with a midrail of 1- by 6-in. lumber (or material providing equivalent protection), and toeboards 4 in. high?			
19. Is wire mesh installed between the toeboard and the guardrail, extending along the entire opening?°			
20. Does such mesh consist of No. 19 gage U.S. Standard wire $1/2$-in. mesh, or the equivalent?			
21. Is the planking used of scaffold-plank quality, even if not officially "graded" as scaffold plank?			
22. Does the span of the planks exceed the maximum allowable depending on the designation light duty, medium duty, or heavy duty? (1926.451(a)(10))			
23. Are your men instructed always to replace guardrails, midrails, and bracing if they have had to be temporarily removed for passing materials and equipment?			
24. Do your sectional scaffolds over 125 ft in height have drawings designed by a registered professional engineer? They must.			
25. Do you obtain safety rules and instructions on use of scaffolds from your scaffolding supplier? You should—they are free.			
Ladders			
26. Are both legs of rung ladders supported at top and bottom?			
27. Is the footing slippery, and if so are safety feet installed?			
28. Can the work height be reached without standing on the top of stepladders or the top three rungs of rung ladders?			
29. Does the extension ladder "fly" section rest on top of the base section and not hang underneath?			
30. Does the wood look crumbly? If so, check by inserting a sharp knife end under a sliver of wood and pry up. If it results in a long splinter, it is O.K.; if it results in "crumbling" of the wood, it is decayed.			
31. Never drop ladders or allow them to fall unless they are thoroughly and minutely inspected for damage before reuse.			

Maintenance of Scaffolds and Ladders   1-185

OSHA SCAFFOLDING CHECKLIST (*Continued*)

Project _____
Inspection area _____ Area supervisor _____
Inspected by _____ Date _____

Ladders and scaffolding	Yes	No	Action/comments
32. Never use a bent aluminum ladder or one which has been bent and restraightened; the material has been overstressed on both occasions and is no longer reliable.			

° Wire mesh required only when employees are required to work or pass underneath the scaffold.

APPLICATION

Let us examine a typical example, and first assume that the exterior windows of a 400-ft-long by 24-ft-high building are to be cleaned. The building front is brick up to 6 ft, with standard glass panes from 8 to 24 ft. One man is assigned to do the job, with occasional additional help if he needs it. Six factors are:
    1. The work is within ladder range.
    2. The window frames afford support.
    3. The area below is asphalted, with some grass sections.
    4. No traffic need pass below—wheeled or pedestrian.
    5. The work ranges both horizontally and vertically.
    6. There is no chance of electrical contact.

    At first consideration, most factors indicate this work could be done by one man with a 28-ft aluminum extension ladder, which has a maximum extended height of 25 ft; if the man does not use the upper three rungs (safe practice), he will stand on a rung at 22 ft and be able to reach 28 ft high. He can move the ladder easily by himself. He can reach 2 ft either side of the ladder and can cover a 5-ft-wide strip of windows in a vertical pass. This is obviously the correct choice, yes? No!

    The man will spend excessive time climbing up and down and relocating the ladder. His *productive* working time will be not more than 50 percent. Remember—a ladder must be installed one-fourth of its extended length away from the building at base; therefore, although he can reach the upper-level windows easily, he will have to lower and reposition the ladder frequently to reach the lower levels. Resorting to overreaching sideways and behind the ladder is unsafe and is the initiating factor in many ladder accidents. A platform stepladder with the platform 18 ft high is a better possibility. The best choice, however, is a 29-in.-wide by 10-ft-long aluminum ladder scaffold, about 19 ft high to top working level, with outboard safety supports and a so-called climb-through wood platform which has a hatch which can be slid out of the way to climb through and thence replaced to make a full platform cover. The reasons for the choice of this equipment are
    1. The man can achieve all heights by varying the levels of the plywood platform for a full working width of 10 ft; from one work level plus his own reach he can clean, say, 60 sq ft, compared with a typical 6 sq ft from one extension-ladder position.
    2. He can reach all heights with one initial placement plus only *two* platform repositionings.
    3. He can cover 10 ft by 24 ft, less 6 ft brick, that is, 180 sq ft, with minor downtime for repositioning.
    4. When a 10-ft vertical pass is completed, he can *roll* the scaffold himself to the next position. With certain youth and agility he could even erect it himself; the aluminum components are extremely light; otherwise, he would require help for initial installation.
    5. For the grass areas (or even at the rear of ornamental-shrub beds), the scaffold can be rolled on leveled planks. Four-inch steel channels can also be used as wheel guides and bridges over minor humps and deviations from level. Level such earth or grass with sand so that there are no "holes" *under* the planks.
    NOTE: Aluminum scaffolds are standardly equipped with 5-in.-diameter casters but are optionally available with 8-in. ones; both have a 24-in. range of screw-leg adjustment for

support from different levels. The 8-in. wheels provide easier rolling over rough surfaces such as old asphalt.

Maintenance Aluminum scaffolds of all types require minimum maintenance. Stairways, ladders, and platforms should be inspected frequently, and any grease or oil should be removed immediately. Make sure the plywood platforms or platform planks are solid, with no splits. Do not store platforms near excessive heat, to avoid drying and warping. Casters should be cleaned and lubricated, and brakes should be checked for satisfactory operation. Threads on extension legs should be cleaned and lubricated periodically for smooth operation. Coupling pins used to join frames vertically should be kept clean so that upper frames slip over the pins easily and freely.

With aluminum scaffolds, slight bends in the tubing due to severe impact or mishandling should be straightened. The spring-lock devices used to fasten the braces should be kept free from dirt to ensure proper operation of the lock. On the more popular types of aluminum scaffolding, this mechanism is exposed and can be cleaned easily with a wire brush.

The steel types of scaffolding should be kept clean by scraping or wire brushing; any rusted spots on frames or braces should be scraped and touched up with quick-drying enamel. Stud threads should be lubricated, and wing nuts run off and on to ensure fast, secure fastening during erection. All frames should be checked frequently for missing vertical-coupling sprockets and the pins that lock them in place, as well as wing nuts. Cross braces should be straightened if bent, and the alignment of the tops of the frames should be checked and braces realigned if necessary.

For maximum safety, swinging scaffolds must be properly maintained. The operating mechanism, or winch, should be kept free of dirt and grit at all times. A wire brush usually is satisfactory for this work. All operating parts should be properly lubricated as outlined in the manufacturer's instructions. The safety devices in the winches should be inspected frequently. Pawls and pawl springs should be checked for proper working condition. Teeth on the drum casting should be inspected, and if broken or worn, the manufacturer should be consulted about replacement. The steel cable should be run off and checked for excessive damage and kinks and then rewound through an oily rag to clean the cable and give it a thin coat of oil. Worm and gear mechanisms should be checked for excessive play, cleaned, and repacked with fiber grease. The stirrup should be checked for alignment and straightened, and all painted surfaces of the machine should be recoated where necessary for rust prevention.

Wooden platforms require careful inspection and maintenance. Grease or oil spilled on the platform should be removed immediately. After a job, the platform should be placed across horses for inspection, overhaul, and repair. Mortar, concrete, and paint should be removed with a wire brush and scraper. Broken rungs, slats, and damaged or missing toeboards should be replaced, as should missing hinges and hooks and eyes. After necessary repairs have been made, the platform should be given a thick coat of quick-drying paint. Finally, clean and examine the S and L hooks from which the scaffold is hung, clean and inspect center stanchions, and replace missing or defective wing nuts and bolts.

Tube and coupler scaffolding should undergo a systematic inspection. Bent tubes should be strengthened and, in case of seriously damaged tubes, discarded or cut into shorter lengths for short bearers. Very dirty or rusty tubes should be cleaned with a wire brush. During this operation, inspect the male and female fittings for damage which would affect their safety and then clean with a wire brush. Remove damaged couplers from stock. Studs should be kept covered with a light film of oil, and catch bolts should be checked for stripped threads.

Ladders and scaffolds are vital accessories for all "off-the-ground" work. It behooves all persons, employers and employees alike, to keep their equipment in first-class operating condition and always use it with safety as the maximum prerequisite.

Chapter **12**

Chain Hoists and Chain Slings

H. E. HOLLIDAY*
Service Manager, CM Hoist Division,
Columbus McKinnon Corporation,
Tonawanda, N.Y.

J. D. ZAJAC
Product Standards and Service, CM Hoist Division,
Columbus McKinnon Corporation,
Tonawanda, N.Y.

GENERAL

Chain hoists, both hand and electric, are a widely used type of material-handling equipment. Their simplicity, dependability, and relatively low cost have made them standard material-handling equipment in manufacturing plants, foundries, mills, repair shops, and garages and in practically every phase of the construction field.

This chapter describes the various types of chain hoists and their relative advantages and usual applications. It provides information on preventive maintenance, inspection, and trouble shooting.

TYPES OF CHAIN HOISTS

Rigger Ratchet Hoist. This is the simplest and least expensive type of chain hoist, with approximately 15:1 mechanical advantage. This lightweight portable tool can be used for pulling horizontally, vertically, or at any angle. A directional level mechanism determines if the load is being applied to or released from the tool by the up-and-down movement of the handle. When loading, a driving pawl engages the ratchet and, by turning the liftwheel, causes tension on the chain. By shifting the direction lever to the unload position, the tension can be released one tooth at a time (Fig. 12-1).

Cyclone and Satellite Spur-Geared Hoists. This type of hoist is more efficient than the rigger ratchet hoist, with mechanical advantages of approximately 22:1 for low-capacity units varying upward with capacity increase, usually obtained by handwheel size and a set of reduction gearing (Fig. 12-2 to 12-4). The initial cost is higher than the ratchet hoist. A self-energizing Weston-type brake is incorporated, which enables the load to be lowered with comparatively little effort, at a very slow rate, and with precise positioning if desired. This type of hoist is an accepted standard for industrial applications requiring high speed and efficiency. The spur-geared hoist is generally used for vertical loading and is available in several model variations. An overload protection device is often incorporated.

*Deceased.

1-188 Basic Maintenance Technology

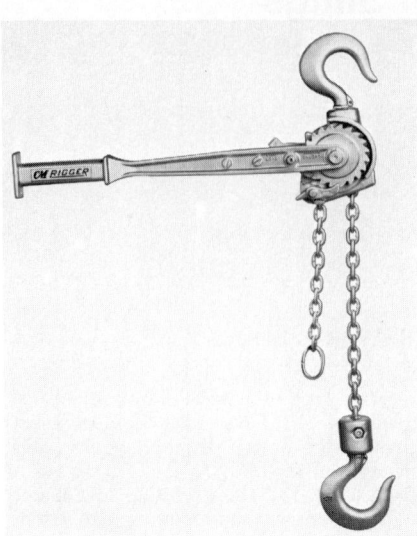

Fig. 12-1 Rigger ratchet hoist.

Fig. 12-2 Cyclone spur-geared hoist.

Fig. 12-3 Satellite spur-geared hoist.

Fig. 12-4 Multiple-reeved cyclone spur-geared hoist.

Modern Spur-Geared Hoists. These hoists utilize more compact design, lightweight alloys, and more antifriction bearings to achieve greater portability through a weight reduction.

Cyclone Spur-Geared Low-Headroom Trolley Hoist. This hoist is built integrally with a trolley for installations where headroom is limited. The 1-ton model requires approximately 13-in. headroom. In heavier capacities, the headroom saving is correspondingly greater. This type of hoist is available with plain or geared trolleys. The geared trolley is moved along the beam by a hand chain drive (Fig. 12-5).

CM Model B Puller (Lever-Operated Hoist). These hoists are lightweight portable tools which are designed for pulling, lifting, or dragging with capacities from ¾ ton to 6 tons. (Figure 12-6 illustrates a ¾-ton model.) A reversing mechanism located in the lever permits short-stroke operation for loading and unloading. This tool has a mechanical advantage of approximately 25:1. An automatic friction-type load brake holds the load

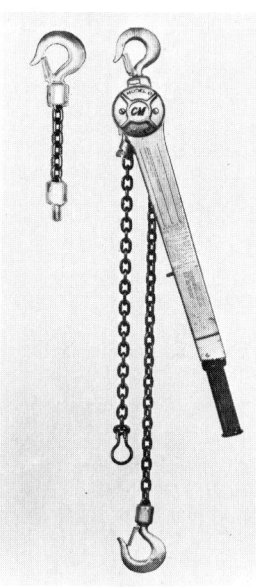

Fig. 12-5 Cyclone spur-geared low-headroom trolley hoist.

Fig. 12-6 Model B puller, lever-operated hoist. At left, an anchor sling attachment.

securely. These pullers are available with a load-sensing device to warn the operator the tool is being loaded beyond its rated capacity. In close headroom operation, an anchor sling is available (shown on left of puller photo, Fig. 12-6).

Lodestar Electric Hoists. These hoists are used for high-speed repetitive duty. They are equipped with either pushbutton or pendant rope controls. Electric hoists are usually equipped with safety limit switches to control both up and down travel. This control prevents the load hook from jamming against the bottom of the hoist or the chain from running out of the hoist (Figs. 12-7 and 12-8).

Electric hoists are available for use with different power supplies. Many are equipped with single-phase 115-volt motors which can be plugged into a conventional three-prong receptacle. Some manufacturers offer three-phase dual-voltage single-speed models. Three-phase single-voltage two-speed models are also available. Both types are designed to operate on 50 or 60 Hz ac 230 or 460 volt power for dual voltage and 230 or 460 volts for two-speed models. Recent electric hoist models are equipped with load protectors to prevent overloading of the hoist (Fig. 12-9).

Electric chain hoists are available in capacities from ⅛ to 5 tons. With lifting speeds

1-190   Basic Maintenance Technology

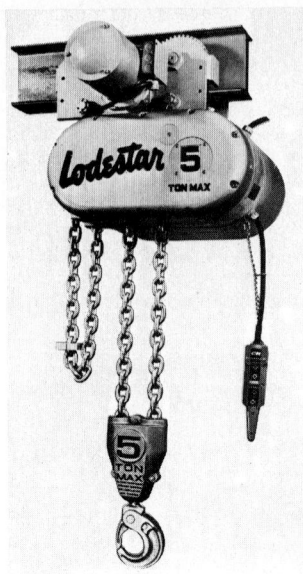

Fig. 12-7 Electric hoist with pendant control station.

Fig. 12-8 Electric hoist with motor-driven trolley and pendant control.

from 2½ to 64 fpm, these are widely used throughout industry because of their convenience, low cost, and durability.

SELECTION OF A CHAIN HOIST

In selecting either hand-operated or electric-powered hoists, certain considerations are basic and common to both types. Figure 12-10 illustrates the importance of performance and physical characteristics of both hand and electric types which must be considered in selecting a hoist for a given use and installation.

Initial cost, frequency of use, labor savings, safety, portability, and maintenance requirements are overall factors in selection of a hoist. Of prime importance is the capacity of the heaviest load to be lifted. Environmental conditions such as moisture, heat, chemicals, and foreign material in the atmosphere, must also be considered. These conditions may require weatherproofing or special protection. Under normal conditions, standard hoists are satisfactory. Other factors which affect the selections of a hoist are headroom, height of lift, location, height of hand chain, type of suspension, and trolley clearances.

PREVENTIVE MAINTENANCE

The design of modern hand and electrical hoists is such that maintenance has been reduced to a minimum. However, correct and adequate maintenance of all types of lifting equipment is of the utmost importance. Many older models still in active service will be included in the maintenance instructions published by the manufac-

Fig. 12-9 Electric-hoist load protector.

turer. Hoist operating mechanisms are subject to wear and should be inspected. A regular periodic inspection should be initiated to ensure that worn or damaged parts are removed from service before they become unsafe.

Individual applications will determine the frequency of inspection. However, the following formula can be used as a guide to the amount of time which should be allowed for general maintenance. The times obtained from this formula do not refer to major items such as gears, frames, and liftwheels but rather to inspection of chain, hooks, limits of travel, brakes, and electrical controls.

$$Hm = \frac{(W \times A) + K + M}{90}$$

where Hm = maintenance hours per month required
W = working hours of hoist per week
A = age of hoist in years
K = humidity
  40% = 10   80% = 52
  50% = 13   90% = 80
  60% = 20   100% = 120
  70% = 38
M = Atmospheric conditions (dust)
  Clean = 10
  Medium dusty = 20
  Foundry dusty = 40

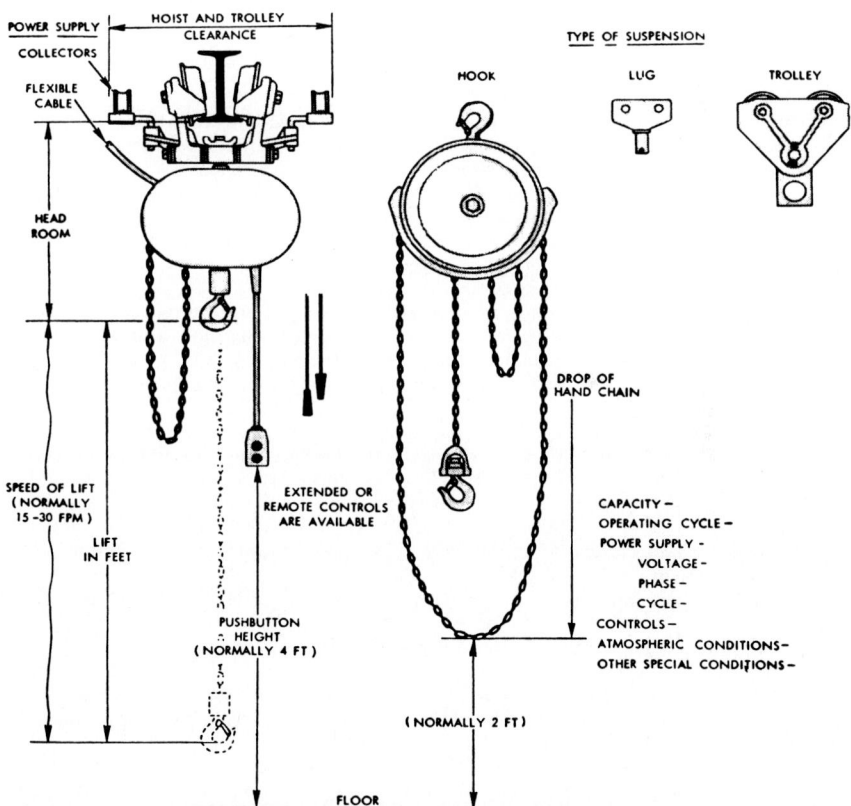

Fig. 12-10 Hoist-instruction check diagram.

Service

Types of Service for hand- and electric-operated chain hoists are defined as follows:

1. *Normal service*   that service which involves operation with randomly distributed loads within capacity or uniform loads up to 65 percent capacity for not more than 15 percent for hand-operated hoists and 25 percent for electric- or air-operated hoists during a single work shift.

2. *Heavy service*   that service within rated capacity which exceeds normal service.

3. *Severe service*   that service which involves normal or heavy service in adverse environmental conditions.

Inspection

Inspection of hand- and electric-operated chain hoists is a critical part of preventive maintenance.

Initial inspection   Prior to use, all new and altered hoists shall be inspected by the user to ensure compliance with the provisions outlined.

Procedure   Inspection procedure for hoists in *regular service* is classified as frequent or periodic based on the intervals at which inspection should be performed. The intervals depend on the nature of the critical components of the hoist and the degree of their exposure to wear, deterioration, or malfunction. Maximum intervals between frequent and periodic inspections are outlined:

1. *Frequent inspection*—visual inspection by the operator or other designated personnel with records not required.
   a. Normal service—monthly.
   b. Heavy service—weekly to monthly.
   c. Severe service—daily to weekly.
   d. Special or infrequent service as authorized by a qualified individual—before and after such special or infrequent period of service with records of operation.

2. *Periodic inspection*—visual inspections by appointed person making records of apparent external conditions to provide the basis for a continuing evaluation.
   a. Normal service—equipment in place; yearly.
   b. Heavy service—as in (*a*) unless external conditions indicate that disassembly should be done to permit detailed inspection; semiannually.
   c. Severe service—as in (*b*); quarterly.
   d. Special or infrequent service as authorized by a qualified individual—before the first such period of service and as directed by the qualified individual for any subsequent special or infrequent period of service.

Frequent inspection   Inspect items such as those in the following list for damage at the intervals outlined (under the preceding heading Procedure) or as specifically indicated. Be sure to include observation during operation for any damage that might appear between the regular inspections. Carefully examine deficiencies and determine whether they constitute a safety hazard. Inspect:

1. Braking mechanism for evidence of slippage under load.
2. Load chain for lubrication, wear, or twists and broken, cracked, or otherwise damaged links. Check chain also for deposits of foreign material that might be carried into the hoist mechanism.
3. Hooks for deformation, chemical damage, or cracks. Hooks damaged from chemicals, deformations, and cracks, or hooks having more than 15 percent in excess of normal throat opening or more than 10° twist from the plane of the unbent hook must be replaced.

NOTE: Any hook that is twisted or has a throat opening in excess of normal indicates abuse or overloading of the unit. Other load-bearing components of the hoist should be inspected for damage.

4. Hooks for proper operation of latch if one is used.

Periodic inspection   Make a complete inspection of the hoist at the intervals outlined (under the previous heading Procedure) depending upon its activity, severity of service, and environment, or as specifically indicated below. Include the requirements of frequent inspection and, in addition, items such as those in the following list. Carefully examine any deficiencies and determine whether they constitute a safety hazard.

1. Check for external evidence of wear of chain, load sprockets, idler sprockets, and handwheel pockets or for chain stretch.
2. Inspect hook-retaining nuts, collars and pins, welds, or riveting used to secure the retaining members.
3. Examine brake mechanism for worn, glazed, or oil-contaminated friction disks, worn pawls, cams, or ratchets. Watch for corroded, stretched, or broken pawl springs.
4. Look for worn, cracked, or distorted parts, such as hook blocks, suspension housing, outriggers, hand chain wheels, chain attachments, clevises, yokes, suspension bolts, shafts, gears, and bearings.
5. Check for loose bolts, nuts, or rivets.
6. Inspect supporting structure and trolley for continued ability to support the imposed loads.
7. Make sure the warning label on proper use of the hoist is attached and legible.

Idle-hoist inspection Hoists not in regular use must be checked carefully before they are put back into service.

1. A hoist that has been idle for a period of 1 month or more but less than 6 months must be inspected by, or under the direction of, a designated person. It must meet the requirements of frequent and periodic inspections before placing it in service.
2. A hoist that has been idle for a period of 6 months shall be given a complete inspection to conform with the requirements of frequent and periodic inspection.

Inspection records Written, dated, and signed inspection reports and records shall be made on critical items such as brakes, hooks, and chains and the time intervals specified in the inspection classification. Records should be readily available.

Testing

Operation tests All new hoists shall be tested by the hoist manufacturer. All altered or repaired hoists or hoists that have not been used within the preceding 12 months shall be tested before use by, or under the direction of, a designated person to ensure compliance with the standard, including the following:

1. All functions of the hoist including hoisting and lowering shall be checked with the hoist suspended in the unloaded state.
2. After testing in the unloaded state, a load of 50 lb times the number of load-supporting parts of the chain shall be applied to the hoist in order to check proper load control.

Load tests All new hoists shall be tested with at least 125 percent rated load. If testing by the manufacturer is impractical, the user shall be notified and the test shall be accomplished at the job site by or under the direction of an appointed person.

All hoists in which load-sustaining parts have been altered, replaced, or repaired shall be tested statically or dynamically by, or under the direction of, an appointed person and a written report prepared and made readily available. The applied test load shall be 125 percent of the rated load. The replacement of normal maintenance items such as chain is specifically excluded from this hoist load test. A functional test of the hoist under a load of at least 50 lb times the number of load-supporting parts of the chain shall be made.

On hoists incorporating load-limiting devices which prevent the application of 125 percent of rated load, a load test shall be conducted with at least 100 percent rated load, after which the function of a load-limiting device shall be tested. The hoist manufacturer should be consulted for the value of the upper load limit of the load-limiting device.

HOIST OPERATOR INSTRUCTIONS

Prior to the start of normal operations, the hoist operator should test all limit switches, brakes, and other safety devices. Any failure should be reported immediately and the equipment should be removed from service until repairs can be made. It should be the responsibility of the operator to perform the frequent inspection.

The hoist operator should be responsible for the safe operation of the equipment. At no time should he leave the control position with a load suspended. Limit switches are for emergency use and should not be used for ordinary operation. During normal operation, if the equipment fails to respond properly, the failure should be reported immediately.

Basic Maintenance Technology

TABLE 12-1 Troubleshooting Guide for Spur-Geared Hoists

Problem	Probable cause	Check/remedy
Hoist is hard to operate in either direction.	Load chain worn long to gauge, thus binding between liftwheel and chain guide.	Check gauge of chain. Replace if worn excessively.
	Load chain rusty, corroded, or clogged up with foreign matter such as cement or mud.	Clean by tumble polishing or solvent. Lubricate with penetrating oil and graphite. (SC-46 and SC-146)
	Load chain damaged.	Check chain for gouges, nicks, bent or twisted links. Replace if damaged.
	Liftwheel clogged with foreign matter or worn excessively, causing binding between the liftwheel and chain guide.	Clean out pockets. Replace if worn excessively.
	Hand-chain worn long to gauge, thus binding between handwheel and cover.	Check gauge of chain.
	Handwheel clogged with foreign matter or worn excessively, causing binding of chain between the handwheel and cover.	Clean out pockets. Replace if worn excessively.
	Liftwheel or gear teeth deformed.	Excessive overload has been applied. Replace damaged parts.
Hoist is hard to operate in the lowering direction.	Brake parts corroded or coated with foreign matter.	Disassemble brake and clean thoroughly. (By wiping with a cloth—not by washing in a solvent.) Replace washers if gummy, visibly worn, or coated with foreign matter. Keep washers and brake surfaces clean and dry.
	Chain binding.	See first three check/remedy items in this table.
Hoist is hard to operate in the hoisting direction.	Chain binding.	See first three check/remedy items in this table.
	Chain twisted. (3-ton capacity and larger)	Rereeve chain, or on 3- and 4-ton unit, if both chains are twisted, capsize hook block through loop in chain until twists are removed. Caution—do not operate unit in hoisting direction with twisted chain or serious damage will result.
	Overload.	Reduce load or use correct capacity unit.
Hoist will not operate in either direction.	Liftwheel gear key or friction hub key missing or sheared.	Install or replace key.
	Gears jammed.	Inspect for foreign material in gear teeth.

TABLE 12-1   Troubleshooting Guide for Spur-Geared Hoists (*Continued*)

Problem	Probable cause	Check/remedy
Hoist will not operate in the lowering direction.	Locked brake due to a suddenly applied load, shock load, or load removed by other means than by operating unit in the lowering direction.	With hoist under load to keep chain taut, pull sharply on hand chain in the lowering direction to loosen brake.
	Chain binding.	See first three check/remedy items in this table.
	Lower hook all the way out. Load chain fully extended.	Chain taut between the lift wheel and loose end screw. Operate unit in hoisting direction only.
Hoist will not operate in the hoisting direction.	Chain binding.	See first three check/remedy items in this table.
Hoist will not hold load in suspension.	Lower hook or load side of chain on wrong side of liftwheel.	Lower hook must be on same side of liftwheel as upper hook. Refer to assembly. Rereeve chain.
	Ratchet assembled in reverse.	Ratchet must be assembled as shown.
	Pawl not engaging with ratchet.	Pawl spring missing or broken pawl binding on pawl stud. Replace spring and clean so pawl operates freely and engages properly with ratchet. Do not oil.
	Ratchet teeth or pawl worn or broken.	Replace pawl and/or ratchet.
	Worn brake parts.	Replace brake parts which are worn.
	Oily, dirty, or corroded brake friction surfaces.	Disassemble brake. Clean thoroughly. (By wiping with a cloth—not by washing in a solvent.) Replace washers if gummy, visibly, worn, or coated with foreign matter. Keep washers and brake surfaces clean and dry.

When operating hoisting equipment, the operator must always use safe material-handling methods. He is responsible for the safe slinging of loads and must know the location and operation of all main electric-power feeder switches and emergency stop buttons.

REPAIRS

Hoist manufacturers have available maintenance and parts manuals for use when repairs become necessary. These manuals give assembly and disassembly instructions, location, and identification of all parts. When replacement parts are needed, name plate information, especially the serial number, should be included with correspondence. Table 12-1 is a troubleshooting guide for the spur-geared hoist. Tables 12-2 and 12-3 are troubleshooting guides for the electric hoist.

Basic Maintenance Technology

TABLE 12-2 Troubleshooting Guide for All Electric Hoists

Problem	Probable cause	Check/remedy
Hook does not respond to the control station.	No voltage at hoist—main line or branch circuit switch open; branch line fuse blown or circuit breaker tripped.	Close switch, replace fuse or reset breaker.
	Phase failure (single phasing, three-phase unit only)—open circuit, grounded or faulty connection in one line of supply system, hoist wiring, reversing contactor, motor leads or windings.	Check for electrical continuity and repair or replace defective part.
	Upper or lower limit switch has opened the motor circuit.	Press the "other" control and the hook should respond. Adjust limit switches.
	Open control circuit—open or shorted winding in transformer, reversing contactor coil or speed selecting relay coil; loose connection or broken wire in circuit; mechanical binding in contactor or relay; control station contacts not closing or opening.	Check electrical continuity and repair or replace defective part.
	Wrong voltage or frequency.	Use the voltage and frequency indicated on hoist identification plate. For three-phase dual-voltage unit, make sure the connections at the conversion terminal board are for the proper voltage.
	Low voltage	Correct low-voltage condition.
	Brake not releasing—open or shorted coil winding; armature binding.	Check electrical continuity and connections. Check that correct coil has been installed. The coil for three-phase dual-voltage unit operates at 230 volts when the hoist is connected for either 230-volt or 460-volt operation. Check brake adjustment.
	Excessive load.	Reduce loading to the capacity limit of hoist as indicated on the identification plate.
Hook moves in the wrong direction.	Wiring connections reversed at either the control station or terminal board (single-phase unit only).	Check connections with the wiring diagram.
	Failure of the motor reversing switch to effect dynamic braking at time of reversal (single-phase unit only).	Check connections to switch. Replace a damaged switch or a faulty capacitor.
	Phase reversal (three-phase unit only).	Refer to installation instructions.

TABLE 12-2 Troubleshooting Guide for All Electric Hoists (*Continued*)

Problem	Probable cause	Check/remedy
Hook lowers but will not raise.	Excessive load.	Reduce loading to capacity limit of hoist as indicated on the identification plate.
	Open hoisting circuit—open or shorted winding in reversing contactor coil or speed selecting relay coil; loose connection or broken wire in circuit; control station contacts not making; upper limit switch contacts open.	Check electrical continuity and repair or replace defective part. Check operation of limit switch.
	Motor reversing switch not operating (single-phase unit only).	Check switch connections and actuating finger and contacts for sticking or damage. Check centrifugal mechanism for loose or damaged components. Replace defective part.
	Phase failure (three-phase unit only).	Check for electrical continuity and repair or replace defective part.
Hook raises but will not lower.	Open lowering circuit—open or shorted winding in reversing contactor coil or speed selecting relay coil; loose connection or broken wire in circuit; control station contacts not making; lower limit switch contacts open.	Check electrical continuity and repair or replace defective part. Check operation of limit switch.
	Motor reversing switch not operating (single-phase unit only).	Check switch connections and actuating finger and contacts for sticking or damage. Check centrifugal mechanism for loose or damaged components. Replace defective part.
Hook lowers when hoisting control is operated.	Phase failure (three-phase unit only).	Check for electrical continuity and repair or replace defective part.
Hook does not stop promptly.	Brake slipping.	Check brake adjustment.
	Excessive load.	Reduce loading to the capacity limit of hoist as indicated on the identification plate.
Hoist operates sluggishly.	Excessive load.	Reduce loading to the capacity limit of hoist as indicated on the identification plate.
	Low voltage.	Correct low voltage condition.
	Phase failure or unbalanced current in the phases (three-phase unit only).	Check for electrical continuity and repair or replace defective part.
	Brake dragging.	Check brake adjustment.

Basic Maintenance Technology

TABLE 12-2 Troubleshooting Guide for All Electric Hoists (Continued)

Problem	Probable cause	Check/remedy
Motor overheats.	Excessive load.	Reduce loading to the capacity limit of hoist as shown on the identification plate.
	Low voltage.	Correct low-voltage condition.
	Extreme external heating.	Above an ambient temperature of 104°F, the frequency of hoist operation must be limited to avoid overheating of motor. Special provisions should be made to ventilate the space or shield the hoist from radiation.
	Frequent starting or reversing.	Avoid excessive inching, jogging, or plugging. This type of operation drastically shortens the motor and contactor life and causes excessive brake wear.
	Phase failure or unbalanced current in the phases (three-phase unit only).	Check for electrical continuity and repair or replace defective part.
	Brake dragging.	Check brake adjustment.
	Motor reversing switch not opening start winding circuit. (Single-phase unit only.)	Check switch connections and actuating finger and contacts for sticking or damage. Check centrifugal mechanism for loose or damaged components. Replace defective part.
Hook fails to stop at either or both ends of travel.	Limit switches not opening circuits.	Check switch connections, electrical continuity, and mechanical operation. Check the switch adjustment. Check for a pinched wire.
	Shaft not rotating.	Check for damaged gears.
	Traveling nuts not moving along shaft—guide plate loose; shaft or nut threads damaged.	Tighten guide plate screws. Replace damaged part.
Hook stopping point varies.	Limit switch not holding adjustment.	Check switch connections, electrical continuity, and mechanical operation. Check switch adjustment. Check for pinched wire. Check for damaged gears. Tighten guide plate screws. Replace damaged part.
	Brake not holding.	Check the brake adjustment.

LUBRICATION

All moving parts of a hoist for which lubrication is specified shall be regularly lubricated. Lubricating methods shall be checked for proper delivery of lubricant. Particular care should be taken to follow the manufacturer's recommendations as to points and frequency of lubrication, as well as to quantity and type of lubricant to be used. Be cautious when substituting lubricant since most manufacturers recommend a specific lubricant only.

TABLE 12-3 Troubleshooting Guide for Two-Speed Electric Hoists

Problem	Probable cause	Check/remedy
Hoist will not operate at slow speed in either direction.	Open circuit.	Open or shorted motor winding loose or broken wire in circuit, speed-selecting contactor stuck in opposite speed mode. Replace motor, repair wire, and/or repair speed-selecting contactor.
	Phase failure.	Check for electrical continuity and repair or replace defective part.
Hoist will not operate at fast speed in either direction.	Open circuit.	Open or shorted motor winding, loose or broken wire in circuit, speed-selecting contactor stuck in opposite speed mode. Replace motor, repair wire, and/or repair speed-selecting contactor.
Hook will not raise at slow speed.	Open speed-selecting circuit.	Open or shorted winding in speed-selecting contactor coil. Loose connection or broken wire in circuit. Mechanical binding in contactor. Control station contacts not making or opening. Replace coil; repair connection, contactor, or control station.
Hook will not lower at slow speed.	Phase failure.	Check for electrical continuity and repair or replace defective part.
Hook will not raise at fast speed.	Excessive load.	Reduce loading to capacity limit of hoist as indicated on the identification plate.
	Phase failure.	Check for electrical continuity and repair or replace defective part.
	Brake not releasing.	Check electrical continuity and connections. Check that correct coil has been installed. The coil for three-phase dual-voltage unit operates at 230 volts when the hoist is connected for either 230- or 460-volt operation. Check brake adjustment.
Hook will not lower at fast speed.	Phase failure.	Check for electrical continuity and repair or replace defective part.
	Brake not releasing.	Check electrical continuity and connections. Check that correct coil has been installed. The coil for three-phase dual-voltage unit operates at 230 volts when the hoist is connected for either 230- or 460-volt operation. Check brake adjustment.
Hook moves in proper direction at one speed—wrong direction at other speed.	Phase reversal.	Wiring reconnected improperly. Interchange two leads of motor winding that is out of phase at the speed-selecting relay.

1-200 Basic Maintenance Technology

Load chain articulates slowly under high bearing pressures and should be lubricated as specified by the hoist manufacturer. In the absence of recommendations, use antiweld or EQ-type lubricant, applied sparingly but frequently as it dissipates during use. Hand chain is lightly loaded and normally needs no lubrication.

WELDED LINK LOAD AND HAND CHAIN

Hoist hand chains and load chains are carefully manufactured. The size of the links and pitch are held to very close tolerances. Pockets of liftwheels, sheaves, and handwheels are accurately formed to ensure a close fit. Hoist manufacturers have charts available for the user of various sizes of chain pitch. Often, this includes information on permissible wear beyond which a chain is no longer safe and should be replaced. Figure 12-11 is an

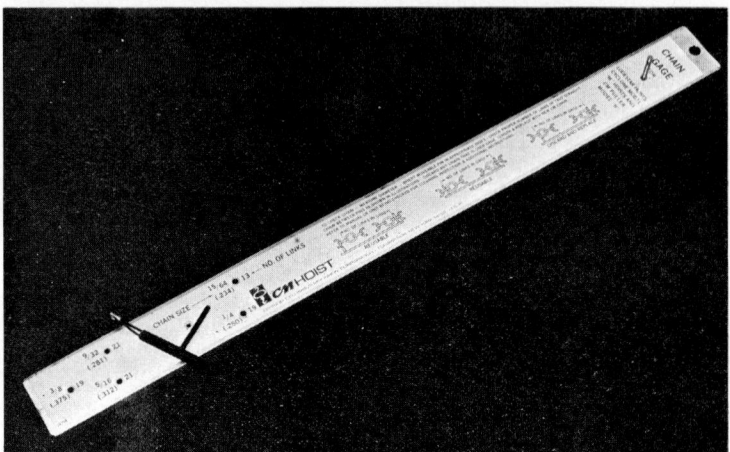

Fig. 12-11 Hoist-load chain gauge.

example of a chain gauge used for determining if the chain is within allowable limits. Lacking manufacturer's data, a fairly accurate check on wear can be made by comparing dimensions of a worn section versus an unused section, such as the loose end. If a used chain is 1½ percent longer than an unused chain for electric and air hoists (2½ percent for hand hoists), replace the chain. Before gauging, the chain must be cleaned in a solvent. After gauging, check for nicks, gouges, and twisted links by a link-by-link inspection. If any of these conditions exists, the load chain must be replaced.

PREVENTIVE MAINTENANCE VERSUS BREAKDOWN MAINTENANCE

Preventive maintenance may be defined differently in scope and intensity in various industries. It has been recognized as extremely important in the reduction of maintenance costs and improvement in the reliability of equipment and production output. Preventive maintenance consists of inspecting equipment and keeping records showing wear or other deterioration of parts. Such records may include repetitive servicing, lubricating, painting, and cleaning. They are the basis for routine inspection and point to the need for major overhaul before breakdown.

Obviously, preventive maintenance reduces the corrective maintenance workload. As preventive maintenance takes over, the timing of the corrective workload is shifted from when you *must* do it back to when you *want* to do it. Thus work can be done more efficiently and at lower cost. However, not every plant can expect to derive equal benefits. For this reason, a preventive maintenance program should be approached with caution.

The cost of a preventive maintenance program is initially high. But the shutdown of equipment for no reason other than periodic inspection may be intolerable from a production standpoint. To help in cost reduction, a good recordkeeping system is essential in planning, scheduling, and training of maintenance personnel. Anyone who expects quick, full benefits from a preventive maintenance program will be disappointed, since a program such as this may take several months before much progress is shown. Breakdown in equipment could result in severe damage and, thus, be far more costly to repair. After inspection, an inspector can report worn parts prior to failure, allowing lead time to order replacement parts. A preventive program must be tailored to fit a company's needs. Such industries as plating companies, chemical plants, and heavy manufacturing plants may require more frequent maintenance.

Some companies prefer a program of replacement units whereby an entire unit can be removed from service and replaced with an identical unit. Such a philosophy can be used for certain types of equipment in a plant where standardization exists.

Sometimes when a centrally administered preventive maintenance program cannot be achieved, qualified mechanics can be assigned to individual pieces of equipment. Because of their familiarity with the equipment, mechanics can effectively reduce cost and breakdown in plants where equipment is not used continually. A complete overhaul of production equipment can be accomplished during a shutdown period similar to the automobile industry changeover for a model year.

In a continuous manufacturing plant, vital inspection and replacement is difficult. Often, these inspections and replacements can be accomplished during the same time it takes to perform primary repairs. However, such methods require reporting and recordkeeping of deficiencies during normal plant operation. Parts must be available and personnel must be familiar with the equipment. A sound preventive maintenance program undoubtedly has advantages over the breakdown approach. Although some plants would prefer breakdown to deliberate shutdown preventive maintenance, we might expect newly created state and federal laws to insist on preventive maintenance.

Chapter **13**

Maintenance of Belt Conveyors and Conveying Equipment

R. A. HOPPENRATH
Product Application Manager—Conveyors
Barber-Greene Company, Aurora, Ill.

Belt conveyors are the key elements in nearly every system for moving bulk materials. They will operate for many years with a minimum of attention but they do require care.

Conveyors appear to be relatively simple mechanisms; therefore they are often neglected. Yet, regardless of the care and ingenuity which have gone into their manufacture, conveyors can serve well only when operated correctly, maintained properly, and lubricated regularly. Maintenance of conveyors is just as important as taking proper care of more complex pieces of equipment. Failure of just one conveyor can shut down an entire operation.

GENERAL

Typically, a conveyor consists of the belting itself, a structure to support it, a drive mechanism, idlers to hold the belt, a method of tensioning the belt, feeding and discharge hoppers, and a variety of accessories which perform such special functions as clearing the belt where sticky or wet material is being handled.

Other accessories include covers that prevent material from blowing off the belt in high winds. Other covers prevent the belt from being blown off the idlers or protect the belt and its contents from sun and weather. Then there are special idlers which cushion the load and keep the belt aligned.

Portable and permanent conveyors Portable conveyors are equipped with pneumatic-tired wheels so they can be moved about readily. Some models have hinged booms to reduce their length for transport. Permanent conveyors have a pair of A-frame supports instead of wheels. Both portable and permanent conveyors are frequently used in the same plant (Figs. 13-1 and 13-2).

Portable conveyors are supplied with screw take-up tail ends (Fig. 13-3) for belt tensioning. Permanent conveyors are provided with either screw take-ups or fixed bearings if the conveyor was ordered with automatic gravity take-up.

Fixed tail ends (Fig. 13-4) are assembled and properly aligned at the factory. No further pulley alignment is necessary. Gravity take-ups are either the vertical or the horizontal type. The latter are generally used where space is limited.

Horizontal gravity take-ups (Fig. 13-5) require a support frame for a movable carriage and pulley. A cable is attached to the movable carriage and to a counterweight to keep the belt automatically at proper tension.

1-204 Basic Maintenance Technology

The maintenance instructions outlined in this chapter apply to both portable and permanent conveyors. Most components are similar on both types of machines.

To neglect maintenance of any conveyor component is to invite trouble. Operating costs will soar, conveyor life will be shortened, and the whole plant operation will be made inefficient by frequent conveyor breakdowns.

Trouble-free belt conveyor operation results from:
1. Selection of quality components
2. Careful installation and initial start-up
3. Proper operating, lubrication, and maintenance procedures
4. Routine inspection

Working around any kind of machinery may be dangerous. Mechanics and operators must be extremely careful while performing maintenance on belt conveyors.

Safety There is no set of safety rules which can cover every conceivable danger. Safety is basically common sense. One simple rule which could prevent thousands of

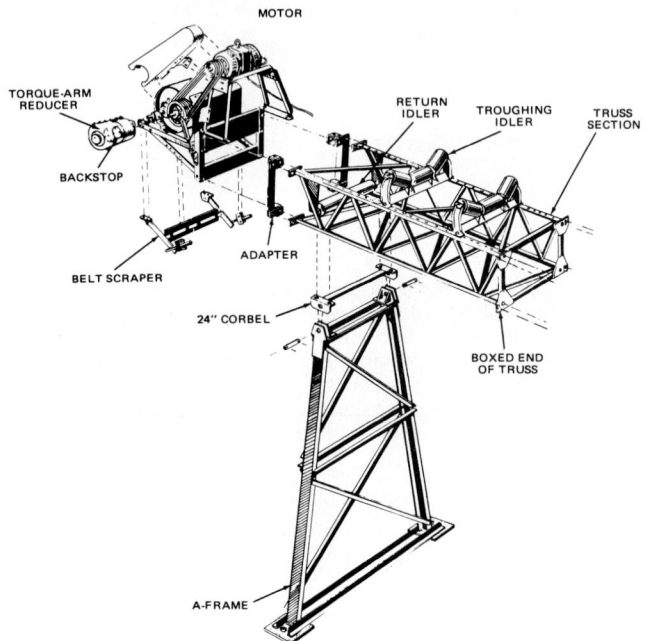

Fig. 13-1 Exploded view of permanent conveyor.

serious accidents every year is: *Never attempt to clean, lubricate, or adjust any machine while it is in motion.*

Owners are advised to consider the use of safety switches and sturdy guards. All personnel who work near equipment powered by electric motors should observe these safety rules:
1. All electrical equipment must be grounded.
2. Always pull the main switch before disconnecting a plug from a receptacle.
3. Never grease, adjust, or repair a machine without removing the plug from its receptacle or pulling the main switch.
4. Use only the proper size and style of fuses; keep a supply on hand and never substitute pieces of metal for proper fuses.
5. Keep electrical cables in good condition at all times and keep them out of water.

Professionalism A true professional is the operator or mechanic who knows the value of proper lubrication. This cuts down lost time. Lost time means lost wages for everyone. Correct lubrication also increases the life of a machine, thus ensuring job security.

Maintenance of Belt Conveyors and Conveying Equipment

The real professional
1. Keeps the grease gun clean and wipes each grease fitting with a clean cloth to prevent grit from being pumped into the bearing
2. Keeps grease and oil containers clean and their covers in place to keep dust and dirt out of the lubricant
3. Keeps each lubricant container well labeled and always uses the correct lubricant in the correct place
4. Studies lubrication charts and does not guess when lubricating a machine
5. Keeps machines as clean as possible and removes excess grease and oil which may accumulate during the day
6. *Lubricates ball or roller bearings carefully so as not to blow seals through overgreasing*

MOST VULNERABLE COMPONENT—THE BELT

The conveyor belt is the most expensive and vulnerable part of the machine and should be cared for accordingly. Most conveyors are designed to provide maximum belt life, provided that proper attention is given to operation and maintenance.

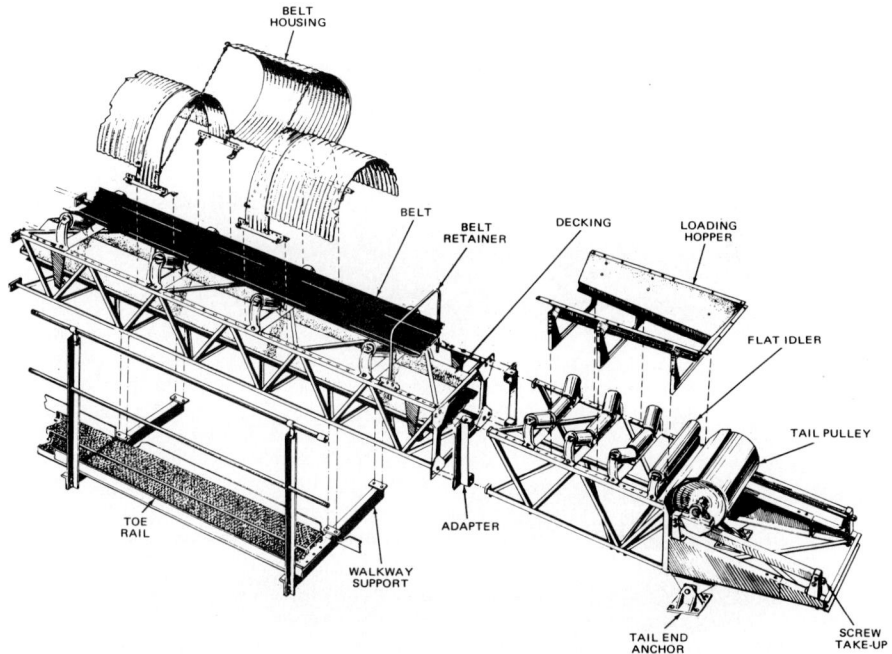

Fig. 13-1 (Continued.)

Good-quality conveyor belting is generally constructed of a high-tensile synthetic carcass with square edges and an extra-heavy cushion layer of rubber between all plies.
1. It has high strength, impact resistance, and great flexibility.
2. It has no shrinkage or excessive elongation.
3. It is rot-proof and mildew-proof.
4. It is resistant to chemicals, alkalis, minerals, and organic acids.

Belt wear The following items are common causes of excess conveyor belt wear and suggestions for their correction:
1. *Improper installation.* The belt must line up so that it runs squarely on the center of pulleys and idlers. If not properly aligned, belts will not give maximum service. Obstructions will damage the belt or permit abrasive material to rub against the moving belt, causing excessive wear and premature failure.

1-206 Basic Maintenance Technology

Fig. 13-2 Portable conveyor.

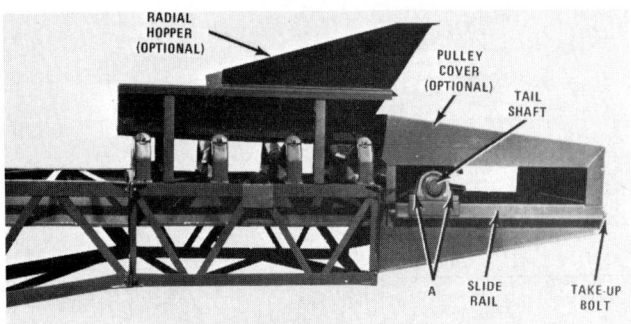

Fig. 13-3 Tail end with take-up screw.

Fig. 13-4 Fixed tail end.

Maintenance of Belt Conveyors and Conveying Equipment   1-207

2. *Belt tension.* Conveyor take-ups should be adjusted to provide just enough tension to operate the belt under load, without slipping. Excess tension places the belt under strain. Slippage between drive pulley and belt obviously will wear the belt.

3. *Starting under load.* The conveyor should never be started under load if it can be avoided. If possible, empty the belt before stopping and start the conveyor before loading.

4. *Moisture.* Cuts or tears should be cleaned with gasoline and filled in with rubber cement. Otherwise, moisture and abrasive particles can enter and start breaking down the carcass of the belt.

5. *Oil and grease.* Petroleum oil or grease is extremely harmful to rubber. Care should be taken while lubricating the conveyor to prevent lubricant from contacting the belt.

6. *Damaged idlers.* Idlers that do not turn freely because of improper lubrication or damaged bearings cause drag and wear on the belt. Hard-turning idlers should be repaired or replaced. Do not allow material to build up on idlers.

Fig. 13-5 Horizontal gravity take-up.

7. *Buildup on idlers.* Some materials tend to build up on idler rollers—especially return idlers and snub pulleys where the carrying side of the belt contacts them.

Buildup will cause difficulty in training a new belt. Damage to the belt can occur when a material builds up to such an extent that it will cause a belt to rub against obstructions. Buildup will be worse in freezing weather.

8. *Friction from external causes.* Hoppers, skirtboards, and chutes should be adjusted so that they do not touch the belt. Flashing on hoppers and skirts should be kept in proper alignment.

9. *Decking.* Decking over the return belt should be kept clean so that material is not carried between belt and pulley.

Inspection Inspect the entire length of the belt as installed on the conveyor to be sure that there are no obstructions which will rub, tear, or cut the belt while it is running. Be sure that no particles of the material handled can catch at some point and damage the belt. Make certain that no stones can drop or bounce onto the return run of the belt and get between the belt and foot pulley. Check belt lacings periodically and replace if they are defective.

Basic Maintenance Technology

Belt Tension Belt tension should be checked at frequent intervals and adjusted when necessary. Adjustment of the conveyor take-up (Fig. 13-3) should provide just enough tension to operate the belt under load without slipping. Excess tension places the belt and drive machinery under strain. Slippage between drive pulley and belt will cause serious belt wear.

If the conveyor is to be shut down for a length of time, it is advisable to remove the belt and store it until ready to resume operations. This is particularly true if the conveyor is exposed to severe weather.

Store the belt in a cool, dry place. Do not lay the roll so that the belt rests on edge. Table 13-1 shows common problems that occur with conveyor belts and recommendations for solutions.

BELT-CUTTING PROCEDURE

Cutting a belt and fitting it to a conveyor require special care. Improperly cut belts cause a number of conveyor problems (as outlined in the preceding section) and can shorten belt lift drastically. The correct procedure is listed:
 1. Measure 8 ft from the end of the belt as shown in Fig. 13-6.
 2. In four places, measure from the outside edge of the belt to the center.

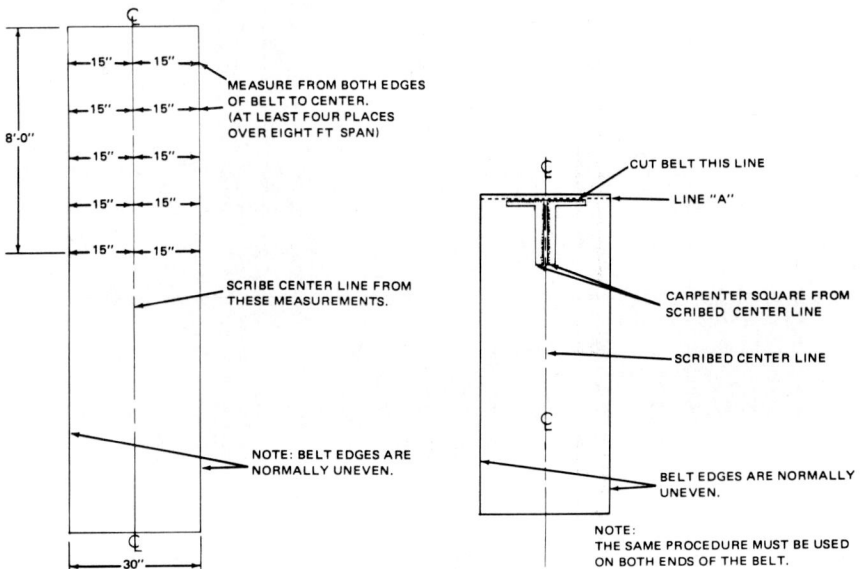

Fig. 13-6 Measurements for cutting conveyor belt.

Fig. 13-7 Scribing a centerline in the conveyor belt.

 3. Scribe the centerline through points determined when measuring from the belt edge (Fig. 13-7).
 4. Locate a carpenter's square on scribed centerline. Scribe a line from center of belt to outside edge in both directions. This line is designated as line A in Fig. 13-7.
 5. Cut the belt along line A.
 6. The same procedure should be used on the other end of the belt.

NOTE: Because of the unevenness of belt edges, it is extremely important to mark off points in four places over an 8-ft span.

A belt tightener (Fig. 13-8) should be used to bring the two ends of the belt together for splicing.

NOTE: It is extremely important that conveyor belting be cut square before beginning installation of lacing.

TABLE 13-1 Troubleshooting Chart

Problem	Probable cause	Recommendation
Making conveyor belts run straight		
Belt runs true when empty, crooked when loaded.	Off-center loading or variations in nature or formation of load.	Adjust chute and other loading devices so that load is delivered to center of belt and in line with direction of belt travel.
Belt climbs sideways on some idlers.	Loose idler.	Return idler to proper position; fasten securely.
	Idler sticks or jams.	Lubricate properly. Replace any sticking idlers having worn spots.
	Idlers out of alignment.	Realign idlers while belt is unloaded.
Part of belt running off idlers is in vicinity of splice.	Improper splice; ends not cut squarely for a mechanical splice.	Resplice. Make sure ends are square for mechanical-type splice.
Same section of belt repeatedly runs off idlers along entire conveyor.	Crooked belt.	Replace with straight belt.
Belt with worn edge becomes crooked.	Worn edge becomes stretched because of high-friction pull or shrinks from moisture absorption.	Eliminate cause of wear. Repair damaged edge.
Top cover gouged or grooved.	Material trapped under skirts.	Prevent jamming by providing an increasing gap under skirts in direction of belt travel. Increase belt tension or space loading-point idlers more closely.
Blisters in cover.	Fine material working into cuts or punctures.	Make spot repair.
Lengthwise strip swelling of bottom cover.	Oil.	Avoid over lubrication and spillage of oil and grease. Use oil resistant belt if necessary.
Bottom cover wear.	Drive pulley.	Increase belt tension if belt rating permits. Lag drive pulley.
	Sticking rollers.	Lubricate properly and replace idlers as required.
	Excessive troughing idler tilt.	Reset not more than 2 deg from upright.
	Corroded troughing idler rolls.	Replace.
Carcass damage		
Carcass breaks.	Impact.	Load lumps between idlers. Decrease height of free fall of lumps.
	Material trapped between belt and pulley.	Use plows ahead of tail pulley on return side. Use deflector over take-up pulley.
	Material buildup on pulleys	Use scrapers.

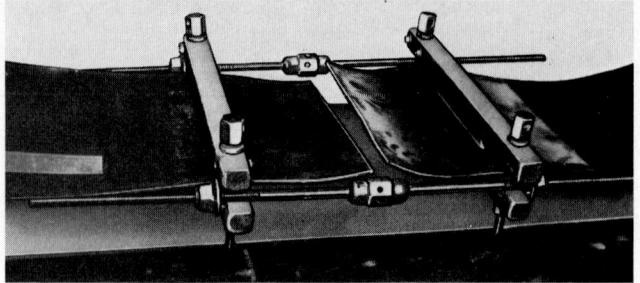

Fig. 13-8 Belt tightening for splicing.

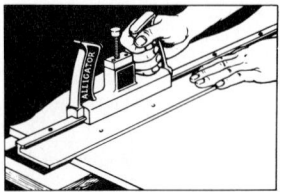

1. Square belt ends and cut to length. To simplify the cutting job, use an Alligator Wide Belt Cutter.

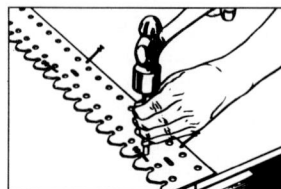

2. Support belt ends with wood plank. Nail Flexco Templet in position with belt ends tight against lugs. Punch or bore bolt holes.

3. An impact tool with Flexco Power Punch or Flexco Power Boring Bit speeds hole boring operation. Remove templet. Leave plank under belt ends for a work surface. All work can be done from the top of the belt.

4. Assemble bolts in bottom plates. Fold one belt end back out of the way. Then insert bolts from under side along one row of holes.

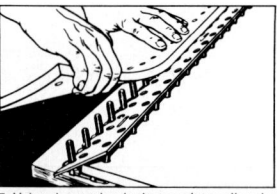

5. Using the notches in the templet to align the opposite row of bolts, place the other end of the belt over the bolts. Press belt onto bolts with hands. Remove templet. Continue to press belt until it is in place.

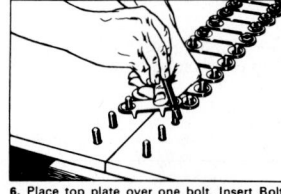

6. Place top plate over one bolt. Insert Bolt horn Tool through the other plate hole and over the second bolt to pry it into place.

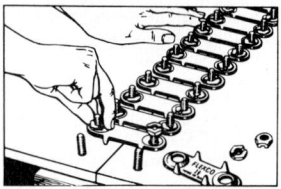

7. Assemble all top plates same way as in Direction No. 6. Start nuts down by hand far enough so that wrench will engage bolts.

8. Before tightening fasteners, cut a piece of Flexco-Lok® Tape three times the width of the belt plus six inches and cut a point on one end. Thread pointed tape between fastener teeth on top of belt, back through the bottom plates, and across the top again.

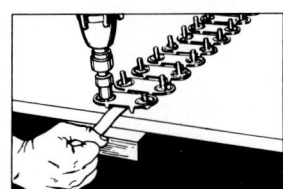

9. Pull tape tight and hold in position by tightening a fastener at each end of the splice. Then snug down all other plates.

10. Tighten all fasteners from edges to center. Tighten all nuts uniformly. A Flexco Power Tool Wrench used with an impact tool will speed this step considerably.

11. Hammer plates in belt with metal or hard wood block as illustrated. Then retighten nuts.

12. Break off excess bolt ends using two bolt breakers. On belts with thick rubber covers, retighten all nuts after a few hours running.

Fig. 13-9 Procedure for applying conveyor belt fasteners.

APPLYING CONVEYOR BELT FASTENERS

Fasteners should be retightened at least once after the first 24 hr of service, especially on belts with thick rubber covers (Fig. 13-9).

INSTALLATION OF BELT

Alignment of the whole conveyor should be checked by instrument or stringline. The alignment must be almost perfect to ensure proper belt operation.

Check all flashings on hoppers and skirt boards. Where necessary, make adjustments so that they do not put too much pressure on the belt. Take-ups, such as screws at tail ends and gravity take-ups, should be brought to minimum position and secured there.

When the truss section is lined up and the other adjustments have been made, the belt may be installed. The roll of belting should be kept level and placed in alignment with the installation. Placing a shaft through the center of the roll and supporting it at each end will allow the belt to feed easily and smoothly off the roll. However, control the roll so that it does not unwind and telescope. Throughout installation the belt must be kept taut.

CAUTION: Most conveyor belts have a thicker rubber cover on the carrying side than on the pulley, or underside. Be sure that this carrying side is uppermost when the belt is placed on the troughing carriers. A clamp and cable may be attached to the end of the belt and the cable pulled over the head pulley. Fasten the belt and thread the cable through the return. Bring the ends of the belt together at a convenient place for splicing. In cases where the belt is too long for one roll, of course, splices must be made at all connecting points.

TRAINING OF BELT

A newly installed belt must be lined up to run on the center of the pulleys and idlers. Misalignment can cause belt training problems. The tail-end take-up pulley and all idlers should be squared up so they are at 90 deg with the centerline of the conveyor. The belt can be aligned only by adjusting the troughing and return idlers. Self-aligning idlers should be adjusted so that the belt will touch the guide rollers before it hits the troughing or return idlers.

If the belt is a reversing type and self-aligners are used, they will be supplied with an actuating shoe. The latter will be bolted to the outside of the idler frame and will tip the idler in the proper attitude to center the belt.

When starting a new belt for the first time, run it slowly and intermittently. Check the entire length to be certain it is not running off at any point to an extent that will damage the belt.

Do *not* attempt to train the belt by adjusting the head-end drive pulley. Drive shaft bearings are set at the factory before shipment. If they are moved, it will throw the drive belts out of line.

The return run should be checked first for belt alignment. When a belt runs off a return roll, the edge may be damaged. To train the return belt, start at the head end. In general, adjustments are made 15 to 20 ft behind the point where the belt runs off.

All troughing idlers are provided with slotted holes for the hold-down bolts to permit shifting the ends for belt alignment. To adjust idlers loosen the hanger bracket bolts slightly and move the idler slightly toward the head shaft on the side opposite the point where the belt runs off. The belt will shift to the side where it touches the roll first.

Do not tighten hold-down bolts for troughing and return idlers until the belt has been trained. Adjust idlers (Fig. 13-10) with the belt running. The effect of idler shifting is not immediate. Wait a few minutes to see if the belt trains properly before making further adjustments.

NOTE: When proper training has been finished, tighten all mounting bolts to prevent shifting of idlers.

When training the belt on the carrying side, start at the tail pulley. Follow the same adjustment procedure on troughing rollers as described previously for return rolls. To repeat, make adjustments behind the point where the belt runs off and wait a few minutes for the belt to train. Tighten all hold-down bolts after the belt is properly trained.

Do not try to correct belt alignment by extreme shifting of one idler but rather by

1-212   Basic Maintenance Technology

slightly changing a number of idlers. Proper alignment of idlers at the head end will train a belt to run on the center of the head pulley.

If the belt does not run central after adjustment of idlers and take-up screws, check to be certain the belt lacing is square.

Permanent stretch of a conveyor belt which has a synthetic carcass will happen fast

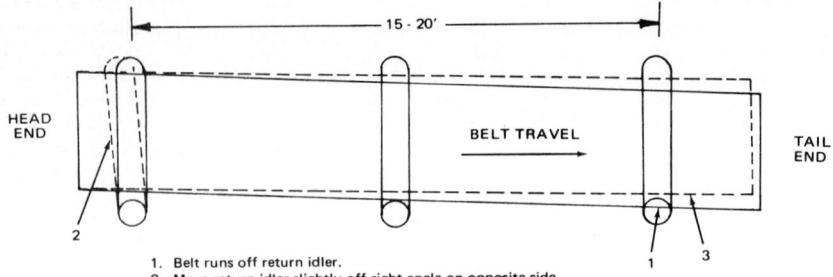

Fig. 13-10   Belt training by idler adjustment (return idler shown).

during the first few hours of operation. It should be checked frequently during the first week. Although permanent stretch occurs more rapidly with synthetic-carcass belts, the total stretch will be no greater than for other types of belting materials.

Conveyor take-up (Fig. 13-11) should be adjusted to provide enough tension for the belt to operate under load without slipping. Correct belt alignment must be maintained while tension adjustments are made. When correctly aligned and adjusted, the belt will be centered on the tail pulley while the conveyor is operating. The amount of belt tension is correct when the belt sags slightly between the return rolls.

Do not apply excessive tension because damage to the belt and to pulley bearings may result. After the belt has been run in, tension should be checked at frequent intervals and take-ups adjusted when necessary.

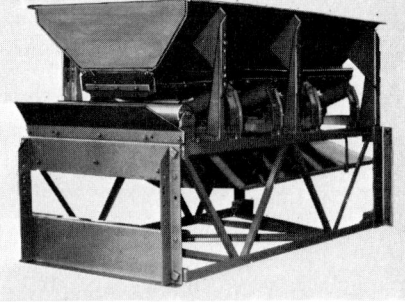

Fig. 13-11   To adjust conveyor belt, turn take-up screws (A) clockwise until desired belt tension is obtained.

Fig. 13-12   Keep the rubber flashing of the loading hopper (shown here) adjusted.

OPERATING PRECAUTIONS

Certain precautions must be observed to ensure safe operation of conveying equipment.

1. Always check belt alignment when starting the conveyor for the first time on a new setup. Be sure that the belt does not rub or catch on any obstruction.

2. When starting operation, be sure that the belt is started and moving freely before

any material is discharged to the hopper. Also, be sure that the discharge to the hopper is such that the hopper is not buried in spilled material.

3. Clean out the hopper end at frequent intervals to remove any accumulation of material. Keep the belt, pulleys, and return idlers clean and free of accumulated material or sticky material.

4. Keep hopper flashing (Fig. 13-12) in proper adjustment.

5. Unless absolutely necessary, the conveyor should not be started under load. If possible, empty the belt before stopping and start the conveyor before loading.

6. If material is fed onto the belt from gravity chutes, it is often possible to overload the conveyor and cause serious damage to the power unit, drive machinery, or belt. The feed of material onto a belt must be regulated not to exceed the conveyor's rated capacity in tons per hour.

7. When handling oily, abrasive materials, such as metal chips, turnings, or oiled coal, use a special Barprene belt. This belt is designed to withstand the corrosive action of oil and grease.

8. To handle hot material, use a high-termperature belt and high-temperature grease for lubrication.

9. Check motor rotation every time electrical leads are changed after moving a conveyor.

Overloading a conveyor As already described, every conveyor is designed to handle a given maximum capacity. If material is fed onto the belt from gravity chutes, it is often possible to overload the conveyor and cause serious damage to motor, drive machinery, or belt. Feed of material onto the belt must be regulated not to exceed the rated capacity in tons per hour for which it was designed.

It is also possible to overload a conveyor by changing operating conditions. For example, if a conveyor is designed to handle a specified capacity in a horizontal position and later is elevated to operate at an 18-deg angle, capacity must be reduced to avoid overloading the conveyor.

Normally, rated capacity is for a conveyor at an 18-deg incline. A belt will not be fully loaded at the rated capacity. Capacity of a conveyor is decreased when the angle of operation is increased. It is advisable to stay below the maximum angles given by the manufacturer. A conveyor loaded with more than rated capacity has extra wear, which reduces the life of the mechanical components and may burn out the motor.

Capacity measurement in the field To determine the actual capacity of a belt, it is necessary to know belt speed. Then, the capacity for any belt width may be determined in the following manner:

1. Measure a length on the loaded belt equal to the length given in Table 13-2 for the corresponding belt speed.

TABLE 13-2 Belt Conveyor Capacity Measurement

Belt speed, fpm	Belt length		Belt speed, fpm	Belt length		Belt speed, fpm	Belt length	
	ft	in.		ft	in.		ft	in.
200	6	0	300	9	0	400	12	0
210	6	3½	310	9	3½	410	12	3½
220	6	7¼	320	9	7¼	420	12	7¼
230	6	10¾	330	9	10¾	430	12	10¾
240	7	2½	340	10	2½	440	13	2½
250	7	6	350	10	6	450	13	6
260	7	9½	360	10	9½	460	13	9½
270	8	1¼	370	11	1¼	470	14	1¼
280	8	4¾	380	11	4¾	480	14	4¾
290	8	8½	390	11	8½	490	14	8½

2. Weigh the material contained in this length of belt.

Each pound of material on this length of belt represents 1 ton per hr that the conveyor is handling.

Feeding Material should be delivered to the hopper with as little drop and impact as possible. Large-size material must be carefully fed to the hopper to prevent damage to the belt.

Basic Maintenance Technology

It is desirable to feed at a uniform rate of speed so that the load on the belt will be evenly distributed along the entire conveyor length. The feed should also be such that the material is carried on the center of the belt so the belt will run true.

Adjusting V-belt tension When they are being mounted, V-belts must not be stretched over the rims of sheave grooves. Reduce the span between sheaves enough to permit the belts to be assembled without stretching.

Tighten new belts about 2 times normal tension. There will be a rapid drop in tension during the run-in period (the first 24 to 48 hr) while the belts seat themselves in grooves (Table 13-3).

TABLE 13-3 V-Belt Tension Adjustment

V-belt cross section	Average Small sheave diam. range, in.	Drive ranges Small sheave rpm range	Speed Ratio range
3V	2.65–3.35	1200–3600	2.00–4.00
3V	4.75–6	900–1800	2.00–4.00
5V	7.1–9	600–1500	2.00–4.00
5V	12.5–16	400–800	2.00–4.00
8V	18–22.4	200–700	2.00–4.00

After the first day or two, check for the correct amount of tension in each belt. If the belt deflection force is over 1.5 times normal, the belts are too tight. If the force is below normal belt tension, they are too loose (Table 13-4).

Adjustment Procedure

STEP 1. Measure span length t of drive (Fig. 13-13).

STEP 2. At center of span t, apply a force perpendicular to the span large enough to deflect one belt on the drive 1/64 in. per inch of span length from its normal position.

TABLE 13-4 Recommended Belt Deflection Force in Pounds per Belt

For normal tension, lb	For 1.5 times normal tension, lb	For 2 times normal tension, lb
3	4½	6
4	6	8
8	12	16
10	15	20
20	30	40

Conveyor hopper flashing The rubber hopper flashing (Fig. 13-12) on conveyors is adjustable. It should be kept adjusted so that it contacts the belt evenly but lightly. Contact that is too hard will cause rapid belt wear.

Hydraulic hoist assembly Whether hand- (Fig. 13-14) or power-actuated (Fig. 13-15), a hydraulic hoist will operate smoothly and with little trouble, free of dirt and foreign matter. All hose connections must be kept tight.

Hand hydraulic pump The hand hydraulic pump used in conjunction with a hydraulic ram for raising a boom is a double-action piston pump. It forces oil from a reservoir into the ram.

If the hand pump fails to raise the boom, either the ram is leaking or the check ball is not seating properly. If the fault is with the check ball, follow this procedure:

1. Remove the valve plug, the spring, and, finally, the ball.
2. Use a magnet to clean out any chips that might be on the valve seat.
3. Replace the check ball and tap lightly with a small hammer and brass rod.
4. Reassemble the valve plug.

Maintenance of Belt Conveyors and Conveying Equipment   1-215

In some instances air will get into the line. Sometimes this air will get into the oil passages in the pump, causing it to function only on one side. To expel this air
 1. Open the release valve.
 2. Give the pump handle a few quick strokes, thus expelling air from oil passages back into the oil reservoir where it will no longer be harmful.

Fig. 13-13 Conveyor-drive V-belts are adjusted by the head shaft torque arm reducer tie-rod turnbuckle.

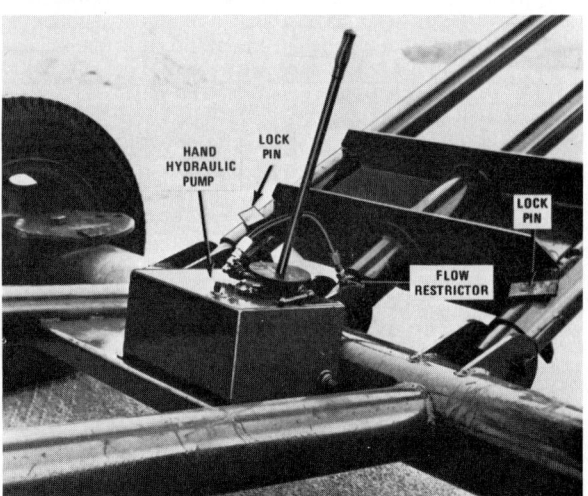

Fig. 13-14 Hand-operated hydraulic boom hoist.

Hydraulic ram The hydraulic ram used on either hand- or power-actuated boom hoists will require little or no maintenance, provided the correct oil is used and it is kept clean and free of dirt. Inspect the ram periodically for leakage around the packing. If leakage is excessive, install a new packing kit.

Basic Maintenance Technology

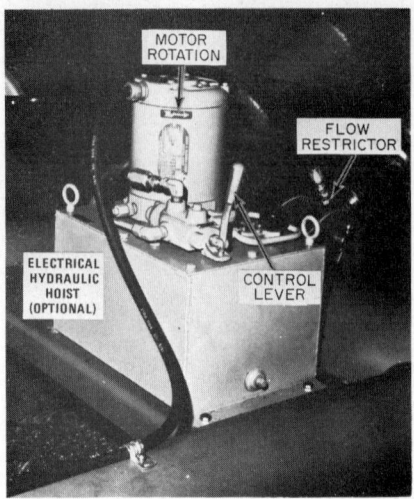

Fig. 13-15 Electrical hydraulic boom hoist.

Tires and wheels Periodically, check tire pressure and keep inflated to correct pressure.

Belt scraper The function of a belt scraper (Fig. 13-16) is to wipe the belt clean of wet, sticky materials. The scraper prevents material buildup on return rolls, eliminating excessive belt tension or failure as well as bearing failure. Return rollers must be kept clean if they are to effectively train a belt onto the belt pulley.

A scraper is mounted so that the scraper blade contacts the belt prior to leaving the head pulley. Compression springs maintain an even scraper pressure against the carrying surface of the belt.

Spring tension is adjustable through use of adjusting nuts. Scraper blades are reversible for maximum wear.

Adjusting belt-scraper tension Tighten nuts increasing the spring tension on the belt by the scraper blade (Fig. 13–17). Proper adjustment is obtained when the rubber blade contacts the belt firmly and evenly. Do not use too much tension because it will cause excessive wear and premature breakdown of the rubber scraper blade.

ADDED MAINTENANCE AND ADJUSTMENTS FOR GASOLINE ENGINE DRIVEN CONVEYORS

Engine Refer to the engine manufacturer's manual for specific maintenance instructions.

Head shaft drive chain adjustment
1. Loosen the drive shaft bearing hold-down bolts (A), four on each side (Fig. 13-18).
2. Loosen lock nuts (B) on retainer clips (C), two on each side.
3. Move shaft to obtain proper chain adjustment. Be sure to move each side the same amount to maintain proper sprocket and sheave alignment.
4. Hold clips (C) firm against bearings and tighten nuts (B).
5. Tighten bolts (A).

Drive shaft V-belt adjustment After adjusting the drive chain, it is necessary to adjust the drive shaft drive belts.
1. Loosen bolts (A) on the countershaft bearing support plates on each side (Fig. 13-19).
2. Loosen lock nuts (B) on adjusting bolts (C), one on each side of the conveyor.
3. Take up on nut (D) to remove excess slack in the V-belts. Be sure to take up equally on both take-up bolts so sheave alignment will remain parallel.
4. Tighten lock nuts (B) and bearing support plate bolts (A). After making this adjustment, it will be necessary to adjust the countershaft drive belts.

Countershaft drive belt adjustment
1. Loosen the two bolts (A) in each of the four-corner power unit sill brackets (Fig. 13-20).
2. Loosen lock nuts (B) on take-up bolts (C).
3. Tighten adjustment nuts on take-up bolts (C) until the drive belts are snug. (Do not run with belts too tight.) Be sure to take up equally on each take-up bolt so sheave alignment will remain parallel.
4. Tighten lock nuts (B) and bolts (A).

LUBRICATION

Nothing can add to the life of a conveyor more than thorough and proper lubrication at correct intervals. When time and availability of a machine are at a premium, a breakdown caused by improper lubrication is absolutely inexcusable since this can so easily be avoided (Fig. 13-21).

Maintenance of Belt Conveyors and Conveying Equipment 1-217

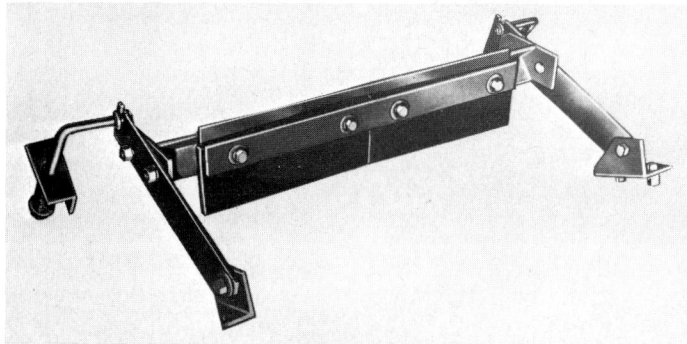

Fig. 13-16 Belt scraper.

Fig. 13-17 Belt scraper adjustment.

Fig. 13-18 Head shaft drive chain adjustment.

Fig. 13-19 Drive shaft V-belt adjustment.

Fig. 13-20 Countershaft drive belt adjustment.

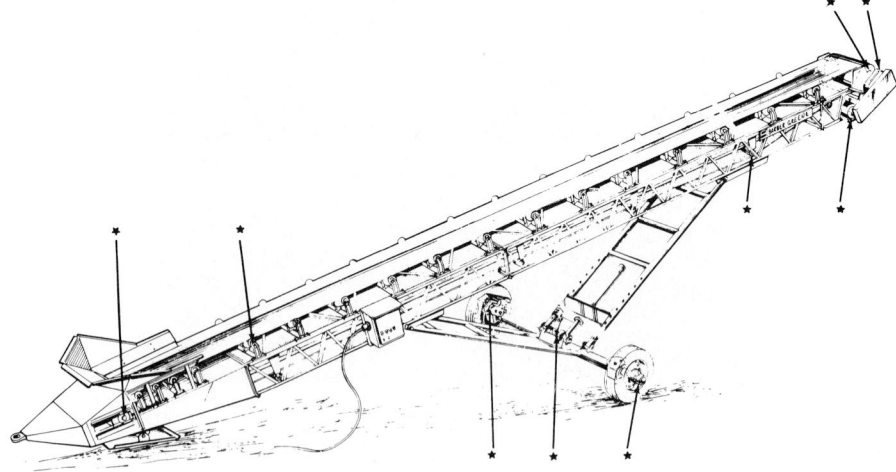

Fig. 13-21 Typical lubrication points.

caused by improper lubrication is absolutely inexcusable since this can so easily be avoided (Fig. 13-21).

NOTE: The number of shots from a grease gun called for in lubrication instructions is based on greasing with a standard 13-oz grease gun. One shot with such a gun equals 1/54 oz. In other words, it takes 54 shots to deliver 1 oz of grease.

The following greases (or others equal to the characteristics of those listed) are recommended for bearings. These greases are suitable for temperature ranges of -10 to $+200°F$ under normal speeds and conditions. For continuous cold weather operation a lighter-grade is advisable.

Consult the factory for recommendations for other temperatures or abnormal conditions. It is important that a grease be used that is suitable for prevailing conditions. Apply lubricant on the basis of lubrication character and temperature.

Characteristics	Temperature, °F	Shots
Penetration unworked	0	165–200
Penetration unworked	77	220–290
Penetration worked	77	235–290
Dropping point	0	300 min.
Mineral oil gravity	—	24–29
Pour	0	0–+15
SSU	100	275–335
SSU	210	50–60

Place safety first Observe the following safety precautions before servicing and lubricating idlers:
1. Wear safety glasses when inspecting idlers while a belt is carrying material.
2. Do not lubricate idlers while a belt is running.
3. Be sure that conveyor drive is locked out before removing an idler or component part.
4. Do not leave tools or parts lying on a belt.

Idlers Keeping idler bearings properly greased is important for the following reasons:
1. To minimize bearing rolling friction and the wear it causes.
2. To avoid the tendency of old grease to thicken with time and lose some of its lubricating quality.
3. To assist seals in preventing dust and moisture from entering bearings and causing wear or corrosion. A good packing of grease not only lubricates but also serves as a barrier against foreign matter.

Usually, a manufacturer will indicate what normal lubrication intervals should be for idler bearings. For some ball or roller bearing idlers, for example, the lubrication interval is 800 to 1200 operating hours under average operating conditions. It is important to note, however, that this is only a guide. The proper lubrication interval may vary from job to job, from conveyor to conveyor, and even from one group of idlers to another on the same

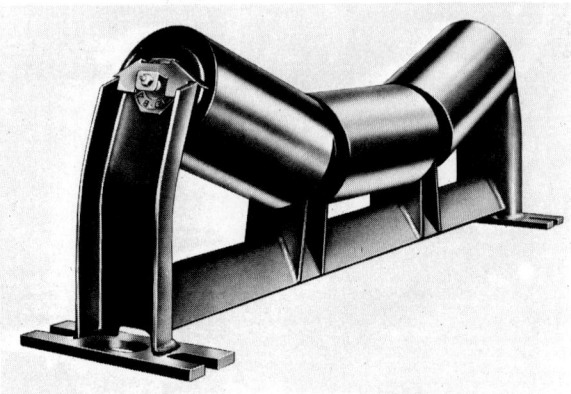

Fig. 13-22 Troughing idler.

1-220 Basic Maintenance Technology

conveyor. The operator must therefore determine the correct lubrication interval for a particular application (Fig. 13-22).

Most conveyor idlers have lubrication fittings on each side. On some types all three rolls of a troughing idler can be lubricated from either side of the belt from either one of the fittings (Fig. 13-23).

Idlers should be lubricated with a high-quality No. 2 or No. 1 grease, either calcium or lithium base. The operating range for this lubricant should be from -10 to $+200°F$. Using a high-quality grease of this type exclusively will help ensure efficient, trouble-free performance and the longest possible service life.

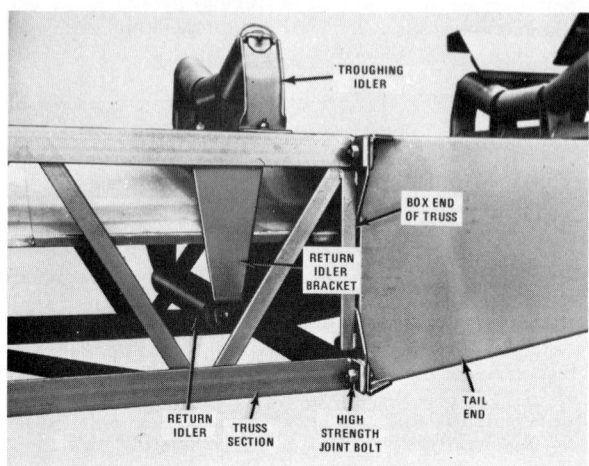

Fig. 13-23 Troughing and return idlers shown in relative working positions.

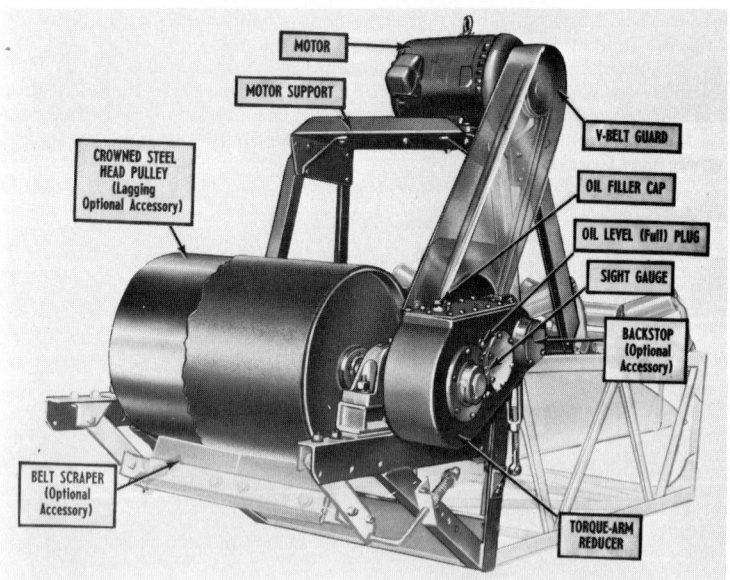

Fig. 13-24 Permanent conveyor head and drive.

Antifriction bearings One factor that determines the lubrication interval is the effectiveness of bearing seals. Another is the set of conditions that prevails at the individual conveyor. If a belt handles much fine, corrosive, or wet material, or if the atmosphere contains a great deal of dust or moisture, idlers should be greased more often than they would be under normal conditions.

With some systems of positive lubrication it is possible to change grease or flush out old grease while replacing it with fresh, clean lubricant. Position a container on the far side to catch old grease so that it does not come in contact with the belt. For normal lubrication, however, the appearance of additional grease at the fitting on the far side of the idler is assurance that all bearings in the rolls have been lubricated. Note that some bearings can be damaged by overgreasing, as explained.

Antifriction bearings used on some conveyors have been lubricated at the factory. The grease cannot be seen since it is concealed within the bearing by the grease retainer seals. Overgreasing distorts and damages these seals, allowing dirt to enter and greatly shorten the life of the bearing.

CAUTION: Never use a power-operated grease gun on antifriction bearings. Lubricate only as directed by the lubrication chart.

Drive Check the oil level in a gear reducer box every 40 to 50 hr of operation. Fill box (Fig. 13-24) to oil level plug, using a good grade of gear lubricant. For temperatures above 32° use SAE 90 gear lubricant. For temperatures below 32° use SAE 80 gear lubricant. Drain, flush out, and refill gear box after 300 hr of operation.

For reducers with bearings requiring external lubrication, use a good grade of grease. For specific grade and type, refer to the operator's manual for the manufacturer's recommendation.

Oil-tight chain guards Fill to oil level plug with a good-quality clean SAE 30 motor oil. SAE 20 motor oil may be used in cold weather. Check oil level once a week *when drive is not operating.* Drain and refill every 3 months.

Plain bearings Plain bearings should be greased every day the conveyor is in operation. Use a good-quality bearing grease.

Electric motors Unless otherwise specified, electric motors are equipped with ball bearings and should be greased only once or twice a year, depending on service. The bearings should be taken out, washed with gasoline, and repacked with electric motor grease, a process that should be performed only by a competent electrician.

Operation of electrically powered equipment can be as safe or as hazardous as you make it. Follow these simple rules for your safety and the safety of others:

1. All electrical equipment must be grounded.
2. Always pull the main switch before disconnecting a plug from its receptacle.
3. Never grease, adjust, or repair a machine without removing the plug from its receptacle or pulling the main switch.
4. Use only the proper size and style of fuse, keep a supply on hand, and never substitute pieces of metal.
5. Keep electrical cables in good condition at all times and keep them out of water.

Gasoline engines Gasoline engines should be lubricated as specified in the operator's manual supplied by the manufacturer.

Chapter **14**

Maintenance of Hydraulic Hose and Fittings

ROBERT L. OLSON, P.E.
Product Application Department, The Gates Rubber Company,
Englewood, Colo.

Power is transmitted through the medium of hydraulics by pressurizing and pumping a liquid at one point and converting it to work at another point. Hydraulic liquids are routed from point to point under pressure through a number of conveyance systems. Usually a combination of conveyance systems is used in a single hydraulic circuit. Pipe, tubing, sandwich channels, drilled holes, hose, couplings, and adapters are among the most popular. When a flexible conveyor is required between two points in a circuit, such as when one point moves relative to the other point, hose is usually used as the conveyor. Wherever hose is used in this type of circuit, couplings are required. This chapter is primarily intended to discuss hose and couplings, but, since couplings and adapters are so closely related, adapters will be included.

HOSE

Generally, hose is designed into a circuit as a flexible connection or vibration dampener. It consists of three major parts: tube, reinforcement, and cover (Fig. 14-1).

Tube

The innermost part of the hose, the tube, is the only part of the hose in contact with the liquid being conveyed. Its purpose is to contain the fluid and keep it from passing into the more porous reinforcement.

There are many tube materials in the hose types available (Table 14-1), having particularly strong resistance to certain degenerative characteristics. For instance, a tube material that is ideal for phosphate ester fluids is not suitable for use with petroleum-base hydraulic oils and vice versa. Most hydraulic-tube materials are compounded using acrylonitrile and butadiene as base polymers. This combination is known in the rubber industry as Buna-N. The chemical and physical properties of this synthetic rubber are subject to a considerable amount of control through compounding with a number of other ingredients necessary to make the material useful.

Chemical compatibility between the tube and the hydraulic fluid is very important. Another factor that can be equally

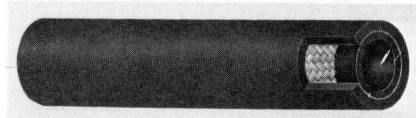

Fig. 14-1 Hose consists of three major parts: tube, reinforcement, and cover.

important is temperature. Almost any hydraulic tube stock is compounded to function properly between −40 to +200°F (−40 to +93°C). At temperatures under −40°F, normal tube material will stiffen and tend to become brittle. At temperatures over 200°F it will harden and sometimes crack. Specially compounded tube material is available for either of these extremes. Certainly there are many other factors a compounder keeps in mind when designing tube materials, but, for a user, the above are the important factors.

Reinforcement

Many reinforcement materials and configurations are used in the manufacture of hydraulic hose. Their purpose is to keep the tube from expanding when its contents are pressurized. Naturally, the greater the pressure requirements of a hose, the greater the strength requirements of the reinforcement.

TABLE 14-1 Characteristics of Hose Stock Types
Choose a hose stock type that best suits application. Chemical and physical properties are subject to a considerable amount of control through compounding; therefore, the characteristics shown for each stock type are generalized to some degree. Tube and cover stocks occasionally may be upgraded to take advantage of improved materials and technology.

Chemical name	Polychloroprene (Neoprene)	Acrylonitrile and butadiene (Buna-N)	Isobutylene and isoprene (Butyl)	Chlorosulfonated polyethylene (Hypalon)	Ethylene propylene diene (EPDM)
ASTM-SAE designation SAE J14 SAE J200	SC BC	SB BG	R AA	TB CE	R AA
Tensile strength	Good	Fair to good	Fair to good	Good	Good
Tearing resistance	Good	Fair to good	Good	Fair	Good
Abrasion resistance	Good to excellent	Fair to good	Fair to good	Good	Good
Flame resistance	Very good	Poor	Poor	Good	Poor
Petroleum oil and commercial gasoline	Good	Good to excellent	Poor	Good	Poor
Resistance to gas permeation	Good	Good	Outstanding	Good to excellent	Fair to good
Weathering	Good to excellent	Poor	Excellent	Very good	Excellent
Ozone	Good to excellent	Poor	Excellent	Very good	Outstanding
Heat	Good	Good	Good to Excellent	Very good	Excellent
Low temperature	Fair to good	Poor to fair	Very good	Poor	Good to excellent
General chemical resistance	Good	Fair to good	Good	Good	Good

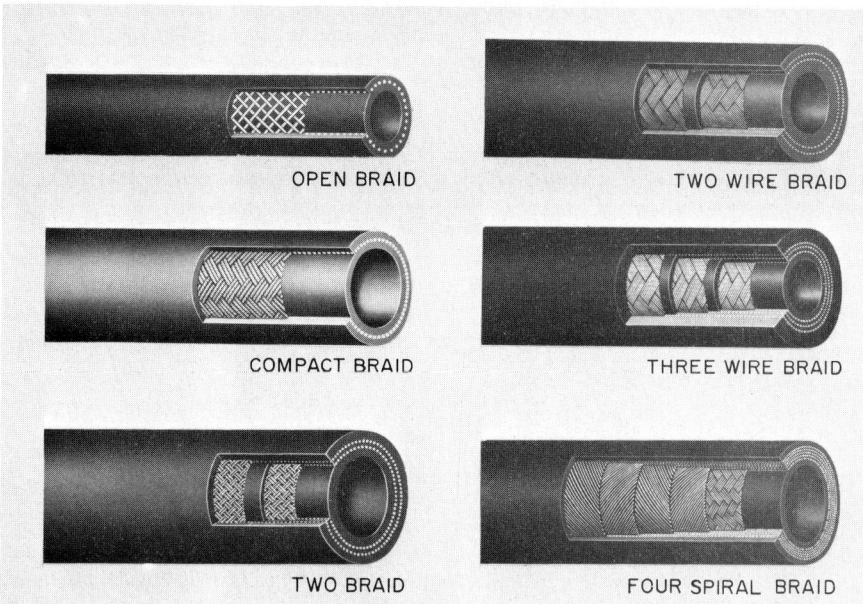

Fig. 14-2 Cutaway drawings of hydraulic hose show reinforcement materials and configurations used to keep a tube from expanding when its contents are pressurized.

The lowest-pressure hydraulic hose generally will have one layer of open-fiber braid. In recent years, the fiber used is synthetic material such as polyester which, because of its added strength over natural fibers, has allowed less dense braiding in some cases or higher pressure ratings for the hose in other cases. As pressure requirements increase, the density of the braid must be increased until a point is reached where additional pack is impractical. Then the designer will use two braids for extra strength. As pressure requirements increase beyond the capability of two-fiber braid, one-wire braid is used. Again, densities of braid and wire diameter may be varied by a designer to meet strength requirements and second- and third-wire braids may be added. Spiral-wire hose has become popular for extremely high pressure applications. It consists of wires lying side by side in a spiral configuration around the hose, each layer spiraling in the opposite direction from that of the preceding layer. The most common spiral hose is *four spiral;* it has four layers or plies of wire. However, six spiral hose is also available (Fig. 14-2).

In designing hose, it has been found that combinations of the above are sometimes advantageous. Fiber braid and wire braid on the same hose and wire braid with spiral wire offer some advantages. The possibilities become limitless for the hose designer, whose knowledge can produce an ideal hose for almost any application.

An interesting observation at this point is that the larger the hose diameter, the less pressure it will accommodate using the same reinforcement construction. For instance, a one-wire-braid ¼-in.-ID hose is rated to burst at 12,000 psi, and a 1-in.-ID hose of the same construction is rated to burst at 3000 psi. This is evident in hose catalogs. The smaller the hose, the higher the pressure rating. The ratio will not be the same as above because hose manufacturers tend to use heavier reinforcement in the larger sizes but the difference is still quite noticeable.

To understand this, visualize a 1-in. length of ¼- and 1-in. hose (Fig. 14-3). Slit each sample on a side parallel to its axis and flatten it; there will be a piece about 1 × ¾ in. or ¾ sq in. from the ¼-in. hose sample and a piece about 1 × 3 in. or 3 sq in. from the 1-in. sample. Now, the pressure in pounds per square inch against ¾ sq in. will apply only about one-quarter of the stress on the wall—where the ¼-in. hose was slit—that it will on the wall of the 1-in. hose of 3 sq in. In other words, the force trying to tear the 1-in.-long section of wall on the ¼-in. hose is only about one-quarter the force trying to tear the 1-in.-long section of wall on the 1-in. hose because the pressure per square inch is pressing against about one-quarter of the square inches in the ¼-in. hose.

Cover

The purpose of cover is to protect reinforcement from damage caused by abrasion or moisture. Hydraulic hose cover is compounded to be resistant to tearing, abrasion, oil,

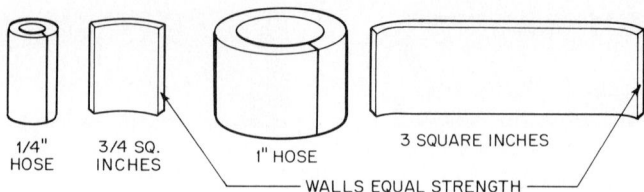

Fig. 14-3 The larger the hose diameter, the less pressure it will accommodate using the same reinforcement construction. In this illustration, the force trying to tear the section of wall on the ¼-in. hose is only about one-quarter the force trying to tear the wall on the 1-in. hose because the pressure per square inch is pressing against about one-quarter the square inches in the ¼-in. hose.

weathering, and ozone. The base polymer is generally polychloroprene (Neoprene). Special compounds are used for extreme temperature, flame resistance, nonconductivity, or other special hose applications. Another variation is called a *cotton cover*. This is an exposed synthetic rubber-impregnated cotton braid which is desirable when the hose is required to slip on a smooth surface or in other applications where the hose is not subject to damage through abrasion. The real bonus in this kind of construction is that the cover contributes to the strength of the hose.

While on the subject of cover, reference to thin-cover wire hose is frequently made. It is necessary to go back in history to fully explain the term. In the early days of wire-braid hose a reasonably heavy cover became standard to protect the costly wire braid. It was quickly determined that for a coupling to stay on the ends of a hose under the pressures the hose was capable of withstanding, the cover had to be removed in the area of the coupling so that the coupling could bite into the steel wire. Later, in an attempt to eliminate this operation of skiving (or buffing) the cover off, a coupling system was developed that would work without removing the cover, provided the cover was not too thick. At this point, thin-cover hydraulic hose was developed, and it has become very popular. Today, regular-cover wire hose is used only where extreme abrasion is evident. Thin-cover hose is easier to couple and is used extensively.

Selecting Hydraulic Hose

To determine the correct hose for an application it is necessary to know the flow velocity of the liquid in the circuit, the maximum operating pressure of the circuit including surges, the temperature range of the fluid, the type of fluid used in the system, and any unusual requirements of a hose such as its resistance to flame or its conductivity.

A nomographic chart (Table 14-2) may be used to select the correct hose size for a given hydraulic system. The velocity of the hydraulic fluid should not exceed the range shown in the right-hand column. When oil velocities are higher than recommended in this chart, the results are turbulent flow with loss of pressure and excessive heating. Higher velocities may be used if the flow of hydraulic fluid is intermittent or only for short periods of time.

The velocity of hydraulic fluid in suction lines should always fall within the range recommended to ensure efficient operation of the pump. The following is an example of the use of this chart: What size hose assembly is recommended to carry 10 gal of oil per minute, and what will be the velocity of the oil through the hose assembly? The hose assembly is to be used in a pressure line and flow is to be continuous.

Locate the flow, 10 gpm (left-hand column), and a velocity, 10 ft/sec (right-hand column), since it is near the center of the recommended range. Lay a straightedge across

TABLE 14-2 Flow Capacity of Hose Assembly at Recommended Flow Velocities

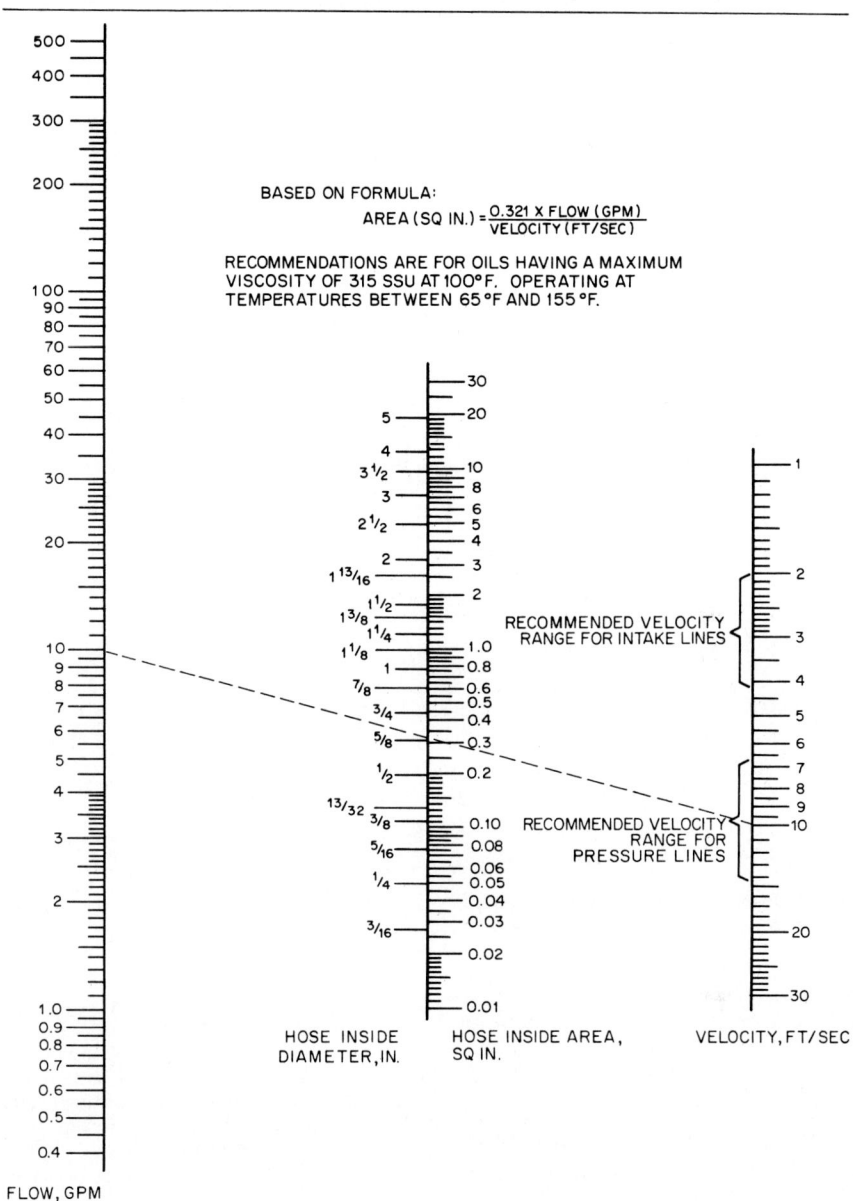

these two points. The straightedge crosses the center column nearest the ⅝-in. hose assembly. Keeping the straightedge on 10 gpm in the left-hand column, cross the center column at the ¾-in. assembly, the ⅝-in. assembly, and the ½-in. assembly. Reading the right-hand column, the straightedge crosses it at 7.5, 10.3, and 16 ft/sec, respectively. Since 7.5 and 10.3 are within the recommended velocity range for pressure lines, either a ¾- or ⅝-in. hose assembly may be used.

Problems concerning suction hoses are solved in a similar manner, except that the recommended velocity range for intake line (right-hand column) and the values for suction hose assemblies (center column) are used. When the hose size has been determined, it is necessary to consult a manufacturer's catalog or the SAE handbook to select the correct hose. These references list the characteristics and dimensions of the most commonly used hose. A complete SAE handbook or individual specifications may be obtained by contacting the Society of Automotive Engineers, Inc., 400 Commonwealth Drive, Warrendale, PA 15096.

If individual specifications are preferred, the following list may aid in a selection:

SAE J30	Fuel and Oil Hoses
SAE J343	Tests and Procedures for SAE 100R Series Hydraulic Hose and Hose Assemblies
SAE J514	Hydraulic Tube Fittings
SAE J515	Hydraulic O-Ring
SAE J516	Hydraulic Hose Fittings
SAE J517	Hydraulic Hose
SAE J518	Hydraulic Flanged Tube Pipe, and Hose Connections, 4-bolt Split Flange Type
SAE J533	Flares for Tubing
SAE J926	Hydraulic Pipe Fittings
SAE J1402	Air Brake Hose—Automotive

A common method of selecting hose for maintenance of existing equipment is to remove the failed hose and judge what the hose type is either by reading the lettering (if it has not been obliterated) or by examining the hose construction and dimensions and duplicating the hose as closely as possible. This procedure is not recommended, but it is done every day and it keeps equipment working.

COUPLINGS

Hydraulic hose assemblies generally consist of a length of hose and two couplings (Fig. 14-4). There are literally thousands of couplings on the market and not too many chances to substitute one for another. Be cautious in selecting the correct stem and ferrule for the hose, and where threads are involved, be sure they are the correct threads, especially in high-pressure applications.

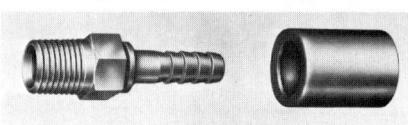

Fig. 14-4 Be sure to select the correct stem (left) and ferrule (right) for hydraulic hose. Correct choice of threads is particularly important in high-pressure applications.

A question often is: Can one manufacturer's coupling successfully be applied to another manufacturer's hose? This is done frequently, but always with some degree of risk. Each manufacturer designs couplings to fit its own hose properly. Even though hose from two manufacturers may comply completely with the same SAE specification, there can be differences in tube hardness and other factors that cause differences in stem and ferrule configurations and dimensions. Each manufacturer tests extensively to be certain its hose and couplings are well matched and compatible. Minute design changes take place in the final design of a coupling. The benefits from this are lost when the coupling is applied to another hose. As mentioned, it is done and may be good first aid in an emergency, but is is not recommended and manufacturers will not guarantee it.

With this knowledge, selecting couplings for an application becomes easy. A hose manufacturer's catalog always recommends couplings for use with the hose. For most hose there is a selection of permanent or reusable couplings. Both have advantages and disadvantages, but the selection is generally determined by the coupling installation equipment the installer has available.

Reusable couplings, as the name implies, can be removed from a worn-out hose and placed on a new hose. This flexibility saves paying for new couplings when a hose is replaced and ensures having the correct couplings on hand. Usually these couplings can

Maintenance of Hydraulic Hose and Fittings 1-229

be changed without special tools. They are available for most hydraulic hose. The disadvantages of reusable couplings are that generally the initial cost is higher and that the time required to install them is greater. Whereas their length of service and reliability have been more than satisfactory, it is generally understood in the industry that permanent couplings are superior. This is substantiated by the fact that five out of seven of the largest hose manufacturers do not list reusable couplings for their highest-pressure hose in their catalogs. Another disadvantage is that reusable couplings are generally larger,

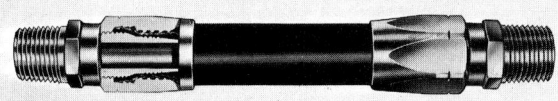

Fig. 14-5 Screw-type reusable coupling has deep, coarse convolutions inside the ferrule that form left-hand threads to assist in getting it over the end of hose. *(Anchor Coupling Co.)*

heavier, and bulkier than the permanent type. Mentioning these disadvantages is not meant to imply reusable couplings do not have a place. They certainly do—in fact, there are a large number of low-pressure hose types for which only reusable couplings are specified.

The most common reusable coupling is the screw type; it has deep, coarse convolutions inside the ferrule that form left-hand threads to assist in getting it over the end of the hose (Fig. 14-5). For thin-cover no-skive wire hose these convolutions are made rather sharp to cut through the cover and grip on the wire. The alternative has flatter-shaped convolutions to avoid cutting the wire on hose when the cover has been skived away in the area of the ferrule. One end of the ferrule has a reduced diameter that does not fit over the hose but is threaded to receive the threads provided on the stem. The end of the stem is tapered ahead of the threads and, as it is rotated into the end of the ferrule, it expands the diameter of the hose, forcing a tight fit onto the convolutions inside the ferrule (Fig. 14-6).

Other reusables are mostly stems similar to crimp-type stems with devices such as two- and four-bolt clamps for higher-pressure applications or stressed steel bands in various configurations for lower-pressure applications to compress the hose onto the stem (Fig. 14-7).

For low-pressure applications, a push-on hose and coupling system is available that

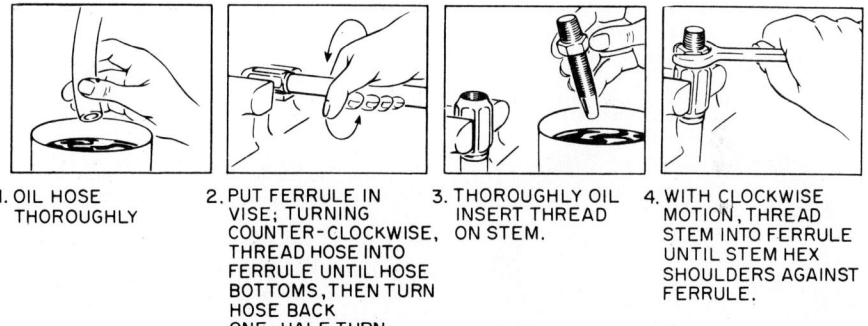

1. OIL HOSE THOROUGHLY
2. PUT FERRULE IN VISE; TURNING COUNTER-CLOCKWISE, THREAD HOSE INTO FERRULE UNTIL HOSE BOTTOMS, THEN TURN HOSE BACK ONE-HALF TURN.
3. THOROUGHLY OIL INSERT THREAD ON STEM.
4. WITH CLOCKWISE MOTION, THREAD STEM INTO FERRULE UNTIL STEM HEX SHOULDERS AGAINST FERRULE.

Fig. 14-6 How to assemble reusable couplings in four easy steps.

simply involves pushing the stem into the end of the hose for coupling. The hose is designed to grip the stem in Chinese finger-puzzle fashion when pressurized.

Permanent Couplings

The term *permanent* is used to indicate that once this coupling is attached to the hose it is not designed to be removed without destroying it (Fig. 14-8). Usually a ferrule of the crimp type is used with a stemmed fitting to constitute a coupling. The primary advantage of the permanent coupling is its reliability; a second advantage is that usually less time is required for installation; third, the permanent coupling and ferrule cost is about one-third to one-half the cost of a reusable; fourth, it is lighter in weight and less bulky.

Crimp-type ferrules vary from thin metal as seen on garden hose to heavy machined metal. Most hydraulic ferrules are machined and have circumferential serrations

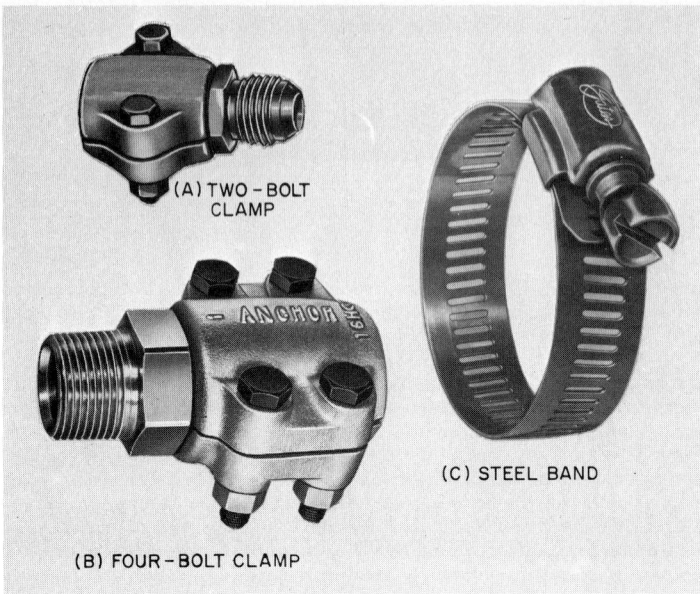

Fig. 14-7 Two- and four-bolt clamps for higher-pressure applications and stressed steel bands in various configurations for lower-pressure applications are other types of reusable stems for hydraulic hose. *(Anchor Coupling Co.)*

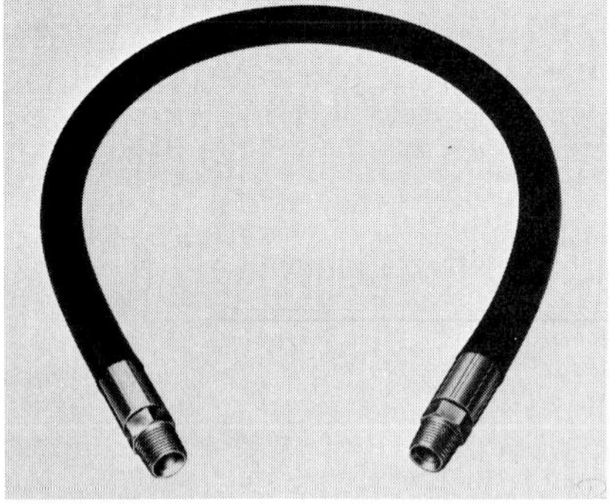

machined on the inside. Again, for no-skive hose these serrations are left sharp to cut through a cover to grip the reinforcement or left flat for skived ends (Fig. 14-9). Stems are designed with saw-toothed circumferential serrations for ease of insertion into the tube and to grip the tube in the other direction. The serrations in the ferrule and stem are designed together so that the hills in the ferrules lie in the valleys of the stems and vice versa to provide optimum gripping onto the hose end.

Crimping the ferrules on permanent couplings is the one thing that cannot be done with tools common around a construction project, although in recent years crimping machines are becoming a part of large contractors' equipment. In addition to having the equipment, it is important to have

Fig. 14-8 Permanent coupling, once attached, cannot be removed without destroying hose. These couplings are very reliable and require less time for installation.

Fig. 14-9 Most hydraulic ferrules have circumferential serrations machined on the inside. Stems are designed with saw-toothed circumferential serrations for ease of insertion into tube and to grip tube in other direction.

Maintenance of Hydraulic Hose and Fittings 1-231

someone available with adequate knowledge to select correctly the hose and fittings for an application and to couple the hose correctly (Fig. 14-10).

A ferrule must be crimped to a specified diameter, usually ±0.005 in. Crimping is usually accomplished with a small hydraulic press having a ram capable of generating from 25 to 100 tons of force. It is not meant to be implied here that all this is in any way difficult to learn; rather, it should not be attempted by an individual who is insufficiently trained. Improperly coupled hose under high pressure can fail, causing undue downtime and possible injury or worse in some cases. Any good hose supplier has the capability and will be more than willing to spend the necessary time to train a customer or his personnel to couple hose properly. In cases where this is not practical, it is urged here that coupled assemblies be purchased complete from a competent hose supplier.

FITTINGS AND ADAPTERS

It was stated earlier that hose is used as a conveyor between two points. The configuration of the connector at each of these points determines what is required on the ends of the

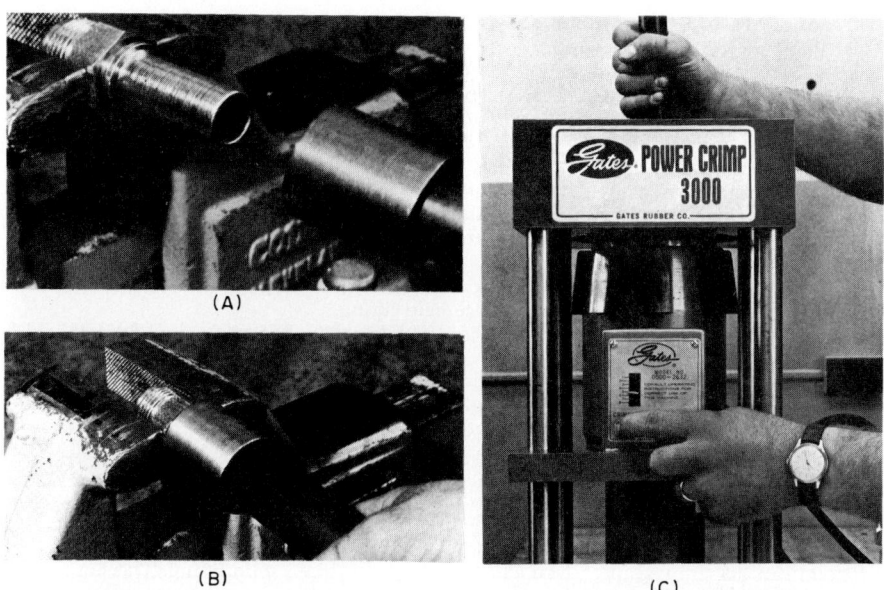

Fig. 14-10 Power crimping machine makes easy crimping of ferrules on permanent couplings. (A) Put stem in vise and place ferrule on hose. (B) Push hose into stem. (C) Insert into machine and push button until pressure is applied and coupling is crimped.

coupling stems. Again, the selection of the correct fitting is extremely important. The following has been prepared to assist you in verbally describing or selecting from a catalog the most frequently used hydraulic fittings. The term *fittings* is being used because it includes couplings and adapters. Frequently both are required to make a hose connection.

Threaded Fittings

There are basically two thread systems used in hydraulics: Iron Pipe Thread and SAE Standard Screw Thread.

The Iron Pipe Thread system generally employs a sealing fit of the mating male and female threads to obtain a leak-proof joint. The thread forms used in the Iron Pipe system are designated by symbols such as NPTF, NPSF, and other combinations using the letters shown below.

Definitions

N	National	F	Fuels
P	Pipe	S	Straight Thread
T	Tapered Thread	M	Mechanical Joint

Descriptions of iron pipe threads

NPTF This is a dryseal thread; the National pipe tapered thread for fuels. This is used for both male and female ends. The sealing fit of the mating threads produces the leak-proof joint.

NPSF The National pipe straight thread for fuels. This is sometimes used for female ends and properly mates with the NPTF male end. However, the SAE recommends the NPTF thread in preference to the NPSF for female ends.

NPSM National pipe straight thread for mechanical joint. This is used on the female swivel nut of iron pipe swivel adapters. The leak-proof joint is not made by the sealing fit of threads, but by a tapered seat in the coupling end.

Descriptions of SAE standard screw threads

Couplings having the SAE standard screw thread usually employ such things as "O" rings, compression sleeves, or tapered seats to obtain a seal.

JIC The Joint Industry Conference specifies a 37-deg angle flare or seat to be used with high-pressure hydraulic tubing. Couplings specified as JIC have 37-deg tapered seats.

SAE A term usually applied to fittings having a 45-deg angle flare or seat. Soft copper tubing is generally used in such applications as it is easily flared to the 45-deg angle. These are for low-pressure applications—such as for fuel lines and refrigerant lines.

Identifying threads A caliper, a thread gauge, and a seat gauge are required to accurately identify threads.

A *caliper* (Fig. 14-11A) measures outside diameter at the largest point on the threads of a male fitting or the inside diameter of a female fitting.

A *thread gauge* (Fig. 14-11B) determines threads per inch. A gauge and fitting should be held toward a light to ensure accuracy.

A *seat gauge* (Fig. 14-12) determines whether seat angles are 37-deg Joint Industry Conference (JIC) or 45-deg SAE. The angles of a coupling and a gauge are matched with the gauge held parallel to the centerline of the coupling.

With the information given above, the charts (Fig. 14-13) may be used to identify the fitting type and nominal size.

Fitting shapes The next thing to determine is the shape of a fitting and, in some cases, whether a swivel is required. The threaded portion of a solid fitting is firmly attached to the hose and cannot rotate unless the hose rotates. When installing an assembly, often the first end can be screwed into place and tightened if it has a solid fitting. The hose then cannot be rotated to attach the second end, so there must be a swivel fitting either on the second end of a hose or on the point to which it is being connected. A swivel fitting allows the thread portion to be rotated independently with respect to a hose or other mounting (Figs. 14-14 and 14-15).

Straight fittings are the most common; however, angle fittings are frequently used. The majority of angle fittings are bent-tube type; however, the block type is not uncommon. Both are used as space savers, and they help to avoid bending the hose too tightly. Angle fittings are found commonly in 45-deg and 90-deg angles (Fig. 14–16 to 14–18).

Flange Fittings

The popular alternatives to threaded fittings are flange fittings. They are primarily used on the larger ID hydraulic hoses because of the difficulties encountered in installing larger-diameter threaded fittings. Flange fittings are identified by a nominal size which is related to the tube diameter on which the flange is brazed.

From the above it can be seen that a nominal 1-in. "O" ring flange is 1¾ in. in diameter and is available for ¾ and 1 in. ID hose, but, in either case, it is called a 1-in. "O" ring flange (Fig. 14-19). "O" ring flange fittings form a seal by compressing the "O" ring against a flat or otherwise prepared surface; generally, two flange halves and four small

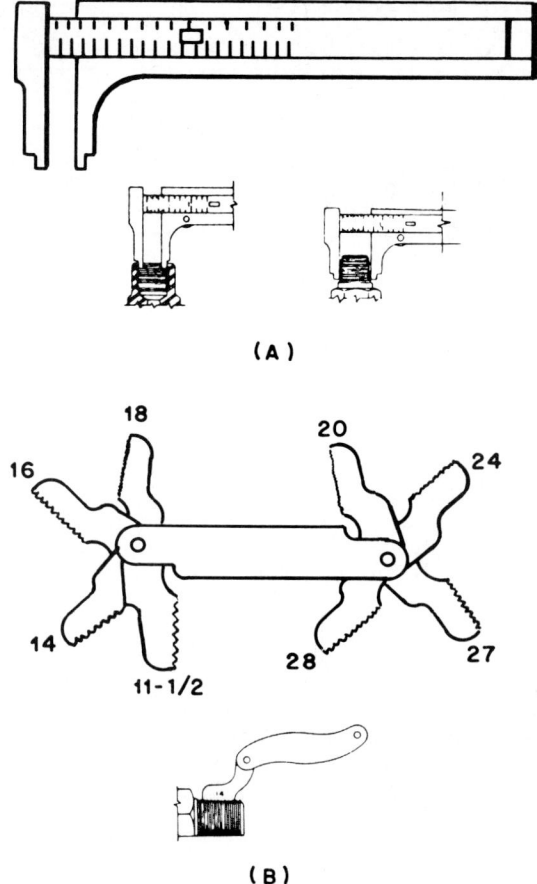

Fig. 14-11 Caliper (A) measures the outside diameter at the largest point on the threads of a male fitting or the inside diameter of a female fitting. Thread gauge (B) is used to determine threads per inch.

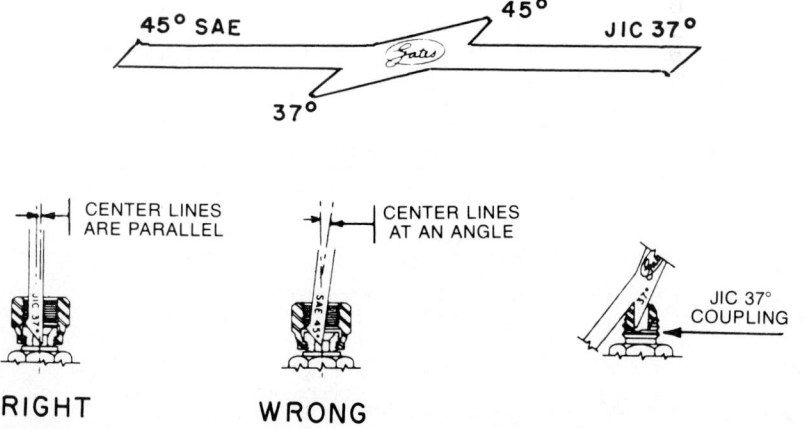

Fig. 14-12 Seat gauge is used to determine whether the seat angles are 37° JIC or 45° SAE.

IRON PIPE THREAD FITTINGS

MALE THREAD			FEMALE THREAD		
NOMINAL THREAD SIZE, IN.	No. THRDS. PER IN.	THREAD O.D., IN.	NOMINAL THREAD SIZE, IN.	No. THRDS. PER IN.	THREAD I.D., IN.
1/8	27	13/32	1/8	27	23/64
1/4	18	35/64	1/4	18	15/32
3/8	18	43/64	3/8	18	19/32
1/2	14	27/32	1/2	14	3/4
3/4	14	1 1/16	3/4	14	61/64
1	11 1/2	1 5/16	1	11 1/2	1 13/64

 NPTF SOLID MALE  NPTF or NPSF SOLID FEMALE   NPTF SWIVEL MALE NPSM SWIVEL FEMALE

JIC FLARED TYPE FITTINGS (37°)

TUBE O.D., IN.	NOMINAL THREAD SIZE, IN.	No. THRDS. PER IN.	THREAD O.D., IN.	NOMINAL THREAD SIZE, IN.	No. THRDS. PER IN.	THREAD I.D., IN.
1/8	5/16	24	5/16	5/16	24	17/64
3/16	3/8	24	3/8	3/8	24	21/64
1/4	7/16	20	7/16	7/16	20	25/64
5/16	1/2	20	1/2	1/2	20	29/64
3/8	9/16	18	9/16	9/16	18	1/2
1/2	3/4	16	3/4	3/4	16	11/16
5/8	7/8	14	7/8	7/8	14	13/16
3/4	1 1/16	12	1 1/16	1 1/16	12	31/32
7/8	1 3/16	12	1 3/16	1 3/16	12	1 7/64
1	1 5/16	12	1 5/16	1 5/16	12	1 15/64
1 1/4	1 5/8	12	1 5/8	1 5/8	12	1 35/64
1 1/2	1 7/8	12	1 7/8	1 7/8	12	1 51/64
2	2 1/2	12	2 1/2	2 1/2	12	2 27/64

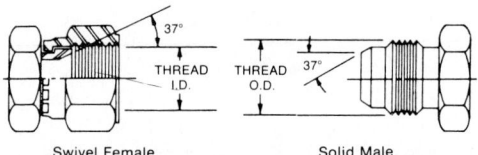

Swivel Female   Solid Male

SAE FLARED TYPE FITTINGS (45°)

NOMINAL THREAD SIZE, IN.	No. THRDS. PER IN.	THREAD O.D., IN.	NOMINAL THREAD SIZE, IN.	No. THRDS. PER IN.	THREAD I.D., IN.
5/16	24	5/16	5/16	24	17/64
3/8	24	3/8	3/8	24	21/64
7/16	20	7/16	7/16	20	25/64
1/2	20	1/2	1/2	20	29/64
5/8	18	5/8	5/8	18	9/16
11/16	16	11/16	11/16	16	5/8
3/4	16	3/4	3/4	16	11/16
7/8	14	7/8	7/8	14	13/16
1 1/16	14	1 1/16	1 1/16	14	63/64

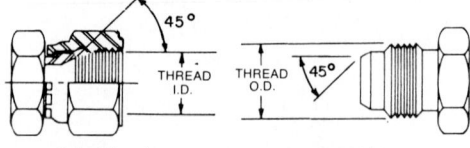

Swivel Female   Solid Male

Fig. 14-13 Charts used to identify fitting type and nominal size.

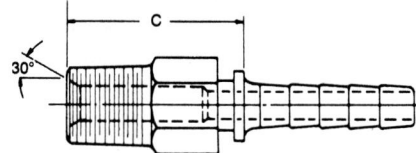
Fig. 14-14 NPTF solid male fitting.

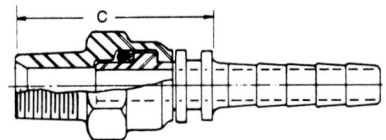
Fig. 14-15 NPTF swivel male fitting.

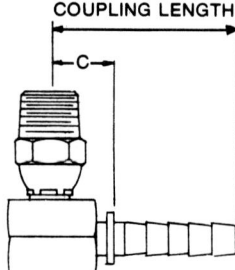
Fig. 14-16 NPTF swivel male—90° (block type).

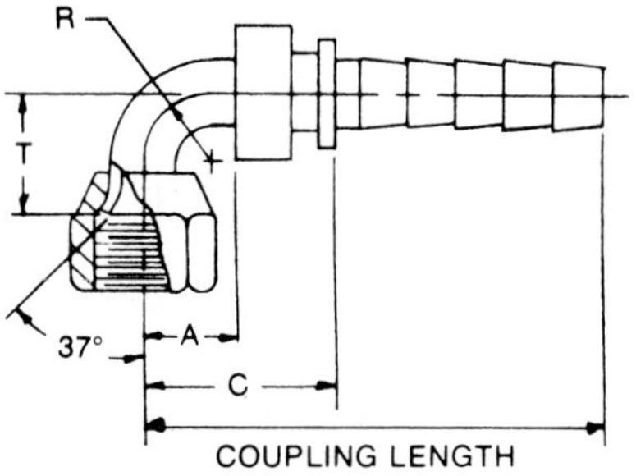

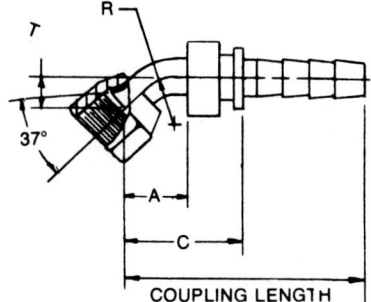
Fig. 14-17 45° JIC swivel female—bent tube.

Fig. 14-18 90° JIC swivel female—bent tube.

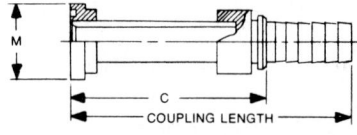

HOSE I.D., IN.	NOM. FLG. SIZE, IN.	M FLG. DIA., IN.	COUPLING LENGTH, IN.	C CUT-OFF, IN.
1/2	1/2	1-3/16	3.29	1.86
1/2	3/4	1-1/2	3.29	1.86
3/4	3/4	1-1/2	3.74	2.04
3/4	1	1-3/4	3.74	2.04
1	1	1-3/4	3.62	1.91
1	1-1/4	2	3.62	1.91
1-1/4	1-1/4	2	4.63	2.33
1-1/4	1-1/2	2-3/8	4.63	2.33
1-1/2	1-1/2	2-3/8	4.94	2.30
1-1/2	2	2-13/16	4.94	2.30
2	2	2-13/16	6.08	2.54

Fig. 14-19 "O" ring flange—straight.

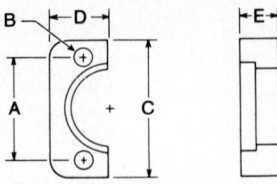

NOM. FLG. SIZE, IN.	A DIM., IN.	B DIM., IN.	C DIM., IN.	D DIM., IN.	E DIM., IN.
1/2	1.50	0.34	2.12	0.91	0.63
3/4	1.88	0.41	2.56	1.03	0.75
1	2.06	0.41	2.75	1.16	0.75
1-1/4	2.31	0.47	3.13	1.44	0.75
1-1/2	2.75	0.53	3.69	1.62	0.81
2	3.06	0.53	4.00	1.91	0.81

Fig. 14-20 "O" ring flange fittings are available with tubes bent to a variety of angles.

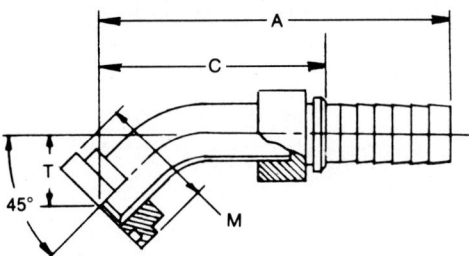

Fig. 14-21 "O" ring flange—45°.

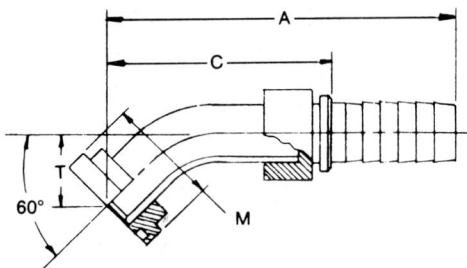

Fig. 14-22 "O" ring flange—60°.

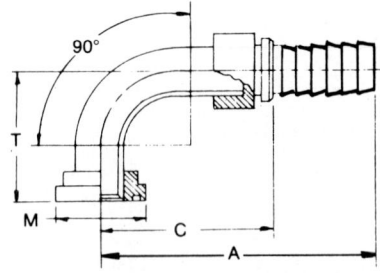

Fig. 14-23 "O" ring flange—90°.

Fig. 14-24 On straight fittings the overall assembly length is measured end to end.

bolts and washers are used. These fittings are available with tubes bent to a variety of angles (Figs. 14-20 to 14-23).

Overall Assembly Length

The method of measuring and identifying the overall assembly length (OAL) has been standardized and, when used properly, can be understood accurately by anyone using it. On straight fittings, OAL is measured from end to end as shown (Fig. 14-24). Angular fittings are measured to the centers of the ends (Fig. 14-25).

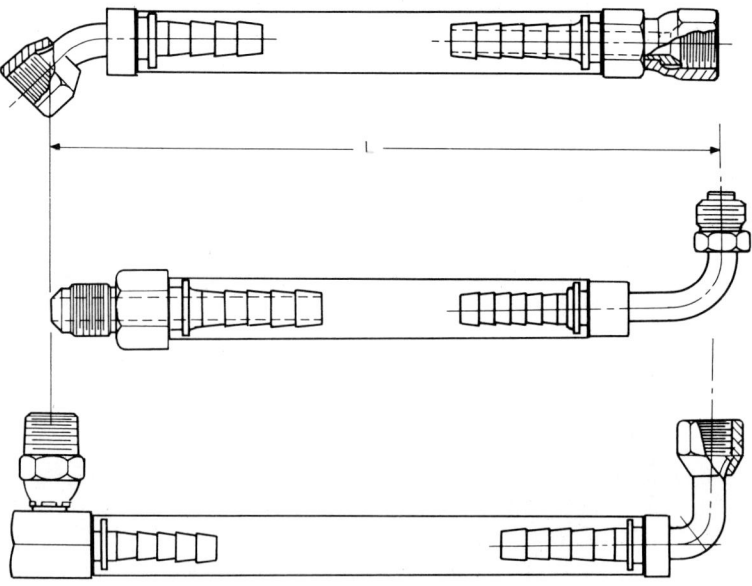

Fig. 14-25 Angular fittings are measured to the centers of the ends.

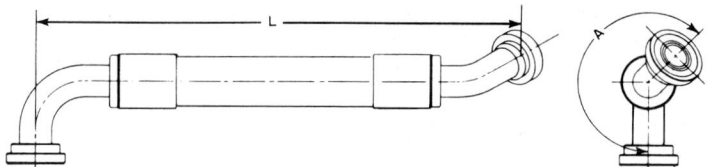

Fig. 14-26 Offset measured with first coupling vertically downward and measured clockwise. Measure from centerline of flanged head to centerline of flanged head for length of assembly.

Another measurement is necessary when two angular couplings are required on the same assembly. It is called the *orientation* and is measured with the first coupling pointing vertically downward; looking at the assembly from the second coupling end, the angle in degrees is measured in the clockwise direction. The length is again measured from centerline to centerline (Fig. 14-26).

Determining the assembly length Hydraulic hose will lengthen or shorten when pressurized and it is important to consider this when determining an assembly length. A hose can shrink up to 4 percent or elongate up to 2 percent, according to SAE standards for higher-pressure hoses, and to 3 percent either way for lower-pressure hoses. Sufficient allowance should be made to permit such changes in length (Fig. 14-27).

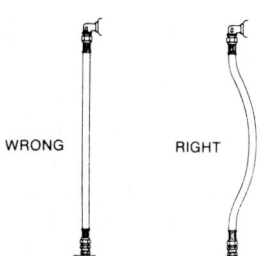

Fig. 14-27 Allow for some slack in hose when determining assembly length.

To cut a hose to the correct length (the so-called cutoff length) to meet an assembly-length specification, coupling lengths must be taken into consideration. In most cases the manufacturer's catalog indicates the amount of hose the coupling replaces. These dimensions must be acquired for each coupling and deducted from the overall length to determine the cutoff length (Figs. 14-28 and 14-29).

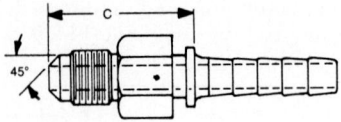

HOSE I.D., IN	THREAD SIZE	HEX SIZE, IN.	BORE SIZE, IN.	COUPLING LENGTH, IN.	C CUT-OFF IN.	STD. PACK (UNITS PER PACK)
1/4	7/16-20	0.50	0.156	2.17	1.08	5
1/4	1/2-20	0.56	0.156	2.23	1.14	5
3/8	5/8-18	0.69	0.232	2.21	1.20	10
3/8	3/4-16	0.75	0.265	2.40	1.39	5
1/2	3/4-16	0.81	0.342	2.88	1.44	5
1/2	7/8-14	0.88	0.406	3.02	1.58	5
3/4	1-1/16-14	1.13	0.625	3.53	1.82	5

Fig. 14-28 SAE solid male. Manufacturer's catalog indicates amount of hose the coupling replaces.

In summary, using the information acquired, a hose assembly may be completely identified as in the following example:

Hose	½ in. 100R1
First coupling	½ in. NPTF solid male
Second coupling	¾-16 JIC swivel female
OAL	9.5 in.

The assembly in the example would resemble Fig. 14-30, which shows permanent couplings, although the example didn't specify them. Either permanent or reusable could have been used.

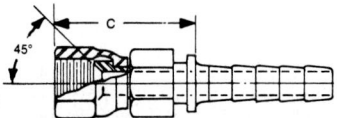

HOSE I.D., IN.	THREAD SIZE	HEX SIZE, IN.	BORE SIZE, IN.	COUPLING LENGTH, IN.	C CUT-OFF IN.	STD. PACK (UNITS PER PACK)
1/4	7/16-20	0.56	0.156	2.31	1.29	5
1/4	1/2-20	0.69	0.156	2.38	1.29	5
1/4	5/8-18	0.75	0.156	2.47	1.38	10
3/8	5/8-18	0.75	0.265	2.39	1.38	10
3/8	3/4-16	0.88	0.265	2.47	1.46	5
1/2	3/4-16	0.88	0.406	2.95	1.52	5
1/2	7/8-14	1.06	0.406	3.04	1.61	5
3/4	1-1/16-14	1.25	0.625	3.39	1.68	5

Fig. 14-29 SAE swivel female.

Maintenance of Hydraulic Hose and Fittings 1-239

Coupling installation equipment Hose and coupling manufacturers provide a variety of tools for coupling installations ranging from small fixtures to high-pressure crimping machines. The reason for the variety is that no one system is right for everyone. A relatively small user might elect to keep an inventory of reusables as opposed to a large user who might find that the saving in permanent coupling inventory cost or value more than offset the cost of the crimping equipment. This can only be determined by working with a hydraulic hose supplier and comparing actual costs.

Cutters. Hydraulic hose can be cut with a standard fine-tooth hacksaw but the high-tensile wire used in medium- and high-pressure hose is tough and hard to cut. The next step up is a converted power handsaw which is generally equipped with a sharp circular blade without teeth. The mounting provides pins to arc the hose, which greatly reduces friction and makes a fast cut possible. Cutters are available specially manufactured for hose; they can get

Fig. 14-30 This completed assembly uses a ½-in. 100R1 hose, with the first coupling a ½-in. NPTF solid male and the second coupling a ¾-16 JIC swivel female. Overall assembly length is 9.5 in.

Fig. 14-31 Hose-cutting saw has sharp circular blade without teeth.

more elaborate, such as the combination cutter and skiver shown in Fig. 14-31. Both cutter and skiver are available as separate machines, and there are skivers available that are less elaborate and less expensive (Fig. 14-32).

Machines are also available to reduce the time and work required to install reusable

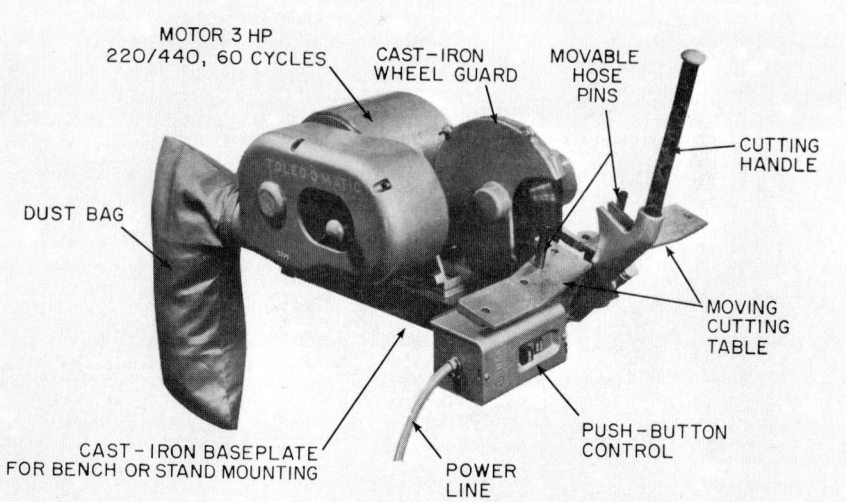

Fig. 14-32 Skiving brush housing equipped with dust opening for connection to exhaust system or dust bag furnished.

1-240　Basic Maintenance Technology

couplings. Some are simply rotating chucks for screw-type couplings. The one shown in Fig. 14-33 has a vise to hold the hose. These machines are a great asset if many reusable assemblies are to be made because so many revolutions are required to complete an assembly and considerable torque is required to make the assembly, particularly on the larger-diameter hoses.

Crimping Machines. For permanent assemblies, a crimping machine is usually required. Crimpers are available from hand-lever-operated types for low-pressure hose to 100-ton hydraulic presses powered by 10,000-psi pumps.

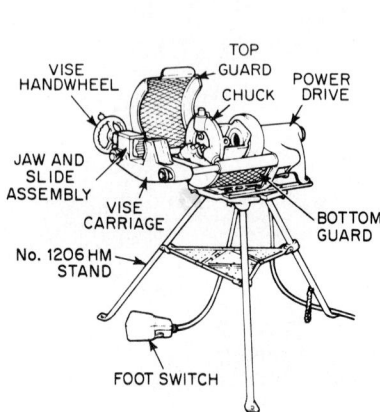

Fig. 14-33 Installing reusable couplings is made easier with machine like this.

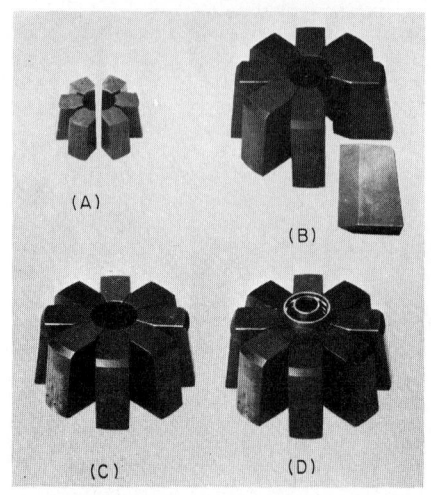

Fig. 14-34 Die fingers for use in crimping machines come in various sizes. (A) Small six-finger set. (B) and (C) Large eight-finger set. (D) Large eight-finger set showing coupling seated inside fingers.

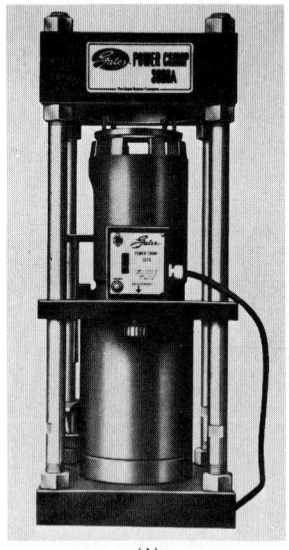

(A)

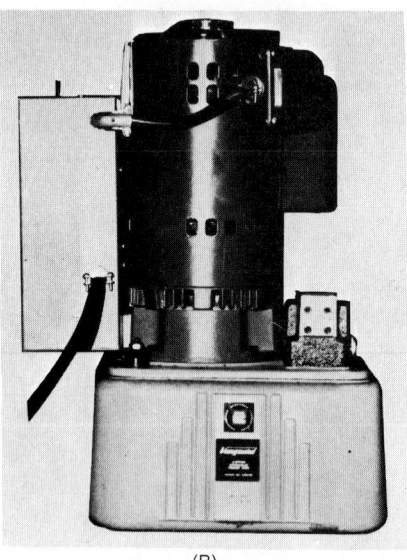

(B)

Fig. 14-35 (A) 100-ton power crimper and (B) 10,000-psi pump.

Maintenance of Hydraulic Hose and Fittings 1-241

Crimping machines usually come equipped with a number of sets of crimping fingers or dies to accommodate a range of hose sizes. Generally, there are six or eight fingers to a set (Fig. 14-34). The crimping faces of the fingers contact the ferrule and squeeze it onto the hose. The surface opposite the finger face is rounded and tapered to fit inside a tapered cone. As the fingers are forced farther into the cone they are forced together uniformly, causing the crimp. The small conical angle provides great mechanical advantage from the press ram through the fingers to the ferrule.

The larger crimping machines shown will crimp hose sizes from 3/16 to 2 in. ID. These machines are generally powered by an electric-driven hydraulic pump. Smaller crimpers are available that will crimp 3/16 to 1 in. ID hose that can be operated using a hand pump, air pump, or electro-hydraulic pump (Fig. 14-35).

A hand pump is often installed with a small crimper on a piece of mobile equipment such as a pickup truck for field crimping. In cases where the truck is equipped with compressed air, an air pump is sometimes used (Fig. 14-36).

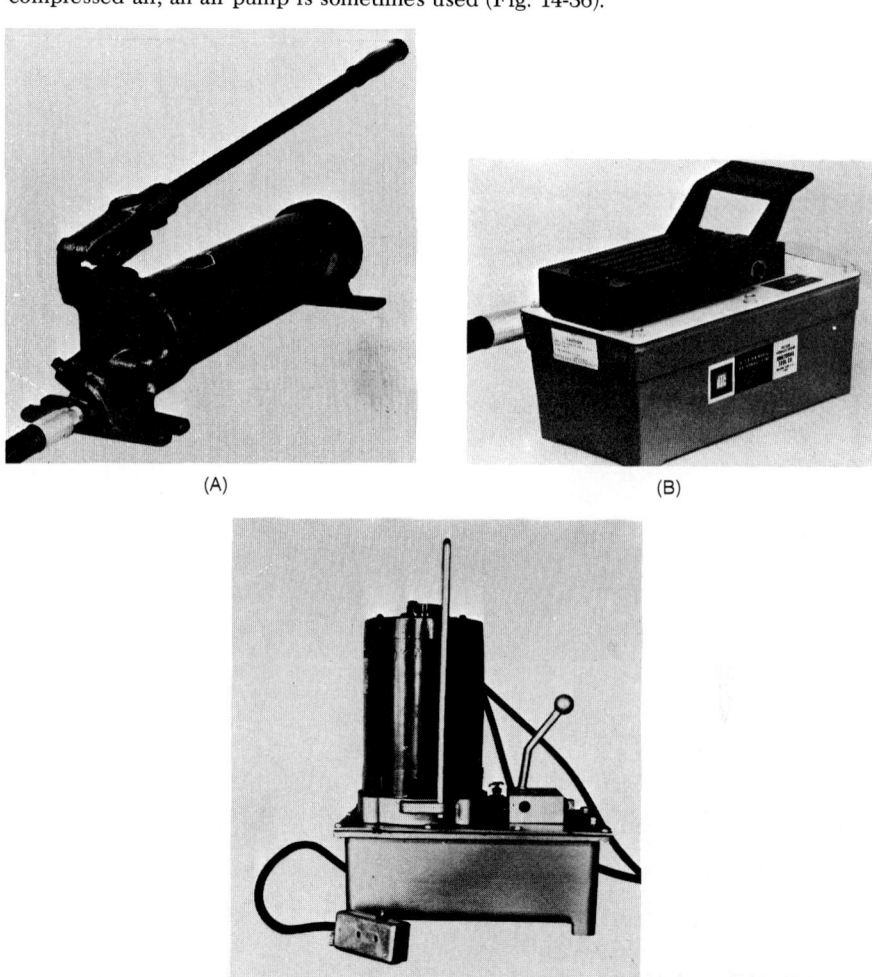

Fig. 14-36 (A) Hand pump—convenient and portable for use anywhere. Makes original equipment manufacturer factory-type assemblies on the spot to minimize downtime. (B) Air hydraulic pump—uses standard air compressor power (90 psi). Convenient and versatile for in-shop use. (C) Electro-hydraulic pump—delivers high oil volume for fast coupling speeds. Perfect for shops doing volume assembly work. Operates on 115 volt ac single-phase current.

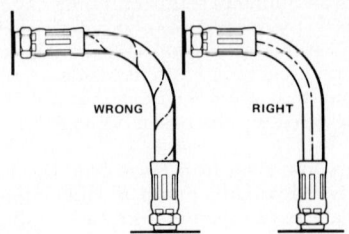
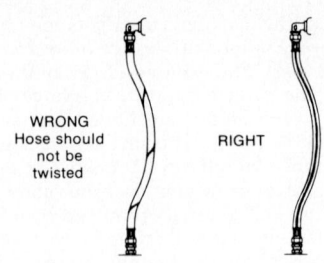

Fig. 14-37 Take care not to twist hose during installation. Correct position can be determined by the printed layline on the hose. Pressure applied to a twisted hose can cause hose failure or loosening of connections. *(Weatherhead Co.)*

Fig. 14-38 Twisting hose will cause threaded fitting to loosen or cause undue stress within hose.

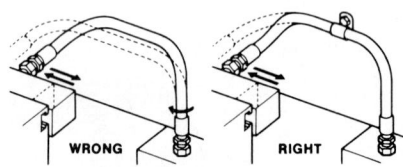
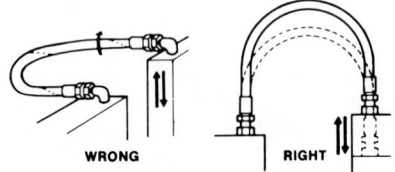

Fig. 14-39 To avoid twisting in hose lines bent in two planes, clamp hose at change of plane (as shown). *(Dayco Corp.)*

Fig. 14-40 To prevent twisting and distortion, hose should be bent in the same plane as the motion of the boss to which the hose is connected. *(Dayco Corp.)*

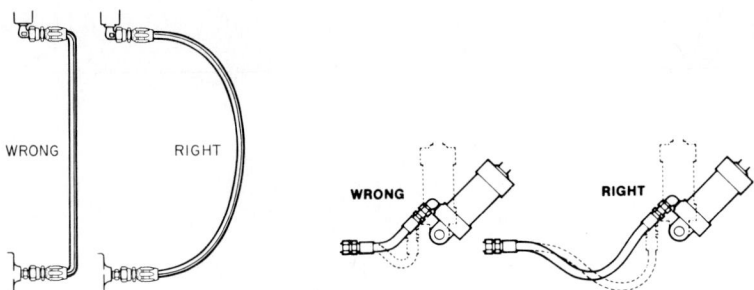

Fig. 14-41 Always avoid bending hose more sharply than the prescribed minimum-bend radius (left). *(Aeroquip Corp.)* Adequate hose length is most important to distribute movement on flexing applications and to avoid abrasion (right). *(Dayco Corp.)*

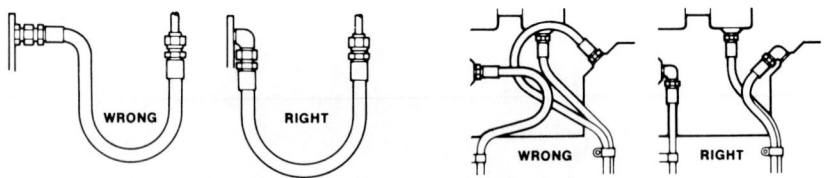

Fig. 14-42 Avoid sharp twist or bend in hose by using proper angle adapters (left). Obtain direct routing of hose through use of 45° and 90° adapters and fittings. Improve appearance by avoiding excessive hose length (right). *(Dayco Corp.)*

Maintenance of Hydraulic Hose and Fittings

Installation of assemblies There are right and wrong ways to install hydraulic assemblies—a wrong installation can loosen threaded fittings and greatly reduce the life of an assembly. The difference between bending and twisting should be recognized, namely, that hydraulic hose can be bent but definitely should not be twisted.

Two basic situations can cause hose to twist: (1) twisting the hose with a wrench during installation, and (2) attaching an assembly end to a moving part in a manner that would cause the hose to twist rather than to bend. Either of these will cause a threaded fitting to loosen or cause undue stress within the hose (Figs. 14-37 and 14-38).

Bending hose into too small a radius is definitely discouraged. Every hydraulic hose has a manufacturer's and/or an SAE minimum-bend radius. This radius is measured to the inside surface of a hose. As an example, a hose having a minimum-bend radius of 5 in. should be bent no tighter than it would be bent if it were wrapped on a drum or pole with a diameter of 10 in. Angle fittings

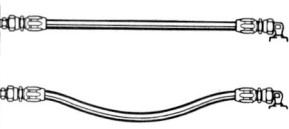

Fig. 14-43 Compensate for length change. *(Aeroquip Corp.)*

and adapters can frequently be used to reduce assembly length and to avoid sharp bends. Such fittings can also be used to dress up the appearance of an installation (Figs. 14-39 to 14-42).

As mentioned earlier, but important enough to repeat, always be sure to leave enough slack in an assembly to allow for the possible +2 to −4 percent change in length when the hose is pressured (Fig. 14-43).

Obviously, good judgment should be used in any installation. Hose must be routed where it will be protected against outside damage as from snagging, rubbing, or contacting hot objects. Aside from this, hose requires little care and no maintenance.

Adapters

Adapters are available for several purposes. The ends of adapters can readily be identified, using the fitting-identification section presented earlier. Most hydraulic hose and coupling catalogs include a selection of the most commonly used adapters and most hose and coupling suppliers stock them.

A common use for adapters is to connect mismatched connections. For example, a hose

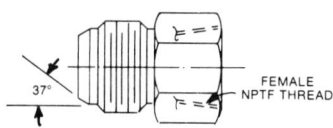

TUBE O.D., IN.	JIC MALE THREAD	NPTF FEMALE THREAD	APPROX. LGTH. HOSE DISPLACED, IN.
1/4	7/16 - 20	1/8 - 27	23/32
1/4	7/16 - 20	1/4 - 18	15/16
5/16	1/2 - 20	1/8 - 27	23/32
5/16	1/2 - 20	1/4 - 18	7/8
3/8	9/16 - 18	1/4 - 18	27/32
3/8	9/16 - 18	3/8 - 18	29/32
1/2	3/4 - 16	3/8 - 18	1
1/2	3/4 - 16	1/2 - 14	1-7/32
5/8	7/8 - 14	1/2 - 14	1-1/8
3/4	1-1/16 - 12	3/4 - 14	1-5/16
1	1-5/16 - 12	1 - 11-1/2	1-15/32

Fig. 14-44 JIC solid male to NPTF solid female.

1-244 Basic Maintenance Technology

with ¾-in. male NPTF thread is to be connected to a tube with ½-in. female JIC thread. An adapter with ¾-in. female NPTF thread on one end and a ½-in. male JIC on the other end would be required (Fig. 14-44).

Another common use is to change the angle of a connection. 45 deg and 90 deg adapters are available and commonly used (Figs. 14-45 and 14-46). A T adapter is used to

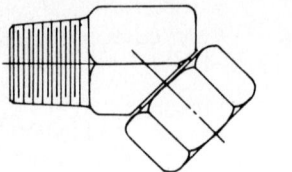

Fig. 14-45 NPTF solid male to NPSM swivel female (45°).

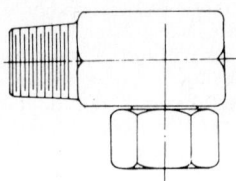

Fig. 14-46 NPTF solid male to NPSM swivel female (90°).

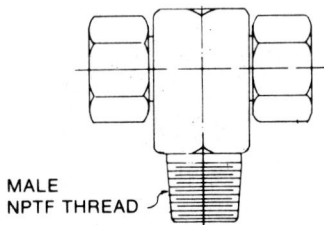

Fig. 14-47 T—NPTF solid male lower end, NPSM swivel female branches.

interconnect an additional line to a circuit. These are manufactured with a variety of end combinations and, usually, the combination required is available (Fig. 14-47).

Innumerable special-purpose adapters can be seen on equipment. They are usually identified by their odd shape, and they are mentioned here only to provide the knowledge that there are nonstandard devices that can be purchased only from an equipment manufacturer or his dealers.

Conclusion

Having read to this point, you know that lack of information can be costly. Fortunately, hose and fitting suppliers have knowledgeable people who are more than willing to assist with any related problem. These people are backed up by factory specialists. Why not take advantage of them?

Chapter **15**

Steam and Hot Water Cleaning

WILLIAM AXELSON
Vice President, The Hotsey Corporation, Englewood, Colo.

The old adage "cleanliness is next to godliness" is an essential principle in the maintenance of construction equipment. Machinery involved in earthmoving, whether building up or tearing down, excavating or lifting, is being used for one of the dirtiest jobs machinery is called on to handle.

Because of this, construction equipment owners and operators have placed tremendous emphasis on proper equipment cleanup. They know that dirt, grime, and grease are some of the tougher opponents of a successful equipment maintenance program.

Experience has shown that
- Dirt can add more than 1000 lb of unneeded weight to an earthmover.
- Dirt and dust can clog an air-cooled power shovel to the point of collapse.
- Dirt can foul a diesel tractor engine into failure.
- Dirt can bring bulldozers to a grinding halt.
- Dirt can ruin hydraulic cylinders, corrode wiring, and destroy components.
- Dirt can make any engine run at excessive temperatures, increasing both fuel consumption and the possibility of cracked blocks.
- Dirt can be harmful to employee morale and can damage a company's reputation.

All these factors contribute to the fact that dirt, grime, and grease can cripple construction equipment performance and dramatically increase operating costs.

HOW TO CLEAN EQUIPMENT

Economics dictates that a construction equipment supervisor should utilize a regularly scheduled machinery cleaning program to spot potential problem areas before they become serious. Also, machinery must be clean before servicing, repairing, or painting.

Statistical evidence has shown that cleaning equipment is a chore that costs time and money. Research indicates that a mechanic in an equipment repair shop spends 20 percent of his time cleaning parts at his work bench or cleaning so that parts can be either removed or replaced. That is 1 hr out of every 5 spent on the job. For this reason, construction equipment operators should implement some type of routine machinery cleaning program.

Pressure Washer and Steam Cleaner

The use of a high-pressure hot water washer or steam cleaner will quickly and efficiently remove crusty, gritty, or corroded buildups of dirt, grease, and grime from transmissions, track and roller assemblies, shovel hoists and crowds, engine blocks, drive trains, and dozer buckets. Once the equipment is clean, the proper repairing, welding, and painting can take place immediately. Hours of wiping, swabbing, and wirebrushing can be reduced to minutes of cleaning when a pressure washer or steam cleaner is used.

In actuality, cleaning construction equipment is similar to washing your hands. You need four basic elements:
- Water
- Pressure
- A chemical agent (if the surface is greasy)
- Heat (150 to 210°F to activate the chemical and lower the viscosity of the oil)

The more there is of each element, the more effective the cleaning job will be. And the pressure washer and steam cleaner provide these basic elements.

There are several important factors to be taken into consideration when setting up a routine cleaning program.

Portability versus permanence An equipment superintendent must evaluate carefully *where* the majority of cleaning will be done. If most of the cleaning will be done in the shop (or adjacent to it), a permanent installation will do the best job. In this situation, the pressure washer's heat and pressure units can be located remotely and multiple cleaning stations located in areas of the yard or shop where cleaning will be done.

If the majority of the cleaning will be done in the field and the construction site, then a portable cleaner mounted on a pickup truck would be preferable.

Ideally, a construction equipment concern should have both systems. As explained below, on-site cleaning is just as important as cleaning at the shop prior to repair.

Size of washer or cleaner Cleaning energy is determined by a simple formula.

$$\text{Amount of water discharged} \times \text{pressure} = \text{cleaning energy}$$

Industry agrees that a cleaner that discharges 120 gph at 500 psi is fine for light-duty cleaning. A machine that delivers 210 gph at 650 psi offers more impact for medium-duty cleaning such as degreasing engines or cleaning farm equipment.

Heavy-duty equipment requires a heavy-duty washer, one that can deliver 240 to 540 gph at 1000 to 2000 psi. A heavy-duty unit that produces 540 gph at 2000 psi generates an astonishing 1 million units of cleaning power.

Experience shows that construction equipment maintenance supervisors should consider cleaning equipment that can deliver at least 240 gph and 1000 psi. This will provide enough cleaning power to blast loose the heavy accumulations of dirt, grime, and grease that can rob a machine of its usefulness.

ESTABLISH A CLEANING PROGRAM

Heavy construction equipment operators and maintenance personnel should adopt a regular cleaning program that will, over the long term, save machinery repair costs, reduce labor time, and save on fuel costs.

Based on in-the-field experience, a guideline to follow includes daily on-site cleaning, on-site inspection, and cleanup before maintenance.

Daily, on-site preventive cleaning When a truck-mounted washer or cleaner is used, equipment should be cleaned each day. A few minutes' cleaning will greatly extend machinery life. It will rinse radiators and oil coolers, free crawler tracks, clear air intakes, unclog power trains, clean cement batchers—in short, cleaning will help assure optimum service from equipment that takes terrible punishment. Daily cleaning also makes it easier for an inspector to spot the more visible trouble areas before they occur.

On-site inspection of machinery Daily cleaning of construction equipment as a part of preventive maintenance will help catch major problems. But all heavy construction equipment should be thoroughly inspected in the field after each 600-hr operational cycle. A few minutes' cleaning of the forklift, backhoe, earthmover, front end loader, crane, or dump truck will prepare it for a careful going over.

Such a cleaning-inspection routine, in the field, can mean the difference between downtime and optimum performance. And it can help keep equipment where it belongs—in the field. Field inspection saves time because the equipment need not be taken back to the shop for inspection. And, using a pressure washer or steam cleaner, an inspector need not fear overlooking a metal fissure, cracked block, hydraulic or oil leak, worn V-belt, frayed cable, corroded battery, or a bent tie rod.

Cleanup before maintenance Should the routine daily cleaning procedure or on-site inspection reveal a problem area, the machinery should be taken to the shop for repairs. Most companies have developed shop procedures to a point where successful repairs are virtually assured. One thing these shops all have in common is a clean machine before repair gets under way.

Pressure washers and steam cleaners quickly and efficiently remove crusty, gritty, and corroded buildups of dirt, grease, and grime from transmissions, track and roller assemblies, shovel hoists and crowds, engine blocks, and drive trains. Once dirt and grime have been removed, repairing, welding, and painting can take place immediately.

Because of the unique flexibility of a cleaning system, an equipment maintenance supervisor can also

1. Add acid to phosphatize parts
2. Clean metal chips out of crevices in parts
3. Sanitize machinery
4. Deodorize equipment
5. Spray insecticides to control insect populations

WASHING TIPS AND TACTICS

As the cleaning program is implemented, several helpful hints can make machinery cleaning more effective.

1. Take heavily greased equipment to a grease pit for washing so that effluent can be properly contained according to environmental considerations.
2. Begin cleaning machinery from the bottom up. Work progressively upward so that soap will not run down dry surfaces.
3. Always rinse from the top down so upper surfaces stay clean.
4. Use a water-soluble oil when rinsing sensitive machine surfaces that are susceptible to rust. As the water evaporates, a thin film of oil is deposited which inhibits the rusting process.
5. Make sure electrical outlets stay dry. The same goes for switch boxes or terminals. Spray engine components such as distributor caps, generator and battery terminals, and spark plug connectors with silicone to promote water repellency.
6. When cleaning parts or assemblies in the shop, spray toward drains. When drains are far apart, remove flushed water with a squeegee.
7. Don't use hot water to remove earth, rock dust, or attached soil—use cold water to save heat and conserve fuel.
8. Clean greasy, oily surfaces with hot water and a detergent.
9. Various cleaning gun nozzles provide different spray patterns for specific cleaning applications. A 0-deg nozzle provides a solid, forceful stream of high-pressure water. When it is necessary to blast loose tough accumulations of dirt and grime, 0 deg is most effective. A broader range (up to 40 deg) is most advantageous for cleaning large, flat surfaces.
10. Hold the cleaning wand 6 to 12 in. from the surface for best results.
11. The trigger shutoff on the pistol grip of the cleaning gun allows for fast water cutoff, thus saving water and energy.
12. A so-called plumbed-in central cleaning system for in-shop use has several advantages. First, the heat and water pressure are generated in a central location. Second, simple hose-connection cleaning stations can be positioned in the most appropriate locations or equipment bays. This will enable the construction equipment operator to fashion a system of the exact specifications required, including volume, temperature, pressure, and additional chemicals.
13. Always remember: Use a simple two-step process when cleaning with a pressure washer or steam cleaner—apply water (and chemicals or detergents if necessary) and rinse the surface clean.

HOW HEAVY-DUTY CLEANERS OPERATE

There are two primary methods of cleaning construction equipment: the steam cleaner and the hot water high-pressure washer.

Steam cleaner This machine produces very wet or saturated steam, with or without chemicals, at the nozzle of a discharge gun. Saturated steam is produced by superheating water to develop pressure in a partially restricted water-heating coil. As the heated, pressurized water leaves the discharge nozzle and enters the atmosphere, it explodes into tiny droplets of saturated steam (Fig. 15-1).

The steam cleaner operates in the following way: Water is pulled from a float tank and is pumped into the water-heating coil. Detergents may be added to the water if desired. A restricting nozzle at the end of the steam gun maintains a small internal back pressure and keeps the coil full of water. When the volume of water being pumped is known, a certain amount of heat can be added to superheat the water in the coil. As heat is applied, the water temperature in the coil rises and the water expands, causing an increase in internal pressure. This expansion of water develops pressure in steam cleaners, not a water pump.

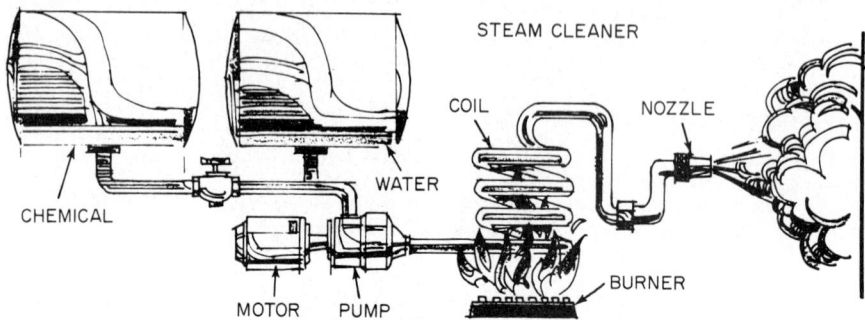

Fig. 15-1 Arrangement of a saturated steam cleaning system.

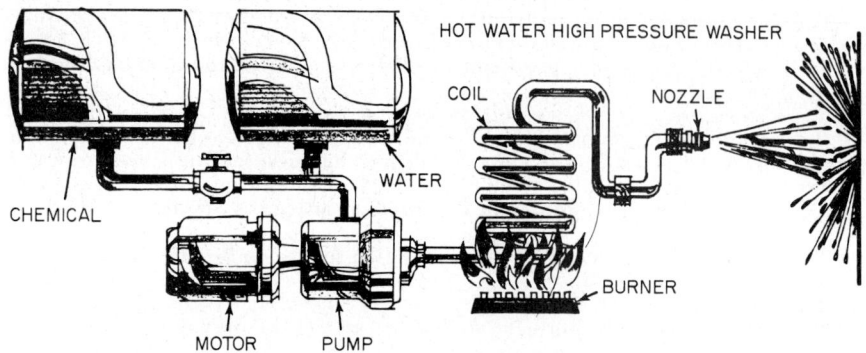

Fig. 15-2 Arrangement of a hot water high-pressure cleaning system.

The boiling point of water rises under pressure, and, when the water is heated to 330°F, the operating pressure is 100 psi. But, as the superheated water hits the atmosphere, it instantly dissipates a tremendous volume of heat or energy. This rapid loss of heat yields clouds of steam vapors (which can inhibit vision in cool weather).

Hot water high-pressure washer This washer uses a water pump to increase pressure tremendously. The pump, the heart of the pressure washer, pushes the volume of water and creates high pressure. The water in the pressure system need not be superheated, but is controlled at temperatures less than 212°F. The discharged water remains in a solid state and flows in a stream with very little surface area. The water temperature leaving the nozzles of both the steam cleaner and the high-pressure washer is less than 212°F since boiling water will not exceed 212°F at atmospheric pressure (Fig. 15-2).

ADDITIONAL CLEANING EQUIPMENT

In addition to a basic cleaning system, one should consider three additional pieces of equipment, the parts washer, the SandJet and the ChemJet, which can further reduce labor costs and save time. Tank cleaners and degassers should be considered if bulk tanks need frequent cleaning.

Parts washer This piece of equipment can greatly ease the job of construction equipment mechanics and repair specialists who deal with dirty or greasy parts. Washers speed the cleaning job and are easy, clean, and safe to use. They make fast work of hard-to-clean parts such as pistons, bearings, carburetors, gears, valves, and seals.

A parts washer should feature large cleaning tanks (18 × 36 in.) and a special safety lid, which will close automatically and shut off the pump in case of fire.

The washer should be operated automatically by a switch which turns on the pump when the lid is raised. The pump should be located in the fluid reservoir to assure a constant supply of solvent.

When the lid is closed, the pump will shut off automatically, assuring that no solvent fumes can escape when the cleaner is not in use. A drain shut-off valve allows the tank to be filled for soaking extra-grimy parts.

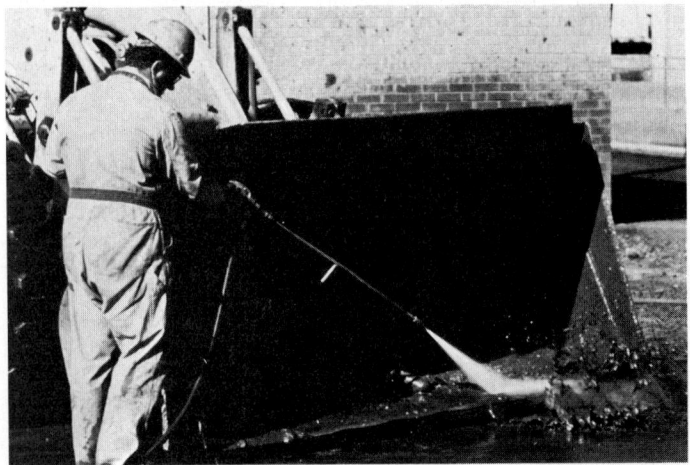

Fig. 15-3 Steam cleaning proves useful in removing dirt or grime from dozer buckets. This cleaning reveals cracks or metal fissures.

SandJet This machine hooks up to a pressure washer and offers an excellent alternative to sandblasting. Because the attachment works with the washer, the sand runs off the surface being cleaned for easy recovery and reuse. There is no dust pollution because of the combination of sand and water and users do not have to screen areas to keep airborne sand under control.

One person can operate the SandJet, and the only protective clothing needed is a pair of goggles. The SandJet can be used to clean steelwork, crane booms, dippers and buckets, masonry, tanks, boilers, engine parts; it can also be used to get any metal ready for painting such as heavy machinery, trucks, tractors, fork lifts, and draglines.

ChemJet This machine also easily attaches to the wand of a pressure washer. A vacuum is created by the washer and it draws any nonviscous fluid and mixes it with the pressure washer's water flow. This chemical applicator has many uses, including phosphatizing and etching raw steel, brightening raw aluminum, stripping paint, and cleaning timbers and supports.

Tank cleaners and degassers These pieces of equipment are useful to construction equipment operators who often need to clean tanks thoroughly (especially fuel cells and bulk storage tanks) before repairs are made. This is an especially important safety procedure when tanks are welded or braised. Use of a tank degasser cuts cleaning time and eliminates a great deal of downtime.

Using a five-phase program involving steam, heat, rinsing, heating, and flushing, a cleaner can completely remove tank contamination as determined by a zero reading on a so-called sniffer instrument.

This specialized cleaning system can be used in degassing tanks for welding and repair

Fig. 15-4 Job-site earth, entrapped in treads and undercarriages of tracked vehicles, adds hundreds of pounds of useless weight to equipment, increasing fuel consumption and reducing engine life. A sound pressure wash relieves this problem.

Fig. 15-5 Tires, like tracks, need to be dirt- and grime-free to enhance life. Consider this in relation to current tire replacement costs.

purposes and in rendering a tank clean for storage or when tank fuel loads must be changed (diesel and gasoline, for instance).

All construction equipment maintenance programs should include a cleaning procedure. This will be an investment with wide-ranging benefits for companies that have to service and maintain expensive equipment. Overwhelming evidence supports the contention that adherence to a cleaning program will help assure long, economical machinery life (Figs. 15-3 to 15-5).

Section **2**

Maintenance of Power Systems

Chapter **1**

Maintenance of Electrical Power Systems

JESS JOHNSON
Kohler Co., Kohler, Wis.

Popular, independent electrical power supply systems at construction sites today are the small, portable gasoline-engine-driven generators in the 500- to 5000-watt range. These lightweight air-cooled units provide ac and/or dc electrical energy for power tools, lighting, floodlighting, warning flashers, and welders at sites remote from commercial power sources. This chapter will be devoted almost entirely to the maintenance of these small portable generator sets.

Generator sets are built to withstand the harsh environment prevalent at most construction sites, but to remain reliable, they require a specific amount of attention at the intervals recommended by their manufacturers (Fig. 1-1). All too often, the unit that performs most reliably does so unnoticed and becomes, unfortunately, the unit most likely to be neglected in terms of service until it's too late to prevent major problems. For example, a simple task such as checking oil level before each start-up may not be done and costly internal damage to the engine may result. Such damage might occur because no one in the crew has been assigned direct responsibility for servicing the unit. A good maintenance program costs very little in time and material, yet lack of proper service leads to costly repairs and, perhaps most important to a contractor, costly downtime.

Maintenance of engine-driven generator sets can be divided into two categories: general routine service and preventive maintenance service.

Routine Service

As can be seen in Table 1-1, the routine service requirements for a typical portable engine-driven generator set are minimal and not difficult to perform. They are important, however, and the person assigned responsibility for servicing a unit should keep an accurate operating hour-service log to record when required services were performed. A sample of a typical service log is shown in Fig. 1-2.

Under the dusty, dirty operating conditions normally encountered at most construction sites, the importance of changing oil and servicing the air cleaner at the prescribed intervals cannot be overemphasized. A dirt-clogged air cleaner contributes to an overrich fuel mixture, leading to formation of harmful sludge in the crankcase of the engine. Unfiltered air entering through loose, missing, or improperly installed air cleaner components quickly causes internal damage—as little as one small teaspoonful of dirt thus introduced can ruin piston rings and the cylinder bore in just a few minutes.

2-2 Maintenance of Power Systems

Even when not contaminated by dirt, lubricating oil absorbs the contaminants which are normal by-products of combustion and which can eventually deteriorate and harm the engine. The best protection is to change oil at specified intervals.

Dust and dirt can also accumulate on air intake screens, cooling fins, and cooling air outlets so that damage could occur because of overheating. To avoid this, a generator set should be operated in an area free of chips, dust, and dirt. As a further precaution, intake screens should be cleaned daily before each start-up, and an operating unit should be stopped and cleaned whenever buildup is noted. Accumulation on brushes and commutator should be blown out at frequent intervals with dry compressed air to allow full output of the generator.

Fig. 1-1 Typical application of engine generator on construction site.

Electrical connections on a generator set should be checked and tightened periodically to prevent shorting and possible damage to electrical devices inside the control box. At the same time, safety guards should be checked to assure that all are in place and securely tightened.

TABLE 1-1 Routine Service Schedule

Frequency	Service
Every day	Check oil level
	Clean air inlets and outlets
	Check fuel filter
	Replenish fuel supply
	Service air cleaner (if dirty)
Every 25 operating hr	Change crankcase oil
	Service air cleaner
	Service fuel filter
Every 50 operating hr	Clean external surfaces
	Clean commutator
	Retighten electrical connections
Every 100 operating hr	Service spark plugs
	Check breaker points
	Replace air cleaner

Preventive Maintenance

In addition to routine services, a generator set should receive preventive maintenance of tuneup services at periodic intervals. Benefits of such services will not only be noted immediately in improved performance but in continued satisfactory performance throughout an extended service life. Such services are often best performed at service centers authorized by a generator manufacturer, but they are not too numerous or too difficult to

OPERATING HOUR - SERVICE LOG

The following is provided to help you keep an accumulative record of operating hours on your generator set and the dates required services were performed. Enter hours to the nearest quarter hour.

DATE RUN	OPERATING HOURS		SERVICE RECORD		DATE RUN	OPERATING HOURS		SERVICE RECORD	
	HOURS RUN	ACCUMULATIVE	DATE	SERVICE		HOURS RUN	ACCUMULATIVE	DATE	SERVICE

Fig. 1-2 Typical service log.

be performed by an experienced member of a maintenance crew. Under normal conditions, a generator set should be tuned up about every 500 hr of operation. A typical tuneup on a small air-cooled engine-powered generator follows.

TUNEUP RECOMMENDATIONS

Exterior: Thoroughly clean, especially cooling fin and air intake areas
Test: Crankcase vacuum and/or compression
Cylinder head: Remove and clean combustion chamber
Air cleaner: Service oil bath or replace element on a dry type
Spark plug: Replace
Breaker points: Replace
Condenser: Replace
Valves: Check valve to tappet clearance and adjust as needed
Breather: Service
Ignition: Check and adjust timing
Carburetor: Restart engine and adjust under load

When a generator set is in for tuneup, this is a good time to look it over closely; there are usually a number of indicators that tell you what condition it's in. Service hints can often be given to the operators as a result of such observations. For example, the condition of the electrodes and spark plugs is an excellent indicator of operating conditions.

2-4 Maintenance of Power Systems

Spark plug analysis When removing spark plugs, always check the firing end as the appearance here gives a very good indication of operating conditions. If abnormal conditions are indicated, always check the number of the removed plug—it may be of the wrong heat range for the engine. If the center electrode is worn round, don't try to square it with a file for reuse—replace the plug to prevent the misfiring often encountered when using worn plugs. Some common firing-end indicators are listed in the following guides for spark plug analysis.

Normal. A plug taken from an engine operating under good conditions will have light- or gray-colored deposits. If the center electrode is not rounded off, a plug in this condition could be regapped and reused.

Worn-Out. On a plug which has been in service too long, the center electrode will be rounded off and the gap will be worn 0.010 in. more than the original setting. Replace worn plugs since they require excessive voltage to fire properly.

Carbon-Fouled. Soft, sooty black deposits indicate incomplete combustion from rich carburetion, weak ignition, retarded timing, or poor compression.

4-CYCLE AIR COOLED ENGINES – TROUBLE SHOOTING GUIDE

PROBLEM	FUEL RELATED CAUSES			IGNITION CAUSES		OTHER CAUSES						
	NO FUEL	IMPROPER FUEL	FUEL MIX. WRONG	NO SPARK	POOR IGNITION	IMPROPER COOLING	IMPROPER LUBRICATION	POOR COMPRESSION	VALVE PROBLEMS	CARBON BUILD-UP	GOVERNOR FAULTY	ENGINE OVERLOADED
WILL NOT START	X			X				X	X			
HARD STARTING		X	X	X	X			X	X			
STOPS SUDDENLY	X			X			X		X			
LACKS POWER		X	X	X	X			X	X	X		X
OPERATES ERRATICALLY		X	X	X							X	
KNOCKS OR PINGS		X	X			X				X		X
"SKIPS" OR MISFIRES				X	X							
BACKFIRES				X	X				X			
OVERHEATS				X	X	X			X			X
IDLES POORLY				X	X							

Fig. 1-3 Typical engine troubleshooting guide.

Wet-Fouled. A wet-fouled plug could be caused by drowning with raw fuel or oil in the combustion chamber. The raw fuel problem may be caused by operating with too much choke. Oil in the combustion chamber area is usually caused by worn rings or valve guides.

Overheated. Overheating is indicated by chalk-white-colored deposits, not burned black as might be expected. This condition is also usually accompanied by excessive gap erosion. Overadvanced timing, lean carburetion, clogged air intake, or blocked cooling fins are some of the causes of overheating.

Tests Crankcase vacuum and compression tests should be made on engines brought in for tuneup (Fig. 1-3). These tests and checks are described as follows.

Crankcase Vacuum Test. A partial vacuum should be present in the crankcase when an engine is operating at normal temperatures. An engine in good condition will have a crankcase vacuum of 5- to 10-in. water column as read on a U-tube water manometer or ½ to 1 in. Hg as calibrated on a mercury vacuum gauge. A crankcase vacuum check is best accomplished with a U-tube manometer. If the vacuum is not in the specified range, consider one or more of the following factors—the condition easiest to remedy should be checked first:

1. Clogged crankcase breather can cause pressures to build up in the crankcase. Disassemble breather assembly and thoroughly clean; then, recheck pressure after reinstalling.

2. Worn oil seals can cause lack of vacuum. Oil leakage is usually evident around worn oil seals.

3. Blow-by, leaky valves can also cause positive pressures. These conditions can be confirmed by making a compression test on an engine.

When using a manometer, place the cork end into an oil fill hole (the other end is open to the atmosphere) and measure the difference between columns. If the water column is higher in the tube connected to the engine, vaccum or negative pressure is indicated. If the higher column is on the atmospheric side of the manometer, positive pressure is present.

Compression Test. The results of a compression check can be used to determine if an engine is in good operating condition or if reconditioning is needed. Low readings can indicate several conditions or a combination of different conditions (Table 1-2).

TABLE 1-2   Low Compression Troubleshooting Guide

Possible cause	Remedy
Cylinder head gasket blown	Remove head, replace gasket, reinstall head, recheck compression
Cylinder head warped or loose	Remove head, check for flatness (see cylinder head service), reinstall and secure in proper sequence to specified torque value
Piston rings worn—blow-by occurring	Recondition engine
Valves leaking	Recondition engine

Higher-than-normal compression can indicate that excessive carbon deposits have built up in the combustion chamber. To check compression, remove spark plugs and run engine up to a speed of at least 800 rpm. Be sure the air cleaner is clean and exhaust is not restricted before checking compression. Set throttle and choke wide open, insert gauge in spark plug hole, and take several readings. Consistent readings in the 110- to 120-psi range indicate good compression. Reconditioning is indicated if readings fall below 100 psi. On two-cylinder engines, take readings on both cylinders.

Cylinder head service   To maintain top operating efficiency and performance, cylinder heads should be removed and carbon cleaned out about every 500 hr of operation. This service should be included in every tuneup. Carbon buildup can be especially heavy in larger-engine models which are often run at reduced load. Constant speed operation also seems to cause increased accumulation of carbon in the combustion chamber. When removing carbon, use a piece of wood or plastic to avoid scratching the aluminum, particularly in the gasket seat area. Always use a new cylinder head gasket and tighten capscrews in the proper sequence and to the torque valve specified in the manufacturers' service manuals when reinstalling cylinder heads.

Air cleaner   Check the condition of the air cleaner on units brought in for tuneup. If poor service is indicated by clogged, dirty elements or improper installation, look for worn-out piston rings or sludge deposits in the oil pan. Service oil bath cleaners or replace dry-type air cleaner elements as part of the tuneup. The service recommendations on some typical air cleaners are as follows:

Dry Air Cleaner.   Elements should be replaced when power loss is noted or after 100 to 200 hr if engine is operated under good clean air conditions—service and replace element more frequently under extremely dusty or dirty conditions. Dry elements should be cleaned after about 50 hr of operating—remove the element and tap lightly on a flat surface to loosen dirt. Replace the element if dirt does not drop off easily. Do not wash dry elements in any liquid or attempt to blow dirt off with an air hose as this will puncture the filter element. Carefully handle a new element—do not use if gasket surfaces are bent or twisted (Fig. 1-4).

Oil Bath Air Cleaner.   If operating under extremely dusty conditions, it may be advantageous to install an oil bath air cleaner in place of a standard cleaner, thus eliminating the need for frequent replacement of the dry element. Normally converting to an oil bath cleaner involves removal of the dry-type cleaner and installation of an elbow and the oil bath unit in its place. The oil bath cleaner should be serviced after every 25 hr of operation; however, if extremely dusty or dirty conditions exist, service the cleaner

2-6 Maintenance of Power Systems

TABLE 1-3 Generator Troubleshooting Guide

Problem	Possible cause	Suggested remedy
No output	Loose terminal connections	Check for loose or bad connections and tighten all brush connections.
	Brushes not seated	Check for loose springs or brushes sticking in holder. Correct any cause for brushes not riding properly.
	Dirty commutator	Poor contact caused by buildup of dirt or oily film on commutator collector rings. Clean with coarse cloth or fine sandpaper or stone (don't use emery paper).
	Residual magnetism lost	After long periods of storage, it may be necessary to "flash" generator to restore magnetism. To do this, lift all brushes off commutator, connect positive (+) battery to positive generator terminal, then momentarily touch negative (−) battery to negative generator brush.
	Short in ac circuit	If engine labors while running or jerks while being cranked, check for short circuit in ac line. If short develops in ac armature, the armature will get very hot.
Low output or excessive drop in voltage	Engine speed too low	Check with tachometer. No-load speed should be about 3750 rpm; speed under load, 3600 rpm. Readjust governor speed.
	Overload	Make sure plant capacity is not being exceeded. Reduce load to 1500 watts; or, at extreme altitudes, derate plant about 3% per 1000 feet of elevation.
	Wrong ac lines	If external leads are too small or too long, excessive resistance is created which reduces output. Shorten lines or use larger-gauge wire.
	Engine in poor condition	Poor compression, excessive carbon, faulty ignition, wrong polarity, or any other condition causing poor performance may show up in reduced output.
	Brush angle wrong	If brush ring shifts or is positioned wrong when installed, brushes will be out of neutral zone, resulting in low output and/or excessive arcing. Position notch in brush ring 57° above horizontal plane (about 5/16 in. above end bracket leg).
Excessive arcing	Brushes sticking	If brushes are too large or too small, they may stick or cock in holder and chatter. Use proper size.
	Brush tension wrong	If spring tension is wrong, brushes may chatter. Adjust tension.
	Wrong brushes	Brush grade and material must be correct—use only specified brushes.

Maintenance of Electrical Power Systems 2-7

more frequently—even every 8 hr or twice daily if conditions warrant. To service normal-capacity oil bath air cleaners, remove the wing nut and remove the air cleaner components as a unit (Fig. 1-5).

Fuel system services Tuneups should include a complete check of the fuel system, including reconditioning of the carburetor and/or fuel pump if needed. Check fuel lines; replace hose in bad condition. Service the fuel filter if an engine is so equipped. Test the

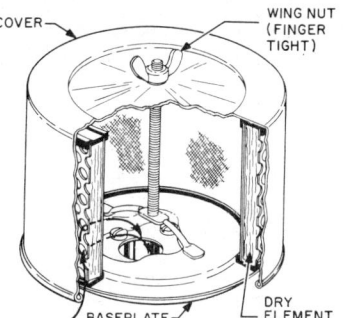

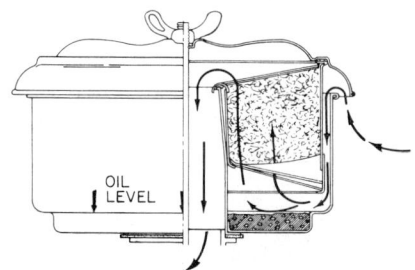

Fig. 1-4 Cutaway view of a typical dry-element air cleaner.

Fig. 1-5 Cutaway view of a typical oil bath air cleaner.

engine under load to check carburetor and pump. If readjustment of carburetor does not restore proper idle or operation, it should be reconditioned.

Exhaust smoke indicators Many people do not realize that even an engine in good condition will consume a certain amount of oil. As engine hours accumulate, ring and bore wear will result in even higher consumption. While blue exhaust smoke could indicate excessive consumption due to wear, it can also be caused by diluted oil or by operating with too much oil in the crankcase.

Generator service Generators do not normally require service on a regular basis. However, it is a good idea to remove the end cover and check the commutator and brushes at least every 6 months or every 50 hr or more often under dusty, dirty conditions. Visually check the commutator first—if a thin skinlike film of uniform thickness is evident on the surface, it is an indication of normal operation. The film acts as a lubricant and promotes longer brush life. If the surface is streaked or has ridges of dirt, clean it with a coarse cloth or, if this doesn't work, use fine sandpaper or a commutator stone—do not use emery cloth. Lift brushes and check their surfaces—replace them if unevenly worn or if worn down to about one-half their original length. Other common causes for rapid brush wear are wrong brush tension, rough commutator surface, high mica on commutator, or brush chatter. Blow dust out with dry compressed air after servicing the generator (Fig. 1-6).

Fig. 1-6 Checking generator brushes.

Some common generator problems that may be easily detected and often corrected without test instruments are stated in Table 1-3, which is a troubleshooting chart. If routine service and suggested corrective action fail to solve a problem, contact a qualified technician to locate and correct the generator malfunction.

Chapter **2**

Maintenance of Diesel Power Systems

VERNON D. HAGELIN[1]
Technical Services, Deere & Company, Moline, Ill.

Need for preventive maintenance of diesel engines cannot be too highly stressed. Regular servicing and prompt attention to warnings of trouble will help prevent costly major repairs, keep necessary repairs from cutting into prime working time, improve operating efficiency, and reduce fuel costs.

Adequate records of filter and oil changes and of other services which should be performed periodically are an important aid to proper servicing and preventive maintenance. Such records should be kept with the machine on which the engine is being used or in the service department.

Many companies make diesel engines, but no two use exactly the same combination or design of cooling and fuel-injection systems, filters, governors, and other components.

This chapter concentrates on the fundamentals of preventive maintenance, common to most diesel engines, which the engine owner or operator should understand. Repairs which involve major engine or component disassembly should be performed only by expert technicians.

Manufacturers' manuals are a must for proper maintenance. If these manuals specify measures more stringent than those recommended here, *follow the manuals.*

The chapter concludes with a general troubleshooting chart. Some sections on various systems also give suggestions for finding malfunctions which are obviously in those systems.

The following are areas of preventive maintenance to be discussed:
- Visual inspection
- Starting engine
- Stopping engine
- Fuel system
- Cooling system
- Lubrication system
- Air intake and exhaust
- Electrical system
- Diagnosing malfunctions

VISUAL INSPECTION

Daily visual inspection of an engine is an important part of adequate preventive maintenance. It is also recommended before any tuneup procedure is started.

[1]The author is now retired.

Maintenance of Power Systems

Oil leaks
An external leak could be the cause of using too much oil.
  Check oil level.
  Look for leaks at the oil pan, drain plugs, front and rear seals, and gaskets.
  Check the coolant for oil contamination.
  Check the oil-cooler top and bottom covers.
  Check oil-filter housings.

Coolant leaks
  Check coolant level.
  Look for coolant leaks at the radiator, water pump, hoses, and oil-cooler inlet and outlet.
  Periodically examine all water hoses for softening, swelling, hardening, and cracking, any of which indicate need for immediate replacement.

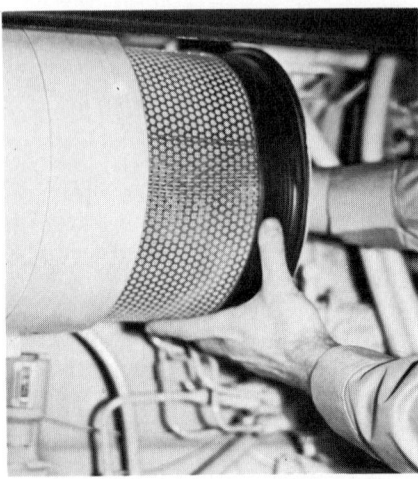 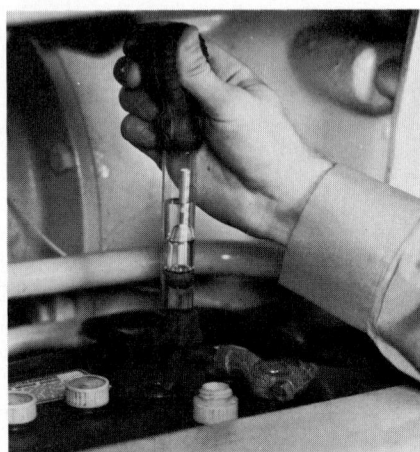

Fig. 2-1 Check the air filter.   **Fig. 2-2** Check battery electrolyte level.

  Check the radiator for trash buildup, bent fins, kinks, dents, fractured seams, and cracked tubes.
  Be sure fan blades are straight and far enough from the radiator to prevent striking the core.
  Be sure the fan belt is in good condition and under proper tension.

Fuel system
  Inspect fuel-tank seams and fuel-pump inlet and outlet connections for leaks. If the fuel-supply pump has a primer level, be sure it is in the lowest position.
  Examine fuel-filter inlet and outlet connections.
  Check high-pressure fuel-supply connections to make sure none is twisted, kinked, broken, or leaking.
  Inspect fuel-delivery and leak-off lines of fuel-injection nozzles for leakage.

Air supply
  Check air filters (Fig. 2-1), air hoses, air-cleaner intake and outlet connections, and the air restriction indicator.

Electrical system
  Check battery electrolyte level and specific gravity (Fig. 2-2); be sure cap vents are open.
  Inspect all connection, especially for corrosion, at battery terminals.
  Look for bare wires which could cause shorts.
  Check for overheated parts (these often have an odor like burned insulation).
  Check alternator drive-belt tension (Fig. 2-3).

Maintenance of Diesel Power Systems 2-11

Turbocharged engines
Check air inlet and outlet connections.
Be sure oil inlet and drain lines are not twisted, kinked, or broken.
Be sure there is no strain on connector between turbocharger and manifold.

STARTING ENGINE

Before starting an engine, make the visual inspection already outlined. Though some of these checks may be made during the engine warmup period, many should not be attempted with the engine running.

Crank the engine no longer than 20 or 30 sec at a time. If it does not start, allow the cranking motor to cool for 2 or 3 min before cranking again. If the engine will not start after several attempts, refer to the troubleshooting section (Diagnosing Malfunctions) at the end of this chapter.

After starting the engine, wait until

Fig. 2-3 Check alternator drive-belt tension and fan-belt tension.

Fig. 2-4 Ether is a common starting aid.

gauges show proper oil-pressure and operating temperature before placing it under load. Run the engine at part load in midspeed range for 5 to 10 min [longer if ambient temperature is near or below 0°F (−18°C)] or until the gauges show proper readings; then apply partial load for a few minutes before going to full load. Do not race the engine during warmup.

Blocking air flow through the radiator will speed warmup, but be sure to remove the blocking material immediately after operating temperature is reached. (If the engine does not warm up properly, check the thermostats.) Partial blocking may also be used if the engine must be kept running in extremely cold weather with no load; however, regardless of ambient temperature, long idling periods must be avoided. An idling engine will not reach optimum operating temperature, resulting in less efficient combustion. Also, oil pressure may not reach its most effective level, preventing some engine parts from receiving adequate lubrication.

To prevent possible damage to an electrical system, the master switch must be in the ON position when the engine is running. The switch should be in the OFF position only when the engine is shut down.

Cold-weather starting Diesel engines depend on heat in combustion chambers for ignition, so starting them in cold weather can be a problem. Some common starting aids are ether (direct-injection systems only; Fig. 2-4), electrical glow plugs which preheat combustion chambers, oil heaters, coolant immersion heaters, battery warmers, and booster batteries.

Maintenance of Power Systems

Ether. Ether has a much-lower ignition point than diesel fuel, and some engines have adapters for injecting it. Heat from this initial ignition warms the fuel-air mixture and normal combustion follows.

Inject ether *only* while the engine is being cranked. If ether or other starting fluid is to be sprayed into the precleaner, use it sparingly and *only* after engine cranking begins. Excessive or improper use of ether can damage an engine.

Never use ether or other starting fluid in conjunction with glow plugs.

Glow Plugs. Glow plugs, which contain heating elements, fit into small turbulence chambers in the engine head. When turned on before using the starting motor, they warm the fuel-air mixture, leading to normal combustion.

Some manufacturers recommend glow-plug heat times of up to 3 min, depending on ambient temperature, before using the starter.

Immersion Heaters. Coolant immersion heaters are available for some diesel engines; they are particularlyuseful if the ambient temperature is expected to drop to $-10°F$ ($-23°C$). These heaters plug into 110-volt outlets (sometimes 220; see instructions), and they usually are left connected for at least 8 hr.

Some diesel engines have excess-fuel buttons to increase fuel delivery when starting; their use is explained in the manufacturer's manual.

Follow the manufacturer's instructions when using oil heaters or battery warmers.

STOPPING ENGINE

Diesel engines develop very high temperatures when working under load; they must be cooled gradually before being stopped.

Operate the engine at half-throttle (no load) for 3 to 5 min before shutting down and then at low idle for 1 or 2 min, to permit the engine and turbocharger to cool from their operating temperatures.

During hot weather, a longer cooling-off period may be desirable.

After stopping the engine, fill the fuel tank to prevent overnight condensation of moisture, and clean trash and dirt from the engine and radiator.

FUEL SYSTEM

Diesel engines burn a wide variety of fuels, but use of clean fuel is imperative, and selection of the right type is important. Check recommendations in the operator's manual. Do not use furnace fuel in modern diesel engines.

The two major grades of diesel fuel are 1-D and 2-D.

Grade 1-D is more volatile and is recommended for modern high-speed engines with variable loads and speeds, as well as for operation in extremely cold weather or at altitudes above 5000 ft.

Grade 2-D is recommended for high-speed engines with relatively high loads and more uniform speeds, as well as for engines not requiring the higher volatility of 1-D.

Ignition qualities of diesel fuels are measured by the cetane number, which is roughly comparable to the octane rating of gasoline. Cetane ratings vary from 33 to 64.

High-cetane fuels permit engines to be started at lower air temperatures and higher altitudes, provide faster engine warmup without misfiring, reduce varnish and carbon deposits, and help eliminate knock caused by slow ignition. However, a too-high cetane rating can cause incomplete combustion.

Ether has a cetane rating of 85 to 96, is highly volatile, and often is used as a cold-weather starting aid. However, it is highly explosive and can damage an engine if too much is sprayed in before cranking. CAUTION: Spray ether into the engine *only* while operating the starter.

Troubleshooting Elsewhere in this chapter is a section devoted to diagnosing most diesel-engine malfunctions. However, the following list may be of value if the malfunction is obviously in the fuel system:

Engine will not start, starts hard, or misfires
    Fuel tank empty
    Water in fuel
    Clogged supply line or filter

Faulty transfer pump
Air lock in injection pump
Fuel-cap vent plugged
Governor linkage to pump loose or broken
Drive shaft or key sheared
Pump plunger or distributor seized
Stuck valves or plugged orifices in nozzles
Excess-fuel control (if any) not activated
Shut-off control not deactivated
Engine will not idle smoothly
Faulty governor idling adjustment
Worn throttle linkage
Stuck plunger or pump rack or sticky or stuck control
Stuck nozzle valve or faulty nozzle-opening pressure
Leaky delivery valve
Pump out of time or calibration; loose control sleeves
Dirty filter
Engine smokes and knocks
Improper fuel
Excessive fuel delivery—faulty fuel-stop setting
Pump out of time
Dirty or fouled nozzles
Nozzle-opening pressure too low
Valves stuck open
Faulty turbocharger
Engine lacks power
Clogged filters
Pump timing retarded
Pump plungers or distributor worn
Faulty nozzles
Governor out of adjustment
Faulty aneroid to control fuel-air mixture
Faulty turbocharger

Filters Diesel fuels tend to be impure, while injection parts are precision made, emphasizing the need for frequent filter attention. Further, filters and their frequent inspection are not intended to compensate for a contaminated fuel supply.

Many diesel engines have three stages of progressive filters: a screen at the tank or transfer pump to remove large particles, a primary filter to remove small particles, and a secondary filter to remove tiny particles. Some filters not only remove suspended matter from fuel but also soluble impurities. Most filters have a water trap where water or heavy sediment can settle to be drained later.

Check all filters frequently. Change filter elements according to instructions in the manufacturer's manual, and keep in mind that it's better to change them too often than not often enough. Change the elements more often when operating in extreme dust or dirt. Check water traps at frequent intervals and drain out water and sediment.

Fuel-transfer pump Analyzers are available for checking fuel-transfer pump delivery and pressure. Usually, however, visual inspection is sufficient.

For a visual check, disconnect the pump-to-filter line at the filter (Fig. 2-5). Set the throttle so the engine will not start, and turn the engine over several times. If fuel spurts from the line, the pump is operating properly.

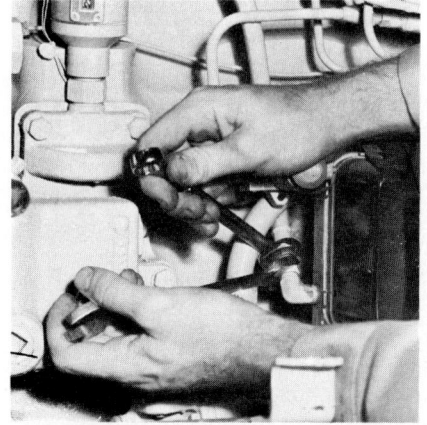

Fig. 2-5 Disconnect the pump-to-pump filter line at filter.

If little or no fuel flows, check for the following:
- Primer lever (if so equipped) left in upward position
- Leaking sediment-bowl gasket
- Plugged sediment-bowl screen
- Loose or damaged connections
- Air leak in inlet line
- Clogged fuel lines
- Loose cover screws on pump

If the trouble is not in these areas, repair or replace the pump. Most pump manufacturers sell repair kits, but it may be more economical to replace the entire unit. Also, some pumps are sealed units and so must be replaced if defective.

If repair is needed, disassemble the pump as outlined in the manufacturer's manual, inspect as follows, and replace parts as necessary:
- If the pump is a diaphragm type, look for punctures or leaks in the diaphragm. Check the slot in the diaphragm pull rod for wear.
- Examine the cover and body assembly for cracked or warped gasket surfaces.
- Examine valve and cage assemblies for worn valves or broken springs.
- Check the diaphragm and rocker-arm spring for proper tension.
- Inspect the rocker-arm link and pin for wear or damage.
- Inspect the filter screen for punctures and clogging.

Fuel lines Diesel engines have three types of fuel lines:

Heavyweight. These are high-pressure lines between injection pump and nozzles.

Medium-Weight. These lines can handle light to medium pressures between tank and injection pump.

Lightweight. These lines are used when there is low or no pressure of leak-off fuel from nozzles to tank or pump.

Inspect fuel lines periodically for leaks which indicate loose connections, breaks, or flaws. Connections should be snug, but not overtightened. (Overtightening could strip threads or damage sealing surfaces.)

Injection pumps Injection pumps are precision instruments, critical in adjustment, and easily damaged by careless handling. Only qualified technicians should disassemble or service these pumps, closely following the manufacturer's manual and using the special tools the manual specifies.

Never steam-clean an injection pump during engine operation, and do not spray a warm pump with cold water; pump seizure could result.

Fuel-injection nozzles Fuel-injection nozzles are comparatively simple devices but so important to proper diesel performance that their faulty operation can cause a variety of engine problems, ranging from hard starting to lack of power.

Various manufacturers use several different types of nozzles. While common features of these nozzles will be discussed here, each operator should consult the manual for his engine to secure specific information. Table 2-1 lists some problems and their possible causes.

The operator or owner should not attempt to remove or test nozzles unless he has the proper equipment and technical skill.

Before removing nozzles for testing and servicing, clean the area around them and remove and cap the injection and leak-off lines. Soak the nozzle assemblies in clean solvent or fuel after discarding the outer seals. Clean carbon and dirt deposits from spray tips and nozzle bodies with a soft-bristle brush (Fig. 2-6); *never* use emery cloth or a steel-wire brush.

When testing nozzles, follow the manufacturer's manual. Use a nozzle tester (Fig. 2-7), a high-pressure hand pump which forces fuel through the nozzle. CAUTION: Fuel comes out of the nozzles at extremely high pressure and can penetrate clothing and skin; point the nozzle away from yourself and any bystander.

Test for three things: spray pattern, opening pressure, and valve leakage.

Spray Pattern. Fuel should be finely atomized and evenly distributed. There should be no stream or large visible drops. If the spray pattern is poor, look for a clogged or eroded orifice or a bent valve.

Opening Pressure. Pressure needed to open the nozzle is checked by pumping the tester steadily until the gauge needle falls rapidly. Check the pressure reading against the manufacturer's manual. If pressure is too low, a weak or broken spring may be the cause,

Maintenance of Diesel Power Systems 2-15

or spring pressure may need adjustment. If pressure is too high, the tip may be plugged, or the valve may be binding in the valve guide.

Valve Leakage. This is the third possible malfunction to check. If these tests show the nozzle is not working properly, it must be disassembled, inspected, cleaned, and, if necessary, repaired. This requires a special tool kit and probably should be entrusted to an experienced technician.

CAUTION: When working on several nozzles, do not mix parts.

TABLE 2-1 Troubleshooting Fuel-Injection Nozzles

Problem	Possible cause
Nozzle opens at wrong pressure	Faulty spring-pressure adjustment
Nozzle will not close	Broken spring
Poor spray pattern	Plugged or chipped orifices (spray tip only); chipped or broken pintle end; deposits on pintle seat; chipped pintle seat
Poor misting of fuel	Plugged or chipped orifices (spray tip only); valve not free; cracked tip
Valve operates erratically	Valve spring misaligned; spring broken; deposits on pintle seat; bent valve; distorted body
Valve will not operate	Valve spring misaligned; spring broken; varnish on valve; deposits in seat area (pintle tip only); bent valve; valve seat eroded or pitted; distorted body
Too much fuel leaks off	Valve guide worn
Too little fuel leaks off	Varnish on valve; insufficient clearance between valve and guide

Inspect the parts for wear, chipped edges, scratches, misalignment, and breakage. Chipped, broken, or bent parts should be replaced, and if damage is extensive, replace the entire nozzle. Some nozzle parts are sold in matched sets, and should be so installed.

If the valve on some types of nozzles is sticking but not bent, a little polishing or use of injector lapping compound around the valve-guide area may be all that is needed. Other nozzle parts often may be salvaged by lapping their surfaces to remove tiny scratches or burrs—improper lapping can result in excess wear.

Before reassembling nozzles, flush each part in clean oil. Then place the entire nozzle in the tester for retesting.

Fig. 2-6 Clean nozzle with soft bristle brush.

Fig. 2-7 Test nozzles with nozzle tester.

2-16 Maintenance of Power Systems

Fuel storage The importance of proper fuel storage cannot be stressed too highly. Many diesel-engine difficulties can be traced to dirty fuel or fuel that has been in storage too long.

Keep all dirt, scale, water, and other foreign matter out of the fuel, and avoid storing fuel for a long period of time. Flush fuel-storage tanks periodically.

Water can condense in partly filled fuel containers. The containers should have provision for draining water which settles in their bottoms.

Governors Diesel-engine governors are speed-sensitive devices which act on the engine throttle to maintain a selected speed, limit slow and fast speeds, and shut down the engine if it threatens to overspeed.

Governors are relatively trouble-free but do need occasional adjustment and servicing. Two principal requirements for proper functioning are a vibration-free drive and clean moving parts. Manufacturer's instructions should be followed in servicing governors.

Here are some suggestions for diagnosing governor malfunctions.

Erratic engine operation, hunting, or misfiring
   Idle spring missing, broken, or wrongly adjusted
   Wrong control spring, or worn or broken spring
   Parts worn, sticking, binding, or improperly assembled
   Faulty high-idle adjustment
   Adjusting screw needs adjustment

Engine idles erratically
   Idle spring missing, broken, or improperly adjusted
   Parts worn, sticking, binding, or improperly adjusted
   Wrong governor spring

Engine not receiving fuel
   Shut-off control not deactivated
   Parts worn, sticking, binding, or improperly adjusted

Engine does not develop full power or speed
   Wrong governor spring
   Faulty fast-idle adjustment
   Adjusting screw needs adjustment
   Parts worn, sticking, binding, or improperly adjusted

Bleeding If the fuel system has been opened (e.g., to replace a fuel filter), air may get in to form an air lock that would keep fuel from reaching or going through the injection pump. Then the engine will not start or will run poorly.

The following procedure is recommended for bleeding the system:

1. Fill the fuel tank with the correct fuel.
2. Loosen the bleed plug on the fuel filter and open the fuel shut-off valve at the tank. If there are dual filters, bleed the filter nearest the tank first.
3. Open the fuel-supply valve. If the system has a manual priming pump or lever, pump the lever until a solid, bubble-free stream of fuel flows from the opening. If the lever will not pump fuel and no resistance is felt at the upper end of the stroke, turn the engine with the starter to change the position of the fuel-pump cam.
4. Tighten the bleed plug.
5. If the engine has dual filters, repeat the bleeding process on the other filter (Fig.2-8).
6. When bleeding is completed, leave the primer lever at the lowest point of its stroke.
7. If an airlock is still present, bleed the injection lines.
8. Using wrenches (two required), loosen the injection-line nuts on at least two lines. CAUTION: Loosen injection-line connectors only one turn to avoid excessive spray.
9. Bleeding half the injection lines usually is sufficient; the others will bleed themselves when the engine starts running.

Fig. 2-8 Bleed air from fuel line at filters.

10. Crank the engine until fuel without foam flows around the connectors. Then tighten the connections carefully until snug and free of leaks. Do not bend the line connections.

COOLING SYSTEM

The intense heat of combustion in modern high-performance engines may cause such components as valves, pistons, and rings to operate near critical temperature limits even when the cooling system is operating normally.

Overheating seriously affects engine lubrication. High metal temperatures may destroy the lubricating film, accelerate oil breakdown, and cause formation of varnish. Cylinder heads and engine blocks often are warped and cracked by the terrific strains set up in the metal by overheating, especially when followed by rapid cooling.

Common causes of cooling malfunctions are clogged systems, low coolant level, and defective water pumps and thermostats.

Clogging Rust clogging, perhaps the most common cause of cooling-system trouble, can be avoided by periodic rustproofing and cleaning when necessary.

Rust forms on walls of the engine water jacket and other metal parts. Particles settle in the jacket and radiator water tubes, cutting down heat transfer until the engine overheats. Overheating stirs up more rust in the block and forces it into the radiator, which eventually gets clogged.

Rust and scale can also build up on the water side of the combustion chambers, causing overheating and eventual engine damage. Rust, scale, and grease can be removed by double-action cleaners, which are harmless to cooling-system metals and connections if used according to directions.

If rust and grease are not completely neutralized and flushed out, they can destroy the corrosion inhibitors in later fills of antifreeze and antirust solutions.

Drain the entire system at least once a year. If the liquid is rusty, use a cooling-system cleaner. Otherwise, use a radiator flush, or flush with plain water. Corrosion inhibitors will not clean out rust already formed.

It is desirable to maintain at least a 25 percent solution of antifreeze containing dependable inhibitors, even during the warm season.

Coolant Even distilled water can cause rust in a system, so if water alone is used as a coolant add a can of rust inhibitor. Keep the system filled to a level midway between the radiator core and the bottom of the filler neck.

If antifreeze coolant is used, ethylene glycol types are recommended, because most diesel engines develop temperatures above the boiling point of alcohol. Add rust inhibitor if the antifreeze does not already contain it.

Use only as much antifreeze as expected temperature extremes require. Because of the nature of ethylene glycol, too strong a cooling-system solution can actually reduce protection against freezing.

Never pour hot water into a cold engine or cold water into a hot engine; a cracked cylinder head or block could result.

Remember that the term *permanent antifreeze* does not mean the solution is good for more than one season; it means the solution will not boil away at normal engine operating temperatures. Adding rust inhibitors or fresh antifreeze to used solutions will not restore full-strength corrosion protection.

Repair radiator or other cooling-system leaks before installing antifreeze coolant. Follow instructions when adding sealing solutions to correct minor leaks. Some of these solutions will react with antifreeze and rust inhibitors and seriously affect coolant performance.

Radiator Though such cooling-system malfunctions as external leaks may be obvious, check the entire system before servicing. Use a pressure tester according to the manufacturer's instructions. Check the radiator (Fig. 2-9), water pump, hoses, drain cocks, and cylinder block for leakage.

Inspect the radiator for bent fins and tubes with cracks, kinks, dents, and fractured seams. Only experienced radiator technicians should make needed repairs.

Radiator cap Most radiator caps have a pressure valve to vent coolant or steam if pressure reaches a certain point and a vacuum valve which opens to prevent vacuum in the cooling system.

2-18 Maintenance of Power Systems

Using a tester available from the dealer, check both valves periodically for proper opening and closing pressures. If either valve malfunctions, use a new radiator cap.

External leaks At best, external leaks will cause loss of costly antifreeze. But they also can cause insufficient engine cooling with possible damage.

Most radiator leakage is due to mechanical failure of soldered joints caused by cooling-system pressure or engine or frame vibration. Other possible sites for external leakage are hose connections, core-hole plugs, gaskets, and stud bolts and capscrews.

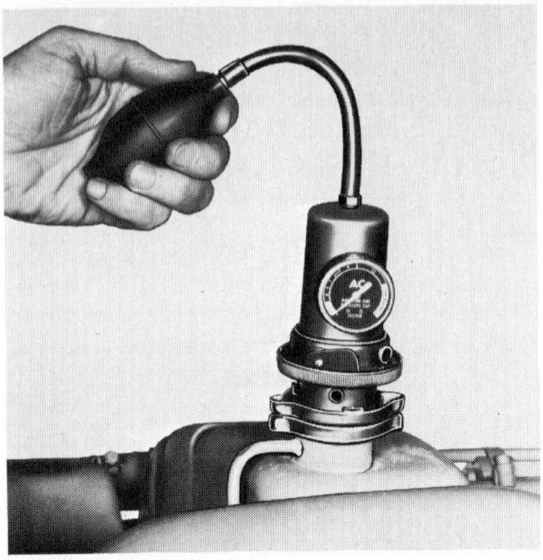

Check the cylinder block for coolant leakage before and after it gets hot and while the engine is running.

Internal leaks Coolant may leak into an engine because of a loose cylinder head or sleeve joint, defective gaskets, a cracked or porous casting, or malfunction in the push-rod compartment.

Water or antifreeze will form sludge when mixed with engine oil, possibly causing lubrication failure (Fig. 2-10), sticking piston rings and pins, sticking valves and valve lifters, and extensive engine damage. If crankcase oil looks milky, the cause may be antifreeze contamination.

If a coolant leak exists in the push-rod compartment, it may be necessary to pressurize the cooling system and tear down part of the upper part of the engine.

Fig. 2-9 Use pressure tester to check radiator.

When replacing cylinder-head gaskets, use only new gaskets designed for the engine in question. Be sure the head and block surfaces are clean, even, and smooth.

Follow the manufacturer's torque specifications and the sequence for tightening cylinder-head bolts.

Thermostats Faulty or improper thermostats can cause engines to warm up too slowly or to operate at the wrong temperatures. Use only the type specified by the engine manufacturer. Never run an engine without thermostat protection (some engines have two or more).

Discard broken, faulty, or corroded thermostats. Do not use bellow-type thermostats in high-pressure cooling systems.

To check a thermostat (Fig. 2-11), suspend the unit and a thermometer in a container of water and, while stirring, heat the water gradually. The thermostat should begin to open at the temperature stamped on it, ±10°, and should be fully open at 22° above the specified temperature. After removing the thermostat from the hot water, observe its closing action.

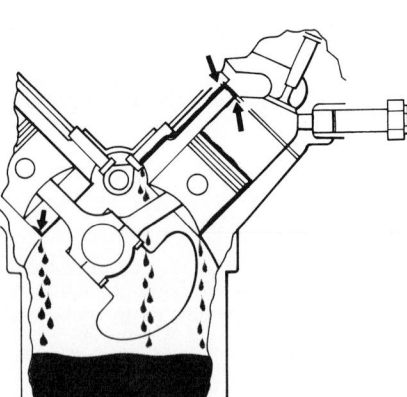

Fig. 2-10 Coolant mixes with engine oil through internal leaks.

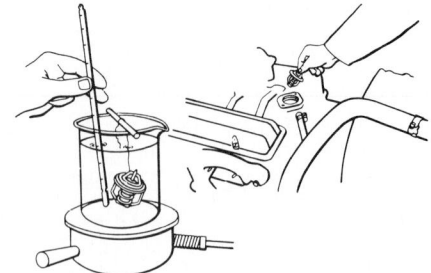

Fig. 2-11 Check condition of thermostat.

Maintenance of Diesel Power Systems 2-19

When installing a thermostat, clean the gasket surfaces and use a new gasket. Position the thermostat with the expansion element toward the engine (the frame must not block water flow).

Hoses Periodically check all hoses for hardening, cracking, softening, and swelling (Fig. 2-12). If hoses must be removed, check any inside reinforcing springs for corrosion.

Fig. 2-12 Inspect hoses for wear.

Replace hoses often enough to be sure they are always pliable and able to pass coolant without leaking or shedding small particles of rubber which could clog the radiator.

Use only the best available hoses, and coat connections with nonhardening sealing compound when installing. Tighten hose clamps securely. A pressurized cooling system can blow off an improperly installed hose.

Water pump Some diesel-engine water pumps turn at 4000 rpm and pump as much as 125 gal (479 liters) of coolant per minute. Overheating occurs quickly if a pump malfunctions.

Pump malfunctions may be caused by leaks in the housing, broken or bent vanes on the impeller, and damaged seals and bearings. If a pump must be removed and disassembled for inspection, replace all damaged or worn parts and use new seals and gaskets when reassembling. Follow the manufacturer's instructions.

Filters Some engines have filters in the cooling system. The filter element and resistor plates often contain chemicals which remove or neutralize corrosives, alkalize the coolant enough to prevent corrosion of metal parts, and form rustproof films on metal surfaces.

Periodic filter servicing should include draining sediment from the lower sump and replacing the filter element when necessary.

Fan and fan belt Effective preventive maintenance must include regular checking of fan-belt condition and tension, similar to the procedure shown in Fig. 2-3. Replace belts when wear warns of early failure.

Adjust fan-belt tension as specified by the manufacturer (Fig. 2-13). Too much tension causes premature failure of belt and fan bearings. Too little permits belt slippage and causes insufficient cooling and excessive belt wear.

Fan service usually consists of making sure the blades are straight and far enough from the radiator so they will not strike the core.

Aeration Aeration in a cooling system, caused by air mixing with the coolant, can speed formation of rust and corrosion. It can also cause foaming, overheating, and loss of coolant through the overflow pipe.

Aeration may result from a leak in the system, turbulence in the top tank, and too-low coolant level.

Use the following steps to check for aeration in the cooling system:
1. Adjust coolant to correct level.
2. Replace pressure cap with plain but airtight cap.

2-20　Maintenance of Power Systems

3. Attach rubber tube to lower end of overflow pipe.
4. With transmission in neutral, run engine at high speed until temperature gauge stops rising and stabilizes.
5. Without changing engine speed or temperature, place end of rubber tube in container of water.
6. A continuous stream of bubbles from the tube will show that air is being drawn into the cooling system.

Exhaust-gas leakage　A cracked head or loose cylinder-head joint can allow hot exhaust gas to be blown into the cooling system under combustion pressures even though the joint may be tight enough to keep liquid from leaking into a cylinder.

Exhaust gases dissolved in the coolant will destroy the inhibitors and form acids which cause corrosion, rust, and clogging. The cylinder-head gasket may burn or corrode because of the gases, and excess pressure also may force coolant out of the overflow pipe.

Fig. 2-13　Adjust fan-belt tension.

Check the cooling system for exhaust-gas leakage if the coolant is rusty or if there are severe rust clogging, corrosion, or overflow losses.
1. Warm up the engine and place it under load.
2. Remove the radiator cap and look for excessive bubbles or an oil film in the coolant.
3. Make this test quickly, before boiling starts, for steam bubbles will be misleading.

Flushing cooling system　Cooling systems should be flushed and thoroughly checked at least once a year and always before installing new antifreeze solution.

Incomplete flushing, such as hosing out the radiator, will close the thermostat and prevent thorough flushing of the water jacket. Either remove the thermostat or, after filling the system with clean water, run the engine long enough to open the thermostat.

After thorough flushing, open all drain points to drain the system completely. Clean out the overflow pipe; remove insects and dirt from radiator air passages, the radiator grille, and screens. Check the thermostat, radiator pressure cap, and the cap seat for dirt and corrosion.

LUBRICATION SYSTEM

Faulty lubrication is perhaps the greatest single cause of premature engine failure. It can result from using the wrong grade or weight of lubricating oil, insufficient oil in the crankcase, deficient oil pressure, lack of proper oil additives, or contaminated oil, any of which can lead to a major and costly engine overhaul long before normal wear would make it necessary.

Oil contamination　The major causes of lubricating-oil contamination include the following.

Improper Storage and Handling. Store lubricants in a clean, enclosed area, and keep all covers and spouts on oil containers when not in use. These practices not only keep dirt out of the lubricant but reduce condensation of water in the containers.

Dust Breathed into Engine with Combustion Air. Regularly clean or replace air filters and the breather on the oil filler.

Cold Engine. A cold engine greatly reduces fuel-burning efficiency and partially burned fuel blows by the piston rings into the crankcase. Oxidation of this fuel in the oil forms a harmful varnish which deposits on engine parts. A misfiring engine will also cause this contamination.

Cold engines also contaminate lubricating oil when water vapor, a normal product of combustion, condenses on cold cylinder walls and is blown past the rings into the crankcase. It then combines with oxidized oil and carbon particles to form sludge, which can effectively block oil screens or passages. Warm up the engine properly before applying full load, making sure the engine reaches operating temperature each time it is started. Use the proper thermostat to warm up the engine as quickly as possible.

Oxidation. This occurs when hydrocarbons in the oil combine with oxygen in the air to produce acids which are highly corrosive and create harmful sludges and varnish deposits.

Carbon Particles. A product of normal engine operation, these particles are created when oil on the upper cylinder walls is burned during combustion. In addition to contaminating the oil, excessive carbon deposits can cause piston rings to stick in their grooves.

Engine Wear. Tiny metal particles are constantly being worn from bearings and other parts. They tend to oxidize and contaminate the oil.

Antifreeze. Should it leak from cooling into lubrication systems, antifreeze can cause sludge formation. Leaking can occur if head gaskets are damaged by improper use of starting fluids, or if head bolts are not torqued to specifications when the head is removed and replaced.

Additives One important way to combat oil pollution is to start with good-quality oil containing proper additives. Remember that these additives eventually wear out, so change oil before they are exhausted.

Anticorrosion additives protect metal surfaces from corrosive attack.

Oxidation-inhibitor additives keep oil from absorbing oxygen, preventing oxidation and varnish and sludge formation.

Antirust additives prevent rusting of metal parts during storage periods, downtime, and even overnight. They also help neutralize harmful acids, and they form a protective coating which repels water droplets and protects the metal.

Detergent additives prevent deposits and help keep metal surfaces clean. They hold carbon particles and oxidized oil in suspension, so these will be eliminated when oil is drained. Black oil is evidence that these detergents are keeping such contaminants in suspension instead of letting them accumulate as sludge.

Oil filters Proper attention to oil filters is essential in guarding against oil contamination, and it is one of the easiest and most important forms of preventive maintenance.

Two types of oil filters are in general use: *Surface filters* have a single surface which stops or removes dirt particles larger than holes or openings in the filter; *depth filters* contain a large volume of filter material which removes particles suspended in the oil as well as some water and water-soluble impurities (Fig. 2-14).

Many depth filters have a relief or bypass valve, which opens as the filter becomes clogged allowing oil to bypass to the bearings. Otherwise, pressure would build up on the inlet side of the filter, causing the pressure-regulating valve in the engine to open completely, sending all oil back to the crankcase, and

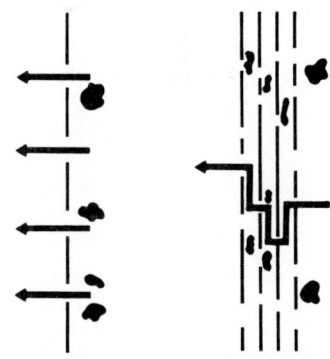

SURFACE FILTER DEPTH FILTER

Fig. 2-14 Filtering methods: surface versus depth.

2-22 Maintenance of Power Systems

seriously damaging the engine. Replace such filters only with an exact duplicate of the original.

Follow the manufacturer's instructions for checking filters. When servicing any filter, use new gaskets and seal rings. Tighten the housing firmly but not too tightly. Then run the engine until the oil pressure registers and check for leaks.

Pressure-regulating valves These valves maintain correct oil pressure in the lubrication system regardless of engine speed or oil temperature. They also bypass oil at filters and oil coolers. Most are adjustable.

Servicing usually is confined to cleaning parts and the valve bore in which the valve slides with a proper solvent. Also check the valve poppet for wear or nicks which might cause it to hang up in the bore.

Check engine oil pressure after servicing any pressure-regulating valve. Causes of too-low oil pressure include low crankcase-oil level, too-thin oil in the crankcase, worn engine bearings, worn oil pump, filter or pump leaks, faulty regulating-valve spring, and improperly adjusted regulating valve.

Causes of too-high oil pressure include too-heavy oil in the crankcase, stuck regulating valve, and improperly adjusted regulating valve. A defective gauge may show deficient or excessive oil pressure.

Oil coolers Many lubrication systems have oil coolers which use engine coolant to dissipate heat from the oil. These coolers may be mounted internally (Fig. 2-15) in the crankcase or externally on the outside of the engine block.

Fig. 2-15 Internal oil cooler.

Normal maintenance of the cooling system usually will keep the oil cooler clean. When cleaning the lubrication system, remove the cooler and use solvent to clean the oil passages.

When replacing or installing an external oil cooler, use new gaskets and be sure the capscrews are properly tightened.

Ventilation All diesel engines have ventilation systems to carry away fuel vapor and water vapor so vapors will not condense into liquids which drain into the crankcase. Manufacturers use a variety of ventilation systems, but these general suggestions will aid in proper servicing.

Begin with the air inlet, which may be the oil-filler cap or the main air cleaner. In either case, service regularly according to instructions in the manufacturer's manual.

If the system has a regulating valve, either clean it with a solvent or replace it, according to the manufacturer's instructions.

If the system has a vent tube, periodically remove it and clean it with a solvent. If the vent tube has a filter, be sure to clean it also.

Oil consumption Some oil consumption is normal during diesel-engine operation. However, if oil consumption seems excessive, check the following possible causes.

Wrong Weight or Grade of Oil. Follow manufacturer's instructions, giving proper consideration to type of engine service and prevailing ambient temperature.

Engine Not Run Long Enough under Load to Seat Rings. Some variation in oil consumption can be expected during break-in, but the level should be stabilized before 250 hr of operation if the engine was broken in properly. If not, assume that a problem exists and must be corrected.

Pressure-Regulating Valve Improperly Adjusted. Rings and valves are flooded with oil.

Crankcase Breather. If plugged, the breather can cause increased crankcase oil pressure.

External Oil Leaks. Even if small, these can add up to loss of quarts of lubricant between changes. Check front and rear seals, all gaskets, and filter-attaching points.

Engine Blow-by. Fumes from the crankcase vent should be barely visible with the engine at fast idle under no load. Excessive blow-by may indicate that piston rings and cylinder liners have worn to the point where the rings cannot seal off the combustion chambers.

Valve-Guide Seals. If defective, these seals permit an excessive amount of oil to enter combustion chambers. To check, warm up the engine and let it idle slowly for 10 min. Then remove the exhaust and intake manifolds. If valve ports and undersides of valve heads are wet with oil, it has been drawn through the valve guides.

If the cylinder head is removed, check the piston heads. If they are wet with oil, it may have been drawn past the piston rings; if excessive blow-by has indicated worn rings, this will serve to confirm it.

Excessive Engine Speeds. Excessive speed is another and common cause of excessive oil consumption. Observe fast-idle limits specified by the manufacturer.

AIR INTAKE AND EXHAUST

Improper or inadequate intake air filtering can lead to early engine failure and a costly overhaul. Enough dust-laden air can pass through an almost invisible crack or leak over a period of time to damage an engine severely.

Intake system Diesel-engine manufacturers use one or more of several types of air filters, but one thing is necessary for all—periodic checking (Fig. 2-1), daily for some types and more often in dusty or dirty conditions.

Rules for servicing air filters depend on type of filter, contamination of air, and type of application. Normal service intervals are specified by the manufacturer, but frequent inspection tells whether they are adequate for conditions under which an engine is operating. Some units have indicator lights which simplify checking.

General Maintenance Rules.
- Keep filter-to-engine connections tight.
- Keep air cleaner properly assembled so all joints are oil- and air-tight.
- Periodically inspect entire air-intake system, including any hoses and the intake manifold.
- Service oil-bath cleaners often enough to prevent oil from becoming thick with sludge. Use specified grade of oil and keep at proper level in cup; oil from an overfilled cup can be drawn into the engine, where it becomes fuel. The engine may overspeed and be damaged by this uncontrolled additional fuel.
- Never wash dry filter elements in fuel oil, gasoline, or solvents; use methods recommended by the manufacturer.
- Periodically inspect the rubber dust-unloading valve (if used), squeezing the rubber end of the valve to be sure it is not clogged.

Servicing Precleaners.
1. Remove and empty bowl (Fig. 2-16).
2. If unit has prescreener, blow or brush off chaff or other foreign matter.

Servicing Oil-Bath Cleaners.
1. Stop engine and remove oil cup.
2. Clean cup if more than ¼ in. of sediment has accumulated. Clean cup if oil has thickened or contains water.

Fig. 2-16 Periodically service the precleaner.

3. Remove caked dirt from bottom of cup, then wash with clean diesel fuel—*never* with gasoline, naphtha, benzene, or other highly flammable solvent.
4. Clean dirt tray (if used).
5. Refill cup to oil-level mark (never above) and replace.

Servicing Dry-Type Cleaners.
1. If cleaner has dust cap, empty it daily. If it has an automatic dust-unloading valve, check this daily to detect any clogging.

2. If unit has restriction indicator, clean whenever the indicator signals a restriction. Some indicators may stick, so watch conditions and total hours of operation.

3. If unit does not have restriction indicator, clean at recommended intervals and more often in dusty conditions.

4. Remove dusty elements and tap gently on heel of hand as element is being rotated. Do not tap on hard surface.

5. If tapping does not remove dust, use a compressed-air cleaning gun (pressure not over 30 psi), blowing up and down pleats and from inside to outside.

6. If element is oily or sooty, use compressed air as described above; then soak and gently agitate element in solution of lukewarm water and commercial filter-element cleaner or equivalent nonsudsing detergent.

7. Rinse element in clean water, shake off excess water, and allow element to dry thoroughly. Do not use compressed air to dry because the wet element can be ruptured easily by air pressure.

8. Inspect element for damage by placing light inside; discard filter if even slight damage is apparent.

9. Discard element after recommended service period (such as 1 year or six washings), if cleaning attempts fail, or if gasket is damaged or missing.

10. Clean the inside of the filter housing with a damp cloth, install the element in the housing with gasket and fin end first, and draw cover tight. Reset restriction indicator if one is used.

Testing Intake System. If air flow into an engine is restricted, the vacuum in the cylinders increases, possibly causing oil to be drawn into the combustion chambers and thereby increasing oil consumption. A vacuum can be detected by an auxiliary vacuum gauge available from the dealer or manufacturer. Proceed as follows:

Warm up the engine.

On engines with restriction indicators, remove the indicator, install a pipe-tee fitting, reinstall the indicator, and connect the gauge to the fitting. (Be sure to remove the pipe-tee after the test.)

On engines without restriction indicators, connect the gauge to the intake manifold.

Run engine at fast idle. Check the gauge reading against the manufacturer's specifications. Too high a reading indicates restriction in the air-intake system.

On engines with restriction indicators, check the indicator by gradually closing the air-intake opening with a board or metal plate, meanwhile watching the gauge. If the indicator is defective, replace it.

Exhaust system Service consists of cleaning carbon buildup from inner passages of the exhaust manifold, replacing defective mufflers, and keeping the entire system free from leaks—especially if the engine is in a machine with a cab or other operator enclosure into which deadly carbon monoxide gas might penetrate.

Turbocharger Turbochargers are exhaust-driven turbines which operate centrifugal compressors to increase air supply to combustion chambers. They run at very high speeds, which can range from 40,000 to more than 100,000 rpm, yet they are relatively simple devices and, if properly serviced, will operate with little or no attention.

Some troubleshooting hints are the following:

Noisy operation or vibration
  Bearings not being lubricated
  Leak in intake or exhaust manifolds
  Improper clearance between turbine wheel and housing
  Rotary vanes out of balance

Engine will not deliver rated power
  Clogged or leaking manifold system
  Foreign matter lodged in compressor, impeller, or turbine
  Excessive carbon buildup behind turbine wheel
  Seized bearing in rotating assembly

Oil-seal leakage
  Faulty seal
  Restriction in air cleaner or air intake creating suction

Inspecting turbochargers
  ▪ Regularly check mounting and connections for prompt detection of oil or air leakage.

- Check crankcase vent to be sure there is no restriction in air flow.
- Operate engine at approximate rated output and listen for unusual turbocharger noise. Abnormal shrill whine could mean bearings are about to fail, but do not confuse this with normal whine heard during so-called rundown as engine speed is reduced.
- Other unusual turbocharger noises could mean improper clearance between turbine wheel and housing.
- Check for unusual turbocharger vibration while engine is operating at rated output; dirty air may have pitted the rotary vanes and caused an imbalance.
- Check exhaust smoke under engine load conditions. Excessive smoke may indicate incorrect fuel-air mixture, which could be due to engine overload, engine malfunction, or turbocharger malfunction. Excessive smoke during acceleration is normal.

ELECTRICAL SYSTEM

Diesel-engine electrical-system maintenance consists largely of battery care, though attention should be given to such components as gauges, the drive belt for the alternator or generator, and the starting motor.

Fig. 2-17 Fill battery with distilled water.

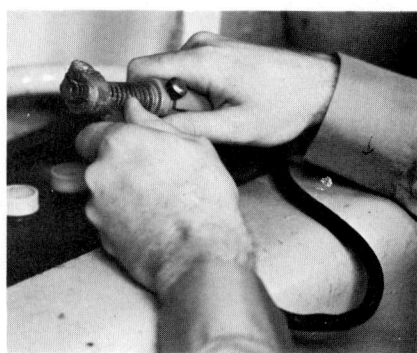

Fig. 2-18 Clean battery terminals.

Batteries

Check electrolyte level every 50 hr, also inspecting vent holes in caps to be sure they are open. Electrolyte level should be at the bottoms of the filler necks and above the tops of the battery plates.

Use distilled water to bring the electrolyte to proper level (Fig. 2-17). Do not use hard water; dissolved minerals will leave deposits on the plates which interfere with chemical action.

Do not overfill. Excess electrolyte will escape through the vent holes and leave deposits on the tops of batteries.

Add only water unless electrolyte has been lost by spilling.

If water must be added to the battery in freezing weather, immediately run the engine long enough to assure proper mixing.

Specific gravity

Before adding water, check electrolyte specific gravity (Fig. 2-2), which should be between 1.215 and 1.270 when electrolyte temperature is 80°F (27°C).

If the hydrometer does not have temperature correction, add four gravity points (0.004) for each 10° above 80, and subtract four points for each 10° under 80.

Battery cells that differ more than 50 specific gravity points indicate an unsatisfactory battery condition caused by an internal defect, short circuit, or deterioration from extended use. The battery should be replaced.

Cleaning batteries

Clean batteries every 250 hr, more often if there are heavy deposits on the case and terminals.

Remove the cable terminals carefully; never pry against the battery case. Clean the terminals (Fig. 2-18) both inside and out, and dip them in a solution of 2 tablespoons of baking soda to 1 pt of water.

Clean the posts (Fig. 2-19) and top of the battery, and brush on a fresh mixture of baking soda and water. Apply this solution until foaming stops, then flush with clean water, making sure neither the solution or flushing water gets through vents in the caps (the electrolyte could be neutralized or contaminated). Dry the battery and posts with a clean cloth.

After replacing the cable terminals, apply a coating of petroleum jelly or light grease to the terminals and clamps to protect them from corrosion.

Be sure each cable terminal is on the proper post, or the generator or alternator may be damaged. Do not pound or force the clamps into place.

Recharging batteries

Cell caps should be removed when recharging batteries, and be sure to follow the charger manufacturer's instructions (Fig. 2-20).

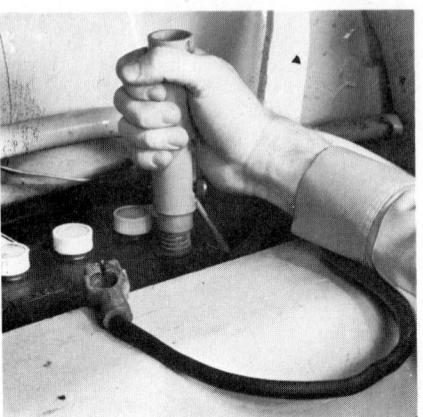
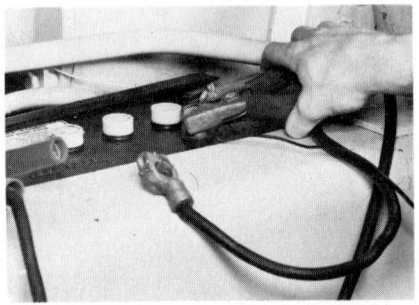

Fig. 2-19 Clean battery posts.   **Fig. 2-20** Recharging a battery.

Badly sulfated batteries will not accept fast charging without danger of damage; they must be recharged slowly. The normal slow-charging period is 12 to 24 hr, but badly sulfated batteries may require 60 to 100 hr for complete recharge.

Booster batteries

If booster batteries are needed for cold-weather starting, be sure all electrical switches and accessories are turned off.

Connect cables first to the machine battery and then to the booster battery. Remove cables from the booster battery first. This guards against explosion of gas from the machine battery.

When attaching booster-battery cables, be sure to connect positive (+) to positive and negative (−) to a grounding point away from the battery. Do not use a 12-volt booster battery with a 6-volt machine battery.

Replacing batteries

Troubleshoot the battery before you replace it; many batteries returned for warranty claim or trade-in have nothing wrong except they are completely discharged.

Choose a replacement battery of an ampere-hour rating at least equal to the original. If accessories have been added, a larger size may be needed.

Remember the cheapest replacement battery is not always the most economical. Consider such factors as ampere-hour rating, construction, and length of warranty.

Safety

Always disconnect the battery ground strap or cable before working on any part of the electrical system of an engine to prevent injury from sparks, short circuits, or the engine accidentally starting.

Do not lay metal tools or other objects across the battery.

Hydrogen gas, released from the electrolyte (more is generated while charging), is extremely flammable, so keep sparks and fires away. Be sure the room where batteries are being charged is well ventilated.

Battery acid is harmful to skin and to most materials. Immediately remove any clothing on which acid is spilled.

Maintenance of Diesel Power Systems 2-27

If acid contacts skin, rinse the affected area 10 to 15 min with running water.

If acid splashes into the eyes, flush with running water 10 to 15 min. Also be sure to force the lids open while washing the eyes. Then, the victim should *immediately go to a doctor* for further treatment.

Acid spilled on the floor or on paint or metal surfaces of a machine can be neutralized by using a mixture of 1 lb of baking soda to 1 gal of water or 1 pt of household ammonia to 1 gal of water.

Alternator Alternators seldom give trouble unless abused, though bearings may fail and insufficient belt tension may reduce output. Check belt condition and tension occasionally, following the manufacturer's belt-tension specifications. If a new belt is installed, check tension after a few hours of operation and compensate for any stretch.

Disconnect the battery ground cable before working on or near the alternator or regulator. If alternator wiring is disconnected, be sure it is properly replaced before the battery is reconnected.

Failure to observe any of the following precautions may result in serious damage to the alternator, regulator, or electrical system:

Never ground a terminal or connect a jumper wire to any alternator terminal.

Never ground the alternator output terminal.

Never ground the alternator field terminal or the field circuit between the alternator and regulator.

Never disconnect or connect any alternator wires with batteries connected or the alternator in operation.

Never attempt to polarize the alternator or regulator.

Generator and starting motor If the engine has a generator instead of an alternator, its service is much the same as for the starting motor.

Most generators and starting motors have sealed bearings which require no lubrication, but some have oil cups for lubrication every 250 hr. Do not overlubricate; excessive amounts of oil can get on the brushes and cause failure.

Occasionally check the condition of the generator belt, and check its tension against manufacturer's specifications every 250 hr. If a new belt is needed, use an exact duplicate and recheck tension after a few hours of operation.

When checking generator brushes (Fig. 2-21), do not pull the brush-connector wire while the brush is under spring tension or allow the tension arm to snap down on a brush. Replace the brushes with new ones if they are worn so tension arms are against the brush holders instead of the brushes or if the brushes are worn to half their original length.

Fig. 2-21 Check generator brushes.

Check new brushes for binding action in the holders. If the brush binds, clean the holder with a cloth or sandpaper and wipe off dirt and grit; do not use a solvent, which could soften insulation on the wires.

Inspect the generator or starting motor for signs of overheating. If solder has been thrown against the cover band or inside of the housing, the unit requires expert care by an experienced technician.

If the commutator is scored or scratched, it must be machined by a competent technician. If merely glazed or dirty, lightly sand it with No. 00 sandpaper, then blow away dust and dirt. Do not use emery cloth, for remaining particles will cause arcing and rapid wear.

Maintenance of Power Systems

If any wire leads were disconnected from the starter or generator while servicing, polarize the generator before starting the engine, using the procedure outlined in the manufacturer's manual.

Gauges Frequently check all gauges when operating a diesel engine. If readings are not normal, stop operation until the cause is found. Keep in mind when looking for the cause that malfunction of the sending or receiving unit of the gauge itself is a possible cause for improper readings.

DIAGNOSING MALFUNCTIONS

Prompt diagnosis and correction of even minor engine malfunctions can, at least, improve operating efficiency and reduce downtime during prime working periods. At best, it may prevent costly major overhauls and permit scheduling corrective action until more convenient times.

Many malfunctions can be corrected in a relatively short time, perhaps during or after the regular work day. Others may require partial engine disassembly or attention from a competent technician, who should check the entire engine as he corrects the immediate trouble.

Here are some symptoms and the malfunctions they suggest.

Engine starts hard or will not start
Fuel system
  Fuel tank empty
  Wrong type of fuel
  Water, dirt, or air in fuel system
  Fuel lines clogged or restricted
  Fuel filter restricted
  Faulty fuel-transfer pump
  Faulty injection pump
  Faulty injection nozzles
  Fuel shut-off is engaged
  Air-intake system restricted
  Air leak in suction side of fuel system
Lubrication
  Wrong oil viscosity for ambient temperature
Electrical system
  Battery weak or dead
  Corroded or loose battery cables
  Cranking speed too slow

Uneven running
Engine
  Faulty valve clearance
  Stuck or burned valves
  Leaking cylinder-head gasket
  Low compression
  Worn or broken compression rings
  Faulty timing
  Engine overheating

Frequent stalling
Engine
  Valves sticking or burned
  Incorrect timing
  Engine overheating

Uneven running or frequent stalling
Fuel system
  Wrong type of fuel
  Water, dirt, or air in fuel system
  Air leak in suction side of fuel system
  Fuel line clogged or restricted
  Fuel filter restricted
  Faulty fuel-transfer pump
  Faulty injection pump
  Faulty injection nozzles
  Injection-nozzle leak-off lines clogged
  Injection pump out of time

Engine misses
Engine
  Weak valve springs
  Faulty valve clearance
  Burned, warped, pitted, or sticking valves
  Low compression
  Worn camshaft lobes
  Engine overheating
Fuel system
  Water, dirt, or air in fuel
  Wrong type of fuel
  Faulty injection nozzles
  Faulty injection pump
  Faulty fuel-transfer pump

Engine lacks power
Engine
  Dirty air-intake system
  Blown cylinder-head gasket
  Worn camshaft lobes
  Burned, warped, pitted, or sticking valves
  Faulty valve clearance
  Faulty valve timing
  Weak valve springs
  Low compression
  Engine overheating
Fuel system
  Wrong type of fuel
  Water, dirt, or air in fuel system
  Air leak in suction side of fuel system
  Fuel line clogged or restricted
  Fuel filter restricted

Speed-control linkage improperly
   adjusted
Faulty fuel-transfer pump
Faulty fuel-injection pump
Faulty fuel-injection nozzles
Injection-nozzle leak-off line clogged
Injection pump out of time
Clogged manifold system

Black or gray exhaust smoke
Engine
   Engine overloaded
   Faulty engine timing
   Restricted air cleaner
   Dirty air-intake system
   Faulty turbocharger
Fuel system
   Wrong type of fuel
   Excessive fuel delivery
   Faulty injection nozzles
   Faulty or improperly adjusted aneroid
   Injection-nozzle leak-off line clogged
   Injection pump out of time

White exhaust smoke
Engine
   Low compression
Fuel system
   Improper fuel
   Faulty injection nozzle
   Injection pump out of time

Slow acceleration
Engine
   Components worn
   Sticky governor
Fuel system
   Improper fuel
   Faulty injection nozzle

Abnormal engine noise
Engine
   Excessive valve clearance
   Worn cam followers
   Bent push rods
   Worn rocker-arm shafts
   Worn main or connecting-rod bearings
   Foreign material in combustion chamber
   Worn piston-pin bushing and pins
   Scored pistons
   Faulty engine timing
   Excessive crankshaft-end play
   Loose main-bearing caps
   Worn timing gears
   Worn oil-pump gears
   Broken oil-pump shaft
   Engine oil level low
   Camshaft oil-pump gear worn or broken
   Gears worn or broken
   Fan striking radiator
   Faulty turbocharger bearings

Excessive oil consumption
Engine
   External oil leaks
   Excessive engine speeds
   Piston rings not seated
   Piston rings worn or broken
   Piston rings sticking in grooves
   Scored pistons
   Faulty piston-ring tension
   Piston-ring gaps not staggered
   Excessive ring-groove wear
   Oil-return slots in pistons clogged
   Worn valve guides or valve stems
   Restricted crankcase breather
   Restricted air-intake system
   Excessive main or connecting-rod
      bearing clearance
   Worn crankshaft thrust bearing
      (misaligned piston and rod)
   Faulty front or rear crankshaft seal
   Excessive loss through vent tube
   Faulty turbocharger oil seal
   Crankcase oil too thin
   Oil level too high

Low oil pressure
Engine
   Oil level low
   Clogged oil-pump inlet screen
   Faulty oil pump
   Oil too thin
   Internal oil leakage
   Faulty regulating-valve operation
   Faulty oil-pressure indicator

High oil pressure
   Oil too heavy
   Faulty regulating-valve operation
   Faulty oil-pressure indicator

Engine overheats
Engine
   Defective head gasket
   Incorrect engine timing
   Scored pistons
   Engine oil level low
   Engine overloaded
Cooling system
   Low coolant level
   Air in coolant
   Faulty thermostat
   Cooling system limed
   Faulty water pump
   Radiator dirty or clogged
   Loose or broken fan belt
   Bent or broken fan blade
   Faulty radiator pressure cap
Fuel system
   Improper fuel
   Excessive fuel delivery
   Faulty injection-pump timing

Maintenance of Power Systems

Engine runs cold
Faulty thermostat
Faulty temperature gauge

Water pump leaks or is noisy
Worn seal and/or shaft or bearing
Worn or broken gasket
Faulty impeller

Oil in coolant or coolant in crankcase
Leaking head gasket
Cracked cylinder-block water jacket
Cracked cylinder liner
Cylinder-liner packings leaking

Turbocharger malfunctions
Abnormal whine
Faulty bearings
Other noise; vibration
Defective bearing lubrication
Leak in intake or exhaust manifolds
Improper clearance between turbine wheel and housing
Engine does not deliver rated power
Clogged manifold system
Foreign matter in compressor, impeller, or turbine
Excessive buildup in compressor
Leak in engine intake or exhaust manifolds
Bearing seizure in rotating assembly
Oil-seal leakage
Faulty seal
Suction caused by restricted air cleaner or air intake

Battery malfunctions
Undercharged battery
Excessive load from added accessories
Excessive engine idling
Low charging-system output
Current leakage
Low battery output
High resistance in circuit
Low electrolyte level
Low specific gravity
Defective cell
Cracked or broken case
Faulty terminal connections
Battery too small
Battery uses too much water
Cracked case
Battery being overcharged
Defective battery

Starting circuit If starting motor does not operate, connect voltmeter to solenoid S terminal and a good ground. With key switch on, press start switch.
Voltmeter registers battery voltage
Defective starting motor
Defective starter solenoid switch

Voltmeter does not register
Defective key switch or start switch
Faulty starting-circuit relay
Faulty wiring between key switch and starting-circuit relay
Faulty wiring between battery and solenoid S terminal
Faulty start safety switch
Starting motor will not spin or engine will not crank
Faulty battery
Burned or faulty solenoid switch contacts
Open, shorted, or grounded solenoid-switch pull-in windings
Poor brush contact or worn-out brushes
Burned commutator
Commutator mica too high
Open or grounded field winding
Faulty brush-spring tension
Grounded brush holder
Solenoid switch chatters
Low battery
Poor connection
Open solenoid hold-in circuit
Starting motor spins but does not crank engine
Faulty overrunning clutch pinion
Broken drive lever
Broken drive-lever pivot bolt
Broken magnetic-switch plunger hook
Faulty overrunning clutch
Engine cranks slowly
Too-heavy transmission oil
Low battery
High resistance in battery cables
Faulty battery-terminal connections
Burned or faulty solenoid-switch contacts
Faulty brush contact or worn-out brushes
Burned commutator
Commutator mica too high
Shorted or grounded armature coil
Faulty brush-spring tension
Armature rubbing pole core
Starting motor keeps running
Defective solenoid
Defective start switch
Shorted wiring
Abnormal noise while cranking
Armature interfering with stationary components
Starting-motor drive gear worn

Charging circuit
Low charging-system voltage
Faulty wiring
Low alternator amperage output
Defective regulator
Defective battery
Slipping drive belt

Low charging-system output
  Slipping drive belt
  Defective battery
  Grounded, shorted, or open field windings
  Defective rectifier bridge
  Defective diode trio
  Defective stator
  Defective regulator
High charging-system voltage
  High resistance at regulator connections
  Defective regulator
  Shorted field windings
  Grounded brush-sleeve clip
Noisy alternator
  Worn or defective bearings
  Defective drive belt
  Loose mounting or drive belt
  Pulley not properly aligned

Gauges
A gauge does not register
  No current to gauge
  Faulty sending or receiving unit
  Poor ground connection
  Connecting wire grounded to unit

A gauge consistently registers too high
  Faulty connection
  Broken connecting wire
  Poor ground at sending unit
  Failure of gauge or sender (usually sender)
Fuel gauge always shows empty
  Poor ground at receiver
  No current to receiver
  Grounded wire between sender and receiver
  Hole in sender float
Fuel gauge always shows full
  Poor ground at sender
  High resistance or open circuit between sender and receiver
  Defective sender or receiver

Fuel shut-off solenoid
High current draw
  Shorted windings
Low or no current draw
  High resistance in internal connection or wire
  Open circuit windings

Chapter **3**

Maintenance of Gasoline Power Systems

C. L. FRICKE
Engine Service Manager, Briggs & Stratton Corp., Milwaukee, Wis.

At almost every engine service meeting, someone asks, "How long should an engine last?" In all probability, the speaker answers by talking about the importance of preventive maintenance. Most small air-cooled engines are capable of operating 1000 hr with only routine maintenance and occasional repair as a result of normal wear. However, only a small percentage of air-cooled engines operate for more than 500 hr, because maintenance is marginal.

No contractor would consider paying two times the regular price of a tool at the time of purchase. Yet, this is exactly what happens when improper maintenance robs half the expected life of that tool. Some years ago, a member of the Outdoor Power Equipment Institute said that the equipment common to his industry was built to deliver 10 years of dependable service. He added that few owners will receive more than 4 years of service because their maintenance procedures were incomplete, improper, or nonexistent.

In the construction field, contractors must be concerned about the cost of equipment downtime. Lost time, resulting from equipment which does not operate, is intangible. The cost of lost time is not easily measured. But it is generally conceded that the cost of lost time far exceeds the cost of a preventive maintenance program.

Manufacturers of small air-cooled engines are particularly concerned about their product, since they realize that the small engine is usually an orphan on construction job sites. Everybody operates the small engines, but it seems no one is really responsible for maintenance.

In the construction industry, a small engine is rarely operated by only one man. Instead, engine-powered portable tools are moved from one job site to another. During any given day, several men may operate the same engine. For this reason, it is usually impractical to attempt to keep a log of the number of hours the engine is operated. Therefore, most contractors find it best to have one man perform preventive maintenance on all small engines on a calendar basis. In this way, preventive maintenance is performed every third or fourth day or perhaps on a certain day of each week.

In discussing the matter of engine life and preventive maintenance with contractors throughout the country, it became obvious that contractors who are pleased with the performance of their small air-cooled engines are those who assigned the responsibility of preventive maintenance to a single person. They charged one individual with performing preventive maintenance as recommended by the engine manufacturer.

Most small engine downtime results from one or more of the following conditions:

    1. Wear, caused by dirt and abrasives which enter the engine because of improper air-cleaner maintenance

Maintenance of Power Systems

2. Scoring, seizures, or parts breakage which occurs when an engine is operated without a sufficient amount of oil in the crankcase or with a lubricant of improper viscosity

3. Damage from abrasives which enter the engine when a spark plug has been improperly cleaned on an abrasive blast-cleaning machine

4. An excessive accumulation of combustion deposits which rob power and shorten valve life

5. Overheating, resulting from plugged cooling fins on the cylinder and cylinder head

6. Damage caused by improper storage during periods when the engine is not being operated

The following pages discuss these areas in greater detail.

AIR CLEANER—FUNCTION AND DESIGN

The average small engine uses about 10,000 gal of air for each gallon of fuel. It is obvious that the air which an engine breathes must be clean. If the air is contaminated with dust and dirt, abrasives eventually find their way into the engine crankcase and contaminate the engine oil. Since most small air-cooled engines do not include an oil filter, abrasives will circulate to internal moving parts again and again.

Briggs & Stratton engines, from 2 to 9 hp, are equipped with an Oil-Foam air cleaner, utilizing a reusable polyurethane foam element housed in a metal cannister. The foam element is oiled so that the millions of microfine cells trap dirt before it can enter the engine. Oil-Foam air cleaners are designed in such a way that sealing lips of the foam element actually become a gasket when the air cleaner is assembled. This design prevents dirt from bypassing the oiled form element.

SERVICING THE OIL-FOAM AIR CLEANER

Oil-Foam air cleaners should be cleaned and reoiled after every 25 hr of operation under normal conditions—more frequently under extremely dusty conditions. Engines of 2 to 5 hp use a rectangular-shaped Oil-Foam air cleaner of the type shown in Fig. 3-1. Service is performed as follows:
1. Remove mounting screw.
2. Remove air cleaner carefully to prevent dirt from entering carburetor.
3. Take air cleaner apart.
4. *a.* Wash foam element in kerosene or liquid detergent and water to remove dirt.
    b. Wrap foam in cloth and squeeze dry.
    c. Saturate foam in engine oil. Squeeze to remove excess oil.
    d. Assemble air-cleaner parts—fasten to carburetor with screw.

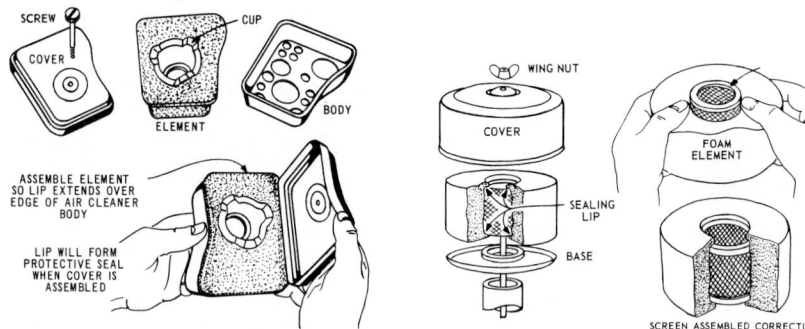

Fig. 3-1 Air-cleaner maintenance procedure, 2 to 5 hp engines.

Fig. 3-2 Air-cleaner maintenance procedure, 6 to 9 hp engines.

Engines of 6 to 9 hp use a cylindrical-shaped Oil-Foam air cleaner of the type shown in Fig. 3-2. Service is performed as follows:
1. Remove wing nut and cover.
2. Lift foam element from base.

3. Push down foam element as shown and pull out screen.
4. *a.* Wash foam element in kerosene or liquid detergent and water to remove dirt.
   b. Wrap foam in cloth and squeeze dry.
   c. Saturate foam in engine oil. Squeeze to remove excess oil.
   d. Put screen inside element. Be sure sealing lip is over end of screen (top and bottom).
5. Reassemble parts as shown. Assemble wing nut finger tight.

Many engines from 10 to 16 hp are equipped with a heavy-duty dual-element air cleaner, utilizing both a washable paper cartridge and an oil-foam precleaner sleeve.

SERVICING THE DUAL-ELEMENT AIR CLEANER

Clean and reoil the foam precleaner sleeve at 3-month intervals or every 25 hr, whichever occurs first (Fig. 3-3).
1. Remove wing nut and cover.
2. Remove foam precleaner sleeve by sliding it off the paper cartridge.
3. *a.* Wash foam in kerosene or liquid detergent and water to remove dirt.
   b. Squeeze foam dry.
   c. Oil foam with 1 oz of engine oil. Squeeze to distribute oil evenly.
4. Install foam precleaner sleeve over paper cartridge.
5. Reassemble cover and wing nut. Assemble wing nut finger tight.

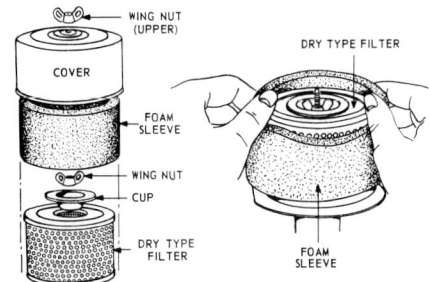

Fig. 3-3 Dual-element air-cleaner maintenance procedure.

Yearly or every 100 hr, whichever occurs first, remove the paper cartridge. Clean the cartridge by tapping gently on a flat surface. If the cartridge is very dirty, replace the cartridge or wash it in liquid detergent and water. Rinse the cartridge until the water remains clear. The cartridge must be air dried *thoroughly* before using. NOTE: Service the air-cleaner assembly more frequently under dusty conditions.

LUBRICATION AND FUEL RECOMMENDATIONS

Many engines are damaged because they are operated without a sufficient amount of oil in the crankcase or because the oil is not of the proper viscosity. The problem generally occurs because the crankcase oil level is not checked at reasonable intervals. It is good practice to check the oil level each time an engine is refueled.

Use a high-quality oil classified "For Service SC, SD, SE, or MS." This is the same type of oil usually recommended for use in truck, automotive, and large stationary engines. Such oils do include a detergent, which helps to keep internal parts cleaner by retarding the formation of gum and varnish deposits. Nothing should be added to the recommended oil.

Select oil viscosity for the season in which an engine will be operated. Briggs & Stratton recommend the following: In summer (above 40°F), use SAE 30; if not available, use SAE 10W-30 or SAE 10-W 40. In winter (under 40°F), use SAE 5W-20 or SAE 5W-30; if not available, use SAE 10W or SAE 10W-30. Below 0°F, use SAE 10W or SAE 10W-30 diluted 10 percent with kerosene.

Use clean, fresh, lead-free or leaded regular-grade automotive gasoline. Purchase gasoline in relatively small quantities, so it can be used within 60 days. In this way, the fuel will remain fresh and the volatility will be correct for the season.

BLAST CLEANING SPARK PLUGS

Cleaning spark plugs on an abrasive blast-cleaning machine has long been an accepted field practice. Many conscientious repairmen and engine owners blast-clean spark plugs frequently, simply as a matter of routine maintenance. They believe that blast cleaning spark plugs is a way to salvage plugs, decrease repair costs, and improve engine performance. Unfortunately, this is often false.

2-36   Maintenance of Power Systems

We investigated complaints on premature wear and short engine life and found that spark plugs had been cleaned on an abrasive blast-cleaning machine. We then used an abrasive blast-cleaning machine to clean spark plugs on several new engines. After only a few hours of operation, extreme wear had occurred on internal engine parts. The wear had been caused by abrasive grit which remained within the plug after blast cleaning. During operation, grit had fallen into the combustion chamber, then worked its way past the piston rings toward the crankcase.

When a spark plug has been cleaned by the abrasive blast-cleaning method, some of the abrasive grit remains tightly packed within the plug after the cleaning operation. Although instructions with spark plug cleaning machines call for solvent cleaning to remove oil deposits before blasting, contacts with local automotive service stations indicate that few, if any, cleaning machine operators degrease the plugs. Further, simply blowing out the plugs with a forceful blast of air after blast cleaning is not adequate to remove all the tightly packed grit. In tests, spark plugs that contained even a small amount of grit caused considerable wear after a few hours of operation.

Fig. 3-4   About 200 mg of abrasive blast-cleaning grit will fill a circle approximately the size of a nickel. This amount can easily remain within a spark plug after blast cleaning.

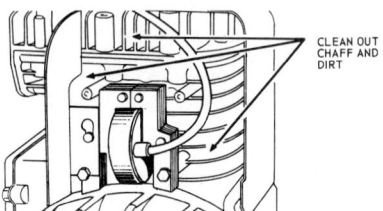

Fig. 3-5   Area for blower housing cleaning.

To get an idea of the way in which spark plugs are being blast-cleaned in the field, we had two plugs cleaned at each of five local automotive service stations. We then washed them out in gasoline to remove any abrasive grit which remained after cleaning. All the spark plugs contained enough grit to cause engine wear. The residue of grit in the 10 plugs ranged between 32 to 372 mg. To give some idea of the quantities involved, 200 mg of grit will fill a circle the size of a nickel (Fig. 3-4).

Engines that have been damaged by abrasive blast-cleaning grit usually have very high piston ring wear, scratched or scored cylinder bores, and extreme amounts of smooth wear on the crankshaft crankpin and bearing journals. The grit, which can be introduced into an engine as a result of improperly cleaned spark plugs, may well account for cases of unexpected wear in which there is no evidence of abuse or neglect by the operator.

In view of our findings, we most urgently recommend that dirty spark plugs be cleaned only in a solvent, or by scraping the electrodes and then regapping. If a dirty spark plug cannot be cleaned in this manner, it should be discarded and a new plug used.

CLEAN OUT COMBUSTION CHAMBER

Industrial engines are often operated at constant speed and at relatively constant load. Under these conditions, the use of leaded automotive fuels results in a gradual buildup of deposits in combustion chambers. When combustion deposits accumulate in excessive amounts, engines lose power. In addition, particles from the deposits may lodge between exhaust valves and seats, causing valves to burn. Removing combustion deposits, after every 100 to 300 hr of operation, will restore power and increase valve life.

Clean out the combustion chamber as follows:

   1. Remove cylinder head screws. Be sure to note if the screws are of different lengths and have steel washers because they must be replaced in their original positions.

2. Turn the crankshaft until the piston is at the top of the cylinder bore and both valves are closed. Scrape and wire-brush the combustion deposits from the cylinder head and combustion chamber.

3. Reuse the cylinder head gasket only if it is in good condition. Reassemble the cylinder head. Turn each screw with a wrench until the screw head is lightly seated.

4. Use a socket wrench with a 6-in. handle and turn all the screws one-quarter turn. Run the engine for about 5 min and retighten all the screws approximately one-quarter turn.

CLEAN COOLING SYSTEM

Air-cooled engines require a generous flow of cooling air, which is directed across cooling fins on the cylinder and cylinder head. Over an extended period of time, dirt and chaff may clog cooling fins, interrupting the flow of cooling air. When this occurs, the engine may overheat severely. Periodically, remove the blower housing and clean out dirt and debris (Fig. 3-5).

STORAGE INSTRUCTIONS

Engines to be stored over 30 days should be completely drained of fuel to prevent gum deposits from forming on essential carburetor parts, the fuel filter, the fuel lines, or the tank.

1. Remove all fuel from the fuel tank. Run the engine until it stops from lack of fuel. Remove the small amount of fuel that remains in the sump of the tank by absorbing it with a clean dry cloth.

2. While the engine is still warm, drain the oil from the crankcase. Refill with fresh oil.

3. Remove a spark plug, pour 1 oz (2 or 3 tablespoons) of engine oil into the cylinder, and crank slowly to distribute the oil. Replace the spark plug.

4. Clean dirt and chaff from the cylinder, the cylinder head fins, and the blower housing.

There is no magic royal road to success with small engines. Engine life, performance, and dependability will be in proportion to maintenance. A preventive maintenance program is a must for contractors who cannot afford unnecessary engine downtime.

Chapter **4**

Principles and Maintenance of Hydraulic Systems

MARVIN SHERMAN AND LINCOLN BURROWS
Sperry Vickers, Division of Sperry Rand Corporation, Troy, Mich.

Hydraulic power has long been one of the most useful and versatile methods of actuating and controlling construction machinery. The brute force and very precise control provided by hydraulic systems have taken much of the work out of operating earthmoving and material-handling equipment and self-propelled vehicles of all kinds. As applications of this form of power grow, so does the need for a better understanding of hydraulic system operation and maintenance.

FUNDAMENTALS

A hydraulic system is used to transfer mechanical energy from one place to another, using pressure energy as the medium. Mechanical energy, driving the hydraulic pump, is converted to pressure energy and kinetic energy in the fluid. This is reconverted to mechanical energy to move a load, such as a bucket, boom, or blade.

The power of the fluid flowing under pressure can be applied directly to the load to produce rotary or linear motion without gears, chains, belts, magnets, or commutators. This direct application of force by an incompressible, nearly frictionless fluid is one of the main factors contributing to the high efficiency of a hydraulic system. Since the fluid also serves as a lubricant, it helps reduce wear and extend the life of hydraulic components.

When hydraulic fluid is confined in a pipe or hose it acts as if it were a steel rod. Power applied at one end of the line causes equal power to be applied at the other end. The power source, that is, the pump, can be connected to the load with relative ease. Hydraulic pipe and hose can be installed around corners, along booms, and through panels. Hydraulic drive costs are competitive with those of mechanical systems and are lower than those for electrical drives on a cost-per-horsepower basis.

The high power-to-weight ratio of hydraulic components is one factor. The economic advantage of hydraulics is particularly evident where a single hydraulic pump is used to drive several hydraulic motors at various locations on a construction vehicle. The high power-to-weight ratio of hydraulic pumps and motors often permits them to be installed in places on a vehicle that would be too small for electrical or mechanical devices having equal capabilities. For example, the cubic space required for a 100-hp electric motor is almost 14 times that required for a 100-hp hydraulic motor.

All hydraulic systems are basically the same regardless of the application. A typical circuit (Fig. 4-1) includes four basic components: (1) a tank (reservoir) to hold the fluid, (2) a pump to force the fluid through the system, (3) valves to control fluid pressure, flow rate,

2-40 Maintenance of Power Systems

and direction, and (4) one or more actuators to convert the energy of fluid movement into mechanical force to perform the work. The actuators are either cylinders for linear motion or motors for rotary motion.

These basic components can be combined in a variety of ways to achieve various motions and, if necessary, at various rates and various locations on a machine. Circuit drawings or diagrams show how the hydraulically operated machine works. These diagrams are also essential for installation of the hydraulic components, for general maintenance work, and for troubleshooting the system.

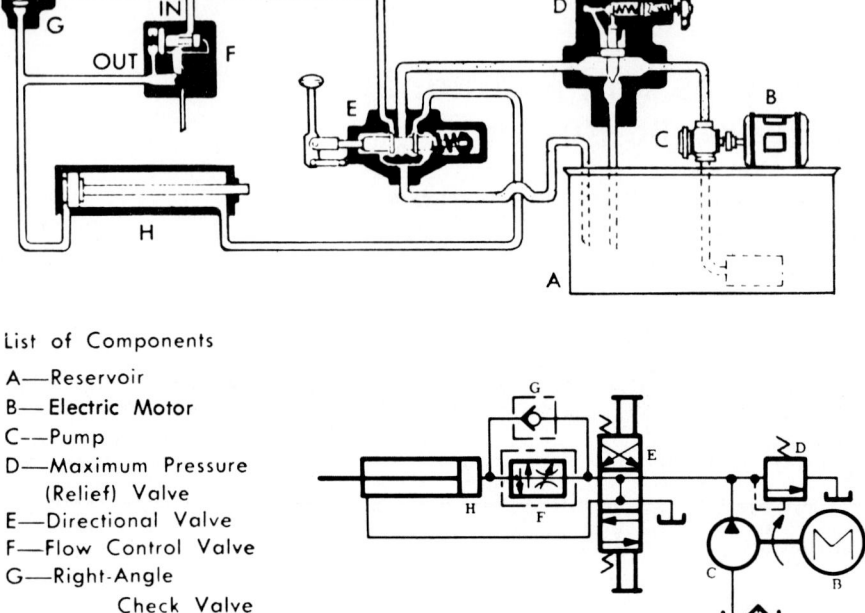

List of Components

A—Reservoir
B—Electric Motor
C—Pump
D—Maximum Pressure
   (Relief) Valve
E—Directional Valve
F—Flow Control Valve
G—Right-Angle
   Check Valve
H—Cylinder

Fig. 4-1 Typical hydraulic circuit is shown here in both pictorial forms and as it would appear using ANSI symbols.

Figure 4-2 shows how graphic symbols are used in most circuit diagrams. These symbols, specified by the American National Standards Institute (ANSI), show the function of system components rather than their shape, internal construction, or operation. The same type of component is often represented by somewhat different symbols, depending on its use in the system.

PUMPS

The pump is the heart of a hydraulic system and, right or wrong, gets most of the attention when a problem develops in a system. Of all the components, it is the most susceptible to failure. It often runs continuously through a work shift, and it is the first component exposed to fluid contaminants when they leave the reservoir. Because of its high performance and relatively small size, a hydraulic pump is more vulnerable to damage from improper operation or maintenance. Still, with proper operation and maintenance practices, the pump can be a most reliable component.

Some equipment operators who stress good maintenance can predict within a few hours of operation when a pump will need rebuilding. They include rebuilding as part of a maintenance program and have very few actual failures.

Classification and rating Practically all pumps used in today's hydraulic systems are positive displacement units. That is, they provide a given amount of fluid for every revolution, stroke, or cycle. Fluid output, except for internal leakage losses, is independent of system pressure.

Pumps are classified as fixed- or variable-displacement units, with displacement depending on the working relationship of internal operating elements. In a *fixed-displacement pump*, the relationship cannot be changed, and pump delivery can be varied only by changing pump speed. The amount of flow going to various parts of the hydraulic circuit is controlled with various valves. In a *variable-displacement pump*, displacement can be varied by an integral controlling device which adjusts the physical relationship of the pump's operating parts.

Pumps are generally rated on the basis of pressure and volumetric output. The *pressure rating* indicates how much pressure the pump can withstand safely for a given time without damage to its parts. This, in turn, determines how much load a system can handle.

Volumetric output, or flow, is given in gallons per minute (gpm) at a designated drive shaft speed. The SAE ratings for vane pumps on construction machinery are designated at 1200 rpm and 100 psi outlet pressure. Other frequently used terms for pump flow rating include *delivery rate, capacity,* and *pump size*. A third means of rating is on the basis of pump *displacement*. This refers to the amount of fluid delivered in one revolution of a pump's rotating members. Displacement is expressed in cubic inches per revolution.

Most hydraulic pumps are the rotary type. That is, a rotating assembly carries the fluid from the pump inlet to the outlet. The type of element that transfers the fluid further establishes another basic approach to pump classification. These elements can be in the form of *vanes, pistons,* or *gears,* which are the three most common types of pumps.

Figure 4-3 shows the configuration of a simple vane pump. A cylindrical rotor with movable vanes in radial slots rotates in a circular housing. As the rotor turns, centrifugal force drives the vanes outward so that they are always in contact with the inner surface of the housing. This action promotes both long life and efficiency since the vanes move automatically to compensate for vane tip wear.

Pumping chambers are formed between succeeding vanes carrying fluid from the inlet to the outlet. A partial vacuum is created at the inlet as the space between vanes increases. Then, fluid is squeezed out at the outlet as the pumping chamber size decreases.

The pump shown in Fig. 4-3 is referred to as *unbalanced* because high pressure is generated on only one side of the rotor shaft. Most of today's vane pumps are of the *balanced* design shown in Fig. 4-4. In this case, an elliptical housing forms two separate pumping chambers on opposite sides of a rotor so that the side loads cancel out, increasing bearing life and permitting higher operating pressures.

Piston pumps A piston-type pump is really a rotary-reciprocating pump. Several pistons—usually seven or nine—reciprocate in rotating cylinder barrels. The pistons retract while passing the inlet port to create a vacuum and permit oil to flow into the pumping chambers. They extend at the outlet to push the oil into the system.

Most piston pumps are classified as *radial* and *axial*. In a radial piston pump (Fig. 4-5) the pistons are arranged like wheel spokes in a short cylindrical block. This cylinder block is rotated by the drive shaft inside a circular housing. The block turns on a stationary pintle that contains the inlet and outlet ports. As the cylinder block turns, centrifugal force slings the pistons outward and they follow the circular housing. The housing centerline is offset from the cylinder block centerline. The amount of eccentricity between the two determines the piston stroke and therefore the pump displacement. Controls are applied to change the housing location and thereby vary the pump delivery from zero to maximum.

In axial piston pumps, the pistons stroke axially, or in the same direction as the cylinder block centerline. Axial piston pumps can be of in-line or angle design.

Gear pumps External and internal gear pumps make up most of the types of pumps classified under gear pumps. *External gear pumps* have the largest application in power transmission. *Internal gear pumps* are used for automatic transmissions and power steering units in automobiles.

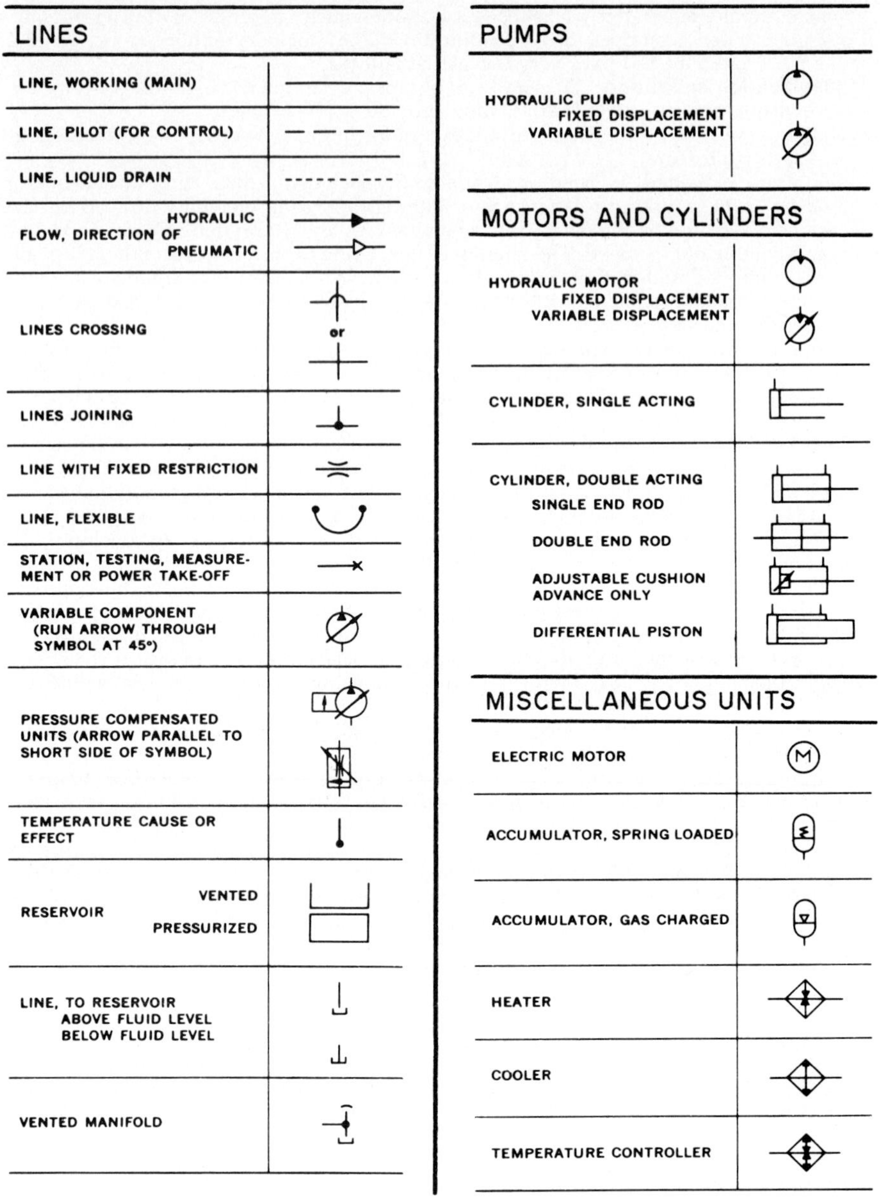

LINES

LINE, WORKING (MAIN)	───────
LINE, PILOT (FOR CONTROL)	─ ─ ─ ─
LINE, LIQUID DRAIN	-------
FLOW, DIRECTION OF HYDRAULIC PNEUMATIC	▶ ▷
LINES CROSSING	┬ or ┼
LINES JOINING	⊥
LINE WITH FIXED RESTRICTION	≍
LINE, FLEXIBLE	⌣
STATION, TESTING, MEASUREMENT OR POWER TAKE-OFF	──✕
VARIABLE COMPONENT (RUN ARROW THROUGH SYMBOL AT 45°)	⌀
PRESSURE COMPENSATED UNITS (ARROW PARALLEL TO SHORT SIDE OF SYMBOL)	
TEMPERATURE CAUSE OR EFFECT	↓
RESERVOIR VENTED PRESSURIZED	⊔ ▭

LINE, TO RESERVOIR
    ABOVE FLUID LEVEL
    BELOW FLUID LEVEL	

VENTED MANIFOLD	

PUMPS

HYDRAULIC PUMP
    FIXED DISPLACEMENT
    VARIABLE DISPLACEMENT
	○
	⌀

MOTORS AND CYLINDERS

HYDRAULIC MOTOR
    FIXED DISPLACEMENT
    VARIABLE DISPLACEMENT
	○
	⌀

CYLINDER, SINGLE ACTING	

CYLINDER, DOUBLE ACTING
    SINGLE END ROD
    DOUBLE END ROD
    ADJUSTABLE CUSHION ADVANCE ONLY
    DIFFERENTIAL PISTON	

MISCELLANEOUS UNITS

ELECTRIC MOTOR	Ⓜ
ACCUMULATOR, SPRING LOADED	
ACCUMULATOR, GAS CHARGED	
HEATER	◆
COOLER	◆
TEMPERATURE CONTROLLER	◆

Fig. 4-2 Symbols for hydraulic-system representation as set down by ANSI.

MISCELLANEOUS UNITS (Cont.)

FILTER, STRAINER	
PRESSURE SWITCH	
PRESSURE INDICATOR	
TEMPERATURE INDICATOR	
COMPONENT ENCLOSURE	
DIRECTION OF SHAFT ROTATION (ASSUME ARROW ON NEAR SIDE OF SHAFT)	

METHODS OF OPERATION

SPRING	
MANUAL	
PUSH BUTTON	
PUSH-PULL LEVER	
PEDAL OR TREADLE	
MECHANICAL	
DETENT	
PRESSURE COMPENSATED	
SOLENOID, SINGLE WINDING	
REVERSING MOTOR	

PILOT PRESSURE REMOTE SUPPLY	
INTERNAL SUPPLY	

VALVES

CHECK	
ON-OFF (MANUAL SHUT-OFF)	
PRESSURE RELIEF	
PRESSURE REDUCING	
FLOW CONTROL, ADJUSTABLE— NON-COMPENSATED	
FLOW CONTROL, ADJUSTABLE (TEMPERATURE AND PRESSURE COMPENSATED)	
TWO POSITION TWO CONNECTION	
TWO POSITION THREE CONNECTION	
TWO POSITION FOUR CONNECTION	
THREE POSITION FOUR CONNECTION	
TWO POSITION IN TRANSITION	
VALVES CAPABLE OF INFINITE POSITIONING (HORIZONTAL BARS INDICATE INFINITE POSITIONING ABILITY)	

2-44 Maintenance of Power Systems

The external gear pump (Fig. 4-6) consists of two meshed gears in a closely fitted housing. The inlet and outlet ports are opposite each other. One gear is powered and, in turning, drives the other. In moving past the inlet, gear teeth create a partial vacuum. Oil is drawn from the inlet and carried to the outlet in the pumping chambers formed between the gear teeth and the housing. The internal gear pump (Fig. 4-7) has an inner

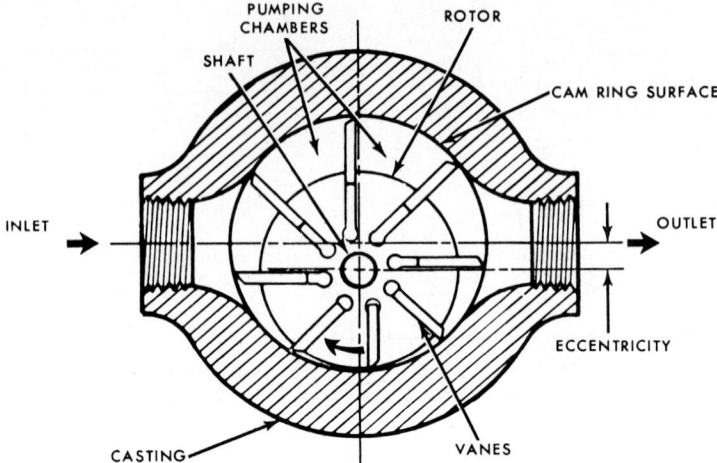

Fig. 4-3 Vane pump—unbalanced design.

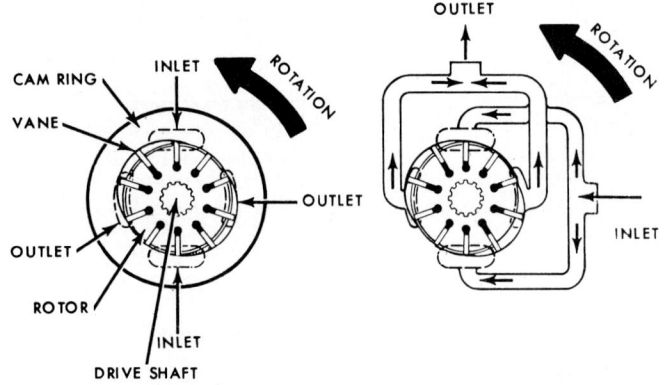

Fig. 4-4 Vane pump—balanced design.

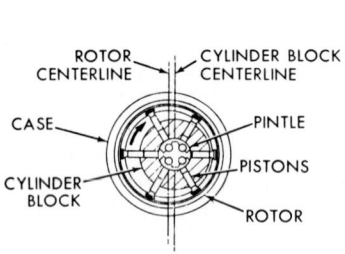

Fig. 4-5 Radial piston pump.

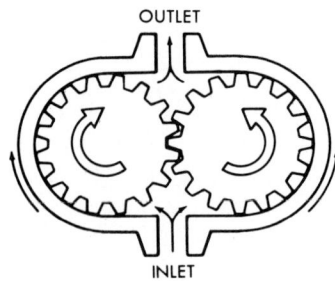

Fig. 4-6 External gear pump.

gear keyed to the drive shaft, a larger external gear, a crescent seal, and a closely fitted housing. The two gears are not concentric; as they rotate, pumping chambers open up between them at the inlet and close off at the outlet. The crescent seals the inlet port from the outlet and both gears carry oil past it.

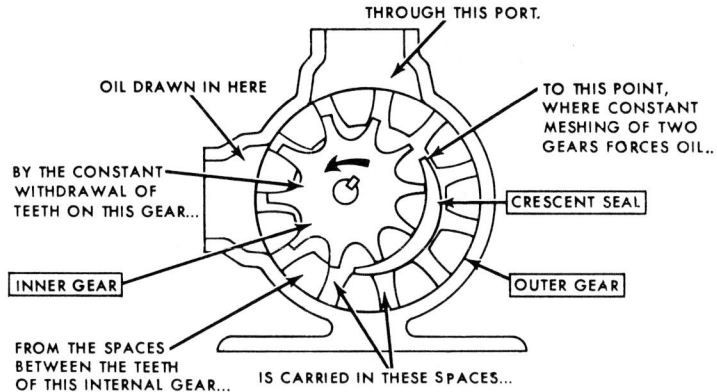

Fig. 4-7 Internal gear pump.

VALVES

Hydraulic valves, which are used to control actuators, fall into three categories: pressure, flow, and directional. As might be expected, some valves have multiple functions that cover more than one category.

All pure *pressure-control* valves operate in a condition approaching hydraulic balance; that is, pressure is effective on one side or end of a ball, poppet, or spool and is opposed by a spring. This is shown in Fig. 4-8, which details a simple check valve of ball and seat placed between two ports. Flow through the seat pushes the ball away to permit free flow, and flow in the other direction pushes the ball against the seat to seal the passage. A check valve could be either a pressure or a directional-control valve or both.

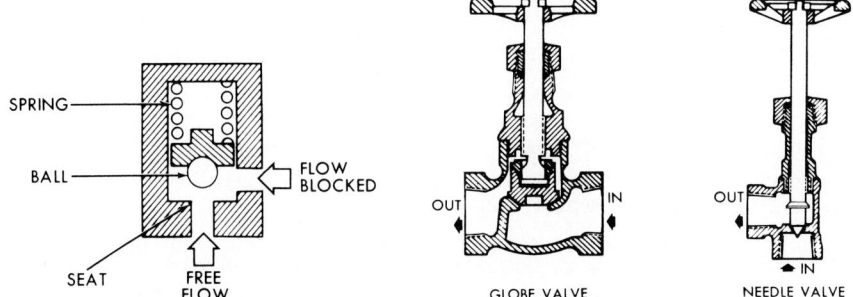

Fig. 4-8 Typical ball-type check-valve arrangement.

Fig. 4-9 Globe- and needle-valve principles and flow pattern.

A *flow-control* valve controls actuator speed by regulating flow. Flow-control valves are rated according to gallon-per-minute capacity and operating pressure. They are classified as *adjustable* or *nonadjustable*. Fixed-size orifices are one form of nonadjustable flow-control valve, but two of the most common are *needle* and *globe* valves. Figure 4-9 provides a schematic of these two valves.

While the basic check valve can and does serve as a direction-control valve, the most common is the reversing directional (or four-way) valve (Fig. 4-10). This valve has at least two finite positions with two possible flowpaths in each extreme position. It must have four ports: P (pump or pressure), T (tank or reservoir), and actuator ports A and B. In one extreme position, the valve has the pump port connected to the B actuator port and the A port to the tank. In the opposite position, flow is reversed; the pump is connected to the A port and the B port to the tank.

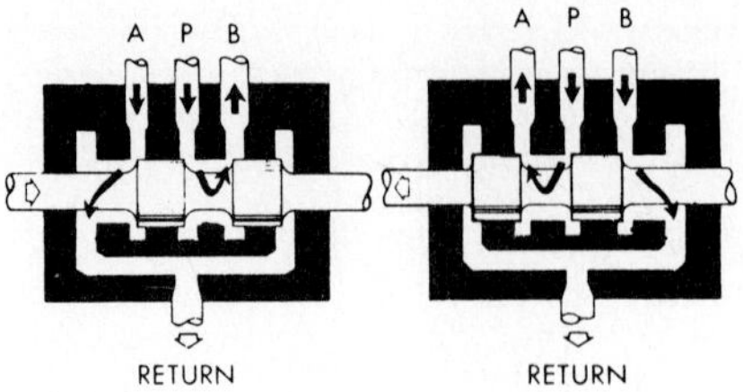

ACTUATORS

Actuators in a hydraulic system are devices that convert pressure energy into mechanical force and motion. They are either *linear* or *rotary*. A linear actuator (cylinder or ram) can give force and motion outputs in a straight line. A rotary actuator (or motor) produces torque and rotating motion.

Fig. 4-10 Reversing directional valves have at least two finite positions, with two possible paths in each extreme position.

The *cylinder* is a piston or plunger operating in a cylindrical housing and the ram is a single-acting plunger-type cylinder. The basic parts of a cylinder are identified in Fig. 4-11. The rotary actuator motor is a pump that is being pushed instead of doing the pushing. The main types of motors are identical to the pumps—vane, piston, and gear. Nearly all are reversible-type actuator motors.

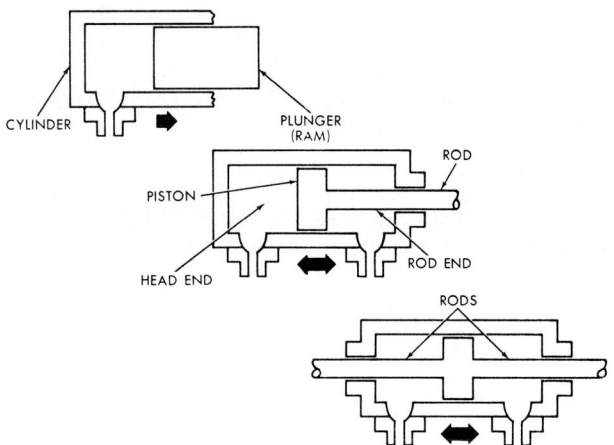

Fig. 4-11 Basic parts and nomenclature of hydraulic cylinders.

MAINTENANCE MORE IMPORTANT

The importance of hydraulic maintenance has gone through a step change. Today's sophisticated construction machinery is designed to meet the demand for increased productivity, meaning bigger payloads delivered faster under more precise control techniques and at higher operating temperatures. Good maintenance practices, most of which are simply a matter of commonsense, can bring out the best from hydraulic components.

Although most of the recommendations made here may have a familiar ring, their importance justifies repetition. The following are problems handled all too frequently by service personnel in working with construction equipment users:

Wrong grade or type of hydraulic fluid
Dirty or clogged fluid filters
Insufficient fluid in the reservoir
Pump driven in wrong direction
Loose pump inlet lines

Any of these situations could be avoided through relatively simple maintenance procedures.

Fluid factors As indicated in the preceding list, and as verified by maintenance records, about three-quarters of all hydraulic service problems can be traced directly or indirectly to hydraulic fluid. It might be a matter of the wrong fluid, or dirty fluid, or perhaps not enough fluid.

The importance of keeping things right with hydraulic fluid cannot be overly stressed. It is the lifeblood of a system, serving not only as the means of transmitting power but also as a lubricant and coolant. Such things as cleanliness, viscosity, and temperature have a great effect on overall system performance.

In the first place, the correct fluid must be used. Selection depends on such characteristics as viscosity index, pour point, acidity, lubricity, usable temperature range, and cost. Operators should keep in mind, however, that the best approach to fluid selection is to follow the recommendations of the machinery manufacturer.

Most fluids have a petroleum base with special additives to provide the qualities required for hydraulic system operation. For severe service such as construction work, antiwear additives help provide reasonable operating life for hydraulic components. Several years ago hydraulic equipment manufacturers were recommending the use of regular crankcase oils with SAE classifications such as SC, SD, and SE. Now, the major oil companies offer specially formulated hydraulic fluids with the antiwear qualities of these crankcase oils.

Viscosity. Fluid viscosity is extremely important. Hydraulic fluid must have sufficient body to provide adequate sealing between working parts of pumps, valves, cylinders, and motors, but not enough to cause pump cavitation or sluggish valve operation. Viscosity is especially critical during start-up at low temperatures. If viscosity is too high, the pump may not get enough fluid and could actually seize up for lack of lubrication.

For extremely wide seasonal variations, the viscosity grade of the fluid should be changed in the spring and fall, as with automobile engines. Original equipment manufacturers' (OEM) operation and maintenance manuals should be checked for the proper grades. Where hydrostatic transmissions are involved, their control mechanisms might require still different viscosity fluids.

Petroleum-base fluids tend to thin out with increasing temperature and to thicken with decreasing temperature. This characteristic is expressed as *viscosity index* (V.I.). A high V.I. number means that viscosity changes at a slow rate with respect to temperature changes.

In general, the V.I. of mobile hydraulic fluids should not be less than 90. Multiple viscosity oils like SAE 10W-30 include additives to improve V.I. These oils generally exhibit both a temporary and a permanent decrease in viscosity due to oil shear that occurs in the operating hydraulic system. Actual operating viscosity can be far less than that shown in the oil specification data. For this reason, such oils should have high shear stability to assure that viscosity remains within recommended limits (Table 4-1).

TABLE 4-1 Typical Oil-Viscosity Recommendations for Mobile Hydraulic Systems

Hydraulic-system operating temperature range (min* to max)	SAE viscosity designation
−10°F to 130°F (−23°C to 54°C)	5W 5W-20 5W-30
0°F to 180°F (−18°C to 83°C)	10W
0°F to 210°F (−18°C to 99°C)	10W-30**
50°F to 210°F (10°C to 99°C)	20-20W

*Ambient start-up temperature.

Hydraulic components have closely fitted mating parts, and fluid lubricity is a must for reasonably long operating life. Crude oil and animal or vegetable oils should never be used. They do not lubricate properly and do not resist rust, corrosion, or foaming.

2-48 Maintenance of Power Systems

Oxidation Resistance. Oxidation resistance has a major effect on the life of hydraulic fluids. Air, heat, and contamination tend to promote oxidation.

All hydraulic fluids will combine with air to a certain degree. High operating temperatures speed up this action. For every 18° rise in temperature the oxidation rate doubles, creating contaminants which then tend to accelerate further contamination. Better grades of fluids usually include inhibitors to retard this action, but operating temperatures should still be watched closely.

To help avoid trouble, bulk fluid temperature measured at the reservoir should be kept in the 100 to 130°F range because localized hot spots may be 100° hotter than the bulk fluid temperature. A thermometer attached to the end of a wire and placed in the reservoir fill pipe can be used to measure fluid temperature. Also, temperature stickers are available for attaching to the side of a reservoir or suspected hot spot. Spots on the sticker change color permanently to indicate the maximum temperature.

When draining a system, try to remove all the used fluid so that none remains to mix with the new fluid. In most cases, bleeding at the lowest point in the system will help. It's also a good idea to drain only after the fluid is warmed up. This helps keep any contaminants in suspension so that they will be removed with the drained fluid. After draining, accumulated deposits should be flushed out with light viscosity fluid. This fluid should have a rust inhibitor to protect metal surfaces against rust after removing the hot fluid.

When a hydraulic pump or motor fails, the system should be considered contaminated, and the unit should be removed for repair. After removal of the pump or motor, the reservoir should be drained and the fluid discarded. Then the reservoir should be flushed to remove all contamination. Lines, cylinders, and valves should also be flushed.

Filtration and handling Fluid contamination must be avoided at every point, both in and out of the system. This includes storage and handling. Drums should be stored on their sides and, if possible, covered to prevent accumulation of dirt and water. Also, they should be kept full to reduce the chance for condensation to build up. Special care is a good idea when transferring fluid from drums into the hydraulic reservoir, and vice versa.

Many equipment builders recommend using a fluid transfer pump equipped with a

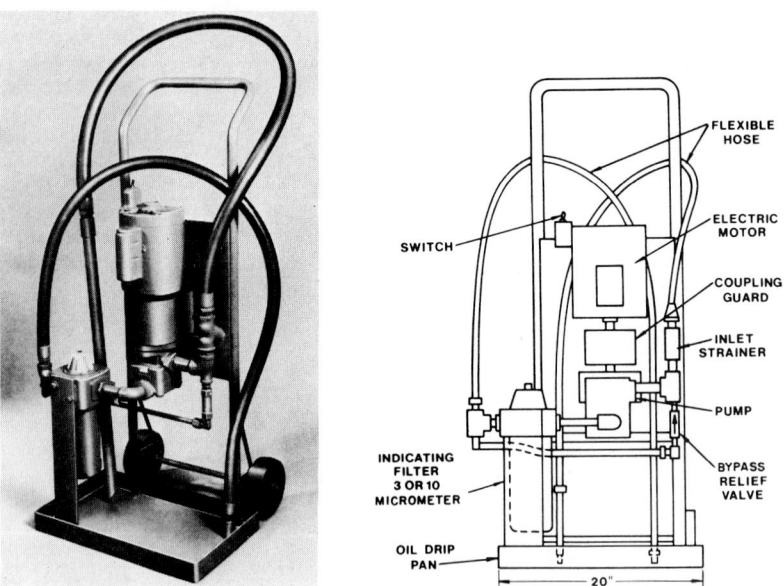

Fig. 4-12 *A* and *B* Portable hydraulic filters can be very helpful in keeping hydraulic systems clean when used on a regularly scheduled basis, following component repair or replacement, or during break-in of new machines.

filter of 25-micron rating, or finer. Portable filters are available for this and other applications (Fig. 4-12A and B). These portable units can be used for periodic filtering of system fluid while the machinery is in operation or for cleaning up the system following repair or replacement of hydraulic components. They are also useful during the initial break-in period on a new machine when so-called built-in contaminants such as metal chips, lint, or paint chips can be dislodged and circulated through the system. At times, service and maintenance procedures require that fluid be drained from a system. Here the portable filter can be used as a unit to transfer fluid into a drum and then subsequently back to the machine.

Establishing a schedule for using this type of auxiliary filtration must be based on experience—preferably documented with good records. The many variables to which each machine is exposed create system-contaminant levels which build up at different time rates in the individual machines. Retaining and analyzing maintenance and production records on these machines provide guidance in setting up schedules for using portable filters.

Hand-carried kits are available which provide on-the-spot fluid-contaminant-level determinations. A sample of fluid is drawn from a system and filtered through a very fine membrane disk. Discoloration of the disk can be compared with that of disks obtained previously. Tying these results into other maintenance records gives a higher degree of confidence that the machine will not have trouble from system contamination.

Typical of filters installed as an integral part of a hydraulic system are micronic-particle units with a replaceable element made of alpha cellulose (Fig. 4-13). The element normally recommended provides two-stage filtration to trap and retain a majority of particles more than 8 to 10 microns in size. For comparison, the smallest particle a sharp eye can see is about 40 microns.

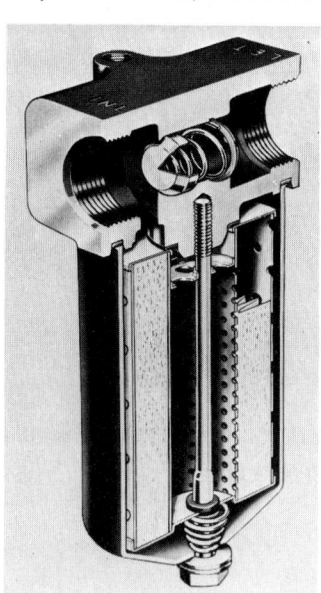

Fig. 4-13 Micronic-particle filters are often used as an integral part of a hydraulic system. The spring-loaded bypass valve at the top opens if the filter element becomes so loaded that it restricts flow excessively.

The main system filter should include a nominal-pressure bypass check valve to protect the system during cold starts and whenever the filter element becomes clogged. When possible, filters should be located in the tank return line where they can trap contaminants before the fluid reenters the reservoir. This location also permits using a low-pressure-type filter. Since filter elements must be changed at regular intervals, the filter should be located in an accessible area and not inside the reservoir.

Reservoir Airborne contaminants—and they're abundant around construction sites—are only one reason for good filtration practice. Another factor is that hydraulic reservoirs have tended to get smaller. This limits the opportunity for dirt particles to settle out before the fluid reaches the pump. It also means that there is a smaller amount of fluid in today's systems so that a given quantity of fluid makes more passes through the system in any given period of time. Contaminants thus have more chance to cause damage.

Inside the reservoir, a strainer is often used on the end of the pump inlet line (Fig. 4-14); this is in addition to the main system filter. This strainer should be cleaned periodically to help assure that sufficient fluid reaches the pump at all times.

The reservoir air breather vent usually includes a filter which also needs attention. This vent provides a means of maintaining atmospheric pressure to force fluid into the pump. A dirty vent filter could be plugged enough to reduce atmospheric pressure at the pump inlet, which, in turn, could produce a vacuum condition that is high enough to cause pump cavitation and, eventually, pump failure.

Cylinder rods and seals are also of concern. A worn or damaged piston rod seal or wiper seal can permit dirt and air to enter a hydraulic system.

2-50 Maintenance of Power Systems

Troubleshooting Improving the service life of hydraulic equipment is basically a matter of maintaining proper conditions within the system. The user should start with a clean fluid, keep it clean, and be on the alert for those symptoms which indicate possible trouble. Routine checks and simple corrective action can solve such difficulties when they

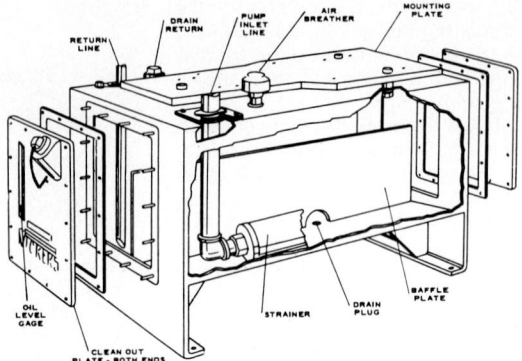

Fig. 4-14 Typical reservoir. The unit here conforms to ANSI specifications.

begin and long before they lead to major problems. The following list provides some helpful troubleshooting pointers:

Noisy Pump.
    Cavitation (pump starving)
- Clean inlet strainer
- Check inlet piping for obstruction
- Fluid viscosity too high
- Operating temperature too low

    Pump picking up air
- Oil level low
- Loose or damaged intake pipe
- Worn or damaged shaft seal
- Aeration of fluid in reservoir (return lines above fluid level)

Other.
- Worn or sticking vanes
- Worn ring
- Worn or damaged gears and housings
- Shaft misalignment
- Worn or faulty bearings

Low or Erratic Pressure.
- Contaminants in fluid
- Worn or sticking relief valve
- Dirt or chip holding valve partially open
- Pressure control setting too low

No Pressure.
- Oil level low
- Pump drive reversed or not running
- Pump shaft broken
- Relief valve stuck open
- Full pump volume bypassing through faulty valve or actuator

Actuator Fails to Move.
    Faulty pump operation (see Noisy Pump)
    Directional control not shifting
- Electrical failure, solenoid, limit switches, etc.

- Insufficient pilot pressure
- Interlock device not actuated

Mechanical bind

Operating pressure too low

Worn or damaged cylinder or hydraulic motor

Slow or Erratic Operation.
- Air in fluid
- Fluid level low
- Viscosity of fluid too high
- Internal leakage through actuators or valving
- Worn pump

Overheating of System.

Continuous operation of relief setting
- Stalling under load, etc.
- Fluid viscosity too high

Excessive slippage or internal leakage
- Check stall leakage past motors and cylinders
- Fluid viscosity too low

Start-up procedures The high speeds and high pressures of hydraulic systems require careful start-up procedures. This is necessary for increased service life, reduction of vehicle downtime, and improved operating efficiencies. By following these recommended practices, you will be taking a big step toward trouble-free hydraulic system operation.

There are three steps that should be followed before the initial vehicle start-up. First, always be sure the system is filled with an oil meeting the vehicle manufacturer's recommendation. Second, check all lines, fittings, and components in the system. Keep them as clean as possible. Third, vent the air from the system. A good job of eliminating air will pay big dividends in efficient operation and longer equipment service life.

1. Fill all pump and motor housings with hydraulic fluid through the inlet ports. If the unit is mounted so that the inlet port cannot be used, fill through the outlet port or drain port. Obviously, this will have to be done prior to installing inlet and outlet connections. The inlet and outlet connections should remain plugged until the connections are ready to be made.

2. When the system is completely plumbed up, fill the reservoir.

3. Break the outlet connection loose on all pumps or use the air bleed valve and hold cracked open until air in the outlet line is purged and solid oil runs out. Bleed air from the highest point in the system and tighten the connection.

4. Loosen the connection at check valves, motor outlet, and heat exchanger outlet (provided the heat exchanger is above the oil level in the reservoir). Turn the engine over on the starter—do not start the engine. Tighten each connection in turn as a solid stream of fluid starts to be expelled.

5. When all connections are tight and the system has expelled as much air as possible, start the engine. Engine speed should be in the low rpm range for the next few minutes when each component in turn is operated.

6. Check fluid level in reservoir and add if necessary.

7. Check system for leaks. Be sure inlets to pumps are tight and not leaking air into the system. Air leaks into the system will cause the pump to be noisy and will damage the unit.

8. Examine the fluid in the reservoir. On initial start-up there may be some air which should come out as bubbles on the surface of the fluid. The fluid in the reservoir should clear up in a short time. (Length of time is dependent on how much air was removed from the system prior to start-up.)

Chapter 5

Maintenance of Electric Motors

R. E. ARNOLD[1]
Former Manager, Product Engineering, General Electric Company, Industrial Motor Division, Schenectady, N.Y.

GENERAL

Electrical equipment will operate better, last longer, and require less repair if it is kept clean and dry and is lubricated properly. This is particularly true of electric motors, which are the driving forces for many operations. The failure of windings and/or bearings means delays from hours to days on vital work.

The electric motor, however, is only one component in an electric drive, and it is extremely important that all system components be correct and properly maintained if unplanned downtime is to be avoided. The power supply must be correctly matched; the control must function well; and particularly, overload, overcurrent, and overtemperature protective devices must be coordinated with the motor and function properly; and the driven equipment must be maintained.

PLAN

Perhaps the most important part of maintenance is to have a well-documented plan that is faithfully carried out. Any good plan will be based on the economics.

If the motor drive is vital to the job, maintenance of the highest order is indicated, but depending upon the physical size, the value, and the complexity of motor, different approaches can be taken. For example, on small, low-cost, more or less standard induction motors, probably the best solution would be to use motor spares or spare parts and keep the ordinary maintenance practices to a minimum.

However, if the motor is a large and/or a high-cost unit, the spare-motor philosophy is not practical because of the investment cost; a high degree of maintenance is justified, with the possibility of carrying spare parts, such as bearings, and form-wound coils if used.

If the motor is small with special features, such as high-temperature bearings to operate in a high-temperature environment, again a high degree of maintenance is indicated with a definite need of carrying critical parts, particularly the bearings.

Motors used in noncritical applications within a relatively dry, clean, normal-temperature environment require only minimum maintenance with the motors lubricated on a schedule suggested by the motor manufacturer. AC wound-rotor and dc armatures (rotors) using commutators or collector rings should have a brush-replacement schedule.

On large motors, of the order of 500 hp and larger for alternating current, and 300 hp and larger for direct current, the minimum maintenance schedule should be based on the manufacturer's recommendations. The value of the equipment is such that it is very

[1]The author is now retired.

important to follow the motor manufacturer's recommendations. There may be other maintenance procedures that can be used, but even these should be checked with the motor manufacturer if not obviously correct. A suggested maintenance schedule for large machines is shown in Table 5-1. Such a schedule would apply to dc motors and ac motors and generators by using the appropriate parts. This schedule could be adjusted to meet the needs of particular installations. Records are important for guidance and should show when more or less attention is justified on particular items.

Advancements have been made in motor inherent protection devices which will protect motor windings from overtemperature as well as overload in most cases and thus prevent motor burnouts.

There are now on the market metal-oxide variable resistors which, when properly placed in the motor circuit, will prevent high-voltage surges from overstressing the insulation. Where repetitive frequent starting, plugging, reversing, and jogging are encountered, such devices should be considered.

The above two items are mentioned because good planning includes upgrading of the equipment and system to prevent failures and in case of failure to take advantage of available improvements in the components and systems. A typical example is the use of a higher-temperature class of insulation for rewind on the stators of ac machines.

The economics of the maintenance plan must include the cost of the loss of production and work-force efficiency as well as the cost of repairs balanced against the cost of the maintenance of the equipment.

In some large operations where many machines are involved, maintenance scheduling can be computerized.

PART I. ALTERNATING-CURRENT INDUCTION MOTORS

The induction motor is one of the oldest types of motors and in its commonest form, the squirrel-cage motor, the simplest. The first commercial installation was in 1889, and the induction motor with distributed primary and secondary windings was developed in 1892. Since the first commercial installation, many designs of polyphase induction motors have been developed and two types have become the recognized standard: (1) the squirrel-cage rotor and (2) the wound-rotor construction.

The characteristic features of the stationary primary element—distributed windings and comparatively small air gaps—are common to both types. The squirrel-cage motor has no external secondary or rotating connection, while the secondary or rotor windings of the wound-rotor motor usually are connected through slip rings and brushes to some form of adjustable resistance.

The modern induction motor, especially the squirrel-cage type, is undoubtedly the most rugged rotating electrical apparatus ever developed. The majority of maintenance requirements, outages, and repair costs would therefore depend to a large extent on the correctness of application. However, the cardinal principle of any electrical maintenance is *keep the apparatus clean and dry*. This immediately points to periodic inspections, which are a highly desirable check on operating conditions.

Before details dealing primarily with maintenance are discussed, a few of the fundamental characteristics of the induction motor which may aid in solving some of the maintenance problems are presented.

Slip At no load an induction motor will run at practically synchronous speed, but when it is loaded, the speed at which the motor will run is below synchronous speed by an amount known as the slip. Thus, if the synchronous speed of a given motor is 1800 rpm and the full-load speed is 1750 rpm, the slip at full load is 50 rpm, or 50/1800, or 2¾ percent. The slip of any induction motor is a function of the losses in the secondary (resistance times the current squared in watts). The higher the secondary resistance, the greater will be the starting torque with any given current; the higher the slip, the higher also are the losses and thus the lower the efficiency.

Torque In analyzing some of the maintenance problems, a consideration of torque characteristics is sometimes important. There are two torque characteristics to consider: starting torque and breakdown torque. Motor torque, in the design state, must be balanced against efficiency and power factor. High starting torque results in lower efficiency and power factor as well as poorer speed regulation. However, if the starting torque is too low,

TABLE 5-1 Suggested Maintenance Schedule

Component	Monthly Inspection or maintenance operation
BEARINGS	Make sure that grease or oil is not leaking out of the bearing housings. If any leakage is present, correct the condition before continuing to operate
Ball and roller	Listen to a few bearings on a sampling basis. Bearings that get progressively noisier will need replacement at next shutdown
Sleeve	Check the oil level. Check the oil color through the sight gauge. Slightly cloudy oil is all right. Black oil is a danger sign
BRUSHES	Check the brush length. Replace when rivet or clip will rub commutator before next inspection. Inspect for worn or shiny brush clips, frayed or loose pigtails, and clipped or broken brushes. However, many dc brushes have no clips. Pigtails are tamped. There is danger that the pigtail of a worn-out brush will cut the commutator. Such brushes have a wear marker on the pigtail. When it gets below the top of the box, the brush should be discarded. Remove a few brushes to check the brush-commutator contact face in the case of dc motors and the brush-collector contact face in the case of ac motors and generators. Burned areas indicate commutation or sliding contact troubles. WARNING: HIGH VOLTAGE. ELECTRIC SHOCK AND ROTATING PARTS CAN CAUSE SERIOUS OR FATAL INJURY. AVOID CONTACT WITH LIVE ELECTRICAL PARTS AND MOVING MECHANICAL PARTS. (IT IS WELL TO NOTE THAT SILICON-CONTROLLED RECTIFIER DRIVES MAY HAVE HIGH VOLTAGES AT THE BRUSHES EVEN WITH THE ARMATURE STATIONARY. LINE SWITCHES SHOULD BE OPEN)
COMMUTATORS	Check the commutator for roughness by carefully feeling the brushes with a fiber stick. Also occasional wiping is recommended using a piece of coarse or nonlinting cloth. Jumping brushes give advance warning that a commutator is going rough. Observe the commutator for signs of threading. If threading is getting worse—take action; threading healed over—all right. Check for excessive commutator wear rate, streaking, copper drag, pitch bar marking, and heavy-slot bar marking. Commutator should have not more than 0.0025 in. total indicator runout or 0.0002 bar-to-bar steps. For high-speed commutators surfacing should meet 0.0010 and 0.0001 limits, respectively
COLLECTORS	Check the collector for roughness, dust, and wear. Ordinarily the rings will require only occasional wiping using a piece of coarse or nonlinting cloth. If the brushes are bouncing up and down on a cycle basis, check for collector-ring concentricity
MECHANICAL	
Air filters	Replace when necessary. Clogged filters cause overheating and lead to premature insulation failure
Bolts	Perform visual observation for loose bolts, loose parts, or loose electrical connections
Noise and vibration	Check for any unusual noise, vibration, or change from previous observations. Loose pole bolts often are the source of magnetic noise when motors are fed from rectifiers

TABLE 5-1 Suggested Maintenance Schedule (*Continued*)

Component	Every 6 Months Inspection or maintenance operation
BEARINGS	
Ball and roller	Listen to all bearings. Pull back bearing cap to inspect grease condition on a few representative machines
Sleeve	Take samples of oil from representative bearings for acidity (neutralization number) test. See manufacturer's sleeve-bearing relubrication recommendations. (Oil acidity is affected most by atmospheric contaminants and temperature—take one sample in each different area.) Change oil if required
COMMUTATORS	Check risers for cracks. If there are cracks, also check end of shaft keyway and shaft fan. (Cracks here mean extreme torsional vibration in system)
INSULATION	Measure 1- and 10-min insulation resistance and calculate the polarization index. Compare with records. Wipe deposits from brush-holder stud insulation and commutator creepage path or collector creepage path. Remove heavy deposits from around field-coil connections on dc machines where grounding might occur. Blow deposits out of the commutator rise area or the collector area with clean dry air. Blow out any blocked ventilation openings in windings. Make visual inspection for signs of overheating (dry, cracked, "roasted-out" insulation and varnish)
MECHANICAL	
Bolts	Check all electrical connections for tightness. Look for signs of poor connections (arching, discoloration, heat). Adjust inspection period to suit experience. Inspect foundation for signs of cracking, displaced foot shims; check foot bolts for tightness. Check frame-split bolts, brush holders, brush-holder studs, and bracket bolts, etc., on sampling basis. Check all coupling bolts
Shaft	Check corners of exposed end of shaft keyway for cracks (due to extreme torsional vibration). If there are cracks, check fan commutator risers on dc machines and the fan and character of the applied load on ac machines
Ventilation	Check for clogged screens, louvers, filters, etc.
Vibration	Check for excessive vibration (more than 0.002 to 0.003 in.) that will indicate change in balance or alignment. If actual operation cannot be seen, check for other signs of vibration (loose parts, chafing, shiny spots, rust deposits)
	Yearly
BEARINGS	
Sleeve bearings	Drain housings. Remove top half of housing. Lift top half of bearing. Inspect bearing surface and rings for wear. If excessive wear or sludge is found in bottom of housing, roll out bottom half of bearing for inspection. Clean if necessary and refill. SEE SLEEVE-BEARING RELUBRICATION RECOMMENDATIONS OF MANUFACTURER

the motor may not be able to start the load. High breakdown or high maximum running torque also results in low power factor and high starting current. If the breakdown torque is too low, the motor may stall on ordinary overloads, which can result in so-called roasted insulation if adequate protection is missing.

Voltage and frequency To obtain optimum results, induction motors should be operated at their normal rated frequency and voltage. Of course, some variation can be tolerated, voltage limits being approximately plus or minus 10 percent from nameplate and frequency plus or minus 5 percent. Both should never be varied at the same time to the extreme allowable limits; neither should they be varied at the same time in opposite direction. The following tabulation shows the effect of variation of voltage and frequency on performance of two-, four-, and six-pole motors of normal design.

	Power factor	Torque	Slip	Full-load efficiency
Voltage high	Decreased	Increased	Decreased	Approx. same
Voltage low	Increased	Decreased	Increased	Slightly lower
Frequency high	Increased	Decreased	Same	Approx. same
Frequency low	Decreased	Increased	Same	Slightly lower

A good rule to follow is that if normal frequency is changed by no more than 10 percent, a corresponding change in voltage should be made proportional to the square root of the ratio of the frequencies. It is not desirable to operate at decreased frequency with less than normal voltage because of increase in current input and temperature unless the load is reduced correspondingly.

Stator windings At first glance, the stator of an induction motor appears to be so rugged and simple that the necessity for maintenance is frequently overlooked. However, an inspection of any electrical-repair-shop records certainly indicates that the stator of the induction motor is a vulnerable item.

Trouble with stators usually can be pinned to one of the following causes: overloading, single-phase operation, moisture, bearing trouble, or insulation failure.

Main contributing factors to stator failures usually are dust and dirt. Some forms of dust or dirt are highly conductive and contribute to insulation breakdown. A good example of this is found in rubber-mill operations where a large amount of lampblack is used in processing the rubber. The type of lampblack used in processing synthetic rubber is more conductive than that used in processing natural rubber, so that failure rates increased when synthetic rubber came in. This is but one of many examples of conductive dirt and dust hazards. Certain steel-mill applications are others.

In addition to insulation failures due to conducting dust, restriction of ventilation can result from dust clogging ventilating ducts, thus causing overheating with possible resultant insulation failure due to excessive temperatures. With the loss of the cooling system or the restriction of ventilation, the failure occurs in many motors because they are inadequately protected by overload relays. Since the heaters are selected on the basis of current, they will not protect the motor when the cooling means fails because the current is affected only slightly to the higher temperatures. Inherent protection based on the temperature of the winding is generally safer. Periodic cleaning with clean, dry compressed air usually will suffice to keep dust accumulations to a minimum. However, some types of dust or dirt have a tendency to stick to windings, and blowing with air does not give a completely satisfactory cleaning. Other methods of cleaning are covered later.

One of the natural enemies of insulation is moisture. Some types of modern insulation have reasonable resistance to moisture, but in general, all types of windings should be kept dry. Many applications make it almost impossible to accomplish this unless special enclosures or other means to keep out moisture are employed. Some success for motors operating in damp locations has been obtained by using special treatment of windings. This will be discussed later.

Vibration frequently hastens winding failures. Vibration during operation may cause coil movement which eventually breaks or wears through insulation. As the motor becomes older, the insulation dries out and loses its flexibility. The mechanical stresses resulting during starting, plugging, and reversing, as well as natural stresses occurring under normal operation, may precipitate short circuits in coils or failures to ground. Periodic varnish and drying treatments properly performed tend to maintain a solid winding, thus minimizing coil movement.

Rotor windings General comments on stator windings here and in subsequent discussion apply equally to windings of wound-rotor motors. However, since the rotor is a moving part, additional maintenance problems are introduced.

Practically all wound rotors have three-phase windings and can therefore have trouble due to single-phase operation. An open circuit in a rotor shows up in lack of torque and slowing down in speed. It usually is accompanied by a grumbling noise and sometimes by failure to start the load. The most logical place to look for an open rotor circuit is in the secondary resistor or the circuit external to the rotor. The stud connections to the slip rings should be checked also, as the open circuit may be found there.

On rotors of greater capacity, where the coils are made of copper strap, clips are used to connect the bottom and top halves of the coil. These connectors should be checked for signs of heating, indicating a partial open. Such end connections, if faulty or improperly made, are a common source of open rotor circuits. Most manufacturers now braze these end connections instead of soldering, which minimizes faulty connections.

A ground in the rotor circuit will not affect the motor performance unless a second ground develops, which may cause the equivalent of a short circuit. This unbalances the rotor electrically and causes reduced torque, excessive vibration, sparking at the collector ring or brushes, or uneven wear of collector-ring brushes.

A reasonably successful method for checking for short circuits in rotor windings is to raise the brushes and energize the stator. If the rotor is free of short circuits, there should be little or no tendency to rotate even when the motor is disconnected from the load. If there is evidence of considerable torque, or if there is a tendency to come up to speed, the rotor should be removed and the winding opened to determine where the fault exists.

Another check which can be made with the rotor in place and with stator energized and brushes raised is to check the voltage across the rings to determine if they are balanced. When this check is made, the rotor should be rotated to several positions and readings taken at each position to be sure that inequality in voltage readings is not due to the relative positions of stator and rotor phases.

Squirrel-cage rotors with die-cast or pressure-cast aluminum rotors comprise the large majority of induction motors. They are very rugged and generally require little maintenance because of the absence of joints and connections. However, they may give trouble because of open circuits (bars) or high-resistance points between the rotor and the end rings. The symptoms of such trouble are reduced torques, higher slip under load, increased heating, and sometimes a noticeable noise.

Open bars associated with cast rotors are seldom visible. Therefore, if open bars are suspected, a method of checking is to apply 10 to 25 percent single-phase voltage to the stator with an ammeter in one line, slowly rotate the shaft, and watch the ammeter for significant changes in current. A significant current change with rotor position indicates a defect in the cast winding cage. Usually in this case, a pulsing sound related to slip under load is the first observation. Of course, if this occurs, a new replacement rotor is indicated.

Large induction motors and some high-resistance rotors are made with the bars brazed to the end rings for technical or economic reasons. A cracked rotor bar on a brazed cage with bar extensions between the iron core and the end ring can usually be seen if one has a reason to look for the defect. If there is trouble in the brazed joints, generally localized heating will cause discoloration. Cracked bars should be replaced, and with high-resistance joints further brazing is necessary, generally in a repair shop.

If cracked rotor bars or die-cast rotors are visible where the bars and end rings meet, the rotor should be replaced, but the manufacturer should be informed so that he can determine if it is a manufacturing defect, a design problem, or an application problem.

Repairs to squirrel-cage rotors, such as replacing bars or brazing broken bars, should be attempted only to competent personnel. Considerable skill is required, and proper care should be exercised in such repairs.

As stated previously, a small air gap is a characteristic of an induction motor. The size of the air gap has a direct bearing on the power factor of the motor, and any alterations that affect the air gap, such as grinding the rotor body, increase magnetizing current and lower the power factor.

Good maintenance on sleeve bearing motors includes a periodic check of motor air gap with feeler gauges to ensure against a worn bearing that would permit a rotor rub. Measurements should be taken on the coupling or driving end of the motor. Four measurements of air gap should be made approximately 90° apart, and one of the points should be on the load side, that is, the point on the rotor corresponding to the load side of the bearing.

On larger motors, a record of air-gap measurements should be kept so that a comparison can be made with previous checks to determine the amount of bearing wear. A rub (rotor on stator) resulting from bearing wear can generate enough heat to cause insulation failure.

Overloading of motors because of increased demands on the driven machine increases the operating temperature, resulting in shortened life of insulation. Momentary overloads within reasonable values usually do no damage; consequently, a thermal-overload device offers best protection. Since the best place to measure the thermal effect of overloads is on the motor, a thermal device can be applied directly to the motor winding. Most manufacturers can supply such a device, which provides effective protection against sustained overtemperature.

The polyphase induction motor is beyond doubt the simplest and most foolproof piece of rotating electrical apparatus. Experience indicates that the most frequent cause of winding failures is probably moisture associated with dirt, dust, chemicals, etc. Since bearings and lubrication of bearings are common to all types of motors, they will be treated in a separate section.

PART II. THE DIRECT-CURRENT MOTOR

The dc motor is more likely to be damaged in operation than the ac motor because a number of current-carrying parts are exposed. This motor comprises two parts: the stationary part, or field, and the rotating part, or armature.

The field of a dc motor consists of a frame and field poles which are fastened to the inner circumference of the frame. The poles are steel, usually laminated, and have mounted on them the field windings which furnish the excitation for the motor. Field windings, in general, are subject to the usual failings of electrical equipment. They can become dirty or oil-soaked, which can interfere with heat being dissipated, causing eventual burnout. Excessive field current caused by malfunctioning of control will cause excessive heating and failure. Field heating can be caused by high voltage, too low speed, brushes being off neutral, overloads, or a partial short in one field coil. An open circuit in a field coil can result in failure to start or excessive speeds at light loads and heavy sparking of the commutator.

The armature of a dc motor consists of two main parts: the windings and the commutator. Cleanliness, while important on all electrical equipment, is particularly so on the dc commutator and brush rigging. Oil, dust, grease, moisture, and corrosive gases should be kept away from commutators and brushes, as they cannot give good performance under adverse conditions such as these.

The armature is the heart, so to speak, of the dc motor. The main line current flows through it, and if the machine is overloaded, the armature shows the first signs of distress. Reasonable attention from the standpoint of cleaning should result in little or no trouble with the armature under normal operating conditions. Repairs to armatures should be done only by competent personnel. Care should be exercised in handling an armature, some of the more important items being:

1. Never roll an armature on the floor; a coil may be injured or a binding band nicked.
2. Support or lift an armature only by its shaft, if possible; otherwise use a wide lifting belt under the core.
3. Never allow the weight of the armature to rest on the commutator or coils.

Windings should be preserved with periodic varnish treatment followed by baking where possible. This will be treated later.

If necessary to renew the banding on an armature, duplicate the banding originally supplied by the manufacturer; in other words, do not change material, diameter, width of band, or location. Band width should not be increased, as to do so may cause restriction of ventilation and also cause heavy current in the bands sufficient to overheat and melt the band solder.

The commutator is probably the most vulnerable part of a dc motor, as it is an exposed current-carrying part rotating at relatively high speeds. The success or failure in operation of a dc machine depends to a large extent on commutation. Regardless of any other excellent feature a dc machine has, if the commutation is unsatisfactory, the machine has little commercial value. It is the intent at this time not to go into the details of manufacture or assembly of commutators but to assume that the manufacturer has produced a device which, under proper conditions, will give trouble-free operation.

Assuming that the design of the machine is such that good commutation may be expected, continued satisfactory operation depends on maintaining the commutator surface in good condition. In general, this means that the surface should be smooth, concentric, and properly undercut. The brush holders should work smoothly and be free of dust and dirt; the brushes should be of the proper grade and manufactured to correct size and tolerance. Brush holders must be spaced equally around the circumference and have correct spacing from the commutator surface (usually 1/16 to 3/32 in.). While conditions under which dc machines are operated vary widely, there are some basic conditions which must be maintained on all such machines to ensure satisfactory commutation, brush life, and a minimum of commutator wear. These conditions and recommended maintenance methods are listed below.

1. The commutator must be concentric. On high-speed machines with peripheral speeds of 9000 fpm or above, the commutator should be concentric within 0.0005 in., which is the practical limit in grinding. For peripheral speeds of between 5000 and 9000 fpm, the concentricity should be within 0.001 in. On slow-speed, large-diameter commutators, this figure can be 0.003 in.

There should be no abrupt change in surface from bar to bar. Variations of over 0.0005 in. are enough to cause bad commutator performance. This bar-to-bar roughness can be detected by using a stick sharpened like a pencil and held on the revolving surface at a slightly inclined angle. If the commutator has negligible bar-to-bar roughness, the stick, so used, will feel as if it is moving over a smooth glass surface.

2. To get a commutator concentric, a grinding rig should be used in all but a few special cases. The grinding rig consists of an abrasive-stone setup similar to a lathe tool in a carriage which can be moved back and forth along the commutator and equipped with a feed to advance the stone into the commutator surface (see Fig. 5-1). The support must be rigged so that the grinding stone is subjected to a minimum of vibration. On most dc machines such a rig can be mounted on a brush arm by removing the brush holders. To be sure of obtaining maximum rigidity, it may be desirable to brace the brush-holder bracket arm during the grinding operation. In some cases, it is possible to support the grinder on parallels fastened to the bedplate, in which case the entire brush rigging can be removed.

Fig. 5-1 Commutator grinding rig in position.

Grinding should be done, if possible, with the armature in its own bearings and, in the case of a constant-speed machine, at rated speed. Low-speed grinding, if there is any evidence of unbalance, will cause the commutator to run eccentrically at rated speed. Care should be exercised to prevent copper and stone dust from getting into the windings. It is important to provide a vacuum cleaner on the grinding rig. In extreme cases where a cleaner is not available, the commutator necks and risers and the coil ends can be covered with paper or cloth to keep the grinding dust out of the machine.

To grind a commutator, the armature must be rotated. Each case must be treated separately, as local conditions will govern. At times it is possible to run the motor with half the brushes out, grinding half the commutator and then repeating with the other half. A driving motor can be used, coupled or belted to the armature on which the work is being performed. On some types of equipment, it is difficult because of space or other considerations to grind the commutator in place. If the machine speed is relatively low, the work should be done in a lathe by taking a very fine cut off the surface and then polishing with a stone. On high-speed machines, a special rig should be used equipped with a grinding

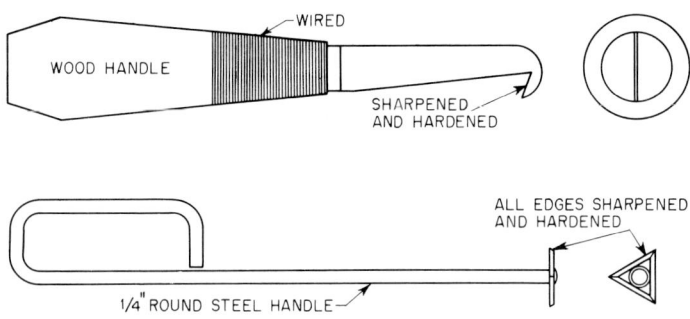

Fig. 5-2 Tools for commutator undercutting.

device, a driving motor, and an adapter for bearings, so that the armature can be run in its own bearings. Grinding should be done at a speed as near to rated maximum speed as possible and still not have the grinding stone affected by vibration.

Three grades of stones are used in grinding commutators: coarse, medium, and fine. The coarse stone has a grit of approximately 80 mesh and takes off large amounts of copper. In general, the use of the coarse stone is not too desirable, because if a great deal of copper has to be removed, it is better to use a lathe tool which can be set up in the grinding rig. The medium stone has a grit of approximately 120 mesh and is used for the bulk of grinding work, finishing up with the fine stone with a grit of approximately 200 mesh for the fine finish. The stone should be contoured or block, and sandpaper should be used to avoid "crowning" the bar.

3. After grinding is completed, all commutator slots should be cleaned out thoroughly and bar edges beveled. Beveling accomplishes two things: burrs, caused by the stone dragging copper over the slots, are removed, and the sharp edge is eliminated at the entering side of the bar under the brush. Bevel as little as possible to just remove "wire edge." This is accomplished with a special beveling tool which should have about $\frac{1}{32}$-in. chamfer at 45° for bars of medium thickness. For thinner or wider bars, the beveling can be changed accordingly.

4. Almost all modern dc machines have undercut mica in the commutator slots. This undercutting should be kept at $\frac{1}{32}$ to $\frac{3}{64}$ in. If, when a commutator is to be ground, sufficient copper is removed so that the undercutting will be shallow, then the commutator should be reundercut before grinding is started. This is done by using a small, circular, high-speed saw about 0.003 in. thicker than the actual thickness of the mica. Where the extent of undercutting is not great, a hacksaw blade mounted in a wooden handle is a popular tool. It is sketched, along with another tool, in Fig. 5-2. Otherwise, one of the undercutting tools put out by any one of a number of manufacturers can be used. Care should be taken when undercutting that a thin sliver of mica is not left against one side of the slot. Slots should be checked, and if slivers are present, hand scarfing will be necessary. If these slivers are left, subsequent operation of the machine raises the mica above the edge of the bar and one of the worst conditions for good commutation develops. As explained previously, the bars should be scarfed after the undercutting is completed. Final polish can best be obtained by using white brush-seater stone on the commutator surface after sanding in the brushes.

5. The bottom of the brush-holder box should be set at the correct angle and the correct distance from the commutator surface. The distance from the bottom of the brush box to the commutator surface on most machines is $1/16$ to $3/32$ in. Failure to maintain the proper spacing results in brushes riding the commutator surface poorly and also has the effect of shifting neutral on machines using inclined brush holders. This shift in neutral contributes to poor commutation. To check this spacing and to ensure uniformity, a piece of hard fiber or similar material of the proper thickness should be used as a gauge.

Spacing of brush holders around the commutator should be maintained evenly, and the spacings should not deviate more than plus or minus $1/32$ in.; also, the holders should be aligned on the brush arm in a straight line parallel to the length of the commutator bars. A popular method of checking brush-arm spacing is to stretch a piece of paper or adding-machine tape around the commutator under or near one brush path and mark the brush position on the paper with a pencil. The paper is then removed, and spacing is checked for uniformity. The brush arms can be shifted to obtain equal spacing. The check should be taken several times to average out any errors. After this work has been done, the brushes should be reinstalled or replaced with the brush holders staggered properly as per manufacturer's recommendations.

Brush and brush-holder tolerances Brushes should slide freely in the brush boxes. The standard tolerances for brush width and thickness are plus 0.000 to minus 0.004 in. up to and including a brush width of $3/4$ in. For brushes over $3/4$ in. wide the tolerance is plus 0.000 to minus 0.015 in. The average industrial-type brush box maintains a tolerance of plus 0.003 to 0.006 in. on the thickness and plus 0.002 to 0.010 in. on the width.

With these close tolerances in the brush boxes and brushes, it can be readily seen that not much dust or dirt is required between brushes and boxes to cause sticking brushes. For this reason a clean machine is essential to good commutator performance. At times brushes will appear to be tight in the holder even after a thorough cleaning. The boxes should then be checked to determine if they have been warped. This occurs at times if the machine has been run with excessive load which caused heating or if the machine has had a severe flash which caused heating of the brush holders. The reaction surface in the box must be flat.

Pressure Brush pressure springs should be set at the manufacturer's recommended values and should be adjusted to be as uniform as possible. Uniform pressures prevent selective action whereby certain brushes tend to take more than their proper share of the load. Tension can be measured with a straight spring scale and should be taken just as a new brush on being lowered touches the current-collecting surface. The pull should be taken in the direction of normal brush motion to avoid setting up friction values differing from those affecting the brush in operation. Dividing the pull obtained with the scale by the cross-sectional area of the brush gives the pressure in pounds per square inch.

Variables affecting commutation The maintenance suggestions outlined above should result in satisfactory commutation, brush life, and commutator conditions. However, there are a large number of variables which can upset the delicate balance between brush and commutator. The resulting unbalance is indicated in a number of ways. Following are listed several of the more common adverse conditions encountered and their remedies, together with what experience indicates as possible causes. These should serve as a guide to the maintenance man in solving his commutation problems but should not be taken as the final analysis on any particular trouble job.

Brush chatter is caused by high friction between the brush and the commutator or by a poor commutator surface. A common cause of high friction between brush and commutator is light-load running. On machines which operate for extended periods with brush densities in the order of 25 amp/sq in. and below, the brushes have a tendency to develop a highly polished glaze on the commutator which invariably causes brush chatter. In order to correct this condition, the brushes in one or more paths around the periphery should be lifted in order to raise the current density in the remaining brushes to approximately 55 amp/sq in. It is always better to operate a brush overloaded for a short period than to operate it lightly loaded for an extended period. If it is not possible to raise the brushes as recommended above, a grade of brush with a slight amount of cleaning action should be used to prevent the formation of the high-friction film, a grade of brush more suitable for low-current density should be installed, or a combination of both can be used.

High friction can also be caused when a commutator is run hot for an extended period.

This condition usually does not develop unless the machine is run well beyond rated capacity and cannot be corrected by any maintenance program. It may be necessary to remove the high-friction film with white seater stone before a normal commutator film can be restored.

Poor commutator surface and high mica also cause brush chatter, which, however, can be distinguished from chatter caused by light load or overload conditions by its lower frequency. It can be corrected by grinding the commutator, trimming out mica fins, scarfing, and polishing.

Brush chipping can result from brush chatter as described above. It can also result from extremely light or extremely heavy spring pressure, high mica, improper fit of brushes in the holders, or severe shock or vibrations set up outside the machine. Correction can be accomplished as described previously.

Threading or Streaking of Commutator Surface. This is caused by the breakdown, in the brush paths, of the film on the commutator surface. The tendency then is for the current to pass through the area where the film has been broken. The surface condition then is further aggravated, and finally threads or streaks are worn around the periphery of the commutator. This condition is caused by particles of copper being embedded in the face of the brush. These particles work-harden and become tools, cutting through the commutator film. Brush faces should be inspected periodically and corrected if necessary.

Selective action, the tendency for one brush or group of brushes to carry more than its share of the load, is also a prime cause for threading or streaking.

Flat spots on the commutator sometimes develop which, if permitted to go uncorrected, will become as much as 0.030 in. deep. The spots may begin singly or at one or two pole pitches apart. Often the flat spots will multiply, with a second, third, and fourth set following the initial set, displaced by several bars. The cause is always a cyclical disturbance, either electrical or mechanical. Electrical sources might be an open or partially shorted armature coil. Mechanical sources would include mechanical unbalance, coupling misalignment, and high or low commutator bars. Flexible coupling should be lined up to about 0.002 in. both transversely and angularly. Atmospheric contamination often aggravates this condition.

Threading or streaking can be reduced by raising the current density above 30 amp/sq in. or by using a brush with some cleaning action. Electrographitic brushes with a mild polishing action are most suitable for this type of situation.

If selective action is causing the streaking or threading, it is necessary to check conditions which could cause unbalance in electrical paths. Electrical resistance in one path different from that in parallel paths causes selective action. Terminal connections, spring pressure, shunt-to-brush connections, brush size, brush-holder spacing, and brush material should be checked. Hardness of brushes ordinarily does not contribute to threading. However, the ash content of a brush may do this, especially if the ash particles are hard, as they can be if ash content is improperly controlled in manufacture.

Sparking. Practically every abnormal condition of the commutator, brush holders, brushes, fields, or armature results in sparking. If the maintenance practices, as outlined previously, have been followed and the machine was correct originally, we must then look for the cause of injurious sparking somewhere in the electrical or magnetic circuit of the machine or in the brush grade. The following are conditions to be looked for: bad connections between armature coils and commutator bars, particularly those which are bad only when the armature is rotating; short-circuited field coils; open-circuited or short-circuited armature coils; unequal air gaps due to worn bearings; improper brush grade; partial short in shunt field; or neutral set-off. The following is a convenient checklist on the most general causes of sparking:

    Rough, eccentric, or dirty commutator
    Brushes sticking in holders
    Incorrect brush tension
    Poor brush fit
    Brushes not parallel with commutator bars
    Unequal brush-holder spacing
    Vibration
    Brushes off neutral
    Unequal spacing of main or commutating poles

Maintenance of Power Systems

Reversed or short-circuited commutating-pole coils
Reversed compensating winding
Open circuit in armature winding
Short circuit in armature winding
Unequal air gap
Grounds
Atmosphere or other commutator contamination (high friction)

Checking Neutral. When a machine is reassembled after it has been dismantled for cleaning or repairs, it is frequently necessary to check and set neutral on the machine. The following is an outline of the kick-neutral method.

This method is based on measurement of voltages induced in the armature coils as the current in the main field on the machine is interrupted. Voltages induced in the conductors located at equal distances to the right and left of the pole centers are equal in magnitude and opposite in direction. If the terminals of a low-range voltmeter are connected to commutator bars corresponding to conductors located midway between poles, no deflection will be caused by breaking the field current. When the brushes are set so that the centerlines of their faces correspond with the centerlines of the commutator bars between which there is no induced voltage, they are on neutral (see Fig. 5-3). A much simpler method is to tie the voltmeter to two studs of opposite polarity.

If the number of commutator bars is not evenly divisible by the number of poles, use the following method: With the machine at standstill, raise all brushes. Replace one of them on each arm by a special brush of the same thickness. This special brush should be beveled to a knife-edge parallel with its longer side and in the center of its face. Connect leads from adjacent brush arms to a dc voltmeter, preferably one having 0.5-, 1.5-, and 15-volt scales. Separately excite the shunt field from a dc source through a quick-break switch. Insert enough external resistance in the excitation circuit to keep the field current small at the beginning. Use the smallest field current that gives a good deflection on the low scale of the voltmeter. When "kick" voltage is read for the first time, begin with the 15-volt scale and change to lower scales only when it is certain that the voltage is within their respective ranges. Before the switch is opened for each reading, wait long enough for the induced voltage caused by closing the circuit to decay. Shift the rocker ring to the point at which the voltage is minimum when the field circuit is opened. If the machine has double brush holders, the center of the brush holder is placed on the neutral mark instead of either of the double holder brushes.

If the number of bars between centerlines of brushes on adjacent arms results in half a bar being included in the commutator pitch (such as 20½ bars between centerlines), this alternative method is used: Raise all brushes. With the voltmeter points on bars 1 and 21 in the approximate neutral zone, open the field circuit as described in the paragraph above and read the deflection. Move the voltmeter points to bars 1 and 22, and read the deflection as the field circuit is opened. Rotate the armature slightly until the two readings are equal but opposite in polarity. This indicates that the correct neutral is exactly on the centerline of bar 1 and on the mica between bars 21 and 22. The rocker ring is shifted until the centerlines of the arc of the brush surfaces are exactly over these positions. The same procedure applies here for double brush holders.

If the number of bars is evenly divisible by the number of poles, it is possible to be off neutral one-half bar with the method just described. If the armature can be rotated, use this alternative method: Raise all brushes. Determine the commutator pitch. Full-pitch windings are *now very rare*. Coil throw (back pitch) is stated in slots. For simplex lap windings, commutator pitch is one segment, or for a simplex wave winding is the number of segments per pairs of poles. With the voltmeter points on bars 1 and 21 in the approximate neutral zone, open the field circuit as described in the above paragraph and read the deflection. Move the voltmeter points to bars 2 and 22, and read the deflection as the field is opened. Rotate the armature in either direction, and repeat these operations until the two readings are equal but opposite. This indicates that the correct neutral is exactly on the mica between bars 1 and 2 or between bars 21 and 22. The rocker ring is shifted to these points, as explained in the preceding paragraph.

If the armature cannot be rotated, the neutral is located by the use of a curve or a calculation (see Fig. 5-4). If the number of bars is divisible by the number of poles, proceed as outlined in the above paragraph. Read the induced voltages on bars 1 and 21, 2 and 22, 3 and 23, etc., until a point is reached at which the polarity of the induced voltage reverses. Then record four readings, two on either side of the reversing point; plot the induced voltages as ordinates and the number of commutator bars as abscissas. Keep in mind that the number indicates the centerline of the end of the bar. After the exact point of reversal has been determined from the curve, mark the relative position on the commutator. This is the correct neutral. Shift the rocker ring as described in one of the above paragraphs. It is possible to calculate the distance from the centerline of a bar on either side of the point of reversal to the neutral without plotting a curve. Only two readings are necessary, one on either side of the point of reversal. Measure the distance between the centerlines of two adjacent bars. The distance from the centerline of one bar to neutral is found by dividing the reading on that bar by the sum of the two readings. This quotient is expressed in percentage of the total distance between centerlines.

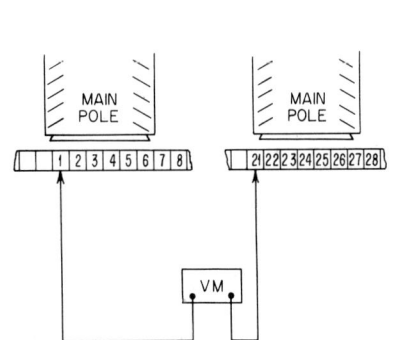

Fig. 5-3 Instrument connections for reading voltage induced in armature.

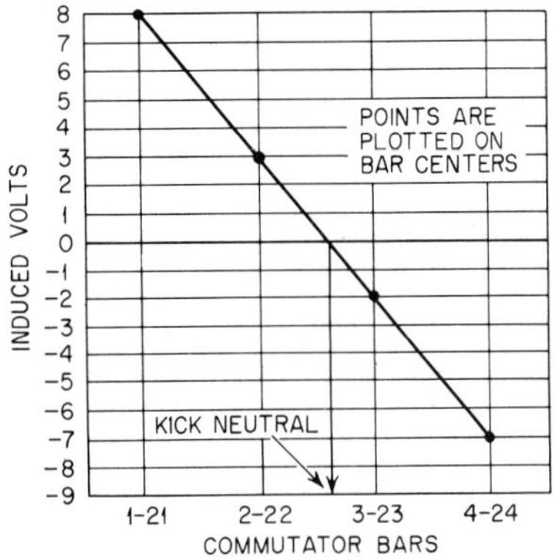

Fig. 5-4 Curve method of determining neutral.

Effect of Setting Brushes off Neutral. At times it is advisable to shift brushes off neutral. This may be done to obtain results that are too expensive to get otherwise or to get better performance curves. When brushes are moved in a direction opposite rotation, the motor becomes undercompounded and compensation increases. Moving brushes in the direction of rotation overcompounds the machine and undercompensates it. Should there be a lack of compensation in a motor which is to run in one direction only and the additional amount of compensation needed is not too great, the machine may be made to commutate properly by moving the brushes slightly against rotation. If the machine is of a reversing type, this cannot be done because poor commutation would be obtained in one direction of rotation. Sometimes a motor-speed curve has a hump in it. Often this can be corrected by moving the brushes off neutral a slight amount as long as commutation is not affected. The effect of shifting brushes off neutral must be observed over an extended period of time. The limit of good commutation determines how far the brushes may be moved. If good commutation is not obtained by moving the brushes, a change should be made in the commutating-pole air gap. This is seldom required but probably should be done by the manufacturer because of the equipment needs and the skills required.

Flashing. A question which has been argued pro and con for years is "Can the brush in itself be the cause for a machine to flash over?" A great deal of study and investigation has been devoted to this. A flash usually is the result of a suddenly charged condition in current or voltage or in the field strength of the machine. Disregarding control failures, the flashing results from a short circuit imposed by brush shunts breaking and touching on the opposite polarity, brushes breaking and causing shorting of commutator bars at the commutator surface, a piece of carbon from broken brushes becoming lodged between opposite polarities of the wiring around frame of the motor circuit, or bar burning on flat spots on the commutator. The large factor of safety in today's standard of commutation is such that a given grade of brush is rarely the cause of flashing.

Commutator Films. The most satisfactory commutator film is evenly colored and is between a light brown or straw color and dark brown. The most important point is that it be a uniform color and not a highly glazed or extremely dull finish.

Undesirable films due to atmosphere conditions are usually quite dark, black or gray. These films can be detected by the fact that the discoloration of the copper is seen on the surface, not in the brush paths.

The amount of contamination in the air that can be injurious to a commutator surface by causing undesirable film is extremely low. Conditions which cannot be detected by smell often cause deleterious results. For example, it is known that 3 parts of sulfur in 1 million parts of air are sufficient to cause heavy formation of undesirable film. When such conditions are found, a cleaner brush should be used with enough abrasive action to prevent the formation of the heavy film or else a brush whose film-forming qualities may develop a satisfactory film of copper oxide on the commutator surface before the injurious film can be established.

Copper Pickup. This undesirable feature is the transfer of copper particles from the commutator surface to the brush faces where they become embedded. Once started, copper pickup is usually progressive to the point where commutation and current collection is impaired. When copper becomes embedded in a brush face, the contact drop is decreased to the point where commutation becomes impossible. The rubbing effect of the copper particles against the commutator surface causes threading. Rubbing work-hardens the particles so that they act as tiny cutting tools. In order to remedy this condition, the proper bar-edge beveling should be maintained, the machine must be kept clean by blowing out thoroughly after working on the commutator, and brush-seating stones should be used sparingly. Also, the brush faces should be sanded to clean out the copper particles.

Brush Wear. This is one of the controversial subjects with operators of machines using brushes. All operators naturally desire longest possible brush life. However, compromises which result in increased commutator wear or poor commutation must not be resorted to in order to obtain long brush wear. Brushes wear away both electrically and mechanically, and these two have a complicated and strange relationship.

Brush life (for the most common electromagnetic brushes) is at its best when the load on the brush is around 55 amp/sq in. Lightly loaded brushes set up friction and chatter conditions, as explained previously.

A newly installed machine frequently has short brush life, caused usually by building and cement dust from the new construction. This condition gradually improves with time, although sometimes over a year may be required to overcome this trouble.

All the possible adverse conditions discussed above have an effect on brush life. Correction of the particular trouble also tends to correct short brush life.

PART III. INSULATION AND ITS CARE

The most important item in the maintenance of electrical equipment is taking care of the insulation. To take care of the insulation, keep it clean, keep it dry, and keep it cool. These three musts are interrelated; for example, if the external surfaces are kept clean, the motor will run cooler because of better heat transfer. A fourth point would be to make sure the power supply and the motor control are basically sound.

In many cases this means that good preventive maintenance can give added years of trouble-free life. It is extremely important to prevent insulation failure, because if the insulation fails, the motor or generator winding most generally has to be removed from the electromagnetic core and a new winding must be inserted and varnish-treated with the proper procedures before that particular motor can be used. This rewinding is time-consuming and varies from 2 days to several weeks depending upon the distance and the availability of the service facilities, the size of the equipment, and the availability of the insulating material as well as the conductors.

Such failures then cannot be considered as temporary shutdown failures, and it is important to have a maintenance plan and program. If the electrical rotating machine is a key to a continuous process, a spare motor is indicated or at least a stator if the machine is an ac squirrel-cage induction motor, or a rotor if the machine is a dc motor.

If the motor is not critical, the rewind time is possibly satisfactory, but in any case, a maintenance plan and program is justifiable.

Let us examine the conditions for long, trouble-free insulation life.

Keep it clean This means the surface of the insulating material shall be free of dirt, dust, salts, oil, cement dust, carbonaceous materials, chemicals, lint, fibers, etc. Quite often, the basic way to do this is to select the right kind of equipment enclosure, for example, a totally enclosed, fan-cooled motor instead of an open, open-dripproof, or splashproof motor, or separate clean air ventilation provided by a blower rather than some form of open motor using the surrounding contaminated air. Many failures would be eliminated, particularly where abrasive dusts such as cement dust, fine conducting dusts such as carbon dust, and strong chemical atmospheres are present.

However, even some dusts that seem innocuous will absorb moisture easily and cause the winding to become wet, which will reduce insulation resistance and dielectric strength; and the dirt, with the absorption of moisture, will affect surface tracking.

Chemical atmospheres can be very bad, because they may react with the insulation materials, causing the insulation to deteriorate.

Even if you enclose the motor, it is necessary to keep the external metal surfaces clean for good heat transfer and the external ventilation paths clean to obtain the maximum air flow for cooling purposes.

For good maintenance practices to give long insulation life, keep the motor clean by (1) removing the source of the contaminant, if possible; (2) selecting the proper enclosure; (3) cleaning the windings; and (4) cleaning the surfaces and keeping the ventilating paths open.

Keep it dry It is extremely important for long life to keep electrical equipment dry. Moisture and humidity can be particularly harmful to insulation, because if the insulation is porous, moisture will be absorbed and the insulation resistance can be reduced drastically. Also the surface of the motor in locations where it is very humid or cold can be completely covered with a thin film of water, which would make the external surface of the winding reasonably conductive. With this condition any pinhole in the insulation can cause a ground or a short between turns because the insulation resistance is lowered drastically, sometimes in the order of several magnitudes, like 1000 to 1.

Dust, dirt, salt, oil, chemicals, and other contaminants generally greatly accelerate the wetting of the surface, emphasizing the fact that the machine should be kept clean as well as dry.

It is important to remember that all insulation is affected to some degree by absorption of moisture or by collection on the surface, but the degree is widely different with various insulations. The higher-temperature insulations are more nonhygroscopic (moisture-resistant) than the lower-temperature insulations, and vacuum-pressure impregnation (not as important on random-wound as on form-wound) is much more moisture-resistant than the normal varnish dips and bakes. Also, more dips and bakes are helpful in decreasing moisture-absorption characteristics of the windings because of the added thickness of varnish and the reduction of pinholes that reduce surface tracking when wet.

In the selection of equipment, proper choice of enclosure can ensure against water in the solid form leaking, dripping, and being blown into the motor when such hazards are present. If recurring flooding of an induction motor is known to happen, encapsulation of the motor windings might be an answer to keep the motor running during such periods and thus keep a continuous process rolling.

If motors are used in very humid or very cold atmospheres, the use of space heaters might be indicated when the motors and generators are not operating (not energized). Space heaters that keep the air temperature higher around the windings by about 3 to 6°C (37½ to 50°F) will prevent surfaces from becoming wet. Such devices are very important in such areas as the Gulf of Mexico coast and Alaska, for standby, storage, and very intermittently operated machines. When machines are in operation, they generate enough heat to drive off moisture.

Keep it cool This means the machine temperatures should be kept within the temperature rating specified on the machine nameplate. Machines are generally designed to meet the ANSI (American National Standards Institute) standards, IEEE (Institute of Electrical and Electronics Engineers) standards, and NEMA (National Electrical Manufacturers Association) standards for insulation and systems. The insulation systems are classified as O, A, B, F, and H with their hottest-spot temperatures of 90, 105, 130, 155, and 180°C, respectively. At this time, O is used only and rarely in small devices; A insulation is used in small motors, particularly fractional-horsepower where flux limitations are met before temperature-rise limitations; B and F insulation are used on integral-horsepower ratings, with B insulation being the NEMA standard on ratings 1- to 200-hp polyphase ac induction motors; and H insulations are used for special motors and special applications, particularly high ambients like 50 and 65°C. The pertinent data on these standards are shown below.

Insulation class	Hottest-spot temp, °C	Permissible rise, °C			
		By thermometer* method	By resistance or embedded detector*		
			TEFC	Max at 1.15 SF	TENV
O	90	35	45		
A	105	50	60	70	65
B	130	70	80	90	85
F	155	105	105	115	110
H	180	130	125		135

*Based on 40°C ambient and suitable hot-spot allowances for the methods of measuring temperatures.

The above table is associated with maximum temperatures, but in actual practice the operating temperatures are generally well below the limits. For example, the ambient temperature of most applications is in the 24 to 30°C range, and usually machines are not fully loaded on a continuous basis.

It is advisable to keep the operating temperature of the machine as low as possible because the thermal aging of the machine is dependent upon the average temperature and the insulation life is generally based on an approximate rule of thumb that the thermal life of insulation is doubled for each 10°C decrease in temperature. Although this number can be different and can vary from 8 to 15°C, it does give a reasonable concept of life within the usual operating-temperature ranges. Therefore, to emphasize our first two points, keeping the motor clean tends to keep it cool, resulting in longer life, and keeping the machine dry makes it easier to keep the motor clean.

In the last few years, in the interest of reducing overall noise, enclosures have been placed around equipment with little regard to the effect of the enclosure on ventilation and the adverse thermal effects on the equipment. Therefore, any time an enclosure is used to surround equipment or the electrical machine, the thermal effects on the equipment and on the machine and the ambient must be checked. The ambient is important because of its relation to overload relays in the control and the ambient of the control center. When the motor ambient is very hot or above normal and the ambient of the control elements, including the overload relay heaters, is normal, the control will not protect the motor from overload. If the ambients are colder than normal, nuisance tripping will occur and control should be ambient-compensated.

Proper power supplies It is important to have the proper power supply to avoid troubles. Low voltage generally results in higher conductor losses and makes the machine run hotter. High voltage results in higher electromagnetic-core losses and, depending upon saturation, can cause the machine to run hotter. Care should be taken to keep the voltages within the standard tolerances allowed in NEMA and ANSI standards.

Many direct-current motors are operated from rectifier power supplies, and the pulsating voltage and current waveforms affect the performance by increasing motor heating and degrading commutation. Because of these effects, it is necessary that the motors be designed or specially selected to suit this type of operation. Since there are many different power supplies, it is important that the proper power supply be matched with the motor. The motor and power-supply manufacturer(s) should be contacted if there is any question of compatibility. The motor nameplate normally specifies by code the type of power supply for which the dc motor is intended.

This is also true of ac adjustable-speed motors supplied by inverters, converters, primary-voltage static power supplies, etc., and it is very important that the motors and power supplies match to prevent undue heating.

Hermetic motors are a special case because they are designed to operate in a hermetically sealed enclosure with one of the freons, gas, or liquids being used as the cooling medium. Such motors generally require special insulating materials to be compatible with a specific freon gas or liquid, and it is therefore important to rewind with the special materials in case of a winding breakdown. Generally, the safest way is to have the rewinding done under the direction of the motor manufacturer or the equipment manufacturer.

Cleaning and Drying Insulation

The operating instructions provided by machine manufacturers always emphasize the importance of keeping electrical apparatus clean, dry, and cool, as mentioned above. Favorable locations, adequate ventilation, application of heaters to prevent condensation of moisture on the motor when the machine is out of service for any length of time, and suitable covers all help to reduce the number of outages and to lessen maintenance costs.

However, motors do get dirty, and when they do, the insulation must be cleaned and restored to as near its original condition as possible and as soon as possible. The amount of cleaning and the method of cleaning are dependent upon the type of dirt, the amount of moisture, how soon the machine is required to go back in service, the availability of a spare machine or component, etc. After cleaning, it is important that the machine is dry. Visual examination is not sufficient, and tests should be made to make sure that the motor windings have an insulation resistance of not less than

$$\frac{\text{Rated voltage}}{(0.075 \times \text{hp rating}) + 1000} \text{ megohms}$$

which would be 0.46 megohm for a 10-hp 460-volt motor and 2.14 megohms for a 1000-hp 2300-volt motor, as two examples. However, a minimum of 1.0 megohm is preferable.

If an electrical machine has been stored in a damp location or has been in a humid environment without being operated for a long time, it should be dried out.

Windings can be dried in a number of ways: (1) Bake the motor in an oven, preferably a circulating-air oven, at a temperature not exceeding 90°C, until the insulation resistance becomes practically constant. (2) Enclose the motor with canvas or similar covering, leaving a hole at the top for moisture to escape, and insert heating units or lamps. (3) Pass a current at one-tenth rated voltage (with rotor locked) through the stator windings, and increase the current gradually until the winding temperature reaches 90°C. Do not exceed this temperature. In a dc machine pass current through the exciting field coils.

As stated above, when harmful dirt accumulations are present, a variety of cleaning techniques are available. First dry dust, dirt, or carbon should be vacuumed without disturbing adjacent areas or redistributing the contamination. Use a small nozzle or tube connected to the vacuum cleaner to enter into narrow openings, as between commutator risers on dc machines. A soft brush on the vacuum nozzle will loosen and allow removal of dirt more firmly attached.

After the initial cleaning with vacuum, high-velocity air may be used to remove the remaining dust and dirt. It is important to vacuum first so that extraneous material such as conducting or other harmful particles will not be driven into the windings or other critical areas. Note the highest pressure for compressed air should be 30 psi, which is the legal limit (OSHA), and during the operation, safety glasses and/or other protective equipment should be used to prevent possible eye injury. Care must be taken to make sure that the air supply is dry.

The presence of oil makes thorough, effective cleaning of machines in service virtually impossible, and repair-shop or service-shop conditioning is recommended. Oil on a surface forms a flypaper which attracts and holds firmly any entrained dust. Neither suction nor compressed air is effective; consequently only accessible areas may be cleaned. First, remove as much of the dirt as possible by wiping with clean, dry rags. For areas not readily accessible, a clean rag is drawn through an opening by means of a hooked wire and the rag is drawn alternately back and forth. This process is continued until a clean cloth thus applied will stay clean. Cloths should be changed frequently in any wiping operation; otherwise, the dirt or contamination picked up by the cloth may simply be transferred to another previously uncontaminated area.

To simplify removal of oily dirt, solvents are commonly prescribed, but in the field on assembled machines, liquid solvents are strongly discouraged, particularly on dc machines and ac machines with commutators or collectors. Generally the solvent application in the field should be by a wiping rag barely moistened (not wet). Where dirt is heavy,

2-70 Maintenance of Power Systems

repeated wipings may be required. Liquid solvent can carry conducting contaminants (metal dust, carbon, etc.) deep into hidden but critical areas to produce shorts and grounds, thus causing machine failure. Without special testing equipment such as surge testers, weaknesses that could result in shorts in an armature, wound rotor, or rotating field cannot be exposed. Grounding weakness may be studied with a megger, but even here, acceptability of the test is questionable because the megger does not develop enough electrical energy to expose all the weak grounding paths.

Solvents may be toxic or flammable. Therefore, adequate ventilation must be provided to minimize fire hazards and health hazards caused by the use of solvents for cleaning purposes. Keep away from sparks, heat, or flame to prevent fire or explosion hazard.

Prior to cleaning and other maintenance activities electric circuits must be deenergized. Also electric circuits should be grounded to discharge capacitors prior to cleaning or maintenance. These precautions are necessary to avoid electric shock which can cause serious or fatal injury.

Certain freons are recommended for cleaning because they are not flammable, have good solvency for grease and oil, are considered safe with most varnishes and insulations, and have a low order of toxicity. Inhibited methyl chloroform is also acceptable. Carbon tetrachloride is effective and nonflammable but is very toxic in confined spaces with repeated usage and is therefore not recommended. The chlorinated solvent is sometimes used in very stubborn cases, but all the safety precautions must be used. Toluene and xylene and Stoddard solvent (hydrocarbon vents) possess good solvency but are not recommended, since they are flammable and attack varnishes quite readily. Steam cleaning is not recommended because, as with liquid solvents, conducting contaminants may be carried deep into inaccessible areas, resulting in shorts or grounds.

Carbon-brush performance may be ruined by absorbed solvents; so brushes should be removed on dc machines, ac wound rotors, and synchronous generators (where used) before solvent wiping.

The above really refers to field-service cleaning. In a repair shop or service shop with a disassembled machine, there are many more options, and generally the shop has more equipment, more instrumentation, and more know-how to perform the necessary operations.

The first step is to get an initial insulation-resistance reading on each machine component. Low readings would be expected with badly contaminated machines, but failure would indicate electrical damage calling for repair, not just cleaning.

Cleaning with water and detergent is a very effective method of cleaning windings when used with a low-pressure steam jenny (maximum steam flow 30 psi and 90°C), but to minimize possible damage to varnish and insulation, a fairly neutral nonconducting type of detergent should be used. A pint of detergent to 20 gal of water is recommended. If a steam jenny (steam-spray machine) is not available, the cleaning solution may be applied with warm water by a spray gun. Tightly adhering dirt will require additional agitation by gentle brushing or wiping. After the cleaning operation, the windings should be rinsed with water or low-pressure steam.

Even plain water can be used to wash out motors that have been plugged with mud or other inert foreign matter by plant operations or floods. The motors can be washed with water from a hose and disassembled so that all parts can be thoroughly cleaned. When water is applied to insulated parts, the pressure should not exceed 25 psi.

After either cleaning operation described above, the surface moisture should be wiped off with a clean cloth and the insulation dried promptly to keep the penetration of water as low as possible. It is advisable to dry the windings further, preferably in a circulating-air oven at a temperature not exceeding 90°C until the insulation resistance becomes practically constant. Other methods such as hot-air heaters are also satisfactory, but care must be taken to keep the temperature within limits and to make sure that local hot spots are not developed.

If an oven cannot be used, the machine's own frame with the addition of some covers will usually make an effective enclosure to contain heat. Some flow of air is desirable to allow the moisture to be carried away. Methods of generating heat include blowing hot air through the machine, heating with heat lamps, or passing a current at low rated voltage with the rotor locked for ac machines and passing a low current through the main-field-coil windings for dc machines.

Solvents can be used in the repair shop or service shops under controlled conditions. If the dirt incrustations are not removable by wiping, blowing, suction, or steam cleaning and the electric-circuit insulation is not restored, solvent cleaning is employed. The two types of solvents are petroleum distillates and other solvents, comprising chlorinated solvents, mixtures of chlorinated and petroleum solvents, and coal-tar solvents. Preferences have been shown for both types.

Actual immersion in trichlorethane or perchlorethylene with air agitation or suspension of the part in a vapor degreaser will soften dirt accumulations, permitting compressed-air removal. Repeated cleanings in this manner may restore the insulated circuits. Under no conditions should carbon tetrachloride or any *chloride* solvent be used because of the highly toxic effect. However, taking the necessary precautions, the *chlorinated* solvents are used effectively.

The petroleum distillates (hydrocarbon solvents) are classed as safety-type solvents and have flash points at about 100°F (38°C). They can be supplied by practically all oil companies under various trade names. Stoddard solution or solvent as described in the National Bureau of Standards Commercial Standard CS 3-40 is a good solvent of this type. It is sometimes modified with perchlorethylene and methylene chloride to give better results. However, all such solvents are flammable, and the vapors form explosive mixtures with air if used at temperatures above their flash point. Since their flash point is higher than that of other types of petroleum distillates, such as gasoline, however, the so-called safety type presents less of a fire hazard. Complete safety precautions should be taken.

An appreciable percent of complex structures (such as dc armatures) may never recover with submerged solvent cleaning, because conducting contaminants become trapped in critical fissures. Also commutators, collectors, etc., should not be cleaned by submerging in a solvent.

While electric-circuit insulation is restored by submerged solvent cleaning, revarnish treatment is required. This also applies to water and detergent cleaning and is recommended for steam cleaning. The varnish used should be what the manufacturer recommends.

One other method of cleaning should be mentioned. Ground-up corncobs, peanut husks, and the like have been employed for mild abrasion and absorption of oil and grease of contaminated electrical components. Where visual contamination is removable in this manner, its use is recommended, as the meal has an affinity for oil and grease and will clean a machine in excellent fashion. When this method is used, the machine should be covered and the dirty meal drawn off by a vacuum. The operators should use a dust mask or respirator.

Silicone high-temperature insulation represents a special case, and the manufacturer should be consulted.

Testing Insulation

The main test in maintenance work is the measurement of insulation resistance. The best guide in making these tests is IEEE 43, "Recommended Guide for Testing Insulation Resistance of Rotating Machiner," published by the Institute of Electrical and Electronics Engineers, 345 East 47th Street, New York, N.Y. 10017. This serves as probably the best indication of whether the machine is in condition for operation and dielectric testing.

The insulation-resistance test gives a good indication of the condition of the insulation, particularly from the standpoint of moisture and dirt. The value of resistance depends upon the type size, voltage rating, etc., of the machine. A good general minimum value is

$$\frac{\text{Rated voltage}}{(0.75 \times \text{hp rating}) + 1000} \text{ megohms}$$

or approximately 1 megohm for each 1000 volts of operating voltage with a minimum value of 1 megohm. Preferably, the resistance corrected to 40°C should measure at least 1.5 megohms. However, don't be fooled; high-insulation-resistance values do not necessarily assure high dielectric strength, although low-insulation resistance may indicate low dielectric strength. The importance of the measured value lies in the relative readings of insulation values under similar conditions at various times. They usually indicate under these conditions the effectiveness of the maintenance work.

A sudden drop or a consistent trend toward low values of insulation resistance gives evidence that the insulation system is deteriorating and that failure may be imminent.

Tests to obtain values of insulation resistance test only the value of resistance to ground and in addition, winding to winding in the case of dc fields. Disconnect all external leads for the test.

The armature and individual field coils in a dc machine can be tested only to ground and not from one circuit to another or along creepage paths from one exposed voltage to another. Superficial cleaning methods may greatly improve resistance to ground but may actually further deteriorate the insulation condition across shorting paths (by washing contaminants into the machine). A machine whose insulation resistance is low could be given an in-place cleaning and later fail because cleaning was not complete and damaging amounts of contamination were left that could not be detected by testing insulation resistance to ground.

Insulation resistance can be measured by a self-contained instrument such as the familiar megger, either hand- or motor-operated; by the electronic type, with a resistance bridge; or with a milliammeter, a voltmeter, and a dc supply. Any of these instruments used in insulation testing must be well maintained and calibrated to be sure that the readings are factual.

As stated above, the insulation resistance of apparatus in service should be checked periodically at approximately the same temperature and under similar conditions of humidity to determine possible insulation deterioration. If such measurements show wide variations, the cause should be determined and corrective measures taken to forestall an insulation failure. On large machines a further refinement is used by obtaining the polarization index. The polarization index is obtained by dividing the 10-min insulation-resistance value by the 1-min insulation-resistance value. The index is a useful means of determining if a machine is suitable for overpotential testing or for further operation. A history of this index is valuable in deciding whether or not the insulation system is deteriorating. If insulation-resistance values are taken for use in calculating a polarization index only, they do not have to be corrected for temperature, since this does not affect the ratio.

The recommended minimum value of polarization index is 2.0 for large form-wound machines. This is subject to somewhat the same limitations that are applied to insulation-resistance measurements.

Dielectric Tests

The purpose of dielectric tests is to determine if the insulation on the machine can withstand the voltage stresses set up during normal and possible abnormal conditions during operation.

The application of the ac high voltage necessary to make a dielectric test presents hazards in that not only can the ac voltage used puncture or break down the insulation, but often severe burning of the machine laminations occurs because the capacity needed to test larger machines is such that in case of breakdown, a large amount of power follows in the arc established. However, in many cases, the risk involved does not outweigh the possible long outage that might occur if the insulation failed while in an operation driving an important load. To diminish that risk, it is important that a satisfactory insulation resistance is obtained because a lot of damage can be done by applying high voltage to a wet and/or dirty winding.

The test voltage applied to new machines or to the winding of machines completely rewound with new coils and insulating material is specified by IEEE and ANSI standards as twice rated voltage plus 1000 volts held for 60 sec, with the exception of field windings of synchronous motors, which are given a test voltage of 10 times the exciter voltage but not less than 1500 volts. For machines in commercial operation or for repaired machines, no standards have been set, but established practice is to use an ac test voltage between 65 and 75 percent of the test voltage for new windings. The lower value should be used for older windings.

Within recent years, high-voltage dc testing has become more and more accepted. It has numerous advantages over ac testing. The capacity used is small, and the test effect in searching out weak insulation is comparable with ac testing. The unit used is considerably smaller physically than the test transformer, the equipment needed to test the largest machine being easily transported in a car, whereas the transformer requires a large truck. The device is electronic and consists essentially of a high-voltage rectification circuit. Instruments measure the current and voltage. Another advantage, which is of prime importance, is that in case of an insulation failure during test, no iron burning results because of the small amount of power used. The test equipment operates from the 60-cycle lighting circuit. Test values have been established whereby a dc test voltage is applied 60 percent greater than the ac test voltage ordinarily used. The cost of the dc test outfits is considerably less, especially when compared with the cost of a test transformer with capacity to test the windings of large machines. The same dc test outfit can be used to test the windings of machines, from the smallest to the largest. This is generally used for large machines 6000 volts or over. See IEEE 95, "Insulation Testing of Large A-C Rotating Machines with High Direct Voltage."

A further test is being used for dielectric testing of high-voltage machines using ac high voltage at very low frequencies (about 0.1 Hz). A working group in IEEE is developing a standard. There are several advantages: the newer equipment is much lighter and more portable than the established ac equipment; it gives a better ac search than high-voltage dc testing; and it does not generate corona during the testing period.

Turn-to-turn insulation can be checked with a surge-comparison tester and is used to locate insulation faults and winding dissymmetries in all classes of equipment regardless of size. It is an electronic device and is portable, so that it can be used for maintenance work as well as shop work. High turn-to-turn voltages are applied without excessive winding to ground stresses, and the testing is nondestructive. Both the surge test and the high-voltage dc testing should be done by experienced, trained operators.

PART IV. BEARINGS AND LUBRICATION

Proper care of bearings, which includes lubrication, is one of the more important maintenance items pertaining to motors. The rotor of a motor, which transfers the electrical energy from the power supply into mechanical energy through the rotating shaft to drive the load, is supported on bearings.

The designer has a choice between sleeve bearings and antifriction ball or roller bearings, but for normal applications, economics will decide the choice of the bearing and lubrication system. Despite the trend in integral-horsepower motors from sleeve to ball bearings in the range of 1 to 500 hp, millions of fractional-horsepower motors are made with wick-oiled sleeve bearings.

In the integral-horsepower range of 1 to 500 hp, almost all normal motors are made with ball bearings or roller bearings except for some of the larger, higher-speed motors. The trend to ball bearings is such that 100 hp and smaller sleeve-bearing motors are considered special.

Over 500 hp, ac and dc motors are normally made with oil-lubricated sleeve bearings for the high surface speeds with large shaft diameters and usually have split bearings for ease in replacement. However, some dc machines as large as 2000 hp are equipped with antifriction bearings.

An analysis of induction-motor failures over a long period shows that bearings are one of the principal causes of trouble and the successful operation of a motor depends on a good preventive-maintenance program. Not only will loss of lubrication fail the bearings, but the failure might result in other problems such as winding or stator-rotor rubbing problems on sleeve-bearing motors.

On small motors with wick-oil sleeve bearings, good maintenance requires that the bearings be reoiled once a year or every 2000 operating hours, whichever comes first. Approximate amounts of oil to be added vary from 30 drops for a 3-in.-diameter rotor to 100 drops for a 9-in. motor. If light turbine oil is unavailable, SAE 10 automotive oil will suffice. During any other motor maintenance work, the yarn or felt packing should be replaced. After the bearing housing is washed, the wick cavity is repacked with clean wool yarn saturated with oil.

For ball bearings and roller bearings, good maintenance requires that the bearings be regreased on a schedule which will vary with motor size, speed, duty, and environment. The larger the motor, the higher the speed, the hotter the temperature or ambient, the greater the vibration, the more severe the duty, and the higher the loading, the more frequent should be the regreasing. The interval of times for relubrication vary from 10 years to 1 month depending upon the above factors. For example, a 1-hp motor used in infrequent operations, such as portable tools, normally used in a clean and reasonable temperature ambient of 10 to 30°C (50 to 86°F), would need to be regreased only once in

10 years while a 200-hp motor used continuously in a dirty, hot environment with severe duty and heavy vibration would need to be regreased every month. This constitutes a range of 84 to 1.

Table 5-2 is an approximate guide for relubrication of electric motors. Note that some manufacturers recommend longer periods and some shorter periods for regreasing. These periods are based on deep-groove conrad ball bearings or roller bearings with open, single-shielded, or double-shielded bearings with a grease reservoir adjacent to the bearings.

TABLE 5-2 Guide for Maximum Relubrication Periods*

Service	Motor horsepower				
	¼–10	15–40	50–150	200–250	Over 250
Easy: infrequent operation, valve operators, door openers, portable tools	7 years	5 years	3 years	2 years	9 months
Standard: one or two shifts, machine tools—air-conditioning conveyors, compressor refrigeration, laundry, textile machinery, woodworking, water pumping, generally Class B insulation	5 years	3 years	1 year	9 months	6 months
Severe: continuous running. Fans, pumps, motor-generator sets, coal and mining machinery, steel mills, some Class F insulation	3 years	1¼ years	6 months	4 months	3 months
Very severe: dirty and vibrating applications, hot pumps and fans, high-ambient Class H insulation	9 months	4 months	3 months	2 months	1½ months

*Some manufacturers recommend longer periods and some recommend shorter periods, based on open and shielded bearings with grease reservoirs adjacent to bearings.

Sealed bearings are not used generally because the grease, and therefore the oil, for lubrication is limited to what is inside the bearing proper and then the life of the bearing becomes the life of the grease. However, on some applications, particularly some in the military field where only 4000 to 5000 hr of life is needed, they are extremely satisfactory because once installed, there is no question concerning the proper grease; there is almost no chance of getting dirt in the bearings; and there is no chance of overgreasing. Sealed and shielded bearings require a stiffer grease to provide greater mechanical stability to minimize churning while closely confined near the rotating balls.

Shielded bearings do allow some feeding of oil from the grease in the housing reservoirs, and open bearings allow the most.

The greases used for ball and roller bearings have constantly been improved over the years, and the modern ball-bearing greases are very stable chemically.

If kept clean and at the intended operating temperatures, these modern greases will lubricate satisfactorily for many years. As a result, some smaller motors are not provided with grease fittings or plugs and are intended to operate without greasing maintenance.

Starting in 1925, the sodium-calcium mixed-base grease replaced calcium cup grease; in 1940, oxidation inhibitors were introduced for general use; in 1953; lithium-soap grease was introduced; and in 1968, the synthetic thickeners in petroleum grease appeared for use in higher bearing temperatures that might be encountered with Class F insulation motors. (The much higher temperatures associated with Class H insulated motors used silicone grease brought out in 1945 or some of the recently developed greases using synthetic thickeners, such as polyurea.) On high-temperature applications always use a grease the same as or compatible with the original grease.

A conscientiously applied program of preventive maintenance will add years of useful life to bearings. Machines with antifriction bearings are shipped from the factory with the bearings packed with grease. If the machines have been stored for a long period of time, it is advisable to regrease the bearings before operation, especially for motors of 150-hp size and larger. The grease used as a lubricant in grease-lubricated antifriction bearings does not lose its lubrication ability suddenly, but the oil bleeds out of the grease over a period of time. For a given bearing and assembly, the loss of lubricating ability of a grease with age depends primarily on the type of grease, the size of the bearing, the speed at which the bearing operates, the temperature and the severity of the loading, and operating conditions. Although it is not possible to predetermine accurately when new grease must be added, good results can be obtained with the following procedure:

1. Wipe all lubrication fillings clean.
2. Remove the relief plug and free the hole of hardened grease.
3. With the machine running, add grease slowly with a hand-operated pressure gun until new grease is expelled through the relief hole. (If the fittings are not safely accessible with the machine running, grease may be added sparingly with the machine at rest.)
4. It is recommended that the motor be regreased at standstill for safety reasons as well as to prevent grease leakage along the shaft-end shield fit and into the internal end shield cavity and the motor windings.
5. Run the machine from 10 to 20 min with the relief plug removed to expel excess grease.
6. Clean and replace the relief (outlet) plug.
7. Make sure the new grease is clean. Dirt and contaminants introduced when adding grease can cause bearing failure, and some foreign particles such as silica or cast-iron dust are almost sure to cause bearing problems.
8. Use the grease (or an equivalent grease) recommended by the motor manufacturer. Some very good greases are not compatible with each other, and if there is any doubt about compatibility with mixed greases, the motor manufacturer should be consulted.
9. Do not overgrease, because this can cause bearings to heat up considerably and cause failure. Sometimes motors are furnished with grease plugs instead of fittings to prevent casual overgreasing.

If a motor has to be disassembled for any reason, such as cleaning or rotor or stator repairs, the bearings and housings should be cleaned of the old grease by washing with a grease-dissolving solvent and the bearing should be repacked at about 50 percent full with the manufacturer's recommended grease or equivalent. Again, cleanliness is the critical word because of the small clearances in ball and roller bearings. Compressive stresses commonly range well over 1,000,000 psi in the minute contact area in the heavily loaded ball-race contacts. It is easily seen that a piece of grit or cast-iron dust could destroy the balls and races.

If a bearing has to be replaced, care must be taken in removing the bearing and reinstalling either the old bearing or a new one. Care must be taken to avoid scratching or nicking critical surfaces. A bearing puller should be used to remove the bearing. It is similar in looks and works like a gear puller or pulley puller. Wherever possible, apply the pulling force on the inner ring or race which is on the shaft. Pulling on the outer race transmits the force through the balls in the bearings, which presents the possibility of damage by brinelling. In case you have to pull on the outer race, the bearing should be replaced. As a matter of fact, any pulled bearing is generally replaced unless it is hard to procure a new one.

Installation of a replacement bearing starts with making sure that it is the correct type. The bearings in many cases are identified on the motor nameplate. If the original bearing called for is not available, it might be wise to check the manufacturer for a suitable substitute. Before reassembling a bearing, all bearing and machined surfaces should be thoroughly cleaned with a suitable solvent. Examine the machined fits of the end shield, cartridge, slinger, and grease caps for burrs. It is important that these surfaces be smooth.

A typical way to reassemble a bearing is as follows:

1. Inspect the bearing housing and related parts for foreign material. Clean, if necessary.
2. The machined fits and critical surfaces of the end shield, bearing cartridge, grease cap, shaft, and bearings should be free of all nicks, scratches, or burrs. If any polishing is

done, care should be taken to avoid a deposit of metal dust in and around the bearing assembly.

3. The internal surface of the bearing cartridge or housing should be coated with a thin film of recommended grease as well as the shaft and shaft fit of the grease cartridge and grease cap. These precautions, although not absolutely essential, will guard against corrosion of the critical surfaces.

4. On smaller bearings, heat the bearing in an oven to 100 to 125°C (but not higher) and place on the shaft while making sure that it is against the locating shoulder. The cooling of the bearing will give a tight fit with the shaft. If no suitable heat is available, a piece of pipe can be used as a pressing fixture to press the bearing tightly against the locating shoulder. The bearing should be gently tapped into place, with pressure applied to the inner race only. Any pressure applied to the outer race is transmitted through the balls to the inner race. Small indentations or impressions in the races (brinelling) will result in making the bearings noisy and will also cause premature bearing failure.

On larger motors with larger bearings, the bearing can be heated in hot oil 50 to 125°C (122 to 257°F) and placed on the shaft. Hold the bearing against the shaft shoulder until the bearing cools.

5. Replace the bearing nut and washer if used.

6. Secure the bearing cap and housing and make sure the shaft turns freely.

If a bearing failure in a specific installation or application occurs too frequently to be considered the result of normal fatigue, then something is wrong and bearings will fail repeatedly until the "something wrong" is corrected. If possible, a bearing should be removed at the first sign of noise or heating so that a bearing specialist or the motor manufacturer can examine the bearing. Much can then be determined by a knowledgeable expert or the manufacturer's bearing specialist. Complete failures, on the other hand, destroy most evidence of the cause of trouble.

Sleeve bearings with oil lubrication are used for larger motors, particularly ones with a large shaft, to avoid the fatigue limits with rolling bearings, to operate above the speed limits for grease lubrication, and for easier bearing replacement. They are commonly used with split-end shields.

Motors with sleeve bearings are normally lubricated with turbine-type mineral oils, which offer the longest life of any petroleum lubricants available for self-contained or circulating oil systems. Their rust inhibitors also minimize corrosion on the shaft, bearing, and housing surfaces.

As a guide for selecting the proper grade of turbine oil, Tables 5-3 and 5-4 show the viscosities for several classes of motor operation. A light oil of 150 SSU (Saybolt Seconds Universal) viscosity at 100°F is recommended for fractional-horsepower motors and large units at 1500 rpm and faster. Lower speeds, high temperatures, and high loads normally call for the more severe requirements of some rolling-contact thrust bearings (such as for vertical pumps) call for heavier oils up to 600 SSU. It can be seen that manufacturers' recommendations should be followed for the best operation.

TABLE 5-3 Suggested Viscosities for Electric-Motor Oils

	Oil-viscosity grade, SSU at 100°F	
Type of bearing system	Below 1500 rpm	1500 rpm and up
Sliding contact		
Pressure-lubricated	300	150
Ring-oiled	300	150
Disk	300	150
Wick-oiled (fractional-hp)	300	150
Plate-type, thrust	150–300	150
Rolling contact	300	300
Ball and cylindrical roller	150	150
Spherical roller	600	600

However, other grades can often be substituted successfully. For example, a 300 SSU oil can give satisfactory overall performance in a high-speed motor, even though the bearing temperatures are higher than desired and power losses slightly greater. On the other hand, motors running below 1500 rpm start more easily with a light oil.

Oils other than the turbine type are normally avoided. Automotive oils not only are more expensive, but their detergents may cause foam, emulsions, and other operating difficulties. Gear oils generally give shorter life, except where their ability to withstand extreme pressures is required, as in gear motors.

TABLE 5-4 Typical Motor-Oil Properties

Characteristic	Grade		
	Light	Medium	Heavy
Viscosity, SSU at 100°F	140–170	270–325	540–700
Viscosity index, min	85	85	85
Pour point, °F, max	0	5	25
Flash point, °F, min	380	400	420
Neutralization number, mg of KOH per gram, max	0.15	0.15	0.8
Aniline point, °F, min	195	195	
Turbine-oil oxidation test, hr, min (ASTM D943)	1000	1000	1000
Rust-prevention test (ASTM D 665)	Pass	Pass	Pass

Synthetic oils intended for aircraft, fire resistance, or other special applications are normally unsuited to electric motors, since most of them attack the paint, rubber, and insulations usually used. If extreme temperatures suggest using a synthetic oil, the complete motor should be considered for compatibility with all components. Under such conditions the best answer might be a circulating oil system, which has supplemental heating or cooling to keep oil temperature at a level which permits use of a turbine oil and avoids the compatibilty problem.

Maintenance with oil lubrication consists mainly of checking the oil every 3 or 4 months for level and appearance. Like ball-bearing relubrication, the ideal interval varies depending on ambient temperature, cleanliness, and severity of service. Level is usually checked at standstill, since rotation of an oil ring or other components may distort the reading. However, the manufacturer sometimes indicates that the oil level should be set while running. If the oil looks clean and not discolored, simply add enough to replenish the level.

Yearly oil changes are common unless ambients or operating conditions call for other intervals. The acidity of the oil can serve as a limiting guide for oil changes. When oxidation raises the neutralization number above 1.0 (ASTM D 974), oil gets darker, varnish deposits can be expected, and bearing corrosion rate increases. Oil should be changed promptly. If an acidity check is impractical, a visual observation is usually adequate. Do not add oil until it drops below full level and flooding should be avoided.

Any cleaning that is needed usually occurs when the oil is changed. This generally flushes water, dirt, and sediment from the bearing housing or oil reservoir. During disassembly for a general motor cleaning, the bearing and housing should be washed with a solvent. Some manufacturers recommend hot kerosene. Coating the drain plug with a sealer such as alkyd-resin compound or No. 3 Permatex before replacing it will prevent leakage.

Extreme care is required in the disassembly of a bearing to prevent nicking or burring of the bearing or machined running surfaces. In addition, the surfaces of the journal and the bearing must be protected from rust and damage when exposed during the process of disassembly and reassembly.

The sleeve bearing, if it must be replaced or rebabbitted, should be ordered from the motor manufacturer or a reliable supplier. If rebabbitting is indicated, the babbitt recommended by the motor manufacturer should be used and no changes should be made except by the motor manufacturer or a knowledgeable expert. There are several types of babbitt, but only two general classes are in use. The tin-base type contains 80 to 90 percent tin with the remainder divided equally between copper and antimony, and the lead-base type runs 75 to 85 percent lead with 5 to 10 percent tin and 5 to 10 percent antimony. Tin-base babbitt is generally used in corrosive atmospheres.

Maintenance of Power Systems

Clearance between the bearing and the journal is important, and the following general rules apply:

1. For shafts 1 in. in diameter and less, the diametrical clearance should be 0.001 in./in. of shaft diameter.
2. For shaft larger than 1 in. in diameter the diametrical clearance should be 0.002 in./in. of shaft diameter.

Since cleanliness is always important, all bearing and machine surfaces should be thoroughly cleaned with a suitable solvent before reassembly. Examine all machine fits for burrs, and remove if present. Remove all oil compound from sealing surfaces.

Prior to actual reassembly, the following precautions should be observed:

1. Inspect the bearing housing and related parts for foreign matter. Clean, if necessary.
2. Inspect the journals and polish them with crocus cloth if any scratches are detected. Do not allow any metal dust to fall into the housing when polishing the journals.
3. Spread a thin coat of oil over the journal and bearing surface before reassembling.
4. Sealing surface of the end shield should be coated with a sealing compound.

A suggested maintenance schedule on sleeve includes a monthly check of the oil level, a check that the oil rings that carry the lubricant to the journal and the bearing are rotating, and a check of the oil color through a sight gauge. Slightly cloudy oil is all right. Black oil is a danger sign. On a 6-month basis, check the oil for acidity and change it if required.

A good maintenance schedule, conscientiously followed with the proper materials and proper procedures, will give long life and service (Table 5-5).

TABLE 5-5 Good Maintenance Practices

Keep motor off line when not needed	Saves unnecessary wear of brushes, commutator, and bearings, saves lubrication
Do not leave field circuit unless motor has been especially designed for this type of duty	Check temperature of shunt fields with themometer to see that it does not exceed 90°C. When field must be excited, caution maintenance men to be sure field circuit is opened before working on the motor. On ac check stator windings
Keep motor clear of metal dust or cuttings that can be drawn into windings and pole pieces	Magnetic attraction will draw metal parts into the air gap and damage windings. Cast-iron dust particularly damaging
Reassembling of motor	Be sure to retain proper air gaps in motor by checking bore of pole faces before removing poles from the frame. Mark shims and poles. Reassemble, replacing poles and liners in original position on dc motors
Note wearing parts and parts frequently replaced to determine anticipated repairs	Carry in proper storeroom stock of replacement parts. Make survey of standard repair parts to avoid duplication of parts to be carried

The main points to remember about bearings are to handle them carefully, remove and install them properly, and lubricate them with the proper lubricants. Keep them clean.

Tables 5-6 to 5-8 summarize maintenance procedures and provide a ready reference.

If the maintenance man or inspector is to do a satisfactory job, proper tools and instruments are necessary. Also, he should have a good knowledge of the electrical and mechanical characteristics of the equipment under his care, together with an understanding of the correct operation of the equipment. The following, as a minimum, should be made available:

1. Tools necessary to disassemble apparatus
2. Extension cords, safety type
3. Flashlights, rubber or molded cases
4. Air-gap gauges
5. Micrometers, inside and outside
6. Dial indicator with assortment of brackets

Maintenance of Electric Motors 2-79

7. Megger, 500-volt
8. Volt-amp-ohmmeter tester such as Simpson, Triplett, or equivalent
9. Thermometers and levels
10. Portable instruments such as ammeters, voltmeters, and graphic meters

Access to instruction books furnished by the manufacturer with equipment purchased should be a must.

Only by having suitably instructed personnel with adequate equipment can motors be given the attention they need for long, trouble-free operation.

TABLE 5-6 AC and DC Motor Check Chart

Trouble	Cause	What to do
Hot bearings—general	Bent or sprung shaft	Straighten or replace shaft
	Excessive belt pull	Decrease belt tension
	Pulley too far away	Move pulley closer to bearing
	Pulley diameter too small	Use larger pulleys
	Misalignment	Correct by realignment of drive
Hot bearings—sleeve	Oil grooving in bearing obstructed by dirt	Remove bracket or pedestal with bearing and clean oil grooves and bearing housing; renew oil
	Bent or damaged oil rings	Repair or replace oil rings
	Oil too heavy	Use a recommended lighter oil
	Oil too light	Use a recommended heavier oil
	Insufficient oil	Fill reservoir to proper level in overflow plug with motor at rest
	Too much end thrust	Reduce thrust induced by driven machine or supply external means to carry thrust
	Badly worn bearing	Replace bearing
Hot bearings—ball	Insufficient grease	Maintain proper quantity of grease in bearing
	Deterioration of grease or lubricant contaminated	Remove old grease; wash bearings thoroughly in kerosene and replace with new grease
	Excess lubricant	Reduce quantity of grease. Bearing should not be more than half filled
	Heat from hot motor or external source	Protect bearing by reducing motor temperature
	Overloaded bearing	Check alignment, side thrust, and end thrust
	Broken ball or rough races	Replace bearing; first clean housing thoroughly
Oil leakage from overflow plugs	Stem of overflow plug not tight	Remove; recement threads; replace and tighten
	Cracked or broken overflow plug	Replace the plug
	Plug cover not tight	Requires cork gasket, or if screw type, may be tightened

2-80 Maintenance of Power Systems

TABLE 5-6 AC and DC Motor Check Chart (*Continued*)

Trouble	Cause	What to do
Motor dirty	Ventilation blocked, end windings filled with fine dust or lint	Clean motor will run 10 to 30°C cooler. Dust may be cement, sawdust, rock dust, grain dust, coal dust, and the like. Dismantle entire motor and clean all windings and parts
	Rotor winding clogged	Clean, grind and undercut commutator, or clean and polish collector. Clean and treat windings with good insulating varnish
	Bearing and brackets coated inside	Dust and wash with cleaning solvent
Motor wet	Subject to dripping	Wipe motor and dry by circulating heated air through motor. Install drip- or canopy-type covers over motor for protection
	Drenched condition	Motor should be covered to retain heat and the rotor position shifted frequently
	Submerged in flood waters	Dismantle and clean parts. Bake windings in oven at 105°C for 24 hr or until resistance to ground is sufficient. First make sure commutator bushing is drained of water and completely dry

TABLE 5-7 DC Motor Check Chart

Trouble	Cause	What to do
Fails to start	Circuit not complete	Switch open, leads broken
	Brushes not down on commutator	Held up by brush springs; need replacement. Brushes worn out
	Brushes stuck in holders	Remove and sand; clean up brush boxes
	Armature locked by frozen bearings in motor or main drive	Remove brackets and replace bearings or recondition old bearings if inspection makes possible
	Power may be off	Check line connections to starter with light. Check contacts in starter
Motor starts, then stops and reverses direction of rotation	Reverse polarity of generator that supplies power	Check generating unit for cause of changing polarity
	Shunt and series fields are bucking each other	Reconnect either the shunt or series field so as to correct the polarity. Then connect armature leads for desired direction of rotation. The fields can be tried separately to determine the direction of rotation individually and connected so both give same rotation
Motor does not come up to rated speed	Overload	Check bearing to see if in first-class condition with correct lubrication. Check driven load for excessive load of friction
	Starting resistance not all out	Check starter to see if mechanically and electrically in correct condition

TABLE 5-7 DC Motor Check Chart (*Continued*)

Trouble	Cause	What to do
	Voltage low	Measure voltage with meter and check with motor nameplate
	Short circuit in armature windings or between bars	For shorted armature inspect commutator for blackened bars and burned adjacent bars. Inspect windings for burned coils or wedges
	Starting heavy load with very weak field	Check full field relay and possibilities of full field setting of the field rheostat
	Motor off neutral	Check for factory setting of brush rigging or test motor for true neutral setting
	Motor cold	Increase load on motor so as to increase its temperature, or add field rheostat to set speed
Motor runs too fast	Voltage above rated	Correct voltage or get recommended change in air gap from manufacturer
	Load too light	Increase load or install fixed resistance in armature circuit
	Shunt field coil shorted	Install new coil
	Shunt field coil reversed	Reconnect coil leads in reverse
	Series coil reversed	Reconnect coil leads in reverse
	Series field coil shorted	Install new or repaired coil
	Neutral setting shifted off neutral	Reset neutral by checking factory setting mark or testing for neutral
	Part of shunt field rheostat or unnecessary resistance in field circuit	Measure voltage across field and check with nameplate rating
	Motor ventilation restricted, causing hot shunt field	Hot field is high in resistance, check causes for hot field, in order restore normal shunt field current. Restore ventilation
Motor gaining speed steadily and increasing load does not slow it down	Unstable speed load regulation	Inspect motor to see if off neutral. Check series field to determine shorted turns. If series field has a shunt around the series circuit, that can be removed
	Reversed field coil shunt or series	Test with compass and reconnect coil
	Too strong a commutating pole or commutating-pole air gap too small	Check with factory for recommended change in coils or air gap
Motor runs too slow continuously	Voltage below rated	Measure voltage and try to correct to value on motor nameplate
	Overload	Check bearings of motors and the drive to see if in first-class condition. Check for excessive friction in drive

Maintenance of Power Systems

TABLE 5-7 DC Motor Check Chart (*Continued*)

Trouble	Cause	What to do
	Motor operates cold	Motor may run 20 percent slow owing to light load. Install smaller motor, increase load, or install partial covers to increase heating
	Neutral setting shifted	Check for factory setting of brush rigging or test for true neutral setting
	Armature has shorted coils or commutator bars	Remove armature to repair shop and put in first-class condition
Motor overheats or runs hot	Overloaded and draws 25 to 50 percent more current than rated	Reduce load by reducing speed or gearing in the drive or loading in the drive
	Voltage above rated	Motor runs drive above rated speed requiring excessive horsepower. Reduce voltage to nameplate rating
	Inadequately ventilated	Location of motor should be changed, or restricted surroundings removed. Covers used for protection are too restricting of ventilating air and should be modified or removed. Open motors cannot be totally enclosed for continuous operation
	Draws excessive current owing to shorted coil	Repair armature coils or install new coil
	Grounds in armature such as two grounds which constitute a short	Locate grounds and repair or rewind with new set of coils
	Armature rubs pole faces owing to off-center rotor causing friction and excessive current	Check brackets or pedestals to center rotor, and determine condition of bearing wear for bearing replacement. Check pole bolts
Hot armature	Core hot in one spot, indicating shorted punchings and high iron loss	Sometimes full slot metal wedges have been used for balancing. These should be removed and other means of balancing be investigated
	Punchings uninsulated Punchings have been turned or band grooves machined in the core Machined slots	No-load running of motor will indicate hot core and drawing high no-load armature current. Replace core and rewind armature. If necessary to add band grooves, grind into core. However, treated glass roving properly varnished and properly processed is the more common method of banding windings in the small and medium sizes. Check temperature on core with thermometer not to exceed 90°C
Hot commutator	Brush tension too high	Limit pressure to 5 psi. Check brush density and limit to density recommended by the brush manufacturer
	High brush friction caused by atmospheric contaminants	Remove cause

TABLE 5-7 DC Motor Check Chart (*Continued*)

Trouble	Cause	What to do
	Brushes off neutral	Reset neutral
	Brush grade too abrasive	Get recommendation from manufacturer
	Shorted bars	Investigate commutator mica and undercutting, and repair
	Hot core and coils that transmit heat to commutator	Check temperature of commutator with thermometer to see that total temperature does not exceed ambient plus 55°C rise, total not to exceed 105°C. Class F and H will be hotter
	Inadequate ventilation	Check as for hot motor
Hot fields	Voltage too high	Check with meter and thermometer and correct voltage to nameplate value
	Shorted turns or grounded turns	Repair, or replace with new coil
	Resistance of each coil not the same	Check each individual coil for equal resistance within 10 percent, and if one coil is too low, replace coil
	Inadequate ventilation	Check as for hot motor
	Coil not large enough to radiate its loss wattage	New coils should replace all coils if room is available in motor
Motor vibrates and indicates unbalance	Armature out of balance	Remove and statically balance, or balance in dynamic balancing machine
	Misalignment	Realign. Alignment of flexible couplings generally must be closer than coupling manufacturer recommends
	Loose or eccentric pulley	Tighten pulley on shaft or correct eccentric pulley
	Belt or chain whip	Adjust belt tension
	Mismating of gear and pinion	Recut, realign, or repair parts
	Unbalance in coupling	Rebalance coupling
	Bent shaft	Replace or straighten shaft
	Foundation inadequate	Stiffen mounting place members
	Motor loosely mounted	Tighten holding-down bolts
	Motor feet uneven	Adds shims under foot pads to mount each foot tight
Motor sparks at brushes or does not commutate well	Brush setting not true neutral	Check and set on factory setting or test for true neutral
	Commutator rough	Grind and roll edge of each bar
	Commutator eccentric	Turn and grind commutator

Maintenance of Power Systems

TABLE 5-7 DC Motor Check Chart (*Continued*)

Trouble	Cause	What to do
	Mica high-hot undercut	Undercut mica
	Commutating pole strength too great, causing overcompensation, or strength too weak, indicating undercompensation	Check with manufacturer for correct change in air gap or new coils for the commutating coils
	Shorted commutating pole turns	Repair coils or install new coils
	Shorted armature coils on commutator bars	Repair armature by putting into first-class condition
	Open-circuited coils	Same as above
	Poor soldered connection to commutator bars	Resolder with proper alloy of tin solder. Current motors use mostly TIG welded
	High bar or loose bar in commutator at high speeds	Inspect commutator nut or bolts and retighten and grind commutator face
	Brush grade wrong type. Brush pressure too light, current density excessive, brushes stuck in holders. Brush shunts loose	See brushes
	Brushes chatter owing to dirty film on commutator	Resurface commutator face and check for change in brushes
	Vibration	Eliminate cause of vibration by checking mounting and balance of rotor
Brush wear excessive	Brushes too soft	Blow dust from motor and replace brushes with a changed grade as recommended by manufacturer
	Commutator rough	Grind commutator face
	Abrasive dust in ventilating air	Reface brushes and correct condition by protecting motor
	Off-neutral setting	Recheck factory neutral or test for true neutral
	Bad commutation	See corrections for commutation
	High, low, or loose bar	Retighten commutator motor bolts and resurface commutator
	Brush tension excessive	Adjust spring pressure not to exceed 2 to 2½ psi
	Electrical wear due to loss of film on commutator face	Resurface brush faces and commutator face
	Threading and grooving	Same as above
	Oil or grease from atmosphere or bearings	Correct oil condition and surface brush faces and commutator
	Weak-acid- and moisture-laden atmosphere	Protect motor by changing ventilating air, or change to enclosed motor

TABLE 5-7 DC Motor Check Chart (*Continued*)

Trouble	Cause	What to do
Motor noisy	Brush singing	Check brush angle and commutator coating; resurface commutator
	Brush chatter	Resurface commutator and brush face
	Motor loosely mounted	Tighten foundation bolts
	Foundation hollow and acts as sounding board	Coat underside with soundproofing material
	Strained frame	Shim motor feet for equal mounting
	Armature punching loose	Replace core on armature
	Armature rubs pole faces	Recenter by replacing bearings or relocating brackets or pedestals
	Magnetic hum	Refer to manufacturer
	Belt slap or pounding	Check condition of belt and change belt tension
	Excessive current load	May not cause overheating, but check chart for correction of shorted or grounded coils
	Mechanical vibration	Check chart for causes of vibration
	Noisy bearings	Check alignment, loading of bearings, lubrication, and get recommendation of manufacturer
	Magnetic noise	Tighten pole bolts. Motors supplied with power from an SCR source will generally have more noise caused by SCR ripple in armature current usually 120, 180, and 360 Hz with harmonics
Wrong rotation	Wrong connection	Consult connection diagram

TABLE 5-8 AC Motor Check Chart

Trouble	Cause	What to do
Motor stalls	Wrong application	Change type or size of motor. Consult manufacturer of driven equipment
	Overloaded motor	Reduce load
	Low motor voltage	See that nameplate voltage is maintained within standard tolerances
	Open circuit	Fuses blown; check overload relay, starter, and push button
	Incorrect control resistance of wound rotor	Check control sequence. Replace broken resistors. Repair open circuits

TABLE 5-8 AC Motor Check Chart (*Continued*)

Trouble	Cause	What to do
Motor connected but does not start	One phase open. Motor may be overloaded	See that no phase is open. Reduce load
	Rotor defective	Look for broken bars or rings
	Poor stator-coil connection	Remove end bells, locate with test lamp, and repair
Motor runs and then dies down	Power failure	Check for loose connections to line, to fuses, and to control
Motor does not come up to speed	Not applied properly	Consult supplier for proper type and size
	Voltage too low at motor terminals because of line drop	Use higher voltage on transformer terminals or reduce load
	If wound rotor, improper control operation of secondary resistance	Correct secondary control
	Starting load too high	Check load motor is supposed to carry at start
	Low pull-in torque of synchronous motor	Change rotor starting resistance or change rotor design. Consult manufacturer
	Check that all brushes are riding on ring	Check secondary connections. Leave no leads poorly connected
	Broken rotor bars	Look for cracks near the rings. A new rotor may be required, as repairs are usually temporary
	Open primary circuit	Locate fault with testing device and repair
Motor takes too long to accelerate	High WK^2. Excess loading	Reduce load. Check moment of inertia with equipment manufacturer
	Poor circuit	Check for high resistance
	Defective squirrel-cage rotor	Replace with new rotor
	Applied voltage too low	Get power company to increase voltage tap
Wrong rotation	Wrong sequence of phases	Reverse connections of motor or at switchboard
Motor overheats while running under load	Check for overload	Reduce load
	Wrong blowers or air shields may be clogged with dirt and prevent proper ventilation of motor	Good ventilation is manifest when a continuous stream of air leaves the motor. If not, check with manufacturer
	Motor may have one phase open	Check to make sure that all leads are well connected

TABLE 5-8 AC Motor Check Chart (*Continued*)

Trouble	Cause	What to do
	Grounded coil	Locate and repair
	Unbalanced terminal voltage	Check for faulty leads, connections, and transformers
	Shorted stator coil	Repair and then check wattmeter reading (form coils)
	Faulty connection	Indicate by high resistance
	High voltage, low voltage	Check terminals of motor with voltmeter
	Rotor rubs stator bore	If not poor machining, replace worn bearings (sleeve bearings)
Motor vibrates after corrections have been made	Motor misaligned	Realign
	Weak foundations	Strengthen base
	Coupling out of balance	Balance coupling
	Driven equipment unbalanced	Balance driven equipment
	Defective ball bearing	Replace bearing
	Bearings not in line	Line up properly
	Balancing weights shifted	Rebalance rotor. Securely fasten blancing means
	Wound rotor coils replaced	Rebalance rotor
	Polyphase motor running single-phase	Check for open circuit in one line or phase
	Excessive end play	Adjust bearing or add washer
Unbalanced line current on polyphase motors during normal operation	Unequal terminal volts	Check leads and connections
	Single-phase operation	Check for open contacts
	Poor rotor contacts to control wound-rotor resistance	Check control devices
	Brushes not in proper position in wound rotor	See that brushes are properly seated and shunts in good condition
Scraping noise	Fan rubbing air shield	Remove interference
	Fan striking insulation	Clear fan
	Loose on bedplate	Tighten holding bolts
Magnetic noise	Air gap not uniform	Check and correct bracket fits or bearing
	Loose bearings	Correct or renew
	Rotor unbalance	Rebalance

REFERENCES

Acknowledgment is made to the following books, pamphlets, and articles:
"Productive Maintenance," General Electric Company.
"Maintenance Hints," Westinghouse Electric Corporation.
Various maintenance articles published in *Power Engineering*.
Booser, E. R.: "Lubrication Plan for Rerated Electric Motors," General Electric Company, June 6, 1967.
Various maintenance and instruction bulletins of motor manufacturers.

Section **3**

Maintenance of Ground Contact Elements

Chapter **1**

Upkeep and Maintenance of Tires for Construction Equipment

LOUIS A. ARBORE, MANAGER

GEORGE ZAMBELAS, ENGINEER
Michelin Tire Corporation, Technical Group, Earthmoving Dept.,
Lake Success, N.Y.

INTRODUCTION

This chapter was composed to help you properly select, use, and maintain tires for construction equipment. Tire costs are a significant part of overall equipment maintenance costs, so it is very important that you know how to get the most out of your tire dollar.

To select a tire properly, you must know as much as possible about the specific application you are considering since each application, even on the same job site, can be very different. Specific job conditions, previous tire performance, and vehicle type and usage are to be thoroughly evaluated and matched to the capabilities of new tires that are being considered.

Tire cost is another factor that must enter into the tire selection procedure. Initial cost as well as expected performance must be considered. Choosing the least expensive tire may not be the answer. Consider also the increased productivity and lower downtime when using tires of higher quality.

After choosing tires, use and maintain them properly to get maximum performance. By familiarizing yourself with the information in this chapter and by contacting tire and vehicle manufacturers for additional information, you will be one step closer to realizing lower tire costs (Fig. 1-1).

SELECTING AN EARTHMOVER TIRE

The following is a list of the main factors to consider before selecting a tire:

Vehicle
- Its original equipment (tire size)
- Its loaded axle weight

3-2 Maintenance of Ground Contact Elements

Site
- Type of surface, condition of haul road
- Type and condition of loading and dumping areas

Use of Vehicle on Site
- Length of round trip (cycle)
- Number of runs (cycles) covered in 1 day's work
- Number of working hours in a 24-hr day
- Average number of miles covered in 1 hr (average speed taken from whole day's travels)
- Maximum loaded speed

Other Considerations
- Behavior of vehicle/tire combination (traction, flotation)
- Previous tire performance
- How tires wear and reasons for final removal
- Any sidewall, tread, bead problems
- The tire
- Cost (original purchase price and cost per mile based on expected performance)

Fig. 1-1 Tire maintenance in construction begins with an evaluation of tires on the scrap pile. Experience in this area reveals the main causes of failure and suggests ways the job site can be changed to improve tire life.

Size and Load Index
   1. The size will normally be the same as that of the tire originally fitted to the vehicle. However, optional sizes available for the vehicle should also be considered.
   2. Load index.

NOTE: It sometimes may be advisable to recommend tire- and rim-size changes to match job conditions.

Type. In general, the choice of tread design and type of tire is essentially determined by

   1. The type of surface the tires are to work on and the problems to be resolved (present and/or future) including muddy conditions, posing grip problems; rocky conditions, posing problems of possible cuts, tread hacking, and shock ruptures (tread or wall); and conditions where both of these problems are encountered (need for grip and resistance to rocky terrain).

NOTE: Each of these problems can be encountered to varying degrees. It is necessary to evaluate them.

2. The average number of miles covered per hour. When this is determined it should be compared with the tire manufacturer's maximum limitations established for each tread/type of tire.

With the above determinations in hand, the proper inflation pressures may now be ascertained (Fig. 1-2).

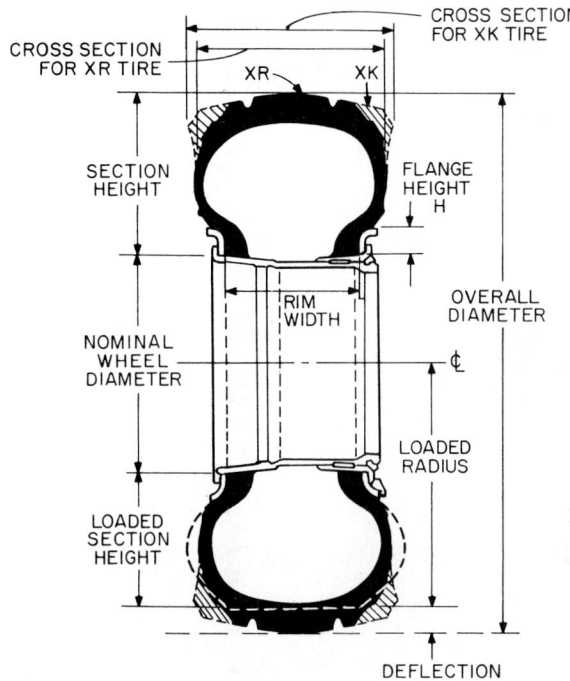

Fig. 1-2 Dimensions to be considered when specifying a tire.

AIDS IN TIRE SELECTION

One help to the earthmover industry is selecting transport tires through the use of the ton-mile-per-hour (tmph) system. Briefly, it consists of establishing the job requirements in terms of mean tire load (tons) and workday average speed (mph). The product of these two figures gives a job tmph value. A tire is then selected which has a tmph value near or above the job value (the tire tmph value or rating is determined by test).

The tmph system was designed for use as a tool to help select the proper tire for any given job. It is important to understand that proper tire application depends on many important factors, some of which are not taken into consideration with the present tmph system. Some of these factors are excessive grades or curves, poor haul-road conditions, wheel position (drive, steer, trail), and heat generation from brakes and planetaries.

A better alternative to the tmph system is to consider loads and speeds (both maximum and average) separately, as well as other important job conditions. First, the tire's load-carrying capacity at various maximum speeds listed in the tire manufacturer's load and inflation tables is compared with the actual tire loads found (or estimated) for a particular job site.

Then, the required work day average speed (WDAS) is determined for each application. (WDAS equals the number of miles traveled in a day, divided by the total shift hours including breaks). This value is then compared to the allowable WDAS listed for each tire as determined by the tire manufacturer.

After a tire is found which has load and speed capabilities that meet or exceed expected job conditions, other job-site factors must also be considered to assure that tire and job are properly mated. As mentioned before, some factors are the existence of severe grades or curves or the conditions of haul roads. Tire manufacturers or their representatives should be consulted to assure that all these conditions are properly evaluated.

CONTACT AREA

On occasion, users or equipment manufacturers request ground-contact-area information from tire manufacturers. This information is used to help predict a tire's flotation and/or ground-contact-pressure characteristics. It must be remembered, however, that flotation/mobility is dependent on more than just contact area. Nevertheless, it is important.

When contact area values are given one must know the specific conditions for which they are valid, especially if they are to be used in comparing tire brands. Actual contact area, that is, rubber-to-ground interface, is a function of many variables, in particular, tire load, inflation pressure, ground penetration, tire casing construction, and tire tread pattern.

Fig. 1-3 Track print left behind in soft ground reveals how radial design and belt-stabilized tread provide uniform ground contact with minimum disturbance of the earth, assuring good traction and flotation.

Usually, the area is specified as a *projected* area on a steel plate for a given load and inflation pressure. (The projected area includes the actual rubber in contact plus the voids within the outline of the contact rubber.)

Although contact area is important in evaluating a tire's flotation/mobility capability, there are other factors to consider. To achieve good flotation, the ground must be disturbed as little as possible. The lower the ground pressure the less the ground will be upset. Theoretically, then, the greater the contact area, the lower the ground pressure (ground pressure equals tire load divided by contact area) and, consequently, the better the flotation. But ground-contact pressure is not uniform throughout the contact patch. Both tire construction and tread design affect ground-pressure uniformity. The radial tire, for example, has a more uniform distribution of ground pressure from the center of the patch to the edge than a bias tire. Under deflection, the center of the patch of a bias tire has relatively low ground pressure, but the edge pressures are relatively high. Therefore, the tire has a tendency to dig in at the edges, thus disturbing the ground. The radial patch, owing in part to its belt-stabilized tread, has more uniform ground pressure, which results in better flotation characteristics.

Moreover, in order to improve flotation, in the field the inflation pressure can be lowered with the approval of the tire manufacturer, who will usually require a reduction of speed. (The contact area is then increased.) This can be done with a radial tire with more success than with a bias tire. Lowering the pressure in a bias tire results in significant tread distortion (hence, ground disturbance), whereas the radial tire with its stabilized tread is less affected and consequently its flotation/mobility capability can be improved (Fig. 1-3).

Of course, in any discussion of flotation, traction, and mobility, there are many other considerations (particularly, those pertaining to the ground or soil itself), which are not included in the discussion above. These, however, are beyond the scope of this chapter.

STORAGE AND HANDLING OF TIRES

Proper tire storage and handling are items which are frequently ignored by people in every phase of the earthmoving industry. To preserve the inherent quality of tires and the

Upkeep and Maintenance of Tires for Construction Equipment   3-5

initial investment in them, there are certain basic procedures to follow when storing and handling tires.

Storage

To avoid all premature aging and degradation of tires during storage, it is necessary to protect them from
   1. Inclemencies, such as changes of temperature, drafts, extreme heat (more than 100°F), and humidity.
   2. Ozone sources, such as arc welders, spark-producing motors, mercury vapor quartz bulbs, battery chargers with mercury rectifiers, and direct exposure to the sun.
   3. Distortion, such as from stacking. Upright positioning of earthmover tires is preferred over stacking to minimize distortion and stress.

If tires are to be stored for any significant length of time, it is recommended that they be placed in closed storage areas which do not contain the conditions just described. In addition, items should be stored to allow the oldest tires to be shipped first. This of course applies to all rubber products, such as tubes, valves, and flaps.

Handling

To eliminate deterioration of beads and the subsequent consequences, follow these rules regarding tire handling:
   1. Do not lift a tire by the beads directly with a crane's hook.
   2. Use flat straps or webbing, not metal slings or chains to lift tires.
   3. Pick tires up under the tread and not by the beads when using a forklift truck.
   4. For all tires shipped with bead protectors, leave this protection on until mounting time. Also, save the protectors since they can be used to cover a tire's beads when it is dismounted for retreading or repair.

The aim of the rules of storage and handling, as outlined here, is to preserve the quality of new tires right up to the time of mounting. Strict adherence to these rules will keep your tires factory fresh and ready for use. Individual tire manufacturers should be consulted for their specific recommendations on storage and handling.

MOUNTING INSTRUCTIONS

   1. Tubeless-type rim:
      Examination of Fig. 1-4 reveals key areas for attention on the tubeless-type rim.
   2. Valve installation:
      a. The valve consists of:
         (1) A sealed base, consisting of a body, a sealing gasket, and a tightening nut (except when the valve hole is in the side of the rim; see Fig. 1-5).
         (2) A large-bore valve stem complete with sealing ring, large-bore (jumbo) core, and a cap.

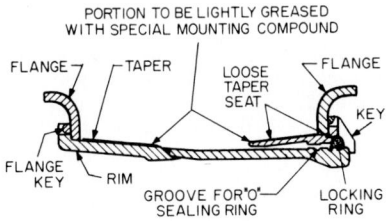

Fig. 1-4 Tubeless-type rim.

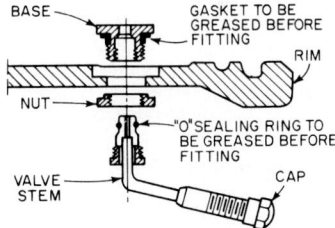

Fig. 1-5 Valve installation procedure.

   b. Before installation of the valve, it is imperative:
      (1) That the parts of the rim and the valve base which come in contact with the gasket be absolutely clean and free of scratches or other defects which might cause a leak.

3-6 Maintenance of Ground Contact Elements

      (2) That the seating surfaces of the "O" sealing ring in the valve stem and the base are clean and free from cracks, scratches, or burns which could cause a leak (Fig. 1-6).
   c. Installing the valve:
      (1) *Thoroughly clean and lightly grease the part of the rim on which the gasket will be placed.*
      (2) Grease the gasket and place it on the valve base. Place the assembly in the valve hole, screw the nut up *Tightly.*
      (3) Clean the valve stem and grease its "O" sealing ring; then, with the stem pointing in the right direction, screw it into the base and tighten, making sure that the stem does not change its position (Fig. 1-7).
      NOTE: When removing the base, *use a new gasket on reinstallation.*

TUBE-TYPE INSTALLATIONS

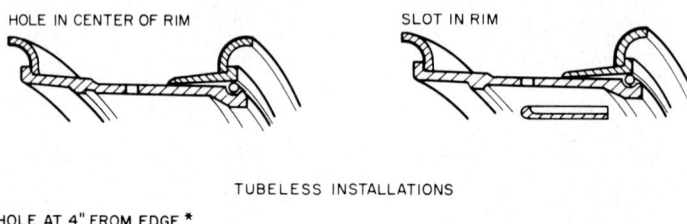

TUBELESS INSTALLATIONS

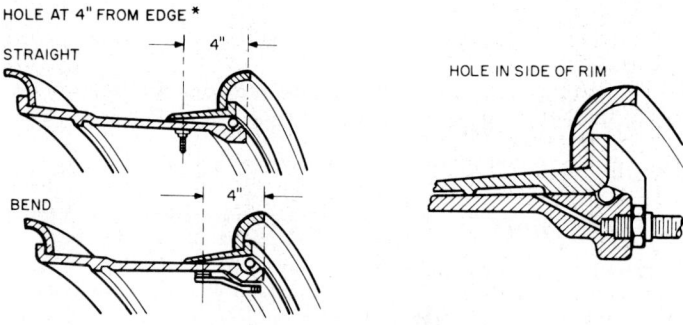

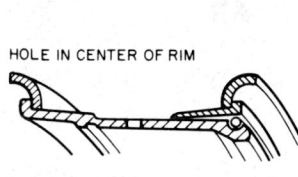

*VERIFY MEASUREMENT, EXCEPTIONS EXIST.

Fig. 1-6 Different valve arrangements are necessary depending on the rim or wheel and on tube-type or tubeless mounting. When selecting valves, make sure the rim or wheel matches the tire (tube type or tubeless) and flap combination, where applicable.

   3. Tire mounting:
    a. General:
      As the tire has to form an airtight assembly with the rim, it is necessary
      (1) To fit it on the correct rim and flange. They must be clean and in **good** condition.
      (2) To carefully follow the instructions and methods described below (Fig. 1-8).
    b. Preparation of the rim:
      (1) If necessary, using a file and wire brush, remove any rough pieces of **metal** or old rubber which might affect air sealing. Then, wipe down the rim and loose bead seat with a dry rag.

Upkeep and Maintenance of Tires for Construction Equipment 3-7

     (2) Those parts of the rim and of the loose taper bead seat on which the tire beads will rest, as well as the groove for the "O" sealing ring, must be perfectly clean and free from burrs.
  c. Preparation of the tire:
     (1) If necessary, remove any accumulation of rubber or grease which might be stuck to the beads. In doing this, be careful not to damage the tire beads.
     (2) Wipe the tire beads with a dry rag.
  d. Mounting operation:
     (1) Put the inner flange into position and check that it is keyed to the rim. (Some rims are manufactured without flange keys.)
     (2) Fit the first flange seal correctly on the rim and push it until it rests against the flange. (Only use flange seals if mounting a used tire and a sealing problem exists.)

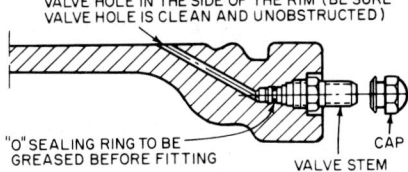

Fig. 1-7 Valve installation when the valve hole is in the side of the rim.

     (3) Lightly lubricate the taper portion as indicated under paragraph 1 (Fig. 1-4).
     (4) Place the tire on the rim and push it back as far as possible.
     (5) Center the tire correctly on the rim.
       NOTE: Bad centering of the tire can make
       (a) Placing of the loose taper seat difficult
       (b) Inflation difficult
       (c) "O" sealing ring come out of its groove during inflation
     (6) Fit the second flange on the loose taper seat; check the keying (Fig. 1-9).
     (7) Fit the second flange seal on the loose taper; push it until it rests against the flange, *if applicable*.

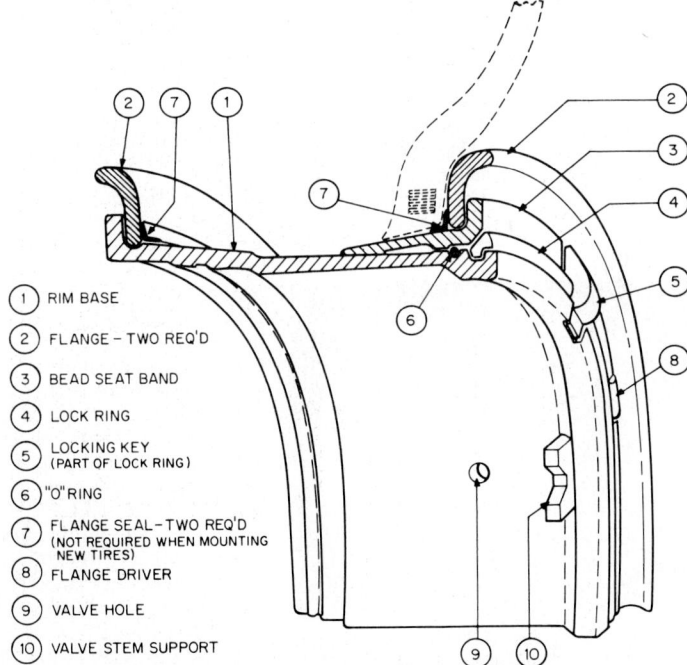

① RIM BASE
② FLANGE – TWO REQ'D
③ BEAD SEAT BAND
④ LOCK RING
⑤ LOCKING KEY (PART OF LOCK RING)
⑥ "O" RING
⑦ FLANGE SEAL – TWO REQ'D (NOT REQUIRED WHEN MOUNTING NEW TIRES)
⑧ FLANGE DRIVER
⑨ VALVE HOLE
⑩ VALVE STEM SUPPORT

Fig. 1-8 Typical five-piece earthmover tubeless-type rim.

3-8 Maintenance of Ground Contact Elements

(8) Lightly lubricate, with mounting compound, the two portions as indicated under paragraph 1 (Fig. 1-4).
(9) Fit the assembly of taper seat on the rim inward so that it clears the groove of the "O" sealing ring (Fig. 1-10).
(10) Place "O" ring in groove, being careful that it is not twisted, and lightly lubricate the exposed portion of the "O" ring. The "O" ring should be clean and dry before placing in the groove.
(11) Fit the locking ring.
NOTE: Before inflation:
(a) Check the keying.
(b) Make sure that the "O" ring is correctly positioned in the groove.
(c) Correct centering of the tire on the rim.

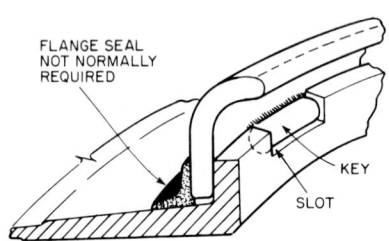

Fig. 1-9 Flange and key seated on taper seat.   **Fig. 1-10** Fitting the taper seat assembly to clear the "O" ring.

e. Inflation:
Remove the valve core and inflate to 90 psi, observing the normal safety precautions for inflation. (Avoid standing in front of the tire.) At this pressure, all parts of the rim assembly should be in place. Check this. Bring the pressure to its working value and screw in the core and cap.
f. Checking for leaks:
Apply soap solution
(1) To valve and valve stem.
(2) To "O" ring (where accessible).
(3) Between tire beads and flanges.
(4) Between flange/bead seat band and lock ring.

Check proper bead seating of earthmover tires. Improper earthmover tire mountings can be the cause of many tire problems. It is important that the mounting instructions on these pages should be followed closely. Flange bead seals should not be used with new tires. They are usually used because of a leak. Air leaks are frequently the result of improper mounting or faulty rim components which may be masked by the use of bead seals. Therefore, before using bead seals, all rim components and the seating of the bead should be checked. Failure to follow rules can prove disastrous. (See Fig. 1-11.)

Fig. 1-11 An example of shoddy workmanship by the tire mounter. Mounting studs are broken or missing. Nuts are missing and the assembly is improperly torqued. All add up to an undesirable and unsafe situation.

It is also important to have the tire as concentric to the rim as possible before inflating. Therefore, the tire should not be resting solely on the top of the rim. It should be resting on the ground with a minimum of weight on it; the vehicle may be partially jacked up or the hoist supporting the tire may be adjusted. If

Upkeep and Maintenance of Tires for Construction Equipment

possible, after inflating to the seating pressure (90 psi), the tire should be run for about a half-hour. After this time, the pressure can be adjusted to the recommended working pressure (after the tire cools down). See Fig. 1-12 for examples of proper and improper bead seating.

Keep in mind that although it is easier to mount a radial earthmover tire, compared with a conventional tire, care must be taken to ensure that the radial is mounted properly. By following the above recommendations and instructions, many tire problems can be avoided.

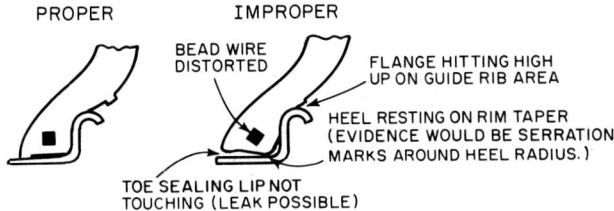

Fig. 1-12 Examples of proper and improper bead seating.

4. Tire demounting:
   After the vehicle has been blocked up to a suitable height
   a. Deflate the tire by unscrewing the valve core.
   b. Push the tire bead away from the seat.
   c. Push the loose seat inward far enough to clear the "O" sealing ring. Remove the locking ring and the "O" sealing ring.
   d. Withdraw the flange and loose seat.
   e. Free the second bead from the rim.
   f. Remove the tire from the rim.
   g. Remove the flange seals from the loose seat and the rim (if applicable).
   CAUTION:
   (1) Never apply heat to the rim when the tire is mounted. Always demount the tire from the rim when heat treatment is required (welding, cleaning).
   (2) With dual fitment, *both* inner and outer tires must be deflated before removing *either* tire from the vehicle.

DETERMINING INFLATION PRESSURE

Any determination of proper tire inflation pressure requires assessment of axle loads. Ideally, equipment should be weighed, both loaded and unloaded. In lieu of this, axle loads can be estimated. Some tire manufacturers have equipment charts which give recommended inflation pressures for most of the more popular earthmover machines. The following paragraphs outline the procedures for determining pressures for different types of machines.

Dumpers and scrapers
- Estimate material density.
- Estimate equipment load capacity (use equipment charts as a guide, but consider any deviation from the standard capacities, sideboards, e.g.).
- Calculate payload.
- Using gross vehicle weight (GVW) distribution as listed in equipment charts, calculate axle loads. Add any significant weight contributed by optional equipment. (Divide the axle load by the number of tires per axle to give the tire load.)
- Determine the top speed of a vehicle loaded (generally, 30 to 49 mph speeds are used).
- Knowing load and speed, use the carrying capacity chart for the tire under consideration to determine the proper inflation pressure.

EXAMPLE:
An empty six-wheel end-dump truck has the following axle weights:

Front (2 tires) = 30,000 lb
Rear (4 tires) = 31,000 lb
Tire size, front and rear: 18.00–33

Operating, it was observed to be carrying heaped loads of earth and rock (approximately 75/25). Its volumetric capacity according to the manufacturer's specifications[1] is 30 cu yd (heaped). The density of material is 2650 lb/cu yd. Therefore, payload is

$$30 \text{ cu yd} \times 2650 \text{ lb/cu yd} = 79{,}500 \text{ lb}$$

The gross vehicle weight is then

Empty weight—front	30,000 lb
Empty weight—rear	31,000 lb
Payload	79,500 lb
Total GVW	140,500 lb

Knowing the GVW distribution[1] to be 33 percent front and 67 percent rear, the loaded axle weights may be calculated

$$\text{Front} = 0.33\,(140{,}500 \text{ lb}) = \frac{46{,}365}{2} = 23{,}183 \text{ lb per tire}$$

$$\text{Rear} = 0.67\,(140{,}500 \text{ lb}) = \frac{94{,}135}{4} = 23{,}534 \text{ lb per tire}$$

The loaded machine was observed to be hitting a top speed of 30 mph. With the foregoing information we can now determine the recommended inflation pressure from the carrying capacity chart. Refer to the chart for the 18.00–33X tire in the *Michelin Earthmover Tire Data Book*.

Front tires At 30 mph we see that the recommended inflation pressure for the load of 23,183 lb would be 90 psi.[2]

Rear tires At 30 mph and 23,534 lb, inflation pressure would also be 90 psi.[2]

At this point other factors which may affect the final recommended pressure should be reviewed. For example, what is the ambient temperature? (See the temperature correction chart.) What are the flotation requirements? What is the condition of the haul road? These factors could require adjusting the inflation pressure.

Loaders

- Estimate material density (refer to Table 1-1, covering load material weights).
- Estimate equipment load capacity. Use equipment charts as a guide, but consider other-than-standard-size buckets.
- Calculate payload.
- Determine axleloads (Fig. 1-13). Of course, this figure will vary with the position of the bucket. Nevertheless, an intermediate position may be used in calculations. The position generally used when determining inflation pressures for normal loading operations is one somewhat between the carry position and the extended position. To complete this axleload calculation, the vehicle wheelbase and empty weights must be known. The equipment charts give the empty axle weights. The wheelbase must be measured. (In lieu of this information, the axle loads may be estimated in a different manner as described below.)
- After axle loads are found, the inflation pressures are obtained from the carrying capacity charts.

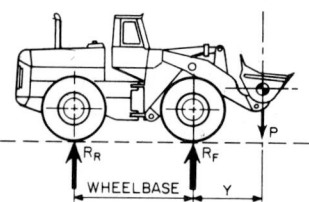

Fig. 1-13 Basic dimensions for axle-load calculations.

Front Axle

A load in the bucket results in a weight increase on the front axle and a decrease on the rear axle. This increase/decrease is a function of the payload, the bucket position, and the wheelbase.

For the front axle the loaded weight could be expressed as

Front axle loaded = unladen front weight + payload + weight transferred from rear

---

[1] This information may be found in equipment charts.

[2] NOTE: The two-star (**) tire meets this pressure requirement.

Upkeep and Maintenance of Tires for Construction Equipment 3-11

An illustration will show how the axle loads may be calculated with the above information.

Material density	2500 lb/cu yd
Bucket capacity	4 cu yd
Weights empty	Front: 20,000 lb
	Rear: 26,000 lb
Wheelbase	80 in.
Distance of bucket CG to front axle centerline (Y):	50 in.
Tire size	23.5-25

The basic formula (see diagram) is

$$\text{Front axle load } (R_F) = \text{empty load} + \text{bucket load} \text{ or } \frac{WB + Y}{WB}$$

TABLE 1-1   Load Material Weights (Approximate)

Material	Weight, lb/cu yd (loose)
Basalt	3300
Bauxite	2400–3200
Caliche	2100–2500
Clay	
Natural bed	2200–2800
Dry	1850–2500
Wet	2500–2900
Clay and Gravel	
Dry	2000
Wet	2200–2800
Coal	
Anthracite	1450–2000
Bituminous	1350–1600
Lignite	1225
Copper ore	2800
Earth	
Dry loam	1550–2100
Moist	2080–2600
Wet	2700–2900
Earth, sand, and gravel	2650
Earth and rock	
25/75	3300
50/50	2900
75/25	2650
Gravel	
Dry	2500
Wet	3400
Iron ore	4150–5500
Limestone	2400–2600
Sand	
Dry	2400–2950
Moist	2800–3100
Wet	3100–3250
Sand and gravel	
Dry	2900–3100
Wet	3400
Sandstone	2600–2950
Shale	2100
Slag	3000
Stone, crushed	2400–2900
Taconite	2900–3850

3-12 Maintenance of Ground Contact Elements

Therefore, for this example

$$\text{Front axle load} = 20{,}000 \text{ lb} + (2{,}500 \text{ lb/cu yd})(4 \text{ cu yd})\frac{80 \text{ in.} + 50 \text{ in.}}{80 \text{ in.}}$$
$$= 20{,}000 \text{ lb} + 16{,}250 \text{ lb}$$
$$= 36{,}250 \text{ lb}$$

Divide this figure by the number of tires per axle to derive the tire load (18,125 lb).

Using the Michelin carrying capacity chart for our 23.5-25X tire, the following inflation pressure is required for the front axle:

$$\text{Front} \quad \frac{36{,}250}{2} = 18{,}125 \text{ lb} \quad \text{per tire at 40 psi}$$

IMPORTANT: When the Y dimension and the wheel base are not known, their effect on the front axle loads as given in the formula can be estimated. The value for the $(WB + Y)/WB$ for most loaders falls between 1.6 and 1.8. This we will call the *leverage factor*. Therefore, in the above example, inserting a leverage factor of 1.6 results in

$$\text{Front axle load } 20{,}000 + (2{,}500)(4)(1.6) = 20{,}000 + 16{,}000$$
$$= 36{,}000 \text{ lb per axle or } 18{,}000 \text{ lb per tire}$$

Rear Axle

The basic formula is

$$\text{Rear axle load }(R_R) = \text{empty load} - \text{bucket load} \frac{Y}{WB}$$

In determining inflation pressures for the rear tires, however, the rear axle load with a loaded bucket is *not* required. The rear axle load is greater with an empty bucket. Therefore, this figure is used when determining inflation pressure.

Again, using the Michelin carrying capacity chart for the 23.5–25X tire, the following inflation pressure is required for the rear axle:

$$\text{Rear} \quad \frac{26{,}000}{2} = 13{,}000 \text{ lb per tire at 30 psi}$$

As with other earthmoving vehicles, other factors affecting the final pressure recommendation should be reviewed. Traction and flotation requirements, for instance, may dictate some pressure adjustments.

Lift and carry If the loader is to be operated in a lift-and-carry application, the recommended inflation pressures could be somewhat different. In general, they are higher. For these applications the individual tire manufacturers should be consulted.

Graders and dozers Inflation pressures are functions of the vehicle axle field weight and vehicle speed. The axle weights for some of these machines may be found in the equipment charts. The speed schedule generally used is 30 mph.

PRESSURE CORRECTIONS FOR HIGH AMBIENT TEMPERATURES

The abundant air mass contained in an earthmover tire is significantly affected by ambient temperatures, and adjustments to the basic inflated pressures are therefore required to avoid harmful underinflation.

To assist in obtaining maximum possible service life, Table 1-2 has been prepared to show pressure corrections for given ambient temperatures.

EXAMPLE: The basic inflation pressure for a Michelin 18.00–25XRB** tire on the rear of an R30 Euclid end-pump truck (maximum speed 30 mph) is 100 psi.

If the tire is to be inflated where the ambient temperature is 100°F, it must be inflated to 108 psi. This is found in Table 1-2 by looking down the left-hand column to the recommended book pressure of 100 psi, then reading across to the 100°F temperature column. At this intersection we read 108 psi.

This chart should also be referred to when the tire's cold inflation pressure is being checked. For instance, if the same machine in the example above was checked early in the morning (after it had been idle all night) when the ambient temperature was 80°F, then, according to the chart, we should find a tire pressure of 104 psi. (Look down the left-hand column to 100 psi, then across to the 80°F temperature column; at the intersection read 104 psi.)

Cold inflation pressure is defined as that pressure at which the temperature of the air inside the tire is equal to the temperature of the air surrounding the tire. This condition is

TABLE 1-2 Adjusted Inflation Pressure (psi)

Recommended inflation pressure, psi	Ambient temperature, °F									
	65	70	75	80	85	90	95	100	105	110	115
30	30	31	31	31	32	32	33	33	34	34	35
35	35	36	36	37	37	37	38	38	39	39	40
40	40	41	41	42	42	43	43	44	44	45	45
45	45	46	46	47	47	48	49	49	50	50	51
50	50	51	51	52	53	53	54	55	55	56	56
55	55	56	57	57	58	58	59	60	61	61	62
60	60	61	62	62	63	64	64	65	66	67	68
65	65	66	67	67	68	69	70	71	72	72	73
70	70	71	72	73	73	74	75	76	77	78	78
75	75	76	77	78	79	79	80	81	82	83	84
80	80	81	82	83	84	85	86	87	88	89	89
85	85	86	87	88	89	90	91	92	93	94	95
90	90	91	92	93	94	95	96	97	99	100	100
95	95	96	97	98	100	100	102	103	104	105	106
100	100	101	103	104	105	106	107	108	109	110	111
105	106	107	108	109	110	111	112	113	115	116	117
110	111	112	113	114	115	116	118	119	120	121	122

found when the tire has been idle for approximately 8 hr. Please note that when a tire has been exposed to the sun for a while its inflation air temperature will be significantly higher than the weather bureau ambient air temperature even though the tire has been idle. This results in a higher pressure reading and, therefore, cannot be considered as the cold pressure.

NOTE: The goal of this adjustment is twofold:
1. To minimize the tire deflection at high ambient temperatures and thereby control excessive temperature/pressure buildup
2. To decrease the chances of operating underinflated owing to a drop in the high ambient temperature

Important

1. Michelin's carrying capacity charts are based on an ambient temperature of 65°F.
2. The foregoing corrections pertain to cold inflation pressures only. Adjustment should not be made on a working tire.

When considering the values of inflation pressures checked while working, anything in excess of 15 percent pressure buildup from cold pressure due to working alone (i.e., excluding the pressure buildup due to ambient temperature increase which could be another 10 percent) would require investigation.

Maintenance of Ground Contact Elements

SOME CAUSES OF PREMATURE DETERIORATION IN EARTHMOVER TIRES

A large number of earthmover tires retire from service prematurely because of:
- Incorrect inflation
- Overloading
- Excessive speed
- Severe shocks
- A combination of the above factors

One particular damage is the separation of certain elements in a tire's construction. This is usually the result of excessive heat generation due to
- Speeds higher than those recommended for the loads and pressure concerned
- Underinflation or overloading
- Heat generated by parts of the vehicle such as brake drums and planetary gears

Separation can be aggravated or caused by mechanical forces such as
- Shocks from badly maintained road surfaces
- Hammering due to road surfacing
- Lateral forces occurring in tight bends

To help avoid the above damage, watch the maintenance of roads and, if possible during their construction, insist on large-radius bends. Table 1-3 gives maximum recommended speeds for flat (unbanked) turns of different radii.

TABLE 1-3 Speed Restrictions in Flat (Unbanked) Curves*

Speed, mph	Minimum turning radius, ft
5	50
6	70
7	90
8	120
9	150
10	190
11	230
12	270
13	320
14	370
15	420
16	480
17	540
18	610
19	680
20	750
21	830
22	910
23	990
24	1080
25	1170
26	1270
27	1370
28	1470
29	1580
30	1690

*EXAMPLE: The radius of a curve in a haul road is 750 ft. Speed in negotiating turn should not exceed 20 mph.

Ply separation can be produced by lateral forces occurring in tight turns. It is important to bank haul-road curves properly and to limit speed in relation to the curve. If sharp bends cannot be avoided, take them at the lowest possible speed.

Eliminating these causes of premature tire deterioration can minimize downtime (loss of production) and the risk of vehicle damage and/or personal injury.

Chapter 2

Wear and Maintenance of the Undercarriage

K. D. ANDERSON
Terex Division, General Motors Corporation, Hudson, Ohio

Owners of crawler tractors have the right to expect optimum performance. New units will deliver that optimum capability, but no make or model can continue such performance without timely and effective inspection, maintenance, and service.

At the outset, the components of a crawler undercarriage are perfectly matched according to dimension. But, as the components work together, they wear. Wear causes mismatch of parts and ultimately accelerates the wear rate of the entire system. This deterioration rate and general repair costs can be controlled. The following should help to lower maintenance costs, increase undercarriage life, and improve job performance and profitability.

WEAR

The undercarriage, the major wear item of a track-laying vehicle, does have a normal wear pattern. Understanding how wear occurs and the wear characteristics of parts can aid in avoiding unscheduled downtime.

Basic to a normal wear pattern is internal pin and bushing wear. Link side wear, accelerated sprocket tooth wear, too much track sag—all these may occur as a result of internal pin and bushing wear.

While a tractor is operating, the chain is in tension. Because the bushing is pressed into one link and the pin is pressed into the adjacent link, the tension in the chain causes contact on only one side of the pin OD and the bushing ID. This contact occurs on the same side whether the tractor is going forward or in reverse. Wear at the pin OD and bushing ID occurs because there is relative motion between the pin and bushing when the chain bends around the sprocket and idler. This internal pin and bushing wear is called *pitch wear* or *track stretch* because it causes the track pitch to increase and the chain actually gets longer (Fig. 2-1).

Normal wear of the sprocket occurs along with pitch wear because the elongation of the chain causes the sprocket to pick up the bushing closer to the tooth tip. Wear occurs as the bushing slides down to the tooth root. This is also called *forward drive side wear*.

In reverse operation the opposite side of the sprocket tooth, usually called the *reverse drive side*, contacts the bushing with the teeth at the top of the sprocket. The effect of reverse operation is more severe to both sprocket and bushing because there is scrubbing action between the sprocket and the bushing as the bushing enters the sprocket. This scrubbing action, or relative motion, between sprocket and bushing does not take place during forward motion of the tractor.

3-16 Maintenance of Ground Contact Elements

In addition to forward and reverse drive side wear, there may be some degree of tip wear on the sprocket. Since the tips of sprocket teeth are left in an as-cast condition it is possible that a tip will get scuffed when the bushings enter the sprocket. This is normal wear and will not affect the performance of the sprocket.

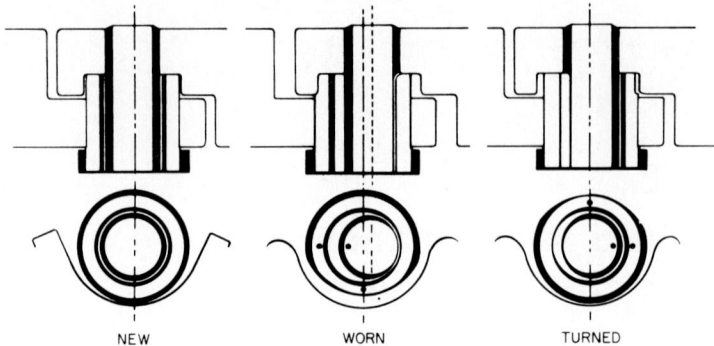

Fig. 2-1 Pin and bushing wear for new and worn conditions and the results of turning.

The rail surface of the link is the side opposite the surface to which the track shoes are bolted. This is the wearing side of the link. It comes in contact with the roller and idler tread surfaces. The rail gets narrower over the pin and bushing bores and consequently wears faster at this point. When the chain travels around the idler, the center of the rail is the only point of contact. This causes the center of the rail to wear faster also. As a result, normal link rail wear becomes wavy as wear progresses (Fig. 2-2).

The normal wear pattern for a roller is directly related to link rail wear. As a link rides over the rollers, the roller tread wears down. Even though this is normal wear, there is a limit which can be tolerated. If the tread wears down too far, the roller flange will hit the pin boss, causing wear of the roller flange and the links. The links may be damaged to the extent that they won't hold the pins (Fig. 2-3).

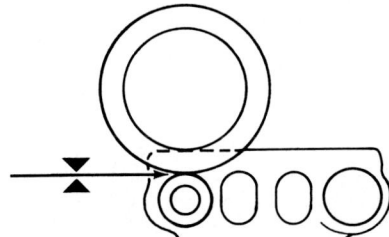

Fig. 2-2 Normal wavy link wear—shaded portion—for the link rail.

Fig. 2-3 Excessive damage means that links cannot hold the pins.

Another basic contributor to wear is front idler and/or track frame misalignment. Track frame mounting-point wear can allow toe-out while the unit is working in the forward direction. This type of misalignment may cause excessive end wear of pins, off-center external wear of bushings, rail side wear and sprocket tooth gouging of the inside of the links, side wear of the sprocket and sprocket teeth, and flange wear of rear rollers.

The front idler assembly may also be misaligned. This misalignment can affect the wear of the front idler flange, the link side rails, and the front track roller flanges. The idler is guided on the roller frame with a series of wear bars and plates which are shimmed to align the idler with the track rollers. There are also side wear plates which serve to guide the idler as it recoils back and forth. Improperly shimmed or unequally worn wear plates will cause the idler to run off center or out of line with the front track rollers. This will allow the links to ride against one flange of the idler and interfere with the front track roller flanges.

The entire alignment problem can be thought of as that of a pulley system with the idler and sprocket as the pulleys. If any part is out of line, the chain will interfere with it and cause wear.

How materials affect wear Nothing affects the life or wear rate of a crawler undercarriage like nature. The normal wear rate of all undercarriage components, especially the chain, is greatly affected by soil conditions. For example, sand greatly accelerates the wear rate of the entire undercarriage but the pins and bushings suffer the most. Sand works into the link counterbore and then into the bushing where it speeds up wear of pins and bushings. Link counterbores can be so badly worn that the links will not be rebuildable. The life of the pin and bushing can be shortened to a fraction of the life of other components.

Rock has its greatest effect on track shoes. Grousers wear fastest. The plate wears thin and beam strength is significantly reduced. If subjected to impact, shoes can bend or crack as they approach the rebuild limit. Track links are subjected to impact and twisting when working in rocky ground, sometimes causing fatigue cracking. These heavy impact loads, when concentrated in one area such as the front or rear rollers, may cause accelerated wear of these parts but normal wear on the rest of the undercarriage.

Wet clay-based soil poses a particular problem to sprockets. The wet clay may pack in the roots of the teeth. As a result, a mismatch of track chain pitch and sprocket pitch can occur which can produce severe wear of the tooth tip on the reverse drive side when the tractor is moving forward. Bushing OD wear is accelerated and in severe cases bushings may even crack.

Generally, coal serves as a lubricant to an undercarriage. The life of an undercarriage working in this material will be much better and longer than that of a comparable unit doing the same work in sandy soil.

The moisture content in soil increases its abrasive effect. A moderately abrasive, dry soil can become very abrasive when water is applied. Tractors working in a moist riverbed, for example, will have shorter pin and bushing life than identical machines a few hundred yards away on higher, drier ground.

The general relationship between different materials and corresponding undercarriage life appears graphically in Fig. 2-4.

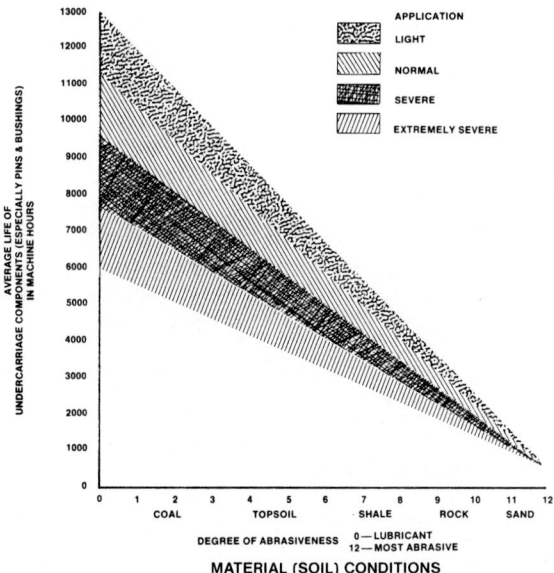

Fig. 2-4 Effect of applications and soil conditions on undercarriage life.

TABLE 2-1 Relationship of Applications to Wear Patterns

Application	Material	Load	Travel pattern	Probable wear characteristics
Ripping or quarrying	Rock, concrete, ore, shale (hard and rough)	High impact, heavy side loads, concentrated, constant	Heavy and frequent cycling and maneuvering	Heavy tread wear on rear rollers
Heavy digging and dozing	Rock, concrete, ore, shale (hard and rough)	High impact, heavy side loads, concentrated, constant	Heavy and frequent cycling and maneuvering	Heavy tread wear on front rollers
Landclearing Logging Digging Dozing	Sand, gravel, stumps (rough and abrasive)	Frequent high impact and side loads, packing	Heavy cycling and maneuvering	Accelerated pin and bushing OD wear; sprocket tip wear due to packing; roller tread wear proportional to soil abrasiveness; link side wear
Leveling Plowing Pushing Pulling Loading	Clay, gravel (rather rough)	Moderate loads—some impact; side loads	Moderate cycling and maneuvering	Pin and bushing wear; sprocket root wear; roller tread wear in proportion to soil abrasiveness
Mining Ditching Grading	Coal, clay, brush (loose and soft)	Uniform loads and little impact	Little maneuvering	Pin and bushing OD wear but not severe; even wear on entire undercarriage
Finish grading Stockpiling Spreading Side sloping	Clay, light gravel, firm earth (smooth and clean)	Little or no impact loads	Very little cycling or maneuvering	Link side wear and retention guard wear; idler wear plate; roller flange wear
Landfill	Trash, garbage, fill dirt	Little impact loads, packing	High cycling and maneuvering	Pin and bushing wear; sprocket tip wear; idler wear plates

Wear and Maintenance of the Undercarriage 3-19

Applications and wear The type of question in which a unit is used will also affect the wear rate and wear pattern of the undercarriage components. The wear chart (Table 2-1) contrasts results for various applications and environments.

Working requirements are closely related to applications. For example, applications which require consistent turning to the right or left will cause faster wear on one track. Side hill work speeds wear of roller flanges and link sides. High speed accelerates undercarriage wear. Over rough ground, high speeds create heavy loads which damage track alignment.

OPERATOR TECHNIQUES AND WEAR

In the same way that the driver of a car affects tire life, the crawler tractor operator can control the wear rate of undercarriage components. Abusive operating techniques can be expected to reduce overall track life, compared with the life conscientious operation would yield. The following are some dos and don'ts for operators:

Don't . . .	*Do* . . .
. . . go too fast unless productivity is worth the increased wear. This is especially true for reverse operation.	. . . ease up on load when the track begins to slip.
. . . spin the tracks. When the track slips, the undercarriage components wear faster and no work gets done.	. . . change operating direction often enough to balance wear.
. . . take such deep cuts with the blade or overuse down pressure.	
. . . park the machine in mud, water, or corrosive environments.	

The operator can also enhance maintenance of an undercarriage with the following helpful hints:

1. Make daily visual inspections of equipment. Check for loose bolts, leaking seals, and abnormal wear. Report all items that need attention to the maintenance team so needed adjustments can be made before extensive damage occurs.

2. Clean mud and debris from undercarriage so rollers can turn properly. Do this as required but always at the end of the day.

3. Do not allow either obvious or subtle problems to go without alerting the maintenance crew.

MAINTENANCE PRACTICES

Without doubt, components of an undercarriage do wear. The rate of wear, however, can be controlled through timely, proper inspection and maintenance. The maintenance cost for a crawler tractor's undercarriage can be as much as half the cost of maintaining the entire unit. Properly maintaining an undercarriage can mean cost savings and less unscheduled downtime.

For maximum service life for all track components, keep the track properly adjusted. The track should be adjusted so that there is about 1½ in. of sag between the front carrier roller and the idler. If the track is adjusted too tightly, a great amount of friction will exist between the pins and bushings as they hinge and travel around the sprocket and idler. This friction causes accelerated wear of pins, bushings, sprocket, idler, and rollers. Severely tight tracks absorb a great amount of horsepower and reduce the amount of power available for work. Also, an extremely tight track can cause severe damage to the final drive hubs, bearings, and gears.

However, if the track is too loose, service life is likewise reduced. Loose tracks fail to stay properly aligned and tend to come off when the tractor is turned. This causes wear to the idler flanges, the roller flanges, and the sides of the sprocket teeth. A loose track will whip at high speeds, resulting in impact loads on carrier rollers.

It is important that tracks be adjusted under actual working conditions of the machine. If the machine works in material which has a packing characteristic, then the material should not be removed when adjusting track for slack.

3-20 Maintenance of Ground Contact Elements

Links normally wear on the rail surface. The case depth of this heat-treated surface will diminish and approach a value beyond which the link is not rebuildable. Beyond this limit the link must be replaced in order to maintain adequate clearance between roller flanges and pin boss. The link is rebuilt by passing layers of weld over the rail surface.

Track rollers and carrier rollers should be checked to ensure free rotation and no leakage of lubricant. All rollers do not wear at the same rate. Uneven wear can be balanced by changing roller positions in the group, by switching them from one side to the other, or by turning them end for end. Changing the position of the rollers, in some cases, distributes the wear and extends the service life of the roller group. If the position of rollers is to be changed it should be done when they are about 50 percent worn.

When rollers are worn up to 75 percent, the maintenance supervisor has to consider the remaining service life of the rollers. They may be used until they are worn out, but this may cause pin boss damage to links and accelerated wear to other parts.

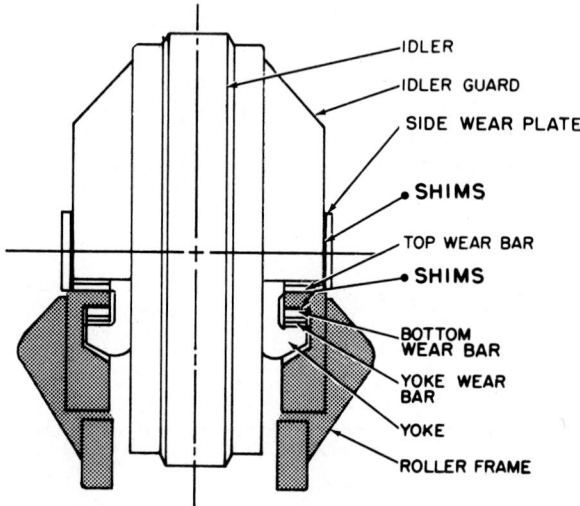

Fig. 2-5 Idler assembly cross section reveals the intricacy of this undercarriage element.

The other alternative is to rebuild the rollers by building up the worn treads and flanges with weld. It is very important that the flanges of the front and rear rollers be in good enough condition to properly guide the track under the middle rollers.

The greatest wear problem with idlers is misalignment. Shims can be added or removed to make up for wear and maintain alignment. The side wear plates which guide the idler as it recoils can be turned end for end to allow the unworn top edge to be used after shimming becomes impractical. The bottom wear bars provide replaceable wear surfaces on which the idler rides during recoil. Worn idlers can be reconditioned by building up the worn rim surface and the sides of the center section with weld (Fig. 2-5).

Sprockets should be checked regularly with a wear gauge to determine when to replace them. If they show more wear on one side of the tooth than on the other, they can be switched from side to side to prolong life and balance wear. Turning pins and bushings will also increase sprocket life. It is not good practice to rebuild worn sprockets. If sprocket teeth are filled with weld, the sprocket will be out-of-round and bushings will wear at a more rapid rate.

Optional equipment helps There are working conditions in which it is not economically advantageous to allow parts to wear normally. Normal wear in such conditions is too fast, too destructive, and too expensive to be compensated by replacement or rebuilding. In such cases, there are usually optional parts available which will prove to wear at a much slower rate and to be more productive than the standard part.

In clay-based mud, a normally equipped tractor will have problems with mud packing between sprocket teeth and causing abnormal, accelerated sprocket wear. Use of a relieved tooth sprocket can materially reduce the mud packing but may also result in accelerated bushing wear caused by the narrow tooth root. These wear rates should be carefully analyzed to produce the most economical balance.

In severe job conditions, extreme-service track shoes usually prove to be tougher and more durable than the standard shoe. They're specially made and proven for high-impact jobs such as rock work.

Track roller guards are optional items which prevent rocks and other debris from entering the track roller area. They, in conjunction with retention guards in the front and rear, also serve to prevent so-called snaking of the tracks. Snaking may occur on uneven ground, side slopes, or in constant maneuvering (turning) applications.

All of these items can affect the life of undercarriage components. It's important to equip a tractor properly to achieve maximum production and service life.

Rebuilding versus replacement As the components of an undercarriage wear, the owner should be concerned about the economic decisions that must be made: Should the parts be replaced, rebuilt, or run to destruction?

Running a component until it is destroyed is most likely to have an adverse affect on the entire undercarriage. For example, if the chain is run to destruction it will affect the sprocket pitch, the rollers, and the idlers by accelerating their wear rates. Normally, when any component is run to destruction, the complete undercarriage will be affected and will have to be replaced.

A more difficult decision is choosing between rebuild and replacement. The decision will probably be based upon the cost savings afforded by each. Rebuilding is the process of welding, or more specifically hard facing, the wear surfaces of eligible parts. Weld is deposited in layers until a wear surface comparable to the new part dimension is acquired. If rebuilding is delayed, the amount of weld that must be applied and the additional labor involved will make the cost of the rebuilt part approach the price of a new part. Conversely, if a part is rebuilt too early, any case hardness remaining will be affected by welding such that, after rebuild, the life of the part may actually be shortened. Therefore, the most important consideration with rebuilding is to rebuild at the proper time. The cost of rebuilding will vary from one area to another. Local sources should be contacted to determine the cost of rebuilding components in a specific area.

When replacing components on an undercarriage, new parts should not be matched with badly worn parts. Mating a worn surface with a new part will only result in accelerated wear of the new part. For example, when a new sprocket is used with a worn chain, the sprocket will wear very rapidly until it matches the chain pitch. Likewise, when new and used links and track rollers are mismatched, the new components have accelerated wear rates.

In deciding whether to rebuild or to replace worn components, the total cost of each alternative and the corresponding cost per hour should be calculated. The following is the cost-per-hour calculation for rebuilding:

Cost of part, new	$_____
Rebuild cost	$_____
Estimated downtime cost	$_____
Removal replacement cost	$_____
Other related costs	$_____
A = total cost with rebuild	$_____
Hours on part before rebuild	_____ hr
Estimated life after rebuild	_____ hr
B = total hours with rebuild	_____ hr

$$\text{Total cost per hour (with rebuild)} = \frac{A}{B}$$

The cost per hour should be calculated because generally rebuilding will give a part additional life. But the additional cost incurred may not be justified by this increased life. The total cost per hour with rebuild should be compared with the cost per hour without rebuild.

$$\frac{\text{Cost of new part}}{\text{Estimated life}}$$

The lower of the two cost-per-hour figures may be the most economical.

3-22 Maintenance of Ground Contact Elements

SUMMARY

This chapter emphasizes the normal wear patterns that can be expected and the wear characteristics produced by various working conditions and styles of operation. It includes maintenance hints for critical components of an undercarriage. The most important consideration in maintenance efforts, including rebuilding and replacement, is timing. Keen observation and analysis of wear characteristics and rates, coupled with timely, effective maintenance, can make the difference between profitable and unprofitable operations.

A crawler tractor owner needs a reliable source for analysis of wear and potential undercarriage life to eliminate unnecessary costs. The original equipment dealer can assist in cutting costs to a minimum and realizing the best possible life for the undercarriage.

Chapter **3**

Dozer Moldboard and Ripper Tooth Maintenance

HERMAN A. HULLMANN
Technical Services Representative, Fleet Services,
Fiat-Allis Construction Machinery Inc.,
Springfield, Ill.

Dozer and ripper manufacturers provide the construction machinery industry with several types of dozers and rippers to suit various job requirements. The steel industry provides alloy steels for cutting edges, end bits, ripper shanks, and points to withstand abrasion and/or the shock loads they are subjected to. It is therefore the responsibility of the user, with assistance from the manufactuer, to select the dozer and ripper best suited for a particular job application. Select the dozer cutting edges, ripper shanks, and points that give optimum performance.

Types of Dozers

There are basically four types of dozer moldboards: the angle moldboard, the straight moldboard, the semi-U moldboard, and the full-U moldboard.

Angle dozer The angle dozer utilizes a C frame and has three positions of the moldboard: straight and 25° angle to the right and to the left. It is a lighter-weight dozer and should be used primarily in dirt or loose-material applications (Fig. 3-1).

Straight moldboard The straight moldboard is the most rugged of the moldboards and is the one best suited for rock or severe applications (Fig. 3-2).

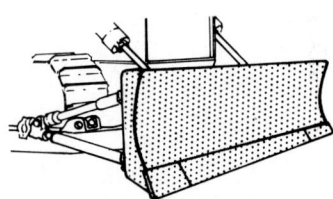

Fig. 3-1 Angle dozer.

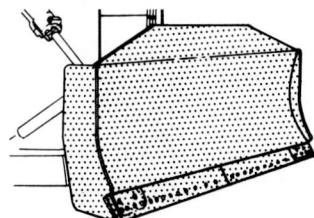

Fig. 3-2 Straight dozer.

Semi-U moldboard The semi-U moldboard, which utilizes the characteristics of both the straight and full-U moldboard, is a rugged, all-purpose moldboard. It works well in all applications and is now the most popular moldboard in the construction machinery industry (Fig. 3-3).

Maintenance of Ground Contact Elements

Full-U moldboard The full-U moldboard is longer than the straight blade with approximately one-quarter of each end of the blade angled forward. The U shape of the moldboard will hold and carry more material in front of the blade, making it ideal for moving loose or ripped material in land reclamation and similar applications. It is also used in severe applications, but is not recommended as the primary rock moldboard. (Fig. 3-4).

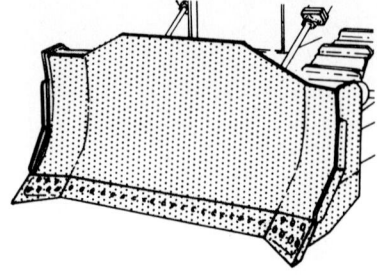 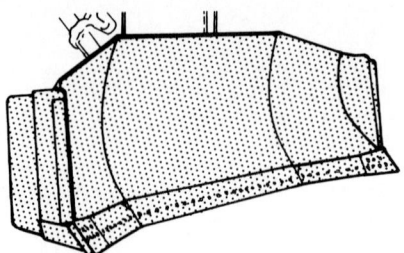

Fig. 3-3 Semi-U moldboard blade.   Fig. 3-4 Full-U moldboard blade.

Types of Rippers

Basically, there are only two types of rippers: the pull or tow type and the integral tractor-mounted ripper. The towed ripper is rapidly disappearing and is no longer considered in most construction applications. The integral mounted ripper, which is mounted directly to the rear housing of the tractor, provides more maneuverability and tractor balance. It is more compact and utilizes hydraulic cylinders and tractor weight for better penetration and ripper depth control. With today's bigger tractors, such as the Fiat-Allis 41B, ripping that previously required blasting is being done economically.

Variations in lifting and control arrangements provide three classes within the integral mounted rippers: the straight bar, the parallelogram, and the radial ripper.

Straight bar ripper The straight bar ripper, which is primarily used on smaller tractors, is a lightweight ripper that utilizes one to five shanks (Fig. 3-5).

Multishank operation is suitable for relatively easy soils, that is, top soils, glacial till (without boulders), chalk, or weak sandstones. It produces relatively high volumes in good conditions but is unsuitable for slabby material or boulders as the distance between the tractor and shank is usually 36 in. or less. Single-shank operation is suitable up to medium-strength or broken limestones or glacial till with small boulders.

Parallelogram ripper The parallelogram ripper is the most practical ripper for all applications. It is heavily constructed and utilizes one to three shanks. Most parallelogram designs have fixed shanks and a constant point angle of penetration as the ripper beam is lowered to provide positive control over ripping depth. Variations in the parallelogram design offer free-swinging or swivel-action shank brackets and a hydraulic pitch adjustment. In difficult materials, swivel-action brackets permit the shank to face into the loads to prevent slewing and to reduce stresses on ripper and tractor. Swivel-action brackets

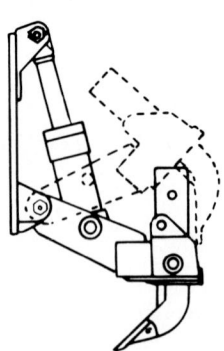 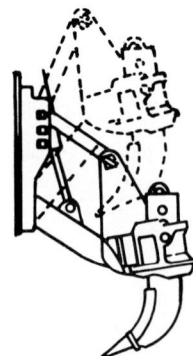

. Fig. 3-5 Straight bar ripper.   Fig. 3-6 Parallelogram ripper.

Dozer Moldboard and Ripper Tooth Maintenance   3-25

also permit steering a tractor during ripping operations without generating side-thrust loads. The advantage of the hydraulic pitch is the ability to change the point angle quickly and easily from the seat of the tractor. A steep point of entry for fast, easy penetration and then a flatter point angle when ripping depth is obtained result in longer point life and increased production (Fig. 3-6).

Multishank operation is widely used for glacial till (without large boulders), medium or broken limestones, or coal, but it is not suitable for work in slabby materials where raking may occur. Single-shank operation of a parallelogram ripper is suitable for the most difficult materials.

Radial ripper   The radial ripper acts in an arc, the beam pivoting to raise and lower the shank. The angle of the point, in relation to the ground, changes as the point is lowered. Some radial ripper designs have one to three shanks on an offset pattern; other designs have only a single shank (Fig. 3-7).

Multishank operation is very much the same as with the multishank parallelogram rippers; but owing to the shank configuration design, it is not as susceptible to raking. Single-shank operation of a radial ripper is suitable for the most difficult materials, including some igneous rocks. The radial shank is considerably longer than the parallelogram shank and will attain a deeper ripping depth. For example, the Fiat-Allis 41B parallelogram ripper has a maximum ripping depth of 42 in., while the 41B radial ripper has a maximum ripping depth of 84 in.

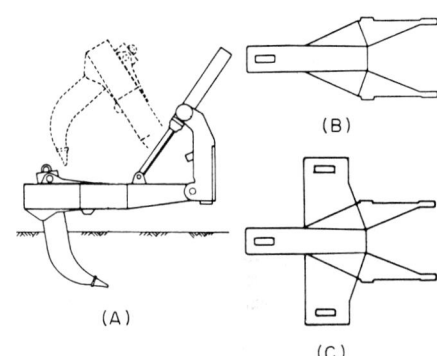

Fig. 3-7   Radial ripper (A), showing a single (B) and multishank (C) arrangement.

Dozer and Ripper Installation

1. When installing a dozer or ripper on a tractor, it is vitally important that all paint and high spots be removed from any matting surfaces to prevent loosening of the attaching part and to obtain proper adjustments.

2. All bolts and capscrews should be torqued to the manufacturer's specifications and not overtorqued. Overtorquing stretches the capscrews beyond the yield point, resulting in broken capscrews when shock loads are applied. Frequent inspections should be maintained to ensure loosening does not occur.

3. If a dozer is to be used to push load scrapers, a push plate should be installed on the face of the moldboard to prevent damage to the moldboard.

Dozer Cutting Edges and End Bits

Construction machinery life is based on hours of operation or use. Hour life of ground-engaging tools is too variable to list in specific hours. Hour life of cutting edges may vary from a few hundred to several thousand hours, depending on the following elements:
   1. Component material and design
      a. Each manufacturer of cutting edges and end bits may use a different alloy steel and heat-treat procedure to obtain optimum performance for their design.
      b. There are usually two classes of cutting edges: standard service and severe service.
         (1) A standard edge is thinner, with a harder heat treatment to resist abrasion, but it will fail in severe shock or rock applications. The thinner edge also has better digging ability in tightly compacted materials.
         (2) A severe service edge is thicker with less brittle heat treatment to withstand the shock loads in severe rock applications, but it is subject to a faster wear rate in highly abrasive materials.
      c. Some of the options for end bits are (Fig. 3-8)
         (1) Standard end bits for average dozing and abrasive materials.
         (2) Severe service end bits for high-impact applications.
         (3) Adapters for end bits to prolong the life of end bits in corner digging applications.

3-26 Maintenance of Ground Contact Elements

    d. The width of the backup plate or cutting-edge bolting plate on the moldboard is vitally important to the support of the cutting edge and end bits. All cutting edges are reversible, to double the life of a cutting edge before replacement is necessary. Inspect frequently for worn or broken edges and end bits, and turn or replace them before the bolting and support area of the moldboard is damaged. Any repair in this area is difficult and costly. End bits usually wear more rapidly and will require replacement before cutting edges.

    e. When turning or replacing cutting edges and end bits, be sure the support area is clean and without rough or high spots before installation. Edges bolted to uneven surfaces will be in stress and will break when shock loads are applied. Plow bolts should be properly seated in the edge and torqued to the correct specification (Fig. 3-9).

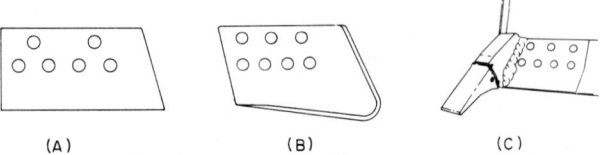

Fig. 3-8 Optional end bits include (A) standard, (B) severe service, and (C) special adapters.

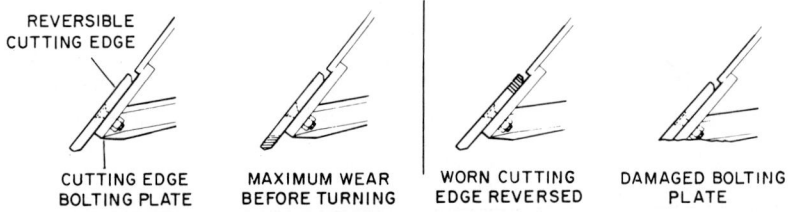

Fig. 3-9 Reversible cutting edges and how they operate.

2. Operator experience
    a. Operator experience is a vital factor in prolonging the hour-life of ground-engaging tools. An experienced operator will move more material with less downtime and fewer costly repairs than the inexperienced operator. Remember operators can make or break a project.
    b. Conduct schools. Let experienced operators share their skills and technology with newer and less experienced operators.
3. Abrasiveness of soil or rock
    a. Material abrasiveness is an uncontrollable but major factor in determining hour life of cutting edges and end bits. Hour-life may vary from a few hundred hours in very highly abrasive material to several thousand hours in less abrasive material.
    b. The highest or hardest heat-treated edges, without excessive breakage, should be used in any application. Keeping proper records is the best method for determining which type is suited for the application to obtain optimum performance.
4. Size of ripped or blasted material
    a. The fragmentation of ripped material is much smaller and more uniform than blasted material and is easier to doze.
    b. Standard cutting edges and end bits work well in ripped material, where as blasted material requires the severe service.
5. Down pressure applied to component. Constant or prolonged down pressure and the abrasiveness of the material are definite factors in the wear rate of cutting edges and end bits. Down pressure can be reduced in most applications by ripping.
6. Cycle time
    a. Hour-life is based on tractor operating hours, and the number of hours the cutting edge is actually engaged in the ground determines its end life.

Dozer Moldboard and Ripper Tooth Maintenance

    b. Long or slow returns have a longer cycle time, less actual engaged time, and/or longer overall operating hours on a cutting edge.
    c. Short, fast cycle time results in increased production and also in fewer hours of life on the cutting edge. Study of production will show correct cycles for application.

Ripper Shanks

Even though there is a wide range of ripper shanks available, there are only three basic designs: the straight shank, the curved shank, and the double-offset shank. Matching the shank design to the duty is of vital importance for economic operation, greater production, better control of fragmentation, and longer component life. Most manufacturers will advise on the correct shank for a specific application.

    Straight shank   Straight shanks are particularly suitable for slabby and blocking materials.
    Curved shank   Curved shanks give a lifting action, resulting in good fracture characteristics in fine-grain, unbroken materials.
    Double-offset shank   Double-offset shanks are most effective in nonabrasive coal and other light, easily fractured, easily penetrated materials. With wear plates, these shanks

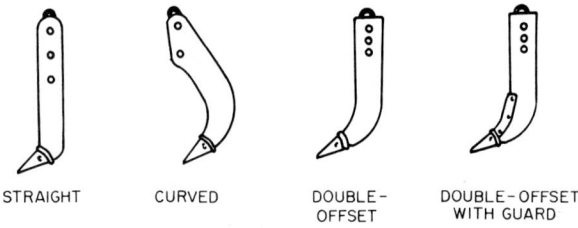

Fig. 3-10 Typical ripper shank forms.

give maximum ripping production in the widest variety of materials. They are the shanks best suited for most general construction and quarry applications. Wear plates, in split segments or one-piece wraparounds, greatly prolong shank life (Fig. 3-10).

    Shank positions   Shank position combinations in the tool bar should be as shown in Fig. 3-11. Improper shank positions cause twisting and side thrusts, resulting in broken ripper frames and shanks.

    Shank repair kits   Worn or broken shank ends can be repaired with a weld-on shank end. Most shank manufacturers offer a shank repair kit with complete installation instructions. As the shank material may vary from one manufacturer to another, it is vitally important to read and follow the instructions provided by each manufacturer.

Ripper Points

Ripper points may also be referred to as ripper teeth or tips, and they constitute the part of the ripper that takes the brunt of the punishment. Different point manufacturers may use different alloy steels and heat-treat procedures to meet specific requirements for their design.

    Ripping capability is determined by the amount of pressure applied per square inch of point surface. A given size tractor can only produce a given amount of pressure on a point. Therefore, the sharper the point, the greater the amount of pressure per square inch of point surface and the greater the ripping capability. The applied pressure per square inch of point surface decreases at a rapid rate as the point wears and becomes dull. In some tough applications, it is

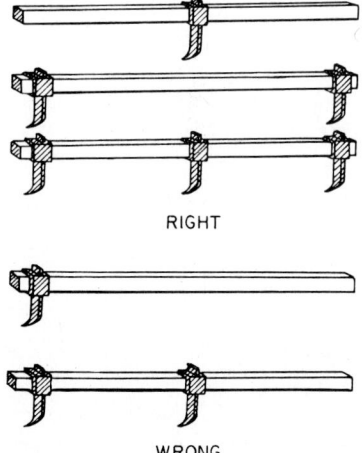

Fig. 3-11 Proper shank position combinations.

3-28 Maintenance of Ground Contact Elements

more economical to change points with only 50 percent wear to maintain a sharp point and an economical ripping operation (Fig. 3-12).

Point selection There are a variety of points to meet every ripping requirement. Selecting the right point for your specific job is vitally important to assure optimum performance. Manufacturers provide points of different lengths and designs, each with a specific alloy steel and heat treatment to meet these requirements. Each manufacturer has a method of identifying various points and will assist in making the right selection. As an example, one manufacturer uses red or blue paint in the point pocket combined with the length of the point as follows:

1. Long *blue* point for normal application with resistance to normal impact and abrasion.
2. Short, sharp *blue* for high impact in severe rocky applications.
3. Long *red* for abrasion and moderate impact.
4. Extra-long *red* for extreme abrasion.

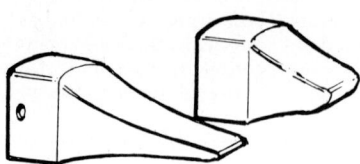

Fig. 3-12 New and half-worn ripper points.

Knowing the different types of points and keeping a proper record is the best method for determining the right point for the application.

Point angle The point angle of penetration is the angle formed by the front or top face of the point and the ground level at point of entry into a material (Fig. 3-13). Before the material can be ripped, it must be penetrated, and the point angle of penetration is an important factor in penetration. Tighter or harder materials require a steeper angle of penetration. In certain rock formations, angles of 45° or more are necessary to obtain penetration. Steeper angles may cause higher-than-average point breakage, but steeper angles may make the difference in whether a material can be ripped. Even with higher point costs, ripping is usually more economical than drilling and blasting.

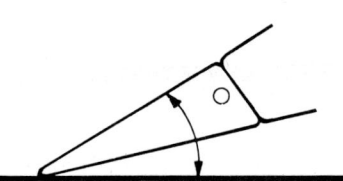

Fig. 3-13 Ripper tooth point of penetration.

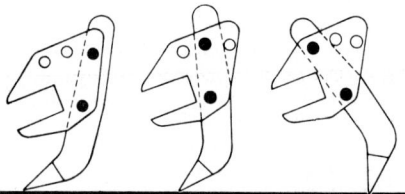

Fig. 3-14 Swivel-action bracket rippers.

Changing the point angle of penetration may be accomplished in three ways:

1. Some swivel-action bracket rippers have a combination of holes in the shank pocket to change the point angle of penetration (Fig. 3-14).
2. With variable parallelogram rippers, the point angle of penetration is changed by hydraulic cylinders or a combination of holes for the top parallel bar (Fig. 3-15).
3. On radial rippers, the angle of point penetration can easily be changed by simply raising or lowering the shank in its retainer (Fig. 3-16).

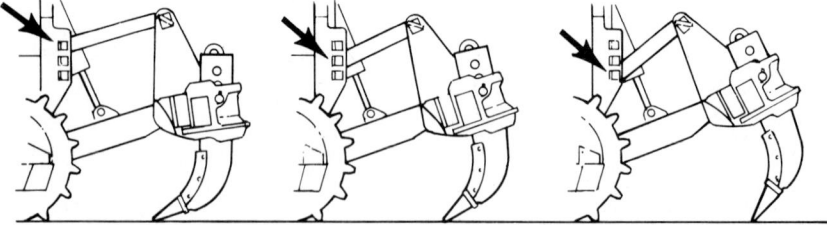

Fig. 3-15 Changing angle of point penetration on variable parallelogram rippers.

Excessive point failures

Some common causes of excessive point failures are
1. Operator inexperience
2. Incorrect point material or point length for an application
3. Worn or broken shank nose
4. Wrong shank angle
5. Backing with point in ground
   a. Operating a ripper is an art requiring the best manpower. Ripping should never be faster than first gear and reduced throttle. If ripping is easy, rip deeper or add more shanks. Speed increases wear and the possibilities of impact damage.
   b. A long and hard point for abrasion, working in rocky material, will result in broken points. A shorter and less brittle point, for impact application and working in abrasive material, will result in points wearing out prematurely. Select the correct point for the application.
   c. Fit of a point on a shank nose is vital to support of the point. A worn or broken shank nose will allow the point to move, and either breakage will occur from a low-support contact area or the point will have too steep a penetration angle. Repair the shank nose by installing a shank repair kit.
   d. Too flat a point penetration angle will result in excessive point wear and lack of penetration. Too steep a point penetration angle will result in excessive point breakage. Adjust the point angle of penetration to obtain a quick penetration and ripping depth without excessive point breakage.

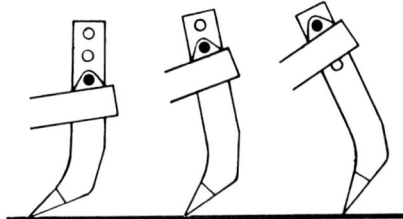

Fig. 3-16 Changing angle of point penetration on radial rippers.

   e. A tractor should never be backed with the point in or near the ground—a rock could catch the back side of the point, resulting in a broken point or loss of the point.

Hard-Facing

Replaceable items such as dozer cutting edges, end bits, ripper points, and shank guards are heat-treated to maximum hardness for best resistance to wear. Rebuilding or hard-facing of these items is seldom economical and can often result in failure. It is therefore not recommended. However, those who still want to rebuild or to hard-face should obtain and follow the instructions of the manufacturer to assure the best possible job and to minimize the danger of failure.

In most cases, the dot method of hard-facing is recommended over the stringer beads since cracking of the base metal adjacent to the stringer beads may result. The dot method consists of a pattern of dots ⅜ to ½ in. in diameter located on approximately 1 to 1 ½ in. centers. The dot method is less expensive because less labor and materials are required, and it is considered as effective as continuous stringer beads. The dots are excellent indicators of extent of wear and should be rewelded before they are completely worn off.

Chapter **4**

Maintenance of Earthmoving Buckets and Bucket Teeth

ENGINEERING STAFF
ESCO Corporation, Portland, Oreg.

Proper selection from the many different types and sizes of earthmoving bucket teeth available and careful maintenance can help increase productivity and at the same time hold down operating costs.

Most equipment owners see to it that expensive machinery is regularly maintained, with engines, hydraulic systems, drive trains, tracks, and tires getting periodic attention. But too often the digging end of that equipment is neglected. And when the bucket doesn't dig or load to capacity, the total machine doesn't do its designed job. Proper teeth can reduce tire wear on front end loaders. And, with the right teeth, regularly maintained, all equipment can operate with less fuel consumption.

The many varieties of earthmoving equipment teeth dictate that each manufacturer's instructions be followed. Tooth locking systems vary; steels used in different points and adapters require different welding techniques and materials. Even digging angles vary for different types of buckets, and teeth must be installed correctly to operate efficiently.

The instructions in this chapter relate to tooth installation and maintenance for the more commonly used buckets in the construction and mining industries: front end loaders, hoe dippers, shovel dippers, clamshell buckets, and dragline buckets. By following these instructions, better performance can be obtained from equipment at lower cost.

Front End Loader Buckets and Teeth

These buckets need teeth to dig into material. They increase penetration of a lip by concentrating force on tooth points rather than on a large lip area.

Some buckets still use solid one-piece weld-on teeth. But most use two-piece (point and adapter) teeth. Solid teeth may originally be inexpensive but maintenance is expensive. When worn, they must be burned off and new ones welded onto the lip. Thus, they are usually run while dull in an effort to get longest life from them.

Two-piece teeth are recommended. They are more economical and versatile. Adapters are bolted or welded to the bucket lip and the points are mechanically attached. When points are worn, they are changed quickly and easily. Also, different point styles can be used to match the material being handled.

Welded adapters are preferred because they are stronger than bolted types. The advantage of bolted adapters is that they can be removed when working, whereas teeth could damage a structure such as in unloading barges.

How to space teeth Too many or too few teeth can adversely affect loader performance. This typical rule of thumb is often used to obtain the correct number of teeth: Adapter nose width, multiplied by 4 and divided into inside lip width of bucket. For

3-32 Maintenance of Ground Contact Elements

example, adapter nose is 3½ in., multiplied by 4 to give 14. Divide this into a lip width of 123 in. to give 8.8 teeth. Round off to 9 teeth. Place a tooth at each end of the bucket and divide the rest evenly along the lip. Divide the remaining 7 teeth into the lip width of 123 in. to get 17.6 in. spacing between teeth.

Adapter types Some manufacturers offer different styles of adapters. A single-leg adapter welds to the top of a lip, leaving the bucket bottom flush for cleanup work.

Others have 1½ or 2 legs which slip over the lip and are welded or bolted to it. These are stronger and provide a wear-resistant runner for the bucket bottom.

Nose types and locking devices Most adapters have a flat wedge-shaped nose which mates with the box section of the replaceable point. These accept head-on loads quite well, but they require a heavy box section on the point and a large pin to resist loads from the side, top, and bottom. They are subject to breakage when vertical or side loads become more than they can bear.

One manufacturer uses a conical nose. It resists loads equally from all directions. A flat rim on the tip of the nose provides holding power so only a small vertical pin is needed to retain the point on the adapter.

A rubber lock fits into a keyslot which squeezes against a corrugated pin, wedging the point onto the nose. The pin can be driven out from top or bottom, making it easy to change points. It is safer to replace points with this system than with side-locked points. Usually there isn't much room to swing a hammer to remove side-locked points.

Point selection Points are offered in a variety of sizes and styles. Some manufacturers provide thicker points for loading rock where penetration is no problem. These give more wear metal to resist abrasion. Long, sharp points are used for penetrating less abrasive material. There are broader points for cutting into coal, hardpan, and clay.

Welding points Points should not be hard-faced or weld-repaired. The alloys and heat treatment make them extremely hard and welding heat can destroy their metallurgical qualities.

Welding two-leg adapters Follow the detailed welding instructions given in "How to Weld Alloy-Steel Bucket Components," later in this section.

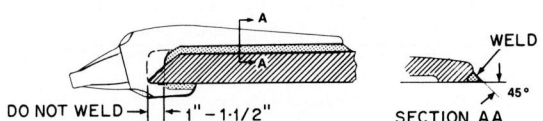

Fig. 4-1 Front end loader two-leg adapters. Fillet weld should cover J groove for effective weld strength through fillet throat. Hold weld angle close to 45°.

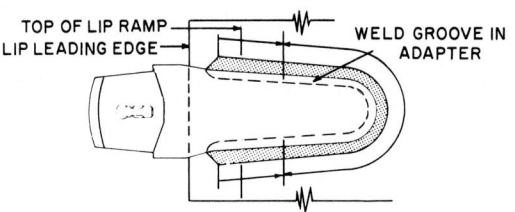

Fig. 4-2 Top view, welding adapters on front end loader buckets.

Use E7016 or E7018 electrodes or E70 wire, all of low-hydrogen content.
1. Locate adapters on bucket lip.
2. Preheat and tack adapters to lip. Use a Tempilstik to check temperature.
3. Weld around both legs but not within 1 to 1½ in. (25 to 38 mm) of the leading edge, as shown in Fig. 4-1. Start weld as shown in Fig. 4-2 at approximate center of leg but behind lip ramp on top leg. Weld one pass around back of leg to center of opposite side. Then backstep with one pass on each side, starting 1 to 1½ in. (25 to 38 mm) back from lip edge to center of leg.

Maintenance of Earthmoving Buckets and Bucket Teeth 3-33

Repeat until weld is complete. Maintain full fillet size to stop off point and tie in the ends.

4. Grind front of weld to smooth contour to relieve stress concentrations (Fig. 4-3).

Welding one-leg adapters

1. Locate adapters on lip.
2. Preheat and tack adapters to lip.
3. Weld completely around adapter legs and along loading edge of lip, as shown in Fig. 4-4.

CAUTION: Welding to lip leading edge may cause cracks in lip if these instructions and good welding techniques are not carefully followed.

Start weld at approximate center of leg behind top lip ramp and weld one pass around back of leg to center on opposite side. Then backstep with one pass from lip leading edge to center of leg, using starter plates, as shown in Fig. 4-5. Do this to both sides. Repeat until weld is finished.

4. Burn off starter plates. Finish welding to lip leading edge, carefully smoothing contours at weld junctions. This weld can be either of two types shown in Fig. 4-6, J groove or fillet. If lightener cavity extends forward of lip leading edge, as shown in Fig. 4-7, weld to edge of cavity. Otherwise, weld across entire width of adapter. Maintain full fillet size over entire weld.

5. Grind off front of weld to maintain smooth contour and avoid stress concentration.

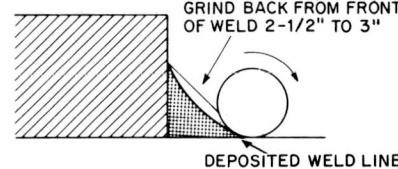

Fig. 4-3 Grind weld back from 2½ to 3 in. to reduce stress concentrations.

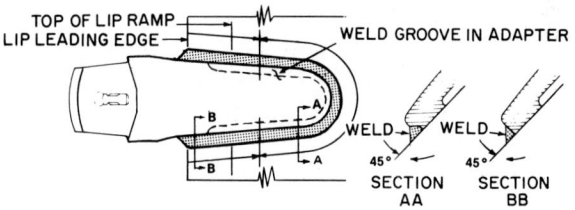

Fig. 4-4 When welding one-leg adapters on front end loader buckets, fillet weld must cover J groove (section *AA*) for effective weld strength through fillet throat. Use fillet weld only in forward area (section *BB*) where there is no J groove.

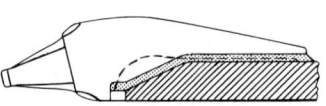

Fig. 4-5 When using a starter plate, backstep from the starter plate. Then remove before welding to lip of leading edge.

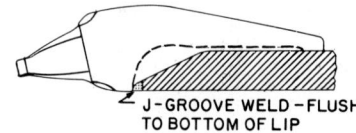

Fig. 4-6 J-groove weld at leading edge is flush with bottom of lip. (Bevel or fillet weld would be equal to lip leading-edge thickness.)

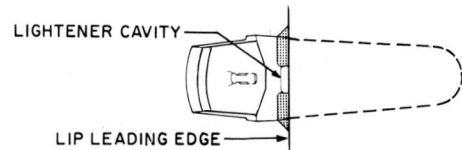

Fig. 4-7 Top view of one-leg adapter, showing lightener cavity.

Maintenance of Ground Contact Elements

Weld sizes

Bucket size		Weld size	
Cubic yards	Cubic meters	Inches	Millimeters
2	1.8	½	13
2¼–5	2–4.6	⅝	16
5½–12	5–11	¾	19
13 and up	12 and up	1	25

Teeth for Shovel Dippers, Hoe Dippers, and Dragline Buckets

There are many types of teeth used on earthmoving buckets. Designs from different manufacturers have similarities as well as differences. Instructions given here are from ESCO. Even so, these instructions do not cover every design variation. Check with the supplier to be sure of proper instructions for maintaining teeth on equipment.

How to install tooth horn adapters Used on hoe dippers, shovel dippers, and dragline buckets, these adapters fit on integrally cast horns on a bucket lip. They are clocked onto the horns by two methods: either fluted spools and wedges or key locks.

Fluted Spool and Wedge
1. Before installing, clean all adapter and lip mating surfaces.
2. Install new adapter and check clearances, as shown in Fig. 4-8. There must be clearances at *A, AA, B,* and *CC*. Bearing surfaces must be uniform at *D* and *DD*. If there are no clearances, rebuild the horns, as shown under "Rebuilding Tooth Horns."

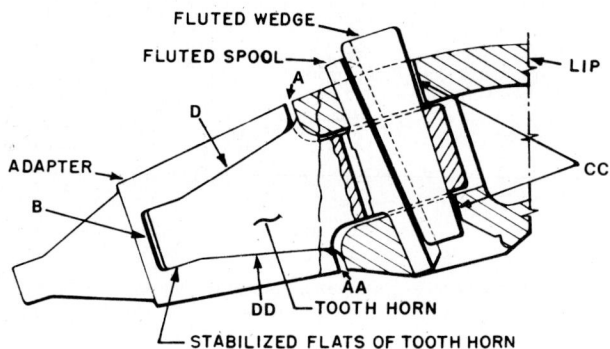

Fig. 4-8 Clearances and bearing areas of tooth horn adapter with fluted spool and wedge lock.

3. Insert fluted spool into cavity with large lug at bottom and facing forward.
4. Insert wedge behind spool, small end down. Drive in with several heavy blows of a hammer until there is no apparent movement—but not to the point of refusal. Use a 4-lb (2-k) hammer.

NOTE: Spool and wedge must be installed properly or it will cause breakage or loss of adapter or breakage of lip tooth horn.

5. Check tightness of wedge regularly. Loose wedges can lead to adapter loss or premature failure. After initial installation, tighten wedges every 2 to 4 hr until no longer needed to prevent looseness. Burn off any part of wedge extending from lip bottom.
6. If wedge drives flush with top of horn, use next oversize wedge.
7. To remove, drive out wedge from bottom upward with drift pin and hammer.

Key Lock
1. Clean adapter and lip mating surfaces.
2. Install new adapter and check clearances *A, AA, B,* and *CC*. There must be uniform bearing surfaces at *D* and *DD*. If no clearances, rebuild tooth horns (see Fig. 4-9).
3. Insert key into cavity and drive with a 4-lb (2-kg) hammer until there is no apparent movement with several heavy blows—but not to point of refusal.

Maintenance of Earthmoving Buckets and Bucket Teeth   3-35

4. Heat and bend tang at an angle (see Fig. 4-10).
5. Check tightness of key regularly. Loose keys can lead to adapter loss or premature failure. After initial installation, tighten keys every 2 to 4 hr until no longer needed to prevent looseness. Burn off key extending past lip bottom. Heat and bend tang as required.
6. If key drives flush with top of horn, use next oversize key.
7. To remove, straighten tang and drive out key from bottom upward with drift pin and hammer.

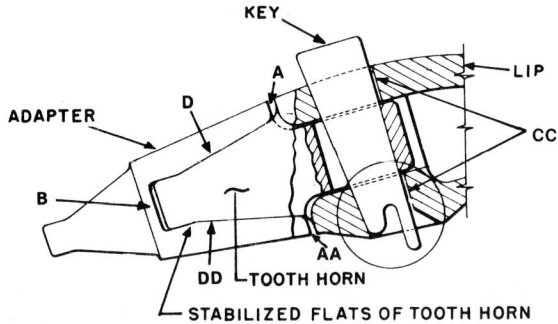

Fig. 4-9 Clearances and bearing areas of tooth horn adapter with key lock.

Maintaining Whistler adapters Whistler adapters are used on shovel dippers and front end loader zipper lips.
1. When lip is new, make a template to locate the leading edge of the keyslot. This will aid in rebuilding the lip and keyslot when they are worn. Cut template to approximate shape of lip bearing pad. Slip over lip and against its leading edge. Mark location of keyslot front edge, both inside and outside the lip.
2. Before installing new adapter, check to see if lip needs rebuilding.
   a. Clean bearing surfaces of lip, adapter, C-clamp, and wedge.
   b. Place new lip template on each bearing pad so it touches lip leading edge. Forward edge of the keyslot must line up with marks on template within ½ in. (13 mm).
   c. Install new adapters but not the clamps or wedges.
3. New adapter must bear against the lip leading edge and both top and bottom legs must contact lip for 3 to 4 in. (76 to 102 mm) back from leading edge (see Fig. 4-11).
4. The rear end of the adapter legs must either contact the lip bearing pad, or the total gap (gaps of both legs added) must not exceed ³⁄₁₆ in. (5 mm) (Fig. 4-11). If any of these conditions are not met, the lip bearing pads are too worn and must be rebuilt.

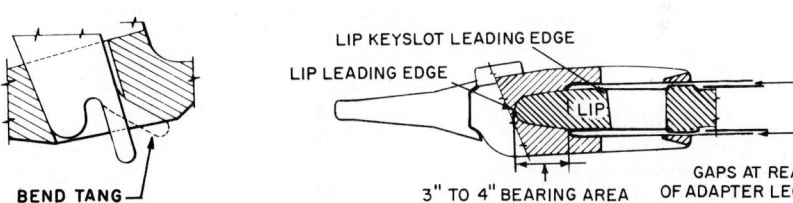

Fig. 4-10 Bend tang on key lock used on tooth horn adapters.

Fig. 4-11 Bearing surfaces and clearances of Whistler adapter.

5. With new adapters in place on lip, install C-clamp, large end first, from inside bucket or dipper.
6. Insert wedge forward of C-clamp, small end first, with flutes engaging C-clamp flutes. Drive wedge with 10 to 20 lb (4.5 to 9 kg) hammer until wedge moves only ⅛ in. (3 mm) per blow, with several successive blows—but not to point of refusal (see Fig. 4-12).

3-36 Maintenance of Ground Contact Elements

7. Check wedge tightness regularly. A loose wedge can lead to adapter loss or premature failure. Tighten every 2 to 4 hr until no longer needed to prevent looseness. Burn off any part of wedge extending below adapter bottom to prevent it from being knocked loose when digging (Fig. 4-12).
8. If wedge drives flush with top of adapter, use next oversize wedge.
9. To remove, drive out wedge from bottom upward with drift pin and hammer.

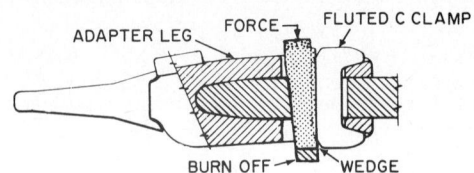

Fig. 4-12 Installing fluted C-clamp and wedge on Whistler adapter.

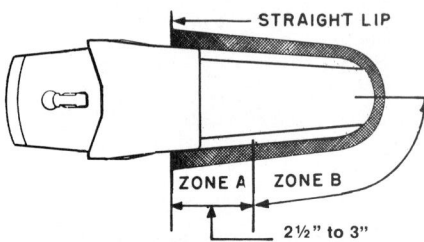

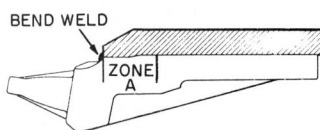

Fig. 4-13 Welding adapters to clamshell buckets; top view shows two welding zones.

Fig. 4-14 Side view of clamshell bucket adapter. Bead weld to leading edge and grind to allow lips to close.

Installing Adapters on Clamshell Buckets

Use E7016 or E7018 electrodes or E70 wire, both of low-hydrogen content. Follow welding details under "Welding ESCO Alloy-12 Series" (see Figs. 4-13 to 4-15).
1. Position adapters on lip.
2. Preheat and tack adapters to lip.
3. Use fillet weld in the two zones as follows:

Bucket size		Fillet weld size			
		Zone A, reduce gradually to zone B			
Cubic yards	Cubic meters	Inches	Millimeters	Inches	Millimeters
½–5	0.5–4.6	⅝	(16)	⅜	(10)
5½–10	5–9	1	(25)	½	(13)
10¼ and up	9.5 and up	1½	(38)	¾	(19)

4. Bead weld to lip leading edge and grind flush to allow lips to close.

Installing Adapters on Hoe Dippers and Dragline Buckets

Use low-hydrogen content E7016 or E7018 electrodes or E70 wire. Follow welding details given under "Welding Structural Components."
1. Position adapters on lip.
2. Preheat and tack adapters to lip.
3. Use same weld sizes in zones A and B as given in "Clamshell Buckets" above (see Figs. 4-15 and 4-16).
4. Weld from rear forward with rotated weld patterns top and bottom and side to side to minimize weld stresses.
5. Grind front ends of welds at least 1½ to 2 in. back to reduce weld stresses.

How to Install Adapters and Teeth on Integral Nose Tooth Bases

There are two types of noses cast into the lips of hoe dippers: shovel dippers and dragline buckets. One has a flat wedge nose with a rounded tip. The other is a patented design with a double conical nose and a flat rim at the tip. These bearing surfaces are different, but the teeth are locked to the noses by the same method.

1. Make a template of the tooth base when it is new to aid in rebuilding when it is worn. Template must fit exactly the tapered bearing surfaces where the adapter or tooth contacts the base. Cut it to fit the base halfway between the edge of the keyslot and the side of the base. Mark on the template the distance from the edge of the keyslot. Cut a clearance over the front face of the base 7/16 in. (11 mm).

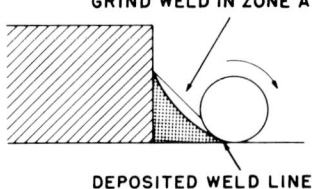

Fig. 4-15 Grind weld in zone A to reduce stress concentration.

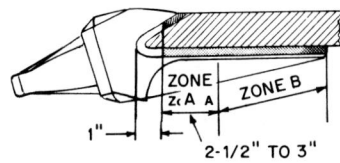

4-16 Hoe and dragline adapters; side view shows weld zones A and B.

2. When installing new box teeth or Kwik Tip adapters, check the fit to see if tooth base needs rebuilding.
   a. Clean all bearing surfaces of tooth base, adapter, spool and wedge, and keyslots of base and adapter (see Fig. 4-17).
   b. Visually check both sides of each base for wear. Severely peened or rounded areas are signs of wear which cause adapters to rock on base. This can cause breakage. Build up these areas as described under "Rebuilding Lip Tooth Bases."
   c. If template of new tooth base is available, place it on each base, both sides of keyslot, at location marked on template.
   d. If template contacts forward edge of base but doesn't contact both tapered surfaces at the same time, tooth base needs rebuilding.
   e. Rebuild any area on tapered bearing surfaces and stabilizing flats on conical tooth bases that is 1/8 in. (3 mm) or more from contacting template when it is located at the marked distance from the keyslot.
   f. Finally, place new adapters on tooth base and insert spool (see Fig. 4-18). Check for bearing of spool against back of adapter keyslot. If there is clearance, tooth bases must be rebuilt. NOTE: Some spools are straight and can be inserted either end first. Most spools are tapered and must be inserted large end first from top of tooth base. Tapered spools are marked on small end with Up or at center with Up and an arrow (see Fig. 4-19).
   g. If spool properly bears against back of adapter keyslot, install wedge. Check for clearance of wedge with front of adapter keyslot and for distance of top of

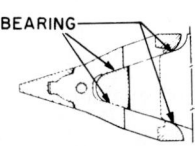

Fig. 4-17 Bearing areas of adapter and integral nose tooth base.

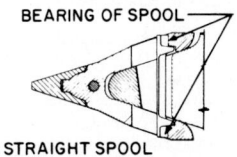

Fig. 4-18 Checking bearing area of integral nose with spool.

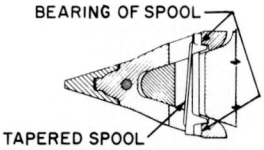

Fig. 4-19 Checking bearing of tapered spool on integral nose and adapter.

wedge above top of adapter. If standard wedge drives flush with top of adapter, use oversize wedge. If oversize wedge drives flush with top of adapter or touches front edge of keyslot, rebuild tooth bases.

NOTE: Insert small end of wedge first from top of tooth base. Spool and wedge must be installed properly or adapters will be lost or broken or tooth base broken.

3. If tooth bases are not worn, install adapters and spools and wedges as described in f and g above.

4. Drive wedge with 10 to 20 lb (4.5 to 9 kg) hammer until wedge moves only ⅛ in. (3 mm) per blow with successive blows. Do not drive to point of refusal.

5. Check wedge tightness regularly. Loose wedges can lead to adapter loss or failure. After initial installation, tighten wedge every 2 to 4 hr until no longer needed to prevent looseness. Burn off wedge extending below adapter bottom to prevent it from being knocked loose when digging (see Fig. 4-20).

6. If wedge drives flush with top of adapter, use next oversize wedge.

7. To remove, drive out wedge from bottom upward with drift pin and hammer.

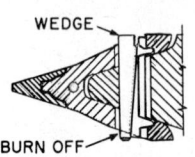

Fig. 4-20 Integral nose with adapter, spool, and wedge.

How to Weld Alloy-Steel Bucket Components

While buckets made of plate steel present no special welding problems, many buckets and components, including tooth and lip mounts, require special techniques. There are many types of alloys used in buckets, adapters, and points.

Varying amounts and combinations of carbon, nickel, chrome, silicon, molybdenum, and manganese give toughness and hardness to bucket parts. Some alloys are weldable, some are not. Some are heat-treated to obtain maximum hardness and welding heat can ruin them.

It is best, of course, to obtain specific welding instructions from the manufacturers, as there is no way to tell what alloy a particular part is made of.

Welding bucket teeth Most manufacturers do not recommend welding or hard-facing replaceable points on two-part teeth. They are cast of alloys and heat-treated to obtain optimum hardness to resist abrasion and breakage.

Welding ESCO alloy-12 series These alloys are used by one manufacturer for bucket parts.

Alloy	Bucket part	Weldability
12C	Front end loader adapters	Excellent
12E; 12F low hardness	Bucket lips and other structural components	Excellent
12 high hardness	Medium-size bolt-on or shank adapters.	Fair
12H	Large bucket adapters, shank, Whistler, and Kwik Tip types	Not recommended
12M low hardness	Bucket components of small to moderate thickness	Excellent
12M high hardness	Bolt-on adapters and wear-protection parts	Not recommended
12S	Bucket and dredge points, ripper points, and bits, and blades	Not recommended
12T, 12F	Large bucket structural components	Excellent

Maintenance of Earthmoving Buckets and Bucket Teeth 3-39

Welding structural components ESCO alloys 12C, 12E, 12F, and 12M low hardness and 12T have excellent weldability. They are used for weld-on adapters and for structural components on dragline bucket lips, cheeks, and arches; shovel dipper beams and most dipper fronts and lips; dredge cutter arms, rings, and hubs.

Structural components of these alloys may be hard-faced. However, since replaceable wear-resistant shrouds are available, hard-facing is not usually required.

These alloys are magnetic and can be distinguished from nonmagnetic manganese steel with a magnet. Since the two different alloys require different welding materials and techniques, it is important to check for magnetism before welding.

Welding electrodes Use E7016 or E7018 low-hydrogen electrodes to weld these alloys to each other or to other low-alloy or carbon-steel castings or forgings.

Controlling moisture in low-hydrogen electrodes Purchase low-hydrogen electrodes in 10-lb hermetically sealed containers. Larger sizes may have too much moisture which can cause underbead cracking. Electrodes pick up moisture when exposed to air. After opening container, keep electrodes dry.

1. Remove only a ½ hr supply a a time. Keep the remainder in a ventilated holding oven at 150°F (66°C).
2. If a partially used container is stored for later use, bake in a 500°F (260°C) ventilated oven for 2 hr before using.
3. If electrodes are taken from a cardboard box or other nonairtight container, do not use until they have been baked for 2 hr in a 500°F (260°C) ventilated oven. Then place in a ventilated holding oven while still warm, removing only a ½-hr supply at a time. Keep holding oven at 150°F (66°C).
4. Rebake electrodes exposed to air for more than ½ hr as in step 2.

Welding techniques Preheat heavy sections to 350 to 400°F (177 to 204°C) with torch. Check temperature with Tempilstik. Remove chill from light sections in cold weather.

Bevel the joints for 100 percent weld penetration. Rebuild worn areas to original contours with E7016 or E7018 before hard-facing. Build up excessively worn areas with low-alloy steel plate to restore original contour.

Weld stringer beads with a slight weave that is not more than three times rod diameter. Remove slag after each pass. Peen each bead to reduce stress concentrations. Maintain interweld temperature of less than 500°F (260°C).

After welding or hard-facing, postheat weld and areas around weld uniformly to 350 to 400°F (177 to 204°C), then air cool.

Dot hard-facing This hard-facing method is preferred to any other. It consists of spots or dots about ⅜ to ½ in. (10 to 13 mm) diameter, welded in rows about 1 to 1½ in. (25 to 40 mm) apart.

It is faster, uses less material, and costs less than continuous stringer bead or solid overlay. It also reduces cracking and weld failure of other methods.

Dots are also good wear indicators. Reweld them before they are completely worn off. Use Stoody 31 or equivalent electrodes for hard facing.

Welding manganese steel For rebuilding structural components such as tooth bases, use lip bearing pads made of ESCO alloy-14 manganese steel.

Check with a magnet to distinguish nonmagnetic manganese from magnetic alloys as it requires different materials and techniques.

Do not preheat—it will destroy toughness and make steel brittle.

Preparation Bevel the joints with a torch, grinder, or arc air to obtain 100 percent weld penetration. Avoid overheating when gouging.

Grind or torch-cut work-hardened surfaces at least $\frac{1}{32}$ in. (1 mm) deep before welding. Check with prick punch to be sure all work-hardened surface is removed. Scarf out cracks completely before welding.

Welding Skip weld and avoid wide weaving beads to distribute heat evenly along joint. Make beads less than 5 in. (125 mm) long. Weld with arc at low current setting (cold arc).

You should be able to put your bare hand on the part within 6 in. (150 mm) of the weld without burning your hand.

Peen each bead immediately to reduce contraction stress.

Rebuild worn areas to original contour with austenitic manganese nickel electrodes before hard facing.

Use these rods Use austenitic chromium nickel stainless-steel electrodes to weld manganese steel.

3-40 Maintenance of Ground Contact Elements

To weld manganese to ESCO 12T, 12M, 12E, and other low-alloy or carbon steel, use stainless-steel electrodes types 307, 308, 309, 310, 312, or 316.

Austenitic manganese nickel rods are sometimes used, but welds are not as durable as stainless welds.

Do not use mild steel or other electrodes to weld manganese to other materials.

Rebuilding tooth horns

1. When horn is new make a template to aid in rebuilding. Scribe it as shown in Fig. 4-21. Use to check keyslot leading-edge bevel and width.
2. If tooth horn does not have stabilizing flats, build them up as shown in Fig. 4-22. Rebuild nose as shown in Fig. 4-23.
3. Grind off work-hardened surfaces if tooth horns are of manganese steel.
4. Use a new adapter, fluted spool and wedge, or key as a gauge.
5. Build up tooth horn with weld until there is bearing at surfaces D and DD with clearance of $1/64$ to $1/32$ in. (0.4 to 0.8 mm) at E and EE. With adapter bearing on surfaces CaD and DD, B must be $1/8$ to $3/8$ in. (3 to 10 mm) maximum. Clearance A must be at least twice that of B. Top of wedge should protrude $3/4$ in. (19 mm) above tooth horn. Use standard feeler gauges (see Fig. 4-24).

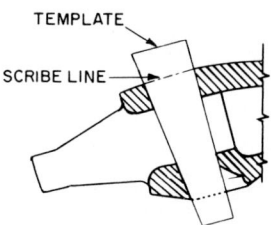

Fig. 4-21 Template for rebuilding tooth horns.

Fig. 4-22 Areas of tooth horn to weld built up and ground.

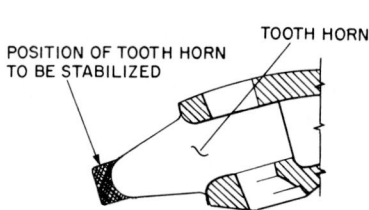

Fig. 4-23 Adding stabilized flats to tooth horn.

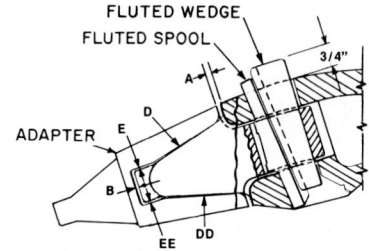

Fig. 4-24 Bearing areas and clearances of tooth horn and adapter.

Rebuilding Whistler lip adapter bearing pads

1. Use template to check wear. If it shows keyslot in bearing pad is too close to lip leading edge, examine keyslot by placing straightedge against its forward edge (see Figs. 4-25 and 4-26).
    a. If forward edge is rounded so straightedge rocks when placed against it, rebuild forward edge and grind smooth so straightedge will not rock. Rebuilt forward edge should line up with marks on template when it is properly positioned in contact with lip leading edge (see Fig. 4-27).
    b. If forward edge of keyslot is straight (the straightedge will not rock more than $3/32$ in. (2.5 mm), then the lip bearing pad can be rebuilt without welding keyslot.
2. Weld buildup and grind leading edge of lip back 3 to 4 in. (76 to 102 mm) to provide bearing against new adapter. Use template as a guide for amount to be added to leading edge by positioning template so marks for keyslot line up with forward edge of keyslot.

Maintenance of Earthmoving Buckets and Bucket Teeth

3. Weld buildup and grind rear adapter bearing pads so the total gap with a new adapter (measuring gap under each leg and adding them) is not more than $\frac{1}{16}$ in. (1.5 mm).
4. Check fit. Use new Whistler adapter, C-clamp, and wedge as gauges.
    a. W-1 wedge should insert through lip and just into lower-leg keyslot of adapter with moderate pounding.
    b. If wedge does not insert far enough, grind buildup from leading edge. If it inserts too far, weld more buildup at lip leading edge (see Fig. 4-27).

Rebuilding lip tooth bases (integral noses) These tooth bases are cast as part of the lip on certain types of hoe dippers, shovel dippers, and dragline buckets. They are also called *integral noses.*

There are two types: conical nose and nonconical nose. Instructions for both are given.

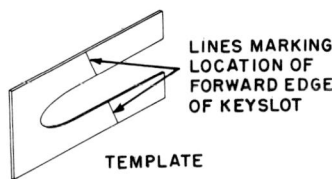
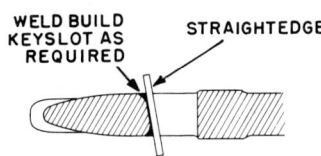

Fig. 4-25 Make a template for rebuilding Whistler lip bearing pads. Scribe lines to mark forward edge of keyslot.

Fig. 4-26 Check keyslot with straightedge. Rebuild keyslot as required.

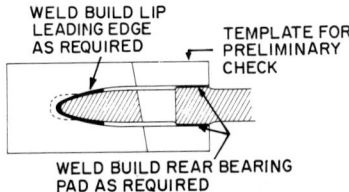
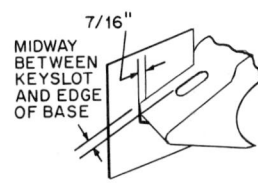

Fig. 4-27 Areas to rebuild on lip leading edge and rear bearing pads; solid areas show weld buildup needed.

Fig. 4-28 Use a template to check wear on lip tooth base.

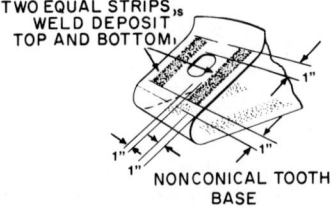
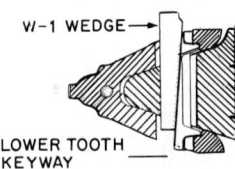

Fig. 4-29 Buildup on worn nose with two equal strips of weld (top and bottom) on nonconical tooth base.

Fig. 4-30 Check the fit of rebuilt tooth base with adapter and wedge.

Nonconical lip tooth bases
1. Make a template as described in "How to Install Adapters and Teeth on Integral Nose Tooth Bases." Use template and an adapter to check for wear (see Fig. 4-28).
2. Rebuild rounded or worn areas back to original contour, using a Kwik Tip adapter or solid tooth as a gauge. Weld two equal strips about $\frac{1}{8}$ in. (3 mm) thick on top and bottom of base. Strips should extend from 1 in. (25 mm) behind keyslot to within 1 in. (25 mm) of leading-edge curve or flat of nose. Keep weld 1 in. (25 mm) away from keyslot and 1 in. (25 mm) away from outside edges (see Fig. 4-29).
3. Check fit, using new Kwik Tip adapter or solid tooth and fluted spool and wedge as gauges.
    a. Fit tooth to base until W-1 wedge will insert through base and adjust into lower tooth keyway with moderate pounding (see Fig. 4-30).

3-42 Maintenance of Ground Contact Elements

b. Check to see if four-point bearing is achieved. Chalk, set tooth or adapter, and grind high spots. Repeat to reach four-point bearing.

Rebuilding conical lip tooth bases Follow above instructions with these exceptions.

1. Use rounded buildup strips to follow contour or conical Kwik Tip adapter or tooth (see Figs. 4-31 and 4-32).

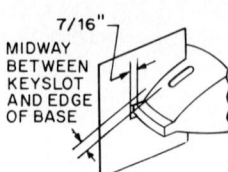

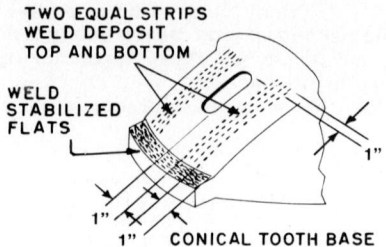

Fig. 4-31 Use a template to check wear on a conical nose lip tooth base.

Fig. 4-32 Areas to build up on a conical lip tooth base. Weld stabilized flats and apply two strips. Weld deposit on top and bottom of nose.

2. Build up stabilizing flats as close as possible to forward end of adapter or tooth cavity. Top and bottom surfaces must be parallel.

Chapter **5**

Maintenance of Vibratory Compactor Drums

ARNOLD DEICHEL
Manager, Product Services
Ingersoll-Rand Company, Compactor Division
Shippensburg, Pa.

Vibratory roller compactors have proven to be profitable in site preparation for a broad range of construction projects ranging from atomic generating sites and airports to shopping areas.

As a result of vibratory compaction of soils ranging from Georgia red clay to Mojave desert sand, contractors have realized site-preparation cost efficiencies and increased profits. The key to the successful operation of the vibratory rollers is a systematic preventive maintenance program as well as routine servicing consistent with the manufacturer's recommendations. Field maintenance starts on receipt of the equipment.

Every piece of construction equipment should have a receipt-condition report attached to record the actual condition of the unit when it arrives at its destination. Fleet managers should check all pertinent data to be certain it corresponds with the equipment ordered—model and serial numbers must match; all keys, parts, service manuals, and auxiliary equipment should arrive at the same time. Visual inspection will determine the condition of tires, gauges, seat, glass, battery and accessories, and the overall appearance of the equipment.

Claims for any transit damage should be immediately filed with the delivering carrier—with a copy of the report sent to the sales representative of the manufacturer. A typical report lists all pertinent condition received data that should be recorded and returned to the factory. The receiver should make a copy for his files and send an additional photofax copy to the carrier (Fig. 5-1).

A predelivery inspection and warranty registration form will help assure trouble-free starting and operation. This comprehensive checklist serves a dual purpose for start-up and routine equipment maintenance. Each unit has such a plaque; it spells out clearly what field maintenance steps must be taken at specific operating intervals.

Periodic service bulletins are issued to advise the latest tips on maintenance as well as auxiliary and product improvement equipment available. Field reports frequently provide hints for service expediency that can save the contractor both time and money.

Maintain master records A master maintenance record card (Fig. 5-2) can be kept up-to-date with all pertinent information included relating to each machine in the field. This file should be maintained throughout the life of the equipment. If any unusual activity is noted on a particular compactor by maintenance mechanics, it signals a need for action.

Spare parts availability is the key to an effective construction equipment support program. Most manufacturers maintain parts inventories throughout the United States

3-43

3-44 Maintenance of Ground Contact Elements

with levels based on experienced need. Naturally, the user should stock certain recommended spare parts for quick field replacement to minimize equipment downtime.

Each unit is supplied with a permanently affixed lubrication schematic diagram (Fig. 5-3) which lists recommended lubrication and maintenance schedules. This same chart usually appears in the compactor's operator manual.

```
┌─────────────────────────────────────────────┐
│              Ingersoll-Rand                 │
│        Machine Receipt Condition            │
├─────────────────────────────────────────────┤
│  MODEL NUMBER                               │
│  SERIAL NUMBER                              │
│  DEALER NAME                                │
│  LOCATION                                   │
│  PARTS & SERVICE MANUALS ☐                  │
│  IGNITION KEYS ☐                            │
│              VISUALLY INSPECT               │
│   TIRES  ☐                  SEAT ☐          │
│   GAUGES ☐                  GLASS ☐         │
│   BATTERY ☐                 CONTROLS ☐      │
│   ACCESSORIES ☐             PAINT ☐         │
│                 COMMENTS                    │
│                                             │
│                                             │
│                                             │
│                                             │
│  File Claim with Carrier for Transit Damage │
│                               Form 8100-10  │
└─────────────────────────────────────────────┘
```

Fig. 5-1 Maintenance begins with a record of machine condition at the time of receipt.

Field Troubleshooting A Key

To keep downtime to a minimum, a systematic, easy-to-read troubleshooting manual is available to locate malfunctions and, where practical, perform field corrections. Simplified fault-logic diagrams help to quickly pinpoint and correct problems so that they can be corrected quickly and efficiently. For example, if vehicle response is sluggish, maintenance personnel should check for a plugged filter, suction line clogged, parking brake engaged, air in the system or, possibly, even a worn pump or motor in the drive circuit. Once the problem is found, its correction is normally simple.

Plugged Filter Change filter.

Suction Line Clogged Inspect suction line from the reservoir to the transmission pump; maximum suction should not be greater than 10 in. of mercury.

Parking Brake Engaged Disengage parking brake lock usually located under the driver's seat by lowering lever to the release position. If brakes fail to release, check the master cylinder for clearance between the brake pedal linkage and the cylinder on air brake systems. Verify gauge pressure and whether the parking control valve is pushed in.

Maintenance of Vibratory Compactor Drums   3-45

Air in System   Check for low fluid level; check inlet filter and suction line for leaks allowing air to enter the system.

Broken or Crimped Control Cable   Inspect the control cable from the console to the transmission for movement and wear, and replace if necessary. Do not attempt to move control unless engine is running.

Pump Drive Disconnected (on Clutch-Equipped Machines Only)   Inspect and adjust clutch assembly between engine and pump drive for proper engagement. Clutch requires tension of 110 ft-lb to operate—measured at the extreme end of the handle.

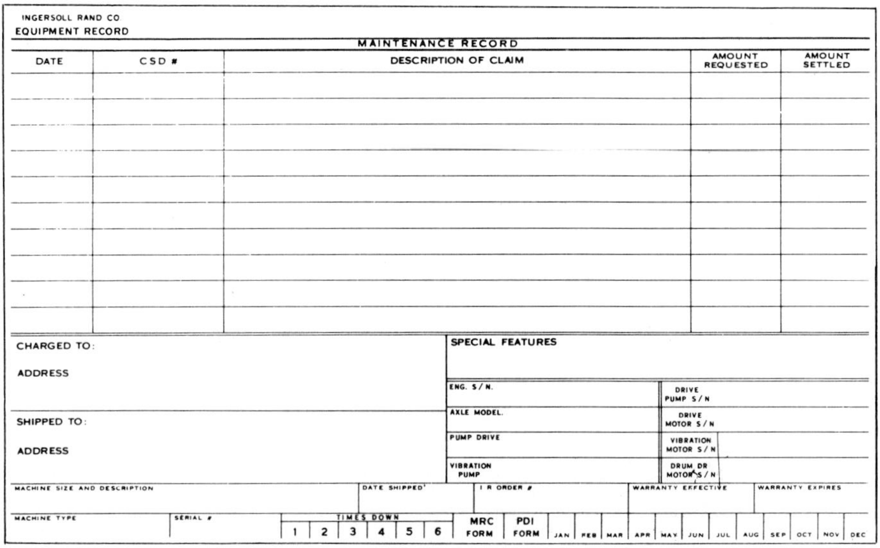

Fig. 5-2   Typical master maintenance record card.

Do Not Disturb Pressure Levels

Before beginning additional troubleshooting, it is important to understand that hydrostatic transmissions must maintain certain pressures to function properly. Variation from the proper pressure levels will damage or render the transmission inoperable.

For hydrostatic heavy-duty transmission, four pressures must be monitored to accurately diagnose a malfunction in the transmission (Fig. 5-4). Pressure gauges should be installed to permit proper troubleshooting and diagnosis (Fig. 5-5).

 1. *Charge pump inlet suction.* The maximum vacuum at the charge pump inlet should not exceed 10 in. of mercury under normal operating conditions. It is normal for the vacuum to be higher during cold start up.

 2. *Charge pressure.* The minimum allowable charge pressure is 130 psi above case pressure. Normal charge pressure is 190 to 210 psi above case pressure when pump is in neutral and 160 to 180 psi above case pressure when pump is in stroke position.

 3. *System or high pressure.* The maximum system pressure obtainable is controlled by the high-pressure relief valves located in the motor manifold. Relief valves have a coded number stamped on the exposed end, stating the valve setting, for example, Sundstrand 50 equals 5000 psi; Eaton 500 equals 5000 psi.

 4. *Case pressure.* Transmission case pressures should not exceed 40 psi under normal operating conditions, except during cold start-up.

Vibration and Steering Start-up

Good start-up procedures can help keep equipment working efficiently and economically. We all require a high degree of reliability, increased service life, and elimination of

3-46 Maintenance of Ground Contact Elements

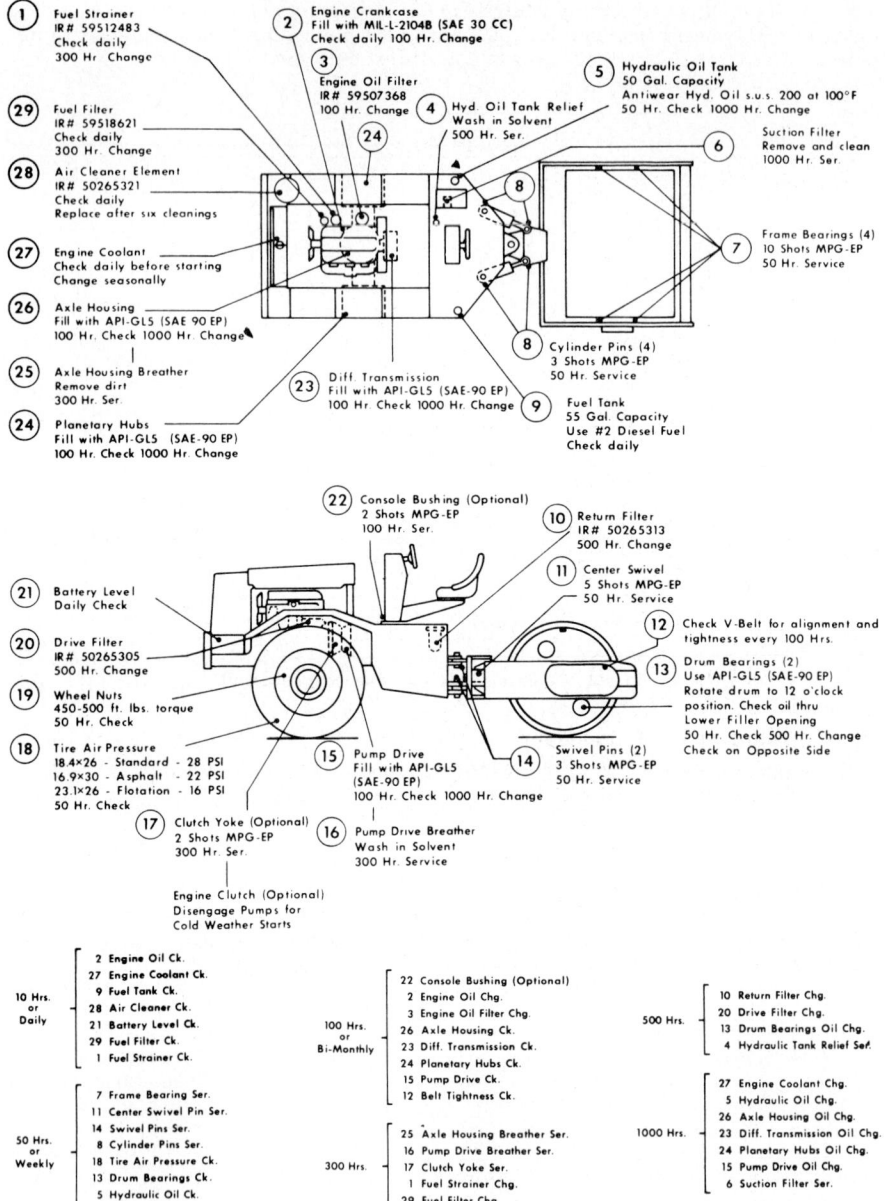

Fig. 5-3 This schematic diagram becomes a sort of portable maintenance manual since it can be affixed like a decal to the side of the unit as a permanent source of information.

downtime for our equipment. Combining good start-up procedures with a good preventive maintenance program, users can be assured of efficient, economical equipment operation resulting in greater profits from each project completed.

Air must be expelled from a hydraulic pump at start-up or it will not prime. Air in the pumping chamber can cause cavitation with resultant pump failure. The following practices are a sure first step toward trouble-free pump operation.

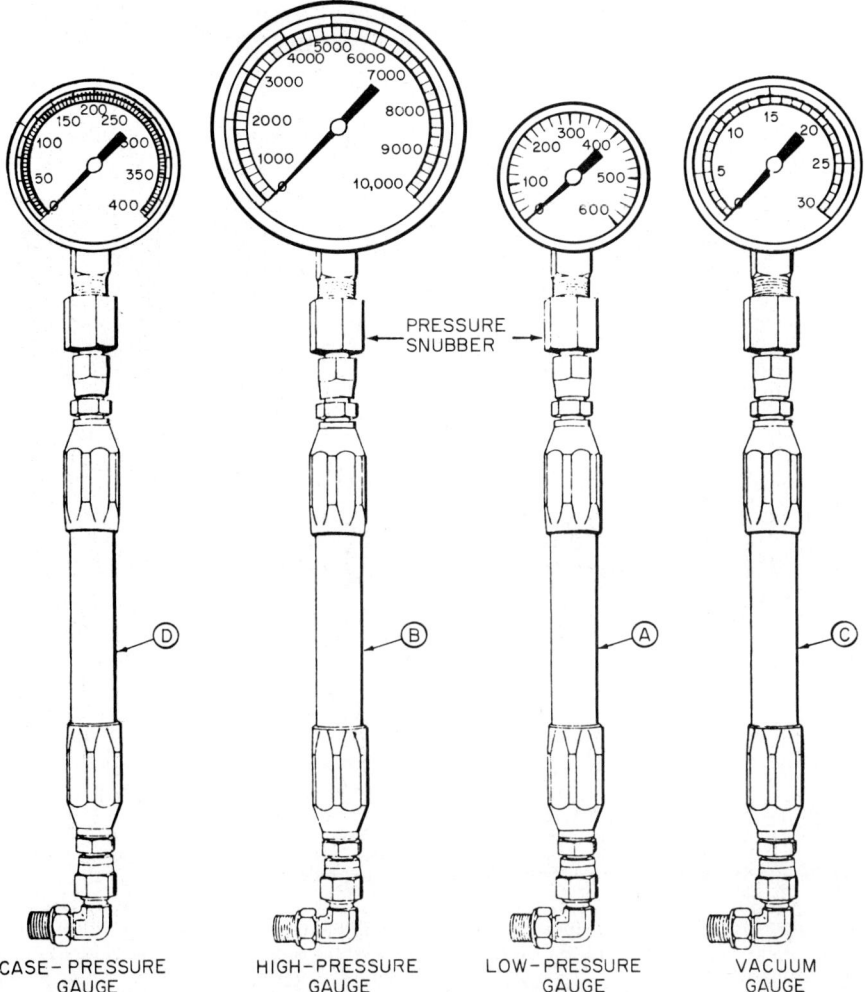

Fig. 5-4 Pressure monitoring gauges.

Remember it only takes 20 sec to burn out a system; therefore, be certain that all components are filled with clean fluid from clean containers before hydraulic lines are installed. It doesn't pay to economize with used fluids; it may cause an extra maintenance job before the next scheduled time. Always refer to the manufacturer's specifications when changing or adding oils.

Flush and clean the reservoir and lines before mounting the system. Fill the reservoir with new, clean fluid as recommended by the manufacturer. With caution, crack open a

3-48　Maintenance of Ground Contact Elements

pressure line fitting at the pump outlet port to bleed air from the system and to prime all newly installed components to assure lubrication at start-up. Now the engine is ready to start.

Turn over the engine by rotating the starter several times—for about 1 min—then start the engine and set the speed between 800 and 900 rpm (avoid high-speed start-up). With

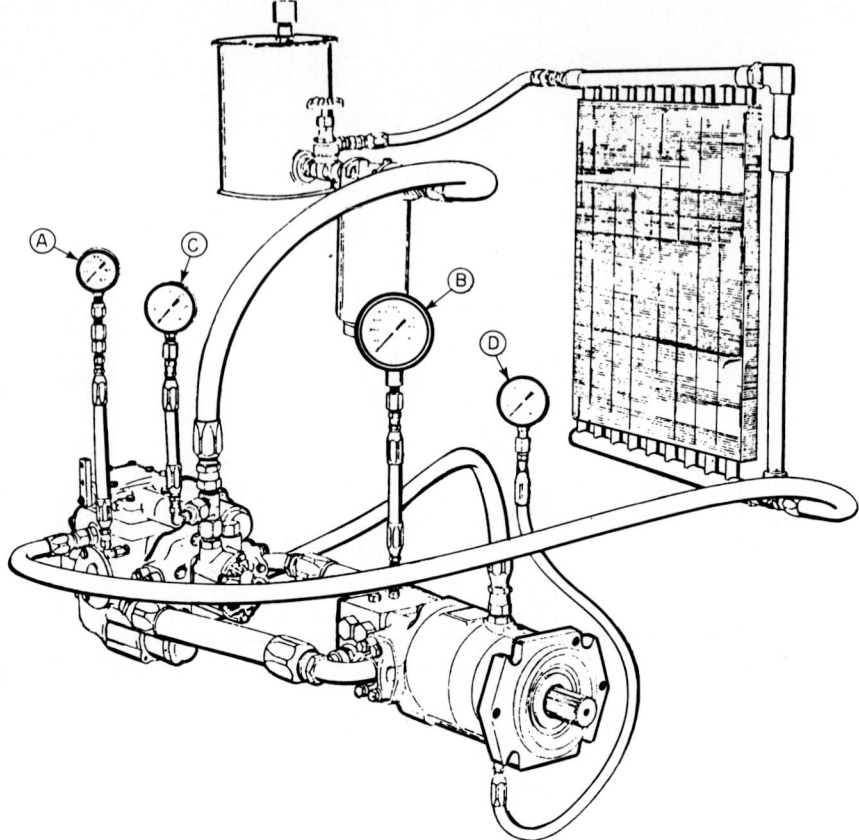

Fig. 5-5　Proper pressure gauge installation. At points A and B, the guage connection is 7/16 in. × 20 SAE "O" ring for all series. At C the gauge connection may also be connected on the suction side of the inlet filter; at D there is a reducer fitting from case port to gauge hose assembly.

the pump primed, air and fluid will bleed through the loose fitting. When a solid stream of fluid begins to flow, all air is bled from the system and the fitting should then be retightened.

Finally, examine the fluid in the reservoir. There may be some bubbles at the top of the fluid—that is the residue of any air which may have been left in the system. This will clear up shortly; then fill the hydraulic oil reservoir to the recommended level.

Cold Weather Start-up Procedures

Consistent daily start-up procedures assure maximum efficiency and service from vibratory compaction equipment.

1. Ensure that oils used in the system conform to manufacturer's specifications for local temperature conditions.
2. Start the engine and allow to idle at a speed between 800 and 900 rpm.

Maintenance of Vibratory Compactor Drums 3-49

3. Do not allow cylinders to travel to the end of their stroke cycle, or permit pressure to build up to the setting on the relief valve. Cold fluid makes relief valves sluggish and can add from 500 to 1000 psi to the maximum setting on the valve.

4. When all components are warm to the touch, the vehicle can be safely placed into service.

Troubleshooting the Vibration Circuit

Whenever difficulties are experienced in the vibration circuit—low frequency, erratic vibration, or no vibration—the following series of checks can be easily performed in the field to ascertain how to correct most malfunctions. Since these are general suggestions, it is always best to refer to the operator's manual for the individual unit for specific guidance before attempting repairs.

Low RPM or No RPM You might find low engine rpm or an inoperative relief valve, worn pump and/or motor, crimped suction line or clogged filter, inoperative control valve, pump not engaged to engine, eccentric shaft unable to rotate, or loose drive belts. Correction for these malfunctions can normally be made in the field.

1. *Relief valve inoperative.* Clean, check, and replace as necessary.

2. *Pump or motor worn.* Disassemble pump assembly and inspect internal components for excessive wear. Where necessary, replace defective components and check for possible contamination of fluid. Always replace contaminated fluid in accordance with recommended procedures.

3. *Inoperative control valve.* Remove control valve. Check valve for internal leakage. Repair or replace valve assembly.

4. *Pump not engaged to engine (where applicable).* Adjust engine clutch assembly to required 110 ft-lb to engage clutch.

5. *Eccentric shaft not rotating.* Check eccentric shaft bearings for resistance to rotation. Replace bearings if necessary, following procedure outlined in service bulletin.

6. *Loose drive belts, where applicable.* Tension as per operator's manual.

Erratic RPM

Erratic rpm can have several causes—air in fluid, relief valve defective, loose vibration drive belt (where applicable), damaged eccentric shaft bearing, or clogged suction line or filter. Follow procedures for insufficient or no rpm for all appropriate causes except:

1. *Air in fluid.* Tighten leaky inlet connections. Fill reservoir to correct level and bleed air from system.

Fig. 5-6 In a typical application, this vibratory drum compactor achieves about 67 percent density while eliminating air pockets and eventually reaching a 75 percent density compaction level through compaction.

3-50 Maintenance of Ground Contact Elements

2. *Relief valve setting below operating pressure.* Reset relief valve to setting as outlined in Ingersoll-Rand service bulletin 8040-25.

3. *Loose vibration drive belts (where applicable).* Adjust V-belt and realign pulleys. Consult operator's manual for proper tension for individual unit.

Major Repairs or Overhauls

Many major repairs can be accomplished in the field. Remember that there is no substitute for consistent service and preventive maintenance to assure maximum machine performance with minimum downtime.

When troubleshooting a vibratory compactor, remember to check for a simple solution first. Often, the difficulty may be solved simply, for example, by replacing a clogged filter.

Should any unusual operational malfunctions occur outside the scope of routine troubleshooting, it is always best to contact the service representative. He should know the equipment and be able to provide the technical backup needed to get the compactor back into operation quickly (Fig. 5-6).

Index

Acids and alkalis for cleaning, 1-5
Air-acetylene soldering (*see* Welding, gas, air-acetylene soldering, heating, and brazing)
Atomic-hydrogen welding, 1-41
Augers, hard-facing, 1-123
Automobile-body soldering, 1-82

Bearings:
  general (*see* Mechanical power transmission equipment, bearings)
  plain, 1-125 to 1-134
    babbitt thickness as a measure of bearing life, figure, 1-127
    bearing temperature as a determinant of life, figure, 1-130
    case tolerances, table, 1-126
    cast-babbitt liners, renewal of, 1-133
    connecting rod tolerances, table, 1-126
    design of, 1-125 to 1-127
      grooving, 1-127
      loads, 1-125
      lubrication, 1-127
      materials, 1-125
      tolerances, 1-125
    housing reclamation, 1-133
    inspection and reconditioning of, 1-129 to 1-130
      bearing replacement, 1-130
      connecting rods, 1-129
      journals, 1-129
    load-carrying capacities for various materials, figure, 1-128
    maintenance and care of, 1-128
      cleanliness, 1-128
      lubricant selection, 1-127

Bearings (*cont.*):
      pressure, 1-128
      temperature, 1-128
    oil-clearance values, recommended, table, 1-127
    reassembly of, 1-131 to 1-133
      bolt torque, 1-131
      crush, 1-131
      end clearance, 1-132
      final checking, 1-133
      free rotation, 1-133
      measure of clearance, 1-132
      oil clearance, 1-132
      preliminary lubrication, 1-133
    shaft tolerances, table, 1-126
  rolling, 1-135 to 1-154
    bearing grease dropping-point temperatures for various materials, 1-154
    boundary dimensions, 1-135
    design and nomenclature for, 1-135
    general considerations, 1-135
    load ratings for, 1-138
    lubrication of, 1-150 to 1-154
      grease lubrication, 1-153
      oil lubrication, 1-151
    mounting and dismounting methods, 1-144 to 1-150
      cold mounting, 1-144
      dismounting procedures, 1-149
      general considerations, 1-144
      hydraulic removal, 1-149
      tapered-bore bearing mounting, 1-146
    mountings of, 1-139 to 1-143
    series breakdown of, 1-136
    shaft and housing fits, 1-139
    temperature mountings, 1-144

2   Index

Bearings (*cont.*):
    hot-oil bath, 1-145
    hot plate, 1-145
    introduction heaters, 1-145
    temperature-controlled ovens, 1-145
    temperature/viscosity relationships, figure, 1-153
Blades:
    dozer (*see* Moldboard, dozer)
Brazing (*see* Welding, gas)
Bucket teeth, 3-38
    (*See also* Buckets and bucket teeth)
Buckets and bucket teeth, 3-31 to 3-42
    components: bucket, alloy steel welding of, 3-38 to 3-40
        bucket teeth, 3-38
        dot hard-facing, 3-39
        electrodes, 3-39
        ESCO alloy-12 series, 3-38
        manganese steel, 3-39
        moisture control, 3-39
        preparation for, 3-39
        structural attachments, 3-39
        techniques for, 3-39
        welding rod choices, 3-39
    dippers, shovel and hoe, 3-34 to 3-36
        adapter installation, 3-36
        clamshell bucket adapter installation, 3-36
        integral nose tooth base adapter installation, 3-37
        tooth horn adapter installation, 3-34
        Whistler adapter maintenance, 3-35
    dragline buckets (*see* dippers, shovel and hoe, *above*)
    front end loaders, 3-31 to 3-34
        adapters: one-leg, welding of, 3-33
        two-leg, welding of, 3-32
        types of, 3-32
        locking devices, 3-32
        nose types, 3-32
        point selection, 3-32
        tooth spacing, 3-31
        weld sizes, 3-34
        welding points, 3-32
    rebuilding operations, 3-40 to 3-42
        conical lip tooth bases, 3-42
        integral noses, 3-41
        lip tooth bases, 3-41
        nonconical lip tooth bases, 3-41
        Whistler lip adapter bearing pads, 3-40

Carbon-arc welding, 1-42

Centralized lubrication systems (*see* Lubrication, centralized systems)
Chain hoists (*see* Hoists and slings, chain)
Chemical cleaning (*see* Cleaning, chemical)
Cleaning:
    chemical, 1-5 to 1-9
        vs. alternative cleaning methods, 1-8
        application methods, 1-6
            fill and empty, 1-6
            flow-through vessels, 1-6
            large hollow vessels, 1-7
            new ideas, 1-7
            pigs, plugs, balls, and jets, 1-7
        data and decisions for planning a job, 1-8
        disposal problems, 1-7
        materials for cleaning, 1-5
            acids and alkalis, 1-5
            organic solvents, 1-6
            sequestrants, 1-5
            specialty cleaning products, 1-6
            synthetic detergents and acid inhibitors, 1-6
        planning a job, data and decisions for, 1-8
        references, 1-9
    steam and hot water, 1-245 to 1-250
        auxiliary equipment for, 1-249
            ChemJet, 1-249
            degasser, 1-249
            parts washer, 1-249
            SandJet, 1-249
            tank cleaner, 1-249
        cleaning unit operation for, 1-247
            hot water high-pressure washer, 1-248
            steam cleaner, 1-248
        portability vs. permanence, 1-246
        pressure washer, 1-245
        process of, 1-245
        program establishment for, 1-246
            daily cleaning, 1-246
            on-site machinery inspection, 1-246
            premaintenance cleanup, 1-247
        size of unit for, 1-246
        steam cleaner, 1-245
        washing techniques for, 1-247
    vibratory drum, 3-43 to 3-50
        cold weather start-up procedures, 3-48
        erratic rpm, causes of, 3-49 to 3-50
            air in fluid, 3-49
            belts loose, 3-50

Cleaning (cont.):
    relief valve setting off, 3-50
    field troubleshooting (see troubleshooting, field below)
    general maintenance considerations for, 3-43
    major repair of, 3-50
    pressure levels, maintaining, 3-45
    record keeping for maintenance of, 3-43
        figures, 3-44, 3-45
    repair, major, 3-50
    schematic guide to, figure, 3-46
    start-up procedures: cold weather, 3-48
        steering system, 3-45 to 3-48
        vibration system, 3-45 to 3-48
    steering system start-up, 3-45 to 3-48
    troubleshooting, field, 3-44 to 3-45
        air in system, 3-45
        broken control cable, 3-45
        filter, plugged, 3-44
        parking brake engaged, 3-44
        plugged filter, 3-44
        pump drive disconnected, 3-45
        suction line clogged, 3-44
    vibration circuit, troubleshooting, 3-49
    vibration system start-up, 3-45 to 3-48
Conveying equipment (see Conveyors)
Conveyors, 1-203 to 1-221
    conveyor belts, 1-205 to 1-212
        cutting, procedure for, 1-208
        fasteners, applying, figure, 1-210
        inspection of, 1-207
        installation of, 1-211
        training of, 1-211
        troubleshooting for, table, 1-209
        wear, causes of, 1-205
    gas-driven unit adjustment, 1-216
        countershaft drive belt, 1-216
        drive shaft V-belts, 1-216
        head shaft drive chain, 1-216
    introduction to, 1-203
    lubrication, 1-216
        bearings: antifriction, 1-221
            plain, 1-221
        chain guards, oil-tight, 1-221
        criteria for, 1-204
        drive, 1-221
        engines, gasoline, 1-221
        idlers, 1-219
        motors, electric, 1-221
        oil-tight chain guards, 1-221
        safety, 1-219

Conveyors (cont.):
    operating precautions (see precautions, operating, below)
    permanent, description of, 1-203
    portable, description of, 1-203
    precautions, operating, 1-212 to 1-216
        belt scraper, 1-216
        belt scraper adjustments, 1-216
        capacity measurement, belt, 1-213
        conveyor hopper flashing, 1-214
        feeding, 1-213
        general, 1-212
        hand hydraulic pump, 1-214
        hydraulic hoist assembly, 1-214
        hydraulic ram, 1-215
        overloading, conveyor, 1-213
        safety considerations, 1-204, 1-219
        tires and wheels, 1-216
        V-belt tension adjustment, 1-214
            figure, 1-215
Corrosion control, 1-1 to 1-4
    inhibitors, 1-3
    methods of stopping corrosion, 1-2 to 1-3
    nonmetallic materials, 1-3
        plastic, 1-3
        rubber and elastomers, 1-3
    protective coatings, 1-3
    references, 1-4
    types of corrosion, 1-1
        crevice corrosion, 1-2
        erosion, 1-1
        exfoliation and selective leaching, 1-2
        galvanic corrosion, 1-1
        intergranular corrosion, 1-2
        pitting, 1-2
        selective leaching and exfoliation, 1-2
        stress-corrosion cracking, 1-2

Depreciation, machinery or equipment (see Accounting practices, fixed costs)
Diesel engines, 2-9 to 2-31
    air intake and exhaust, 2-23 to 2-25
        exhaust system, 2-24
        intake system, 2-23
        turbocharger, 2-24
    cold-weather starting, 2-11
    cooling system, 2-17 to 2-20
        aeration, 2-19
        clogging, 2-17
        coolant, 2-17
        exhaust gas leakage, 2-20
        external leaks, 2-18
        fan and belts, 2-19
        filters, 2-19

Diesel engines (*cont.*):
   flushing cooling system, 2-20
   hoses, 2-19
   internal leaks, 2-18
   radiator, 2-17
   radiator cap, 2-17
   thermostats, 2-18
   water pump, 2-19
  electrical system, 2-25 to 2-28
   alternator, 2-27
   batteries, 2-25
   gauges, 2-28
   generator, 2-27
   starting motor, 2-27
  engine stopping precautions, 2-12
  fuel system, 2-12 to 2-17
   bleeding, 2-16
   filters, 2-13
   fuel-injection nozzles, 2-14
    troubleshooting, table, 2-15
   fuel lines, 2-14
   fuel storage, 2-16
   fuel-transfer pump, 2-13
   governors, 2-16
   injection pumps, 2-14
   troubleshooting, 2-12
  lubrication of, 2-20 to 2-23
   additives, 2-21
   oil consumption, 2-22 to 2-23
   oil contamination, 2-20 to 2-21
   oil coolers, 2-22
   oil filters, 2-21 to 2-22
   pressure-regulating valve, 2-22
   ventilation, 2-22
  start-up considerations for, 2-11 to 2-12
  troubleshooting, by malfunction, 2-28 to 2-31
  visual inspection of, 2-9 to 2-11
   air supply, 2-10
   coolant leaks, 2-10
   electric systems, 2-10
   fuel systems, 2-10
   oil leaks, 2-10
   turbocharged engine checks, 2-11
Dozers (*see* Moldboard, dozer)

Earthmoving buckets and bucket teeth (*see* Buckets and bucket teeth)
Electric motors (*see* Motors, electric)
Electrical power syastems (*see* Power systems, electrical)
Electron-beam welding, 1-43
Electroslag welding, 1-41
Engines (*see* Diesel engines; Gasoline engines)
Explosive welding, 1-43

Flash welding, 1-42
Flow welding, 1-43
Friction welding, 1-43

Gas welding (*see* Welding, gas)
Gasoline engines, 2-33 to 2-37
  air cleaner: dual-element, servicing, 2-35
   function and design, 2-34
   oil-foam, servicing, 2-34
  combustion chamber cleanout, 2-36
  cooling system cleaning, 2-37
  fuel for, 2-35
  general considerations for maintenance of, 2-33
  lubrication of, 2-35
  spark plug blast cleaning, 2-35
  storage of, 2-37
Generators, electric, 2-1 to 2-7
Graphite (*see* Lubricants, solids, characteristics of, 1-17)
Grease (*see* Lubricants, grease)

Hammer welding, 1-43
Hoists and slings, chain, 1-187 to 1-201
  electric, 1-189
  general considerations, 1-187
  instruction check diagram, 1-191
  instructions for operators, 1-193
  lever-operated, 1-189
  lubrication of, 1-198, 1-200
  preventive maintenance of 1-190 to 1-193
   frequency, formula for, 1-191
   inspection, 1-192
    frequent, 1-192
    idle-hoist, 1-193
    initial, 1-192
    periodic, 1-192
    procedure for, 1-192
    records, 1-192
  preventive maintenance vs. breakdown maintenance, 1-200
  rigger ratchet, 1-187
  selection of, 1-190
  service, 1-192
  spur-geared (*see* spur-geared, *under* types of *below*)
  testing, 1-193
   load, 1-193
   operating, 1-193
  troubleshooting on: electric, all, table, 1-196 to 1-198
   electric, two-speed, table, 1-199
   spur-geared, table, 1-194 to 1-195
  types of, 1-187 to 1-190

Hoists and slings, chain, (*cont.*):
   electric, 1-189
   lever-operated, 1-189
   rigger ratchet, 1-187
   spur-geared, 1-187, 1-189
     cyclone and satellite, 1-187
     low-headroom trolley, 1-187
     modern, 1-189
   welded link load and hand chain, 1-200
Hose, hydraulic, 1-223 to 1-244
   characteristics of, by stock type, table, 1-224
   components of, 1-223
     cover, 1-226
     reinforcement, 1-224
     tube, 1-223
   couplings, 1-228 to 1-231
     general, 1-228
     permanent, 1-229
   fittings and adapters, 1-231 to 1-244
     adapters, 1-243
     assemblies: assembly length, overall, 1-237
       coupling installation equipment, 1-239
       determining, 1-237
       installing, 1-243
     flanged fittings, 1-232
       definitions for, 1-232
       iron pipe, 1-232
       shapes, 1-232
       standard screw, 1-232
       thread identification, 1-232
     flow capacity of, table, 1-227
     selection of, 1-236
     threaded fittings, 1-231
Hose fittings (*see* Hose, hydraulic)
Hot water cleaning (*see* Cleaning, steam and hot water)
Hydraulic hose (*see* Hose, hydraulic)
Hydraulic systems, 2-39 to 2-51
   actuators for, 2-46
   ball check valve, figure, 2-45
   circuitry for, figure, 2-40
   cylinder, nomenclature and parts of, figure, 2-46
   external gear pump, figure, 2-44
   filters, micronic-particle, figure, 2-49
   fluid, viscosity of, table, 2-47
   fundamentals of, 2-39
   globe valve, figure, 2-45
   internal gear pump, figure, 2-45
   maintenance considerations, 2-46 to 2-51
     filtration, 2-48

Hydraulic systems (*cont.*):
     fluid factors, 2-47
     reservoir, 2-49
     start-up, 2-51
     troubleshooting, 2-50
   micronic-particle filters, figure, 2-49
   needle valve, figure, 2-45
   nomenclature and parts of cylinder, figure, 2-46
   pumps for, 2-40
     classification of, 2-41
     external gear pump, figure, 2-44
     gear types, 2-41
     internal gear pump, figure, 2-45
     piston types, 2-41
     radial pump, figure, 2-44
     vane pump, figures, 2-44
   radial pump, figure, 2-44
   reversing directional (four-way) valve, figure, 2-46
   symbols for, figure, 2-42 to 2-43
   valves for, 2-45
     ball check valve, figure, 2-45
     globe valve, figure, 2-45
     needle valve, figure, 2-45
   vane pump: balanced, figure, 2-44
     unbalanced, figure, 2-44
   viscosity of fluid, table, 2-47

Induction welding, 1-43
Inhibitors, corrosion, 1-3

Jets in cleaning:
   ChemJet, 1-249
   SandJet, 1-249
   water, 1-7
Journals, bearing, 1-129

Ladders (*see* Scaffolds and ladders)
Lubricants, 1-11 to 1-29
   additives for, 1-15
     anticorrosion additives and rust preventives, 1-18
     antioxidants (inhibitors), 1-15
     engine-cleanliness, 1-18
     extreme-pressure, 1-18
     foam depressants, 1-18
     pour-point depressants, 1-18
     viscosity-index improvers, 1-15
   endurance value factors for, 1-10
     adhesions, 1-19
     emulsification, 1-19
     interfacial tension, 1-19
     saponification, 1-19
     surface tension, 1-19
     wetting ability, 1-19

Lubricants (*cont.*):
  grease, 1-12, 1-15
    characteristics of, 1-12
      base, 1-12
      dropping point, 1-12
      melting point, 1-12
      penetration, 1-12
    definition and discussion of, 1-15
    protection of, 1-18
      against fire, 1-19
      need for proper, 1-18
      storage in, 1-19
  solids, characteristics of, table, 1-17
    bentones, table, 1-17
    boron nitride, table, 1-17
    fullers earth, table, 1-17
    graphite, table, 1-17
    mica, table, 1-17
    molybdenum disulfide, table, 1-17
    talc, table, 1-17
    zinc oxide, table, 1-17
  synthetics, characteristics of, table, 1-16
    esters, table, 1-16
    fluorocarbons, table, 1-16
    glycols (polyethers), table, 1-16
    hydrocarbons, table, 1-16
    polyalkylene oxides, table, 1-16
    polyethers (glycols), table, 1-16
    silicones, table, 1-16
  tests of, 1-11
    carbon-residue content, 1-11
    demulsibility, 1-12
    emulsification, 1-12
    flash point, 1-11
    neutralization number, 1-12
    pour-point, 1-11
    saponification number, 1-12
    viscosity, 1-11
  types of, 1-12
    circulating oils, 1-13
    gear oils, 1-13
    greases (*see* grease, *above*)
    machine (engine) oils, 1-14
    steam-cylinder oils, 1-14
    synthetics and solids, 1-15
Lubrication:
  centralized systems, 1-21 to 1-25
    advantages of, 1-21
    basic system of, 1-21
    front end loader example, 1-22
    large drill or shovel example, 1-23
    savings through use of, 1-24
  conveyors (*see* Conveyors, conveyor belts, lubrication)

Lubrication (*cont.*):
  diesel engines (*see* Diesel engines, lubrication of)
  gasoline engines, 2-35
  general considerations for, 1-11 to 1-20
    approach to, 1-18
    failures attributed to, 1-20
    lubricant types (*see* Lubricants)
    personnel selection for, 1-20
    procedures for, 1-20
    timing of, 1-20
    tools of, 1-20
      hand-pressure grease guns, 1-20
      power guns, 1-20
  mechanical power transmission equipment: babbitted and bronze sleeve bearings, 1-158
    chain drive, 1-164
    dry fluid drives and couplings, 1-169
    mounted bearings, 1-157
    shaft-mounted speed reducers, 1-163
  motors, electric (*see* Motors, electric, bearings and lubrication)

Mechanical power transmission equipment, 1-155 to 1-169
  bearings, babbitted and bronze-sleeve type, 1-158
    adverse operations of, 1-159
    alignment of, 1-158
    correct load direction of, 1-158
    load rating limits of, 1-158
    lubrication of, 1-158
    shaft journal surface finish of, 1-158
    temperature limits on, 1-159
    wear inspection of, 1-159
  bearings, mounted, 1-158
    alignment of, 1-157
    flingers for, 1-157
    load direction for, 1-157
    lubrication of, 1-157
    mounting of, 1-156
      adapter type, 1-156
      eccentric collar type, 1-156
      setscrew type, 1-156
    seals for, 1-157
    shaft tolerances for, table, 1-157
    shafting for, 1-157
    troubleshooting on, 1-158
  bushings, tapered, 1-159
    cleanliness for, 1-159
    size recommendations of, 1-159
    tightening of mounting screws, 1-159
    wall thickness considerations for, 1-159

Mechanical power transmission (*cont.*):
  chain drives, 1-163
    installation of, 1-163
      alignment of, 1-163
      chain installation, 1-163
      chain tensioning, 1-163
      component cleaning, 1-163
      sprocket mounting, 1-163
    maintenance of, 1-164
      cleaning, 1-164
      inspecting, 1-164
      lubricating, 1-164
      replacing, 1-165
      troubleshooting, 1-165
        table, 1-166 to 1-167
  dry fluid drives and couplings, 1-165
    adverse operating conditions of, 1-168
    changing characteristics of, 1-168
    erratic acceleration of, 1-169
    frequent starting, steps to take, 1-169
    high-speed operation, 1-168
    lubrication, 1-169
    overload protection, 1-168
    slippage, 1-169
  flexible couplings, 1-165
    alignment of, 1-165
    clamping bolt tightening for, 1-165
    visual inspection of, 1-165
  speed reducers, shaft-mounted, 1-161
    installation of, 1-161
      belt alignment in, 1-163
      driven shaft check for, 1-161
      input sheave mounting for, 1-163
      lubrication, initial, for, 1-163
      positioning and tightening of, 1-161
      tape removal for, 1-161
    operational maintenance of, 1-163
      regular oil changes, 1-163
      routine inspection, 1-163
      seasonal oil changes, 1-163
      troubleshooting, 1-163
  V-belt drives, 1-160
    installation of, 1-160
      alignment checking for, 1-160
      belt mounting during, 1-160
      belt selection for, 1-160
      belt tensioning for, 1-160
      sheave inspection for, 1-160
      sheave mounting during, 1-160
    maintenance of, 1-161
      belt dressing not proper in, 1-161
      dirt accumulation prevention, 1-161
      oil and grease buildup prevention, 1-161

Mechanical power transmission (*cont.*):
    troubleshooting, 1-161
      table, 1-162
Metal resurfacing, 1-107 to 1-114
  alloy selection criteria, 1-110
    forms of alloy, 1-113
    types of alloy, 1-113
      build-up, Group 1, 1-111
      high-alloy ferrous, Group 3, 1-112
      low-alloy ferrous, Group 2, 1-112
      nonferrous, Group 5, 1-113
      tungsten-carbide, Group 4, 1-112
    types of wear, 1-110
      abrasion, 1-110
      corrosion, 1-110
      heat, 1-111
      impact, 1-110
      stress-related, 1-111
  applications of hard-facing, 1-120
    augers, 1-123
    baffle plates, 1-121
    cable sheaves, 1-121
    chutes, 1-121
    conveyor screws, 1-123
    crusher roll shells, 1-123
    engine valves, 1-122
    shafts, 1-121
    swing hammers, 1-121
    teeth, 1-122
  functions of rebuilding and hard-facing, 1-108
  methods of rebuilding and hard-facing, 1-114
    base metal use, 1-114
    preheating, 1-115
    welding preparations, 1-114
    welding procedures, 1-115
      automatic rebuilding, 1-117
      manual arc hard-facing, 1-116
      oxyacetylene hard-facing, 1-115
      semiautomatic hard-facing, 1-117
      thermal spraying, 1-118
  original equipment surface hardening, 1-107
    diffusion alloying, 1-107
    flame hardening, 1-107
    hard chrome plating, 1-108
  patterns of hard-facing, 1-120
  processes of rebuilding and hard-facing, 1-109
    automatic welding, 1-110
    manual arc welding, 1-110
    oxyacetylene welding, 1-109
    semiautomatic welding, 1-110
  surface checks, 1-119
  welded overlays, 1-108

Millwright, apprenticeship schedules for, 1-22
Moldboard, dozer, 3-23 to 3-27
  cutting edges of, 3-25 to 3-27
  end bits, 3-25 to 3-27
  installation of, 3-25
  types of, 3-23 to 3-24
    angle, 3-23
    full-U, 3-24
    semi-U, 3-23
    straight, 3-23
Motors, electric, 2-53 to 2-88
  ac induction type, 2-54 to 2-59
    rotor windings, 2-77
    slip, 2-54
    stator windings, 2-57
    torque, 2-54
    voltage and frequency, 2-56
  bearings and lubrication, 2-73 to 2-79
    adding lubricant, 2-75
    bearing and journal clearance setting, procedure for, 2-78
    bearing and journal reassembly, procedures for, 2-78
    bearing reassembly, procedures for, 2-75
    motor-oil properties, table, 2-77
    relubrication periods, guide for, table, 2-74
    viscosities of motor oils, table, 2-76
  check charts, 2-79 to 2-87
    ac and dc motors, 2-79
    ac motors, 2-85 to 2-87
    dc motors, 2-80 to 2-85
  dc type, 2-59 to 2-66
    armature handling, 2-59
    brush and brush holder, 2-62
    brush pressure, 2-62
    commutation, variables affecting, 2-62 to 2-66
      brush chatter, 2-62
      brush life, 2-66
      brush wear, 2-66
      commutator films, 2-65
      copper pickup, 2-66
      flashing, 2-65
      neutral brush setting, effect of, 2-65
      sparking, 2-63
      streaking of commutator surface, 2-63
      threading of commutator surface, 2-63
    commutator, 2-59
      conditions for sound operation, 2-59 to 2-62

Motors, electric (*cont.*):
    importance of, 2-59
  general approach to maintenance of, 2-53
  good maintenance practices, table, 2-78
  insulation and its care, 2-66 to 2-73
    cleaning of, 2-69
    cleanliness requirements, 2-67
    coolness requirements, 2-67
    drying of, 2-69
    dryness requirements, 2-67
    power supply for, determining, 2-68
    testing, dielectric, 2-71
      general, 2-72
  plan for maintaining, 2-53
  references, 2-88
  suggested schedule, table, 2-56

Organic solvents for cleaning, 1-6
Oxyacetylene welding (*see* Welding, gas, oxyacetylene welding, cutting, gouging, and hard-facing)

Percussion welding, 1-42
Pipefitter, apprenticeship schedules for, 1-23
Plasma-arc welding, 1-42
Power shovels (*see* Shovels, power)
Power systems, electrical, 2-1 to 2-7
  air cleaner servicing, 2-5
  cylinder head servicing, 2-5
  engine troubleshooting, figure, 2-4
  exhaust smoke indicators, 2-7
  fuel system servicing, 2-7
  generator switching, 2-7
  generator troubleshooting, table, 2-6
  low compression troubleshooting, table, 2-5
  preventive maintenance on, 2-3
  routine servicing of, 2-1
  scheduling routine service, table, 2-2
  spark plug analysis, 2-4
  tests on, 2-4
    compression, 2-5
    crankcase vacuum, 2-4
  tuneup practices, 2-3
Power transmission equipment (*see* Mechanical power transmission equipment)
Pressure welding, 1-43
Projection welding, 1-42
Protective coatings, corrosion, 1-3

Resistance welding, 1-41
Resurfacing (*see* Metal resurfacing)

Rigger, apprenticeship schedules for, 1-24
Ripper tooth and dozer moldboard, 3-24 to 3-29
  dozer moldboard (see Moldboard, dozer)
  hard-facing, 3-29
  installation of, 3-25
  ripper points, 3-27 to 3-28
    failure of, 3-28
    point angle, 3-28
    selection of, 3-28
  ripper shanks, 3-27
    curved, 3-27
    double-offset, 3-27
    position of, 3-27
    repair kits for, 3-27
    straight, 3-27
  types of, 3-24 to 3-25
    parallelogram, 3-24
    radial, 3-25
    straight bar, 3-24

Scaffolds and ladders, 1-171 to 1-186
  aluminum tube and coupler scaffolds, 1-181
  general considerations, 1-171
  OSHA scaffolding checklist, 1-183 to 1-185
  safety roles for ladders, 1-175 to 1-178
  scaffolding applications, 1-185
  scaffolding maintenance, 1-186
  scaffolds, 1-178 to 1-183
    safe use and safety rules for, 1-183
    safety requirements for, 1-181
    safety swinging, 1-181
    special-design, 1-181
    tube and coupler (steel and aluminum), 1-181
    welded aluminum, 1-178
    welded sectional steel, 1-179
  steel tube and coupler scaffolds, 1-181
  stepladders and extension ladders, 1-172 to 1-175
  basic ladder groups, 1-173
  choice of materials for, 1-172
  extension, 1-173
  metal, 1-175
  precautionary measures for, 1-175
  single, 1-173
  special-purpose, 1-174
  stepladders, 1-173
  tube and coupler scaffolds (steel and aluminum), 1-181
Seam welding, 1-42
Shears and nibblers, 2-23

Sheet-metal welding, 1-54, 1-56
Shielded-metal-arc welding (see Welding, arc, shielded-metal-arc)
Slings (see Hoists and slings, chain)
Soldering (see Welding, gas, air-acetylene soldering, heating, and brazing)
Solid lubricants (see Lubricants, solids, characteristics of)
Solvents, organic, for cleaning, 1-6
Steam cleaning (see Cleaning, steam and hot water)
Stud welding, 1-42
Submerged-arc welding, 1-53 to 1-54
Synthetic lubricants (see Lubricants, synthetics, characteristics of)

Teeth, bucket (see Buckets and bucket teeth)
Thermit welding, 1-43
Tires, maintenance and upkeep of, 3-1 to 3-14
  adjusted inflation pressure, table, 3-13
  flotation, 3-4
  ground-contact area, 3-4
  handling rules, 3-5
  inflation pressure, adjusted, table, 3-13
  inflation pressure determination, 3-9 to 3-12
    dozers, 3-12
    dumpers, 3-9
    front axle, 3-10
    front tires, 3-10
    graders, 3-12
    lift and carry, loaders, 3-12
    loaders, 3-10
    rear axle, 3-12
    rear tire, 3-10
    scrapers, 3-9
  load material weights, table, 3-11
  mounting, 3-5 to 3-9
  premature deterioration, causes of, 3-14
  pressure corrections, high temperature, 3-12
  selection, 3-1 to 3-3
    miscellaneous considerations, 3-2
    site factors, 3-2
    size and load index factors, 3-2
    surface type, 3-2
    vehicular conditions, 3-1
  selection aids, 3-3
  speed restrictions, unbanked curves, table, 3-14

Tires, maintenance and upkeep of, (*cont.*):
  storage considerations, 3-5
Tooth, ripper (*see* Ripper tooth and dozer moldboard)
Track maintenance (*see* Undercarriage, wear and maintenance)

Undercarriage, wear and maintenance, 3-15 to 3-22
  maintenance, 3-20 to 3-22
    general factors for, 3-20
    optional equipment for, 3-20
    rebuilding vs. replacement, 3-21
  operator actions to reduce wear, 3-19
  soil conditions as a wear factor, figure, 3-17
  wear, 3-15 to 3-19
    forward drive side, 3-15
    front idler, 3-16
    link, 3-16
    pin, 3-15, 3-16
    pitch, 3-15
    reverse drive side, 3-15
    roller, 3-16
    sprocket, tooth tip, 3-16
    track frame misalignment as a cause of, 3-16
    track stretch, 3-15
    wear-causing factors, 3-17
    wear patterns related to use, table, 3-18
Upset welding, 1-42

Vibratory drum compactors (*see* Compactors, vibratory drum)
Viscosity:
  hydraulic fluids, table, 2-47
  lubricants, tests of, 2-11
  motor oils, table, 2-76
  viscosity-index improvers, 1-15

Welding:
  arc, 1-27 to 1-78
    abrasive wear resistance, 1-56
    alloy steels, 1-45 to 1-47
      chromium, 1-47
      high-manganese, 1-47
      high-tensile low-alloy, 1-45
      stainless, 1-46
      stainless clad, 1-47
    carbon arc: average welding conditions for, table, 1-64
      hand, maximum currents for, table, 1-64

Welding (*cont.*):
    carbon-arc welding, 1-63
      torch for, 1-64
    carbon steels, 1-44 to 1-45
      cast iron, 1-45
      high-carbon, 1-45
      low-carbon, 1-44
      medium-carbon, 1-45
      preferred analysis range of, table, 1-44
    condensers, 1-75
    delay relays, 1-75
    distortion control, 1-48 to 1-53
      buildup and manganese-steel electrodes, 1-52
      causes and remedies of, 1-48
      cellulose-coated electrodes (EXX10 and EXXII), 1-49
      lime-covered low-hydrogen electrodes, 1-52
      shielded-metal-arc welding, 1-48
      titania-coated electrodes (EXX12 and EXX13), 1-50
      type(s) E6010 and E6011, 1-50
      type(s) E6012 and E6013, 1-51
      type E6027, 1-52
      type E7014, 1-52
      type E7015, 1-53
      type E7016, 1-53
      type E7018, 1-53
      type E7024, 1-52
      type E7028, 1-53
    equipment handling, operational practices for, 1-71 to 1-75
      abuse avoidance, 1-72
      cleanliness of, 1-71
      coolness of, 1-71
      maintenance of, 1-73
    equipment installation, 1-69 to 1-71
    equipment selection and maintenance, 1-67 to 1-69
      accessories, 1-68
      machines, 1-67
    general considerations, 1-27
    hard-facing: guide to, figure, 1-60 to 1-61
      selecting material for, 1-57
    hard-surfacing, 1-56
      with submerged-arc process, 1-62
    impact wear resistance, 1-57
    input cable wire sizes: for ac/dc unit, table, 1-71
      for motor-generator units, table, 1-71
    metal cutting with welding equipment, 1-64

Welding (*cont.*):
  nonferrous metals, **1**-47
    aluminum, **1**-47
    copper and copper alloys, **1**-47
  partial wear surface checks, **1**-59
  processes of, **1**-41 to **1**-43
    atomic-hydrogen, **1**-41
    carbon-arc, **1**-42
    electron-beam, **1**-43
    electroslag, **1**-41
    explosive, **1**-43
    flash, **1**-42
    flow, **1**-43
    friction, **1**-43
    hammer, **1**-43
    induction, **1**-43
    percussion, **1**-42
    plasma-arc, **1**-42
    pressure, **1**-43
    projection, **1**-42
    seam, **1**-42
    stud, **1**-42
    thermit, **1**-43
    upset, **1**-42
  references, **1**-78
  resistance welding, **1**-41
  sheet-metal electrode sizes, table, **1**-57
  sheet-metal welding, **1**-54
    currents for, table, **1**-56
  shielded-metal-arc, **1**-30 to **1**-41
    distortion control, **1**-48
    fully automatic flux-cored, **1**-32
    general, gas-shielded, **1**-33
    manual shielded metal, **1**-32
    metal, gas-shielded, **1**-39
      CO_2, **1**-40
    self-shielded metal, **1**-31
    semiautomatic flux-cored, **1**-32
    spot, gas-shielded, **1**-40
    submerged, **1**-32
    tungsten: equipment for, **1**-37
      gas-shielded, **1**-33
    spot welding, **1**-41
  steel-electrode, classification, characteistics, and uses of, table, **1**-49
  submerged-arc welding, **1**-53
    electrodes, **1**-54
    equipment for, **1**-53
    fluxes, **1**-54
  surfacing electrodes, types of, **1**-57
  torch for carbon-arc welding, **1**-64
  troubleshooting for, table, **1**-76 to **1**-78

Welding (*cont.*):
  weldability of metals, **1**-43
  welding cable sizes: for ac/dc units, table, **1**-72
    for motor-generator units, table, **1**-72
  welding procedures, checks on, **1**-59
  welding processes: electric-arc, **1**-28
    general, **1**-28
    manual and automatic, **1**-29
  gas, **1**-79 to **1**-106
    air-acetylene soldering, heating, and brazing, **1**-79 to **1**-86
      automobile-body soldering, **1**-82
      electrical connections, **1**-82
      miscellaneous applications, **1**-85
      paint burning, **1**-84
      precautions and safe practices, **1**-86
      sheet-metal working, **1**-81
      soft solders, common, table, **1**-81
      soldering, **1**-79
      soldering fluxes, **1**-81
      sweat-type fitting installation, **1**-83
    oxyacetylene welding, cutting, gouging, and hard-facing, **1**-87 to **1**-106
      blowpipe motion, **1**-90
      braze welding, **1**-88
      codes, specifications, and standards, **1**-98 to **1**-102
      cutting, **1**-87, **1**-95
        preparation for, **1**-86
      flame adjustment, **1**-88
      fusion welding: cast iron, **1**-82
        general, **1**-89
      gouging, **1**-94
      hard-facing, **1**-97
      hard-facing cast iron, **1**-104
      hard-facing deposit finishing, figure, **1**-105
      hard-facing rods, characteristics of, tables, **1**-103
      hard-facing steel, **1**-97
      heavy braze welding, **1**-88
      outfit setup, **1**-87
      oxygen cutting, **1**-92
      oxygen cutting equipment, **1**-94
      references, **1**-106
      weld making, **1**-90
      welding and brazing, **1**-86
      welding methods: ferrous, table, **1**-93
        nonferrous, table, **1**-94